Functional Materials from Lipids and Proteins

Smart Materials

Series editors:
Hans-Jörg Schneider, *Saarland University, Germany*
Mohsen Shahinpoor, *University of Maine, USA*

Titles in this series:

1: Janus Particle Synthesis, Self-Assembly and Applications
2: Smart Materials for Drug Delivery: Volume 1
3: Smart Materials for Drug Delivery: Volume 2
4: Materials Design Inspired by Nature: Function Through Inner Architecture
5: Responsive Photonic Nanostructures: Smart Nanoscale Optical Materials
6: Magnetorheology: Advances and Applications
7: Functional Nanometer-Sized Clusters of Transition Metals: Synthesis, Properties and Applications
8: Mechanochromic Fluorescent Materials: Phenomena, Materials and Applications
9: Cell Surface Engineering: Fabrication of Functional Nanoshells
10: Biointerfaces: Where Material Meets Biology
11: Semiconductor Nanowires: From Next-Generation Electronics to Sustainable Energy
12: Supramolecular Materials for Opto-Electronics
13: Photocured Materials
14: Chemoresponsive Materials: Stimulation by Chemical and Biological Signals
15: Functional Metallosupramolecular Materials
16: Bio-Synthetic Hybrid Materials and Bionanoparticles: A Biological Chemical Approach Towards Material Science
17: Ionic Polymer Metal Composites (IPMCs): Smart Multi-Functional Materials and Artificial Muscles Volume 1
18: Ionic Polymer Metal Composites (IPMCs): Smart Multi-Functional Materials and Artificial Muscles Volume 2
19: Conducting Polymers: Bioinspired Intelligent Materials and Devices
20: Smart Materials for Advanced Environmental Applications
21: Self-Cleaning Coatings: Structure, Fabrication and Application
22: Functional Polymer Composites with Nanoclays
23: Bioactive Glasses: Fundamentals, Technology and Applications
24: Smart Materials for Tissue Engineering: Fundamental Principles
25: Smart Materials for Tissue Engineering: Applications
26: Magnetic Nanomaterials: Applications in Catalysis and Life Sciences
27: Biobased Smart Polyurethane Nanocomposites: From Synthesis to Applications
28: Inorganic Two-Dimensional Nanomaterials: Fundamental Understanding, Characterizations and Energy Applications

29: Ionic Liquid Devices
30: Polymerized Ionic Liquids
31: Nanogels for Biomedical Applications
32: Reactive Inkjet Printing: A Chemical Synthesis Tool
33: Electrochromic Smart Materials: Fabrication and Applications
34: Layered Materials for Energy Storage and Conversion
35: Smart Membranes
36: Cucurbituril-Based Functional Materials
37: Ambipolar Materials and Devices
38: Applications of Porphyrinoids as Functional Materials
39: Hybrid Metal–Organic Framework and Covalent Organic Framework Polymers
40: Chemoresponsive Materials: Smart Materials for Chemical and Biological Stimulation: 2nd Edition
41: Functional Materials from Lipids and Proteins

How to obtain future titles on publication:
A standing order plan is available for this series. A standing order will bring delivery of each new volume immediately on publication.

For further information please contact:
Book Sales Department, Royal Society of Chemistry, Thomas Graham House, Science Park, Milton Road, Cambridge, CB4 0WF, UK
Telephone: +44 (0)1223 420066, Fax: +44 (0)1223 420247
Email: booksales@rsc.org

Visit our website at books.rsc.org

Functional Materials from Lipids and Proteins

Edited by

Nandika Bandara
University of Manitoba, Canada
Email: Nandika.Bandara@umanitoba.ca

and

Aman Ullah
University of Alberta, Canada
Email: ullah2@ualberta.ca

ROYAL SOCIETY OF **CHEMISTRY**

Smart Materials No. 41

Print ISBN: 978-1-83916-417-0
PDF ISBN: 978-1-83916-798-0
EPUB ISBN: 978-1-83916-799-7
Print ISSN: 2046-0066
Electronic ISSN: 2046-0074

A catalogue record for this book is available from the British Library

The Royal Society of Chemistry is a charity, registered in England and Wales, Number 207890, and a company incorporated in England by Royal Charter (Registered No. RC000524), registered office: Burlington House, Piccadilly, London W1J 0BA, UK, Telephone: +44 (0) 20 7437 8656.

For further information see our website at www.rsc.org

Printed in the United Kingdom by CPI Group (UK) Ltd, Croydon, CR0 4YY, UK

Preface

In the present and forthcoming years, the imperatives of environmental preservation, sustainable energy, and public health unequivocally stand out as central challenges confronting societies worldwide. Functional materials derived from lipids and proteins emerge as pioneering solutions, offering opportunities for advancing sustainability in energy, the environment, and breakthroughs in sectors such as food, packaging, and medical and biomedical sciences. This journey commences with a discerning exploration into the fundamental properties of lipids and proteins, laying the groundwork for a profound comprehension of their pivotal roles in the synthesis and application of functional materials. Within this framework, this book synthesizes the collective expertise of numerous specialists, providing a comprehensive overview of the prevailing trends in current research in this dynamic field.

This book comprises 15 chapters and endeavors to present contemporary trends and research achievements in the realms of materials, cosmetics, food, food packaging, water purification, and biomedical sciences. In Chapter 1, an exploration of lipid- and protein-based materials is undertaken, delving into their classification, properties, and applications. Chapter 2 presents recent advances in the biosynthesis of lipids and proteins, paving the way for chapters 3 and 4, which respectively focus on the packaging applications of protein- and lipid-based materials.

Chapters 5 to 10 delve into the specialized applications of lipids and proteins as bioactive compounds, highlighting their pivotal role as additives to enhance various food systems. This includes an exploration of nanotechnology in chapters 7 and 8, where nanoliposomes emerge as a promising avenue for the encapsulation and delivery of potential nutraceutical ingredients and protein- and lipid-based nanostructures for food bioactive delivery.

The subsequent chapters intricately explore specialized applications, ranging from the utilization of protein-based materials in cosmetics to the

Smart Materials No. 41
Functional Materials from Lipids and Proteins
Edited by Nandika Bandara and Aman Ullah

Published by the Royal Society of Chemistry, www.rsc.org

development of advanced materials for water purification. Additionally, the discussions extend to include insights into lipids and protein-based adhesives, presenting a comprehensive perspective on the diverse applications and versatility of these biomaterials. This panoramic view sheds light on the potential roles these biomolecules may assume in enhancing our daily lives.

In Chapter 14, a dedicated examination of sustainable biomaterials unfolds, introducing the innovative concept of a green biorefinery that employs proteins and lipids as sustainable biomass. This chapter serves as a compelling showcase of the remarkable potential for these biomolecules to take on a pivotal role in eco-friendly applications.

Chapter 15 delves into the realm of functional protein-based biomaterials and explores the application of protein-based materials in wound healing. The profound impact of these materials on the biomedical field is underscored by discussions on promising advancements in health and well-being.

Each chapter stands as a testament to the collaborative efforts of leading experts who generously share their valuable insights and research findings.

We express our heartfelt appreciation to the contributors for their invaluable contributions, which have played a pivotal role in molding this volume into a comprehensive reference suitable for researchers, educators, and students alike.

We warmly invite readers to embark on an enlightening journey through the pages of **Functional Materials from Lipids and Proteins**. Here, the convergence of biological molecules and materials science holds the promise of unveiling new vistas of discovery and fostering innovation.

Aman Ullah and Punita Upadhyay
Edmonton, AB, Canada

Contents

Smart Materials No. 41
Functional Materials from Lipids and Proteins
Edited by Nandika Bandara and Aman Ullah

Published by the Royal Society of Chemistry, www.rsc.org

CHAPTER 1

Introduction to Lipid- and Protein-based Functional Materials

AMAN ULLAH*[a] AND NANDIKA BANDARA*[b]

[a]Lipid Utilization Labs – Lipids/Materials Chemistry Group, Department of Agricultural, Food and Nutritional Science, University of Alberta, Edmonton, Canada; [b]Department of Food and Human Nutritional Sciences, Faculty of Agricultural and Food Sciences, University of Manitoba, Winnipeg, Manitoba, R3T 2N2, Canada
*E-mail: ullah2@ualberta.ca, Nandika.Bandara@umanitoba.ca

1.1 Lipid-based Functional Materials and Their Applications

Lipids are a diverse group of hydrophobic organic compounds, such as oils, fats, waxes, steroids, specific membranes, and hormones, which are mostly insoluble in water but have shown high solubility in organic solvents. Being renewable, economical, and abundantly available, lipids and their derivatives with various possible structural and functional attributes can be used as a promising basic raw material for the advancement of novel functional materials and the development of chemicals/polymers. In addition to animals, lipids are majorly available from plant sources, which are composed of heterogeneous compounds (groups of atoms), such as hydrocarbons, and also other heteroatoms, such as oxygen, nitrogen, sulfur, and

Smart Materials No. 41
Functional Materials from Lipids and Proteins
Edited by Nandika Bandara and Aman Ullah

Published by the Royal Society of Chemistry, www.rsc.org

phosphorus.[1] Due to these characteristics, lipids are found to be very attractive as a raw/starting material for the synthesis of different functional materials for various industrial applications. Generally, lipids are classified into three main categories, which include simple, compound, and derived lipids (Figure 1.1).[1] Simple lipids are further subcategorized into fats, oils, and waxes. Fats and oils are mainly a mixture of triglycerides, which can be converted into glycerol and free fatty acids by the hydrolysis process.[2,3] They are commonly found in food and nutritional processes.[4–6] Waxes are solid at ambient temperatures, generally nonpolar, mostly soluble in organic solvents, and insoluble in water.[7] Similarly, compound lipids are also subcategorized into two classes, *i.e.*, glycolipids and phospholipids (glycerophospholipids). They are the main components of the cell membranes in all animal and plant cells. On the other hand, derived lipids, which are the third class of lipids, are subdivided into three categories, which include carotenoids, terpenes, and steroids (Figure 1.1).

Among these three classes of lipids, simple lipids (fats and oils) have been explored extensively, offering a broad range of applications in the development of functional materials for chemicals and polymers, and in the food industry.[1,8] Structural representations of some commonly found fatty acids in plants/animals and their derivatives,[1] which are mostly being used in the development of various functional materials at an industrial scale, are shown in Figure 1.2. The structure of triglycerides with a variable hydrocarbon chain length containing various functional attributes, such as internal unsaturated bonds, alcohols, epoxides, and esters, makes them

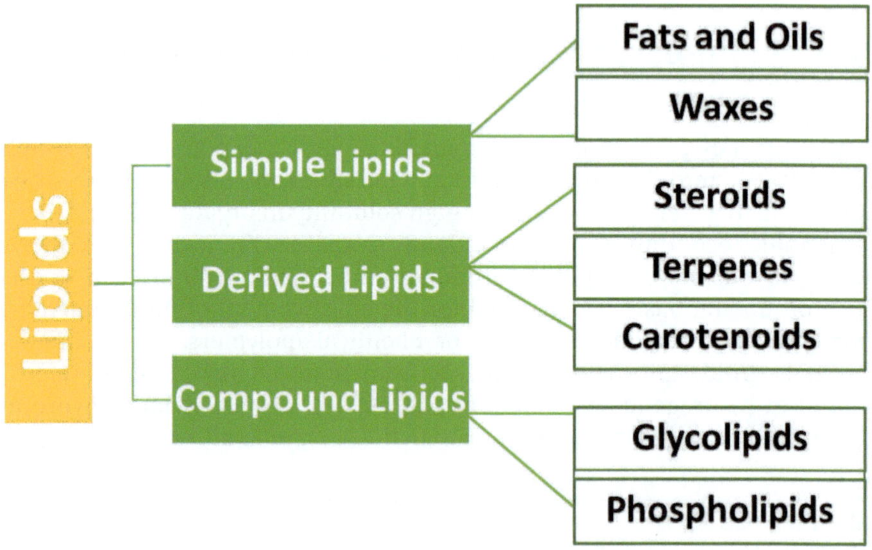

Figure 1.1 Classification of lipids into various classes. Adapted from ref. 1 with permission from John Wiley & Sons, Copyright © 2021 Wiley-VCH GmbH.

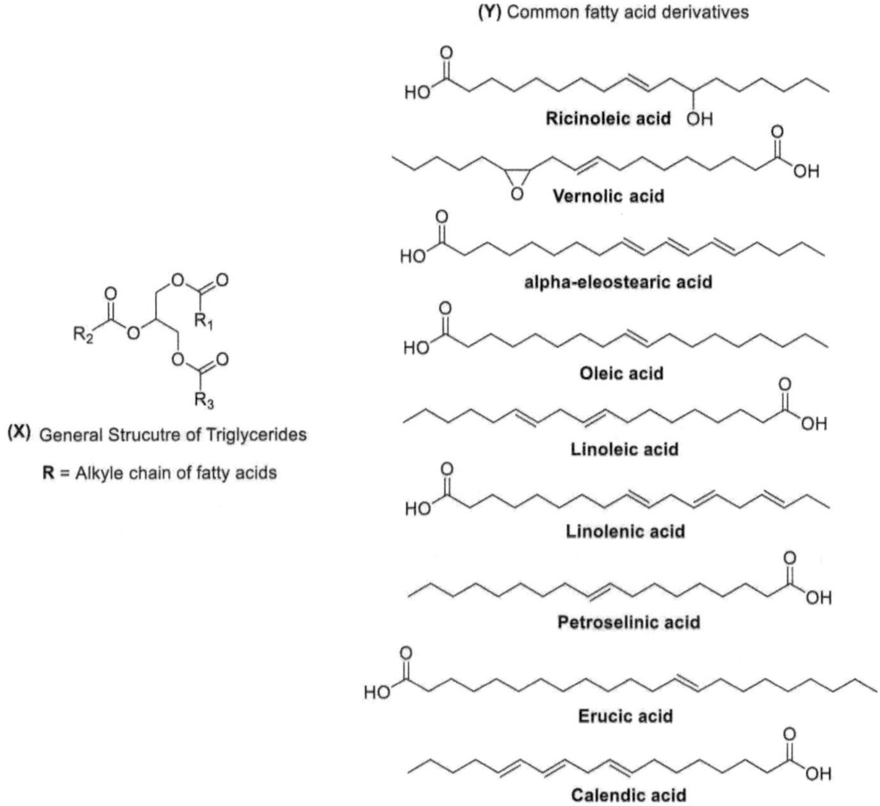

Figure 1.2 A general structural representation of triglycerides (X) and other common fatty acid derivatives (Y) used in the development of novel materials. Adapted from ref. 1 with permission from John Wiley & Sons, Copyright © 2021 Wiley-VCH GmbH.

a promising candidate to be transformed directly into useful polymeric materials. In addition, they can also be modified through various chemical processes, such as epoxidation,[9] hydroxylation,[10] hydrolysis,[11] hydroformylation,[12] ozonolysis,[13] ethenolysis,[14,15] transesterification,[16] hydrogenation,[17] and thiol-ene reaction[18]. These processes make them more reactive or useful monomers or polymer precursors for various industrial applications, such as coatings, polymers, nanocarriers, nanomedicines, biocomposites, lubricants, surfactants, pesticides or herbicides, pharmaceuticals, cosmetics, printing inks, leather processing, hydraulic fluids, solvents, glycerol, and fuels, providing a renewable and sustainable alternative to the existing petroleum-derived materials, such as fuels and plastics.[19–21]

Bio-based plasticizers, specifically lipids, play an important role as a plasticizer, which are great alternatives to phthalates and their derivatives because of their inherent advantages of being nontoxic and eco-friendly. These lipid-based plasticizers are either long-chain alkyl-based plasticizers obtained from triglycerides or phenolic lipid components (cardanol). For food packaging applications, polyvinyl chloride (PVC) is widely used because of its several advantages, such as excellent cost/performance ratio, barrier properties, and easy printability. For this purpose, it has been modified by the addition of several petroleum-based and bio-based plasticizers. Epoxidized soybean oil (ESO) is used as a plasticizer to improve the degradability (thermal stability) and processability of PVC. The effect of ESO on the thermal stability of PVC was investigated by Bueno-Ferrer *et al.*[22] They observed that with the increase of ESO, the thermal stability of PVC also increases. Additionally, ESO offers other benefits, including low migration and low viscosity during the processing of food packaging materials. The combined effect of tri-ethylene glycol ester of gum rosin (TEGR) and epoxidized linseed oil (ELO) on PVC was also reported by Arrieta *et al.*[23] They claimed that ELO caused a reduction in mechanical performance but exerted a plasticization effect and thermal stability on PVC. On the other hand, the elastic modulus was increased with the addition of TEGR. Li *et al.* observed that the thermal stability of tung oil-derived epoxidized diester (TOED) as a plasticizer for PVC was high.[24] They also found that the tensile strength of PVC plasticized with TOED was significantly higher than that of other petroleum-based plasticized PVC. Carbonell-Verdu *et al.*[25] evaluated PVC plasticized with epoxidized cotton-seed oil to gain an insight into the effect of time and curing temperature, which demonstrated good plasticizing effects. Waste cooking oil (WCO) was also found to be a promising candidate as a plasticizer, as proclaimed by Suzuki *et al.*[26] Cardanol (phenolic lipid), which is generally acquired by the distillation process of the liquid of cashew nut shell, has also been used as a plasticizer for PVC.[27,28] Two different cardanol-based plasticizers exist, cardanol acetate (CA), which was formed by the esterification process of cardanol oil, and epoxidized cardanol acetate (ECA), which was obtained by epoxidizing the double bonds in cardanol acetate. These two types of cardanol-based plasticizers were evaluated for their miscibility in PVC by Greco *et al.*[28] They claimed that CA offered limited miscibility, while ECA as a secondary plasticizer, along with diethyl hexyl phthalate as a commercially used primary plasticizer, further augmented the performance. Cardanol can also be converted into resins for anti-corrosive applications.[29] They have successfully converted a phenolic lipid polyamine resin *via* the one-pot green co-condensed method and claimed that these coatings are hydrophobic, mechanically robust, nontoxic, environmentally friendly, and chemically and thermally stable, depicting excellent resistance to corrosive ions up to 21 days.[29]

Furthermore, Rahman *et al.*[30] reported the synthesis of novel, self-healed, completely re-processable, three-dimensional, and printable partly bio-based elastomers by cross-linking canola oil-derived diacrylate monomers with partly oxidized silicon-derived copolymers bearing disulfide bonds and free thiol groups. The fabricated canola oil-derived elastomer displayed a great self-healing property with enhanced thermal and restored mechanical properties, which makes it a potential candidate to serve as an eco-friendly alternative to fossil fuel-based elastomers, with a wide range of applications in soft robotics, strain sensors, flexible wearable devices, healthcare, and next-generation energy-harvesting storage devices.[30] Furthermore, the fabrication of lipid-derived biocomposites, nanocomposites, polyesters, copolymers, and monomers after a slight chemical modification of triglycerides has also been reported, making them potential materials in chemical and polymer industries.[19,31–33]

Several nanomedicines have been endorsed by the Food and Drug Administration and the European Medicines Agency[34] for various disease treatments, and among them, lipid-based nanomedicines (LBMs) are found to be the most clinically admissible nano-formulations as they account for approximately 30% (slightly higher) of the commercial nano-drugs.[35] Based on the source of nanocarriers and structures, LBMs can be classified into three subclasses, which include lipid nanoparticles (LNPs), liposomes, and naturally occurring extracellular vehicles (EVs).[36,37] Liposomes are membrane-encircled vesicles, which consist of phospholipids, cholesterol, and other lipids. They were first discovered as a nanomedicine for cancer treatment and hence studied widely in the literature (more than 50%).[38,39] Currently, LNPs are considered at the forefront for the efficient distribution of ribonucleic acid drugs (for cancer therapy and vaccines, *e.g.*, COVID-19); however, EVs are still in trials.[36,40] In other words, based on a historical perspective, liposomes are regarded as the past, LNPs as the present, and EVs as the future.[40]

Furthermore, lipid nanotubes (LNTs) are open hollow cylindrical structures comprising rolling lipid bilayer walls/membranes and have distinctive properties that are different from those of other nanotubes. These properties include the provision of asymmetrical outer and inner surfaces with identical hydrophilic membrane surfaces having the precise control of diameters ranging from several nanometers to 100 nm, which make them best candidates for extensive applications, such as chemical reaction, drug delivery (DD), and selective nanomaterial fillings. The precise molecular recognition, highly ordered structure, and the possibility of substance transport provide a great system for exploring their large applications in medicine and biological fields, such as the detection of DNA, separation of proteins, and drug delivery systems.[41,42] This lipid assembly (liposome) is an outstanding candidate to be a drug container due to its numerous merits, which include high affinity to bio-environments, functionalized modification, controllable size, response to external stimuli, and biocompatibility.[43–45] Lipid particles and spherical lipid carriers have been widely reported in DD[46–49] and its system design.[49,50] Rod-shaped or tubular drug carriers give longer blood circulation time compared to structures of similar size, *e.g.*, spherical.[51] Greater penetrability into tumors shows better drug delivery applications and potential for tubular nanomaterials.[52–54]

Various synthesis methods have been reported in the literature for the formulation of LNTs, such as the micropipette method,[55] rearrangement/recombination,[56–58] microextrusion,[59] electric field method,[60,61] shear by microfluids,[62] and the method using a combination of external force, electric field, and accessory molecules.[63] In addition, incorporating functional groups (desired) into lipid membranes permits manipulation, modification, and fabrication for further studies and applications.[45,64]

In the rearrangement method, the lipids are rearranged into LNTs (diameter: ~500 nm and length: 1–200 μm) by assembling the vesicles followed by reducing their phase transition temperatures.[56] By repeating the above-mentioned cooling procedure, approximately 100% conversion efficiency of LNTs can be achieved.[65] However, it is highly required to have an accurate and uniform temperature control for the cooling process. The solvent change method (recombination) is another method for the formation of LNTs, where a change in solvent in a system having a mixture of water, lipid, and alcohol takes place at a constant temperature. By adjusting the ratio of water and alcohol, various sizes of LNTs can be produced. Rearrangement/recombination is a simple and classic preparation method for LNTs and is used to date but with the limitation of the type of lipids used, *i.e.*, alkyne lipids.[66] Another developed method is the shear flow in which membrane tubes (cylindrical structures) are formed by stretching of lipid membranes.[67–69] Natural lipid-derived LNTs with a hollow structure of several centimeter lengths are produced by Brazhnik *et al.*[62] Syntheses of LNTs are affected by several factors, which include lipid concentration, pH of the aqueous solution, hydration temperature, and shear flow rate. Numerous LNTs are easily produced by this method. Shear force helps in regulating the length and diameter of the LNTs. However, it is difficult to control the precision and uniformity of the products.

Micropipette (optical tweezer) is one of the micromanipulation techniques and another method for LNT preparation. This method needs luring of the local membrane of lipid vesicles with an external force in the range of tens of pico-newtons, which deforms it into tubes by using a micropipette.[70] This method provides a great way to study membrane properties, which include membrane elasticity, rigidity, remodeling, curvature, and tension. However, this method needs high operation requirements for LNT fabrication. Another relatively modern method for LNT synthesis is the induction or actuation of the electric field.[71,72] In the LNT formation process, from the real-time observation, it is observed that the growth of LNTs is aligned with the direction of the electric field. In addition, the essential reason for LNT formation is the distribution of a special electric field. Based on the combination of accessory molecules, external force, and electric field, some new methods are also reported in the literature.[63] Additionally, the modification of LNTs by different materials (proteins, peptides, metals, and nanoparticles) also enhances their potential applications. The LNTs obtained from these modifications can make more attractive shapes, structures, and functions.

Additionally, polymers are also used to modify and upgrade lipids to improve the stability, functionality, and dispersibility of LNTs. The most used polymer is polyethylene glycol. The polymer-based method gives a quick, simple, and chiral separation (on-site) in medical and biological fields. Silicon deposition on lipids shows a lot of potential applications in the fields of medicine, agriculture, green industry, and environment due to their excellent biocompatibility. LNTs consist of several hollow internal spaces by which they can hold nanoparticles (NPs) (well-defined shapes and controllable size dimensions), which help to fabricate the unique nanomaterials (metal nano-composites) or different conjugates that are beneficial in biomedicine and other important fields.[73,74]

Modification of LNTs has also been carried out by incorporating amino acids[75] and other biomolecules, including DNA,[76] drugs,[77] and proteins,[78] to enhance their unique properties. LNTs, prepared using amino acid-based lipids with hydrophobic channels, could serve as reactors for the formation of imine polymers.[75] Modification of LNTs by metals is one of the earliest developed modification technologies compared with other modification methods, which mainly focused on the surface. The thickness of the metal-coated layer on the outer and inner surfaces of the LNTs can be adjusted by plating time and plating solution concentration. Thin-layer deposition of copper depends on the surface of LNTs and can be used in several potential applications, including biosensors and microelectronic components.[57]

Functionalized LNTs were fabricated for gene delivery *via* the co-assembly process of arginine–lipid, bipolar glycolipid, and polyethylene glycol (PEG)–lipid by Ding *et al.*[76] They observed that the aforementioned LNTs effectively formed complexes with plasmid DNA; as a result, they protected DNA from enzymatic degradation and improved dispersibility. Silver ion and LNT complexes show great antimicrobial activities toward broad-spectrum bacteria. They prevent the propagation of microorganisms by interrelating with electron-donating groups, which include hydroxyls, carboxylates, thiols, and amides.[79] Composites made from biocompatible lipids along with Ag having antibacterial properties (Ag–LNTs) can be used in implants, catheters, medical devices, dental materials, and other clinical treatments.[41,80] The antimicrobial activity of the Ag–LNTs exhibits a bacterial inhibition ratio of over 90% on the *E. coli* bacteria.[81]

The fabrication of lipid-stabilized microbubbles (LSMs) along with their functions (drug loading and long circulation) and therapeutic applications, including sonoporation for tumor-targeted therapy, immunotherapy, and brain-targeted therapy, has also been discussed.[82] LSMs can be synthesized by different methods. In one method, these microbubbles are produced by dispersing lipids with perfluoropropane (at a molar ratio of 94 : 6) in a vial (5 ml) filled, clinched, and supercharged with additional perfluoropropane to pressurize up to 2 atm, followed by sonication for a few minutes in a sonicator (bath-type).[83,84] In an alternate method, the vials containing lipids are shaken with high intensity in the presence of a poorly soluble gas (acts as a facilitator), which then homogenizes the gas and liquid to form microbubbles. Microbubbles are unstable in nature, and they can be preserved or synthesized before use in a suitable manner, *e.g.*, freeze-drying. The physicochemical properties of the microbubbles are mainly dependent on the lipid covering the bubbles and the gas inside the bubbles. In ultrasound applications, microbubbles generally contain a shell of lipids, proteins, surfactants, or polymers and a poorly water-soluble gas.[82]

1.2 Protein-based Functional Materials and Their Applications

Protein as a versatile biopolymer is gaining significant interest in food and non-food industries owing to its remarkable attributes, including biodegradability, biocompatibility, and exceptional functional properties.[85] Beyond their physiological functions in food systems, proteins exhibit potential applications in bio-based nanomaterials, particularly in bioplastics and packaging, encapsulation, adhesives, scaffolds, blends, and biocomposites.[86] The protein sources used for these applications can be broadly categorized into two main groups: animal and plant proteins, with a third emerging group of bacterial proteins as well.[85,86] Plant proteins, notably obtained from agro-industrial by-products of cereals, legumes, oilseeds, nuts, edible seeds, and tubers, offer the dual advantage of being cost-effective and sustainable.[87] Among animal proteins, whey, casein, albumin, keratin, collagen, gelatin, and silk are the most commonly used proteins, each with its unique set of properties and applications.[86]

The rich diversity of proteins is rooted in the molecular structure, characterized by the presence of 20 distinct amino acids.[85] Figure 1.3 shows the structural diversity of the key amino acids. The structural diversity of proteins relies on their diverse functional groups, ranging from anionic to cationic and polar to non-polar. It forms the basis for their functional properties,[88] including gelling, water holding, oil holding, foaming, emulsification, thickening, antimicrobial, and antioxidant functions.[85] Also, the huge diversity of functional groups allows the interaction with diverse compounds and other biopolymers. As a building block, amino acids contribute to the formation of peptide chains and make four protein structural levels: primary, secondary, tertiary, and quaternary. This structural complexity creates a web of interactions, including peptide bonds, hydrogen bonds, electrostatic interactions, hydrophobic interactions, and disulfide bonds. In addition, they define the properties of proteins and underpin their remarkable adaptability for diverse material applications.[2]

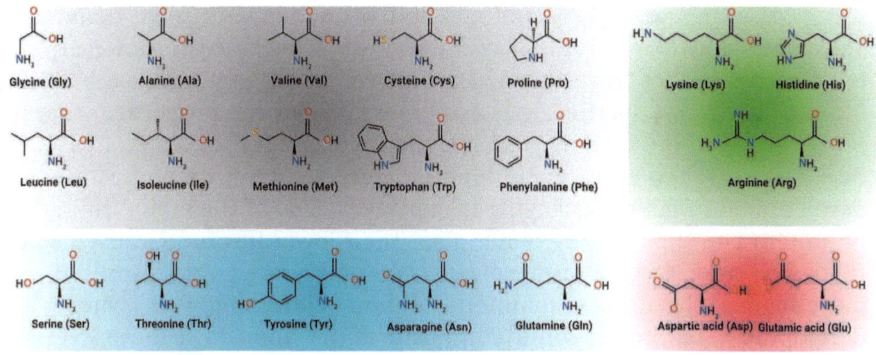

Figure 1.3 Chemical structures of amino acids: the structural diversity. Created with BioRender.com.

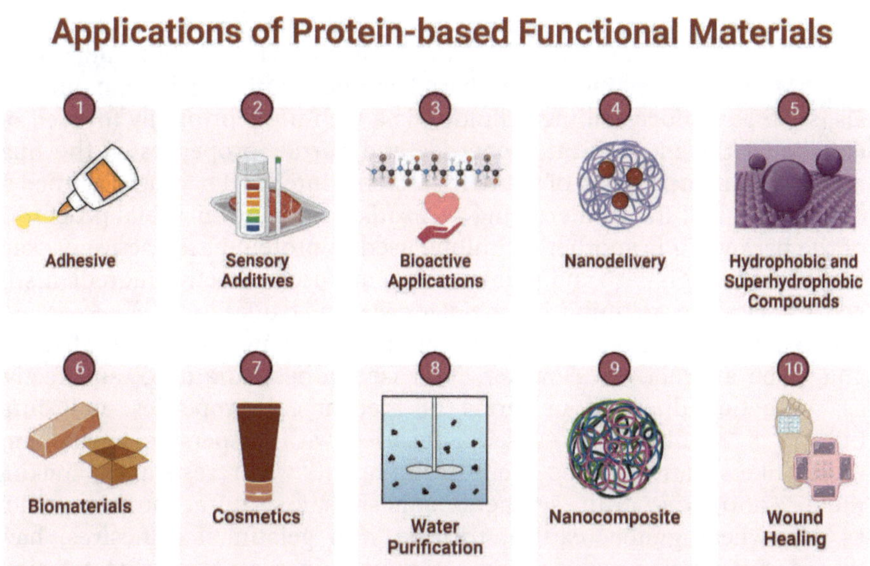

Figure 1.4 Applications of protein-based functional materials. Created with BioRender.com.

These functional material applications could include packaging, adhesive, nanodelivery, wound healing, bioactive applications, sensory additives, hydrophobic interactions, and composite applications. Figure 1.4 shows some of these major protein-based functional material applications.

The development of functional food packaging materials using biopolymers such as proteins can be introduced as an excellent solution to overcome the environmental pollution caused by synthetic plastics while also ensuring economic sustainability.[89] Proteins extracted from by-products and wastes, including but not limited to canola,[90,91] soy,[92] lentil,[93,94] zein,[95] sesami,[96] gelatin,[97] keratin,[98] and whey,[99] have a huge demand in packaging due to their high availability at a lower cost. However, proteins alone cannot be used to develop packaging materials with desirable mechanical properties, gas and moisture barrier properties, thermal properties, and antioxidant and antimicrobial properties. In that case, the development of composite and blend materials, protein modification, and development of active packaging take place.[100] One of the major limitations of protein-based packaging materials is their lower flexibility caused by the rigid structure. Incorporating plasticizers, such as glycerol, sorbitol, and polyethylene glycol, is a common solution addressing this issue.[101] The development of protein–lipid composite films aims to achieve excellent barrier properties, effectively shielding food from moisture, oxygen, and UV radiation. Several research studies have used diverse lipids and oils in protein-based packaging aimed at improving barrier properties. Some of the examples are whey protein isolate–sunflower oil,[102] canola protein–nanocrystalline cellulose modified by oleic acid,[91] soy protein–flaxseed oil,[103] soy protein–rapeseed oil,[92] and whey protein–almond and walnut oil.[104] Moreover, creating protein-based composite films using nanomaterials, such as nanocrystalline cellulose, is a technique primarily focused on enhancing the mechanical properties and barrier properties of the final material.[90] Incorporation of active ingredients into food packaging aimed at enhancing shelf life by increasing antioxidant and antimicrobial properties of the packaging is another technique used in protein-based active packaging. Essential oils[105,106] and plant extracts are used as active ingredients in active packaging to inhibit microbial growth and oxidation.[107,108]

Bio-based adhesives are gaining attention as environmentally friendly and green alternatives. However, bio-based adhesives are not as effective as synthetic adhesives in terms of mechanical properties and durability.[109–116] Modifying bio-based adhesives with dispersing agents and cross-linkers can enhance their strength and water resistance, making them competitive with synthetic adhesives.[117] Natural proteins, such as soy, wheat gluten, canola, triticale, and gelatin as adhesives, have gained significant attention and recognition in recent years as sustainable and versatile alternatives to traditional synthetic adhesives. Potential modifications of the properties of proteins, such as pH, ionic strength, and temperature, and different modifications make them an excellent raw material for developing adhesives.[109,114,116,118] Proteins tend to have a lower environmental impact, as they decompose more readily and release fewer harmful chemicals during their life cycle.[26] Many protein-based adhesives are inherently biocompatible and nontoxic, making them suitable for medical devices and pharmaceutical applications.[119] In construction, they can be used as eco-friendly alternatives to traditional adhesives and sealants, contributing to greener adhesives.[120] In the food industry, protein-based adhesives are used in packaging materials, enhancing food safety and reducing waste.[121]

Protein-based biomedical materials, such as hydrogels and scaffolds, can be applied in wound dressings and tissue engineering.[122] Hydrogels formulated using natural proteins, such as collagen, gelatin, keratin, elastin, and whey protein, have gained substantial attention as protein sources due to their high availability, closely mimicking the properties of soft tissues, self-assembling under physiological conditions, biocompatibility, and low cost.[123] Collagen, a common protein used in these dressings, is a fundamental component of the extracellular matrix and provides a structural framework for cells to migrate and proliferate. On the other hand, collagen can act as a drug delivery system due to its binding capacity and network-forming capacity, which promote tissue regeneration and wound healing.[124] Another important attribute of hydrogels is the structural durability of tissues. Keratin is an excellent biomaterial for this application due to its high sulfur content and cross-linking capacity.[124] Moreover, flexibility and water permeability are important factors of wound dressings that directly impact the hydrating and dehydrating wound, maintaining desired moisture levels and preventing infections. When the protein source lacks all the desirable properties, hybrid materials are developed to acquire desirable properties.[125]

Protein-based delivery systems/encapsulation involve entrapping bioactive compounds, such as drugs and nutraceuticals, in a protein matrix to protect their bioactivity and bioavailability. Recent studies have developed several delivery systems to deliver phenolic compounds, fatty acids, vitamins, minerals, probiotics, and phytosterols to protect them from chemical and physical barriers.[126] A vast range of proteins, including but not limited to pea, soy, canola, gliadin, silk, gelatin, casein, albumin, collagen, and whey, are used as carrier materials in developing delivery systems.[1,43] The potential reduction in the size of proteins to the nanoscale allows them to transfer through cells *via* endocytosis, and non-cytotoxicity, biodegradability, nutritional value, stability, targeted delivery, and biocompatibility make proteins an excellent material to use in encapsulation.[127,128]

Overall, protein-based functional materials represent a dynamic and innovative field at the intersection of biology, chemistry, and materials science. These materials are derived from natural proteins with specific properties and functions, making them highly versatile and applicable to various industries. They have gained significant attention due to their potential to revolutionize various sectors, including biomedicine, food technology, and environmental sustainability. With the potential to impact various industries, protein-based materials offer a sustainable and versatile solution to many modern challenges. Ongoing research in this field promises to unlock new opportunities and lead to groundbreaking innovations in the future.

References

1. M. Zubair, R. A. Pradhan, M. Arshad and A. Ullah, Recent advances in lipid derived bio-based materials for food packaging applications, *Macromol. Mater. Eng.*, 2021, **306**(7), 2000799.
2. U. Schuchardt, R. Sercheli and R. M. Vargas, Transesterification of vegetable oils: a review, *J. Braz. Chem. Soc.*, 1998, **9**, 199–210.
3. Y. Y. Lee, T. K. Tang and O. M. Lai, Health benefits, enzymatic production, and application of medium- and long-chain triacylglycerol (MLCT) in food industries: a review, *J. Food Sci.*, 2012, **77**(8), R137–R144.
4. T.-C. Kuo and Y. J. Tseng, LipidPedia: a comprehensive lipid knowledgebase, *Bioinformatics*, 2018, **34**(17), 2982–2987.
5. M. Oresic, Metabolomics, a novel tool for studies of nutrition, metabolism and lipid dysfunction, *Nutr., Metab. Cardiovasc. Dis.*, 2009, **19**(11), 816–824.
6. B. Rodrigo Valenzuela and B. Alfonso Valenzuela, Overview About Lipid Structure, in *Lipid Metabolism*, ed. B. Rodrigo Valenzuela, IntechOpen, Rijeka, 2013, ch. 1.
7. W. W. Christie, 13-Phenyltridec-9-enoic and 15-phenylpentadec-9-enoic acids in *Arum maculatum* seed oil, *Eur. J. Lipid Sci. Technol.*, 2003, **105**(12), 779–780.
8. M. Arshad, A. K. Mohanty, R. Van Acker, R. Riddle, J. Todd, H. Khalil and M. Misra, Valorization of camelina oil to biobased materials and biofuels for new industrial uses: a review, *RSC Adv.*, 2022, **12**(42), 27230–27245.
9. X. Zhang, J. Burchell and N. S. Mosier, Enzymatic epoxidation of high oleic soybean oil, *ACS Sustainable Chem. Eng.*, 2018, **6**(7), 8578–8583.
10. L. Daniel, A. R. Ardiyanti, B. Schuur, R. Manurung, A. A. Broekhuis and H. J. Heeres, Synthesis and properties of highly branched *Jatropha curcas* L. oil derivatives, *Eur. J. Lipid Sci. Technol.*, 2011, **113**(1), 18–30.
11. F. Shakirova, A. Shishov and A. Bulatov, Hydrolysis of triglycerides in milk to provide fatty acids as precursors in the formation of deep eutectic solvent for extraction of polycyclic aromatic hydrocarbons, *Talanta*, 2022, **237**, 122968.
12. V. Sharma and P. P. Kundu, Condensation polymers from natural oils, *Prog. Polym. Sci.*, 2008, **33**(12), 1199–1215.
13. Z. Zhang, Q. Yang, H. Chen, K. Chen, X. Lu, P. Ouyang, J. Fu and J. G. Chen, In situ hydrogenation and decarboxylation of oleic acid into heptadecane over a Cu–Ni alloy catalyst using methanol as a hydrogen carrier, *Green Chem.*, 2018, **20**(1), 197–205.
14. A. Ullah and M. Arshad, Remarkably Efficient Microwave-Assisted Cross-Metathesis of Lipids under Solvent-Free Conditions, *ChemSusChem*, 2017, **10**(10), 2167–2174.

15. R. A. Pradhan, M. Arshad and A. Ullah, Solvent-free rapid ethenolysis of fatty esters from spent hen and other lipidic feedstock with high turnover numbers, *J. Ind. Eng. Chem.*, 2020, **84**, 42–45.
16. M. Arshad, S. Saied and A. Ullah, PEG–lipid telechelics incorporating fatty acids from canola oil: synthesis, characterization and solution self-assembly, *RSC Adv.*, 2014, **4**(50), 26439–26446.
17. D. P. Pfister, Y. Xia and R. C. Larock, Recent advances in vegetable oil-based polyurethanes, *ChemSusChem*, 2011, **4**(6), 703–717.
18. H. Mutlu, A. N. Parvulescu, P. C. Bruijnincx, B. M. Weckhuysen and M. A. Meier, On the Polymerization Behavior of Telomers: Metathesis versus Thiol–Ene Chemistry, *Macromolecules*, 2012, **45**(4), 1866–1878.
19. M. Arshad, L. Huang and A. Ullah, Lipid-derived monomer and corresponding bio-based nanocomposites, *Polym. Int.*, 2016, **65**(6), 653–660.
20. M. Arshad, R. A. Pradhan, M. Zubair and A. Ullah, Lipid-derived renewable amphiphilic nanocarriers for drug delivery, biopolymer-based formulations: Biomedical and food applications, in *Biopolymer-Based Formulations*, Elsevier, 2020, pp. 283–310.
21. L. Yu, Z. Deng, L. Liu, W. Zhang and C. Wang, Plant-derived nanovesicles: a novel form of nanomedicine, *Front. Bioeng. Biotechnol.*, 2020, **8**, 584391.
22. C. Bueno-Ferrer, M. Garrigós and A. Jiménez, Characterization and thermal stability of poly (vinyl chloride) plasticized with epoxidized soybean oil for food packaging, *Polym. Degrad. Stab.*, 2010, **95**(11), 2207–2212.
23. M. P. Arrieta, M. Samper, M. Jiménez-López, M. Aldas and J. López, Combined effect of linseed oil and gum rosin as natural additives for PVC, *Ind. Crops Prod.*, 2017, **99**, 196–204.
24. M. Li, S. Li, J. Xia, C. Ding, M. Wang, L. Xu, X. Yang and K. Huang, Tung oil based plasticizer and auxiliary stabilizer for poly (vinyl chloride), *Mater. Des.*, 2017, **122**, 366–375.
25. A. Carbonell-Verdu, D. Garcia-Sanoguera, A. Jordá-Vilaplana, L. Sanchez-Nacher and R. Balart, A new biobased plasticizer for poly (vinyl chloride) based on epoxidized cottonseed oil, *J. Appl. Polym. Sci.*, 2016, **133**(27), 43642.
26. A. H. Suzuki, B. G. Botelho, L. S. Oliveira and A. S. Franca, Sustainable synthesis of epoxidized waste cooking oil and its application as a plasticizer for polyvinyl chloride films, *Eur. Polym. J.*, 2018, **99**, 142–149.
27. M. Bocqué, C. Voirin, V. Lapinte, S. Caillol and J. J. Robin, Petro-based and bio-based plasticizers: Chemical structures to plasticizing properties, *J. Polym. Sci., Part A: Polym. Chem.*, 2016, **54**(1), 11–33.
28. A. Greco, D. Brunetti, G. Renna, G. Mele and A. Maffezzoli, Plasticizer for poly (vinyl chloride) from cardanol as a renewable resource material, *Polym. Degrad. Stab.*, 2010, **95**(11), 2169–2174.

29. S. Masood, S. Khan, A. Ghosal, M. Alam, D. Rana, F. Zafar and N. Nishat, Fabrication of cardanol (a phenolic lipid) based polyamine coatings for anti-corrosive applications, *Prog. Org. Coat.*, 2023, **174**, 107304.
30. S. S. Rahman, M. Arshad, A. Qureshi and A. Ullah, Fabrication of a self-healing, 3D printable, and reprocessable biobased elastomer, *ACS Appl. Mater. Interfaces*, 2020, **12**(46), 51927–51939.
31. M. Arshad, M. Kaur and A. Ullah, Green Biocomposites from Nanoengineered Hybrid Natural Fiber and Biopolymer, *ACS Sustainable Chem. Eng.*, 2016, **4**(3), 1785–1793.
32. L. Jin, K. Geng, M. Arshad, R. Ahmadi and A. Ullah, Synthesis of fully biobased polyesters from plant oil, *ACS Sustainable Chem. Eng.*, 2017, **5**(11), 9793–9801.
33. M. Safder, M. Arshad, F. Temelli and A. Ullah, Bio-composites from spent hen derived lipids grafted on CNC and reinforced with nanoclay, *Carbohydr. Polym.*, 2022, **281**, 119082.
34. A. A. Halwani, Development of pharmaceutical nanomedicines: from the bench to the market, *Pharmaceutics*, 2022, **14**(1), 106.
35. T. T. H. Thi, E. J. Suys, J. S. Lee, D. H. Nguyen, K. D. Park and N. P. Truong, Lipid-based nanoparticles in the clinic and clinical trials: from cancer nanomedicine to COVID-19 vaccines, *Vaccines*, 2021, **9**(4), 359.
36. L. Xu, X. Wang, Y. Liu, G. Yang, R. J. Falconer and C.-X. Zhao, Lipid nanoparticles for drug delivery, *Adv. NanoBiomed Res.*, 2022, **2**(2), 2100109.
37. I. K. Herrmann, M. J. A. Wood and G. Fuhrmann, Extracellular vesicles as a next-generation drug delivery platform, *Nat. Nanotechnol.*, 2021, **16**(7), 748–759.
38. A. D. Bangham, M. M. Standish and J. C. Watkins, Diffusion of univalent ions across the lamellae of swollen phospholipids, *J. Mol. Biol.*, 1965, **13**, 238–252.
39. Y. C. Barenholz, Doxil®—the first FDA-approved nano-drug: Lessons learned,*J. Controlled Release*, 2012, **160**(2), 117–134.
40. L. Cheng and A. F. Hill, Therapeutically harnessing extracellular vesicles, *Nat. Rev. Drug Discovery*, 2022, **21**(5), 379–399.
41. Y. Zhou and T. Shimizu, Lipid nanotubes: a unique template to create diverse one-dimensional nanostructures, *Chem. Mater.*, 2008, **20**(3), 625–633.
42. Y. Wang, J. Zhang, H. Gao, Y. Sun and L. Wang, Lipid nanotubes: Formation and applications, *Colloids Surf., B*, 2022, **212**, 112362.
43. P. Davoodi, L. Y. Lee, Q. Xu, V. Sunil, Y. Sun, S. Soh and C.-H. Wang, Drug delivery systems for programmed and on-demand release, *Adv. Drug Delivery Rev.*, 2018, **132**, 104–138.

44. B. Spargo, R. Cliff, F. Rollwagen and A. Rudolph, Controlled release of transforming growth factor- from lipid-based microcylinders, *J. Microencapsulation*, 1995, **12**(3), 247–254.
45. N. Kameta, S. J. Lee, M. Masuda and T. Shimizu, Biologically responsive, sustainable release from metallo-drug coordinated 1D nanostructures, *J. Mater. Chem. B*, 2013, **1**(3), 276–283.
46. H. Bi and X. Han, Magnetic field triggered drug release from lipid microcapsule containing lipid-coated magnetic nanoparticles, *Chem. Phys. Lett.*, 2018, **706**, 455–460.
47. W. L. Odette, C. D. Hennecker, A. K. Mittermaier and J. Mauzeroll, EDTA-Gradient Loading of Doxorubicin into Ferrocene-Containing Liposomes: Effect of Lipid Composition and Visualization of Triggered Release by Cryo-TEM, *Langmuir*, 2021, **37**(38), 11222–11232.
48. R. Sagar, J. Lou, A. J. Watson and M. D. Best, Zinc triggered release of encapsulated cargo from liposomes via a synthetic lipid switch, *Bioconjugate Chem.*, 2021, **32**(12), 2485–2496.
49. H. Bi, Z. Chen and J. Qiu, Drug release and magneto-calorific analysis of magnetic lipid microcapsules for potential cancer therapeutics, *Des. Monomers Polym.*, 2021, **24**(1), 158–163.
50. L. He, D. Fan, W. Liang, Q. Wang and J. Fang, Matrix metalloproteinase-responsive PEGylated lipid nanoparticles for controlled drug delivery in the treatment of rheumatoid arthritis, *ACS Appl. Bio Mater.*, 2020, **3**(5), 3276–3284.
51. Y. Geng, P. Dalhaimer, S. Cai, R. Tsai, M. Tewari, T. Minko and D. E. Discher, Shape effects of filaments versus spherical particles in flow and drug delivery, *Nat. Nanotechnol.*, 2007, **2**(4), 249–255.
52. T. Shimizu, N. Kameta, W. Ding and M. Masuda, Supramolecular self-assembly into biofunctional soft nanotubes: from bilayers to monolayers, *Langmuir*, 2016, **32**(47), 12242–12264.
53. T. Räthel, H. Mannell, J. Pircher, B. Gleich, U. Pohl and F. Krötz, Magnetic stents retain nanoparticle-bound antirestenotic drugs transported by lipid microbubbles, *Pharm. Res.*, 2012, **29**, 1295–1307.
54. K. L. Lee, L. C. Hubbard, S. Hern, I. Yildiz, M. Gratzl and N. F. Steinmetz, Shape matters: the diffusion rates of TMV rods and CPMV icosahedrons in a spheroid model of extracellular matrix are distinct, *Biomater. Sci.*, 2013, **1**(6), 581–588.
55. E. Evans, H. Bowman, A. Leung, D. Needham and D. Tirrell, Biomembrane templates for nanoscale conduits and networks, *Science*, 1996, **273**(5277), 933–935.
56. P. Yager and P. E. Schoen, Formation of tubules by a polymerizable surfactant, *Mol. Cryst. Liq. Cryst.*, 1984, **106**(3–4), 371–381.
57. J. H. Georger, A. Singh, R. R. Price, J. M. Schnur, P. Yager and P. E. Schoen, Helical and tubular microstructures formed by polymerizable phosphatidylcholines, *J. Am. Chem. Soc.*, 1987, **109**(20), 6169–6175.

58. J. S. Chappell and P. Yager, Formation of mineral microstructures with a high aspect ratio from phospholipid bilayer tubules, *J. Mater. Sci. Lett.*, 1992, **11**, 633–636.
59. H. Frusawa, A. Fukagawa, Y. Ikeda, J. Araki, K. Ito, G. John and T. Shimizu, Aligning a single-lipid nanotube with moderate stiffness, *Angew. Chem., Int. Ed.*, 2003, **42**(1), 72–74.
60. S. E. Ghellab and X. Han, Lipid tubes formation induced by electroosmotic flow, *Chem. Phys. Lett.*, 2018, **706**, 515–519.
61. H. Bi, D. Fu, L. Wang and X. Han, Lipid nanotube formation using space-regulated electric field above interdigitated electrodes, *ACS Nano*, 2014, **8**(4), 3961–3969.
62. K. P. Brazhnik, W. N. Vreeland, J. B. Hutchison, R. Kishore, J. Wells, K. Helmerson and L. E. Locascio, Directed growth of pure phosphatidylcholine nanotubes in microfluidic channels, *Langmuir*, 2005, **21**(23), 10814–10817.
63. Y. Zhao, L. An and J. Fang, Buckling of lipid tubules in shrinking liquid droplets, *Nano Lett.*, 2007, **7**(5), 1360–1363.
64. W. B. Han, R. Kwak, J. Y. Kang and T. S. Kim, Generation of Solvent-Free 3D Lipid Structure Arrays on High Aspect Ratio Si Microwell Substrate, *Adv. Mater. Interfaces*, 2019, **6**(5), 1801554.
65. T. G. Burke, A. S. Rudolph, R. R. Price, J. P. Sheridan, A. W. Dalziel, A. Singh and P. E. Schoen, Differential scanning calorimetric study of the thermotropic phase behavior of a polymerizable, tubule-forming lipid, *Chem. Phys. Lipids*, 1988, **48**(3–4), 215–230.
66. H. Bi, Z. Chen, L. Guo, Y. Zhang, X. Zeng and L. Xu, Fabrication, modification and application of lipid nanotubes, *Chem. Phys. Lipids*, 2022, **248**, 105242.
67. K. Sugihara, M. Chami, I. Derényi, J. n. Vörös and T. Zambelli, Directed self-assembly of lipid nanotubes from inverted hexagonal structures, *ACS Nano*, 2012, **6**(8), 6626–6632.
68. H. Liu, G. D. Bachand, H. Kim, C. C. Hayden, E. A. Abate and D. Y. Sasaki, Lipid nanotube formation from streptavidin-membrane binding, *Langmuir*, 2008, **24**(8), 3686–3689.
69. C. Lor, J. D. Lopes, M. K. Mattson-Hoss, J. Xu and L. S. Hirst, A simple experimental model to investigate force range for membrane nanotube formation, *Front. Mater.*, 2016, **3**, 6.
70. T. Bhatia, J. Agudo-Canalejo, R. Dimova and R. Lipowsky, Membrane nanotubes increase the robustness of giant vesicles, *ACS Nano*, 2018, **12**(5), 4478–4485.
71. K. Prathyusha, I. Pagonabarraga and P. S. Kumar, Modification of lipid membrane compressibility induced by an electric field, *Phys. Rev. E*, 2020, **102**(6), 062413.
72. K. Antonova, V. Vitkova and C. Meyer, Membrane tubulation from giant lipid vesicles in alternating electric fields, *Phys. Rev. E*, January 2016, **93**(1), 012413.

73. M. Tanaka, K. Critchley, T. Matsunaga, S. D. Evans and S. S. Staniland, Fabrication of lipid tubules with embedded quantum dots by membrane tubulation protein, *Small*, 2012, **8**(10), 1590–1595.
74. C. Wang, L. Fan, Z. Wang, H. Liu, Y. Li, S. Yang and Y. Li, Directed assembly of hierarchical CdS nanotube arrays from CdS nanoparticles: enhanced solid state electro-chemiluminescence in H_2O_2 solution, *Adv. Mater.*, 2007, **19**(21), 3677–3681.
75. N. Kameta and W. Ding, Supramolecular nanotube reactors for production of imine polymers with controlled conformation, size, and chirality, *Small*, 2019, **15**(19), 1900682.
76. W. Ding, M. Wada, N. Kameta, H. Minamikawa, T. Shimizu and M. Masuda, Functionalized organic nanotubes as tubular nonviral gene transfer vector, *J. Controlled Release*, 2011, **156**(1), 70–75.
77. Q. Wang, L. He, D. Fan, W. Liang, X. Wang and J. Fang, PLA2-triggered release of drugs from self-assembled lipid tubules for arthritis treatments, *ACS Appl. Bio Mater.*, 2020, **3**(9), 6488–6496.
78. N. Kameta, H. Minamikawa, Y. Someya, H. Yui, M. Masuda and T. Shimizu, Confinement effect of organic nanotubes toward green fluorescent protein (GFP) depending on the inner diameter size, *Chem. - Eur. J.*, 2010, **16**(14), 4217–4223.
79. T. Shimizu, H. Minamikawa, M. Kogiso, M. Aoyagi, N. Kameta, W. Ding and M. Masuda, Self-organized nanotube materials and their application in bioengineering, *Polym. J.*, 2014, **46**(12), 831–858.
80. S. Eckhardt, P. S. Brunetto, J. Gagnon, M. Priebe, B. Giese and K. M. Fromm, Nanobio silver: its interactions with peptides and bacteria, and its uses in medicine, *Chem. Rev.*, 2013, **113**(7), 4708–4754.
81. Y. Zhou, M. Kogiso, M. Asakawa, S. Dong, R. Kiyama and T. Shimizu, Antimicrobial Nanotubes Consisting of Ag-Embedded Peptidic Lipid-Bilayer Membranes as Delivery Vehicles, *Adv. Mater.*, 2009, **21**(17), 1742–1745.
82. D. Omata, J. Unga, R. Suzuki and K. Maruyama, Lipid-based microbubbles and ultrasound for therapeutic application, *Adv. Drug Delivery Rev.*, 2020, **154**, 236–244.
83. R. Suzuki, T. Takizawa, Y. Negishi, N. Utoguchi and K. Maruyama, Effective gene delivery with novel liposomal bubbles and ultrasonic destruction technology, *Int. J. Pharm.*, 2008, **354**(1–2), 49–55.
84. Y. Endo-Takahashi, Y. Negishi, Y. Kato, R. Suzuki, K. Maruyama and Y. Aramaki, Efficient siRNA delivery using novel siRNA-loaded Bubble liposomes and ultrasound, *Int. J. Pharm.*, 2012, **422**(1–2), 504–509.
85. T. Dissanayake, X. Sun, L. Abbey and N. Bandara, Recent advances in lipid-protein conjugate-based delivery systems in nutraceutical, drug, and gene delivery, *Food Hydrocolloids for Health*, 2022, **2**, 100054.

86. N. H. C. S. Silva, C. Vilela, I. M. Marrucho, C. S. R. Freire, C. P. Neto and A. J. D. Silvestre, Protein-based materials: from sources to innovative sustainable materials for biomedical applications, *J. Mater. Chem. B*, 2014, **2**(24), 3715–3740.
87. M. Nikbakht Nasrabadi, A. Sedaghat Doost and R. Mezzenga, Modification approaches of plant-based proteins to improve their techno-functionality and use in food products, *Food Hydrocolloids*, 2021, **118**, 106789.
88. C. Li, L. Chen, D. J. McClements, X. Peng, Z. Xu, M. Meng, H. Ji, C. Qiu, J. Long and Z. Jin, Encapsulation of polyphenols in protein-based nanoparticles: Preparation, properties, and applications, *Crit. Rev. Food Sci. Nutr.*, 2023, **0**, 1–15.
89. H. Chen, J. Wang, Y. Cheng, C. Wang, H. Liu, H. Bian, Y. Pan, J. Sun and W. Han, Application of Protein-Based Films and Coatings for Food Packaging: A Review, *Polymers*, 2019, **11**(12), 2039.
90. T. Dissanayake, B. Peng Chang, T. H. Mekonnen, C. Senaka Ranadheera, C. Narvaez-Bravo and N. Bandara, Reinforcing canola protein matrix with chemically tailored nanocrystalline cellulose improves the functionality of canola protein-based packaging materials, *Food. Chem.*, 2022, **383**, 132618.
91. T. Dissanayake, B. M. Trinh, T. H. Mekonnen, P. Sarkar, R. E. Aluko and N. Bandara, Improving properties of canola protein-based nanocomposite films by hydrophobically modified nanocrystalline cellulose, *Food Packag. Shelf Life*, 2023, **35**, 101018.
92. S. Galus, Functional properties of soy protein isolate edible films as affected by rapeseed oil concentration, *Food Hydrocolloids*, 2018, **85**, 233–241.
93. E. D. d. Apodaca, A. Montanari, L. F. -d. Castro, E. Umilta, L. Arroyo, C. Zurlini and M. C. Villaran, Lentil by-products as a source of protein for food packaging applications, *Am. J. Food Technol.*, 2020, **15**(1), 1–10.
94. H. Nadeeshani, T. Dissanayake and N. Bandara, Nonfood Applications of Lentils and Their Processing By-products, in *Lentils*, John Wiley & Sons, Ltd, 2023, pp. 261–278.
95. H. Cui, D. Surendhiran, C. Li and L. Lin, Biodegradable zein active film containing chitosan nanoparticle encapsulated with pomegranate peel extract for food packaging, *Food Packag. Shelf Life*, 2020, **24**, 100511.
96. L. Sharma and C. Singh, Sesame protein based edible films: Development and characterization, *Food Hydrocolloids*, 2016, **61**, 139–147.
97. Z. A. M. Adilah and Z. A. N. Hanani, Active packaging of fish gelatin films with Morinda citrifolia oil, *Food Biosci.*, 2016, **16**, 66–71.
98. D. O. S. Ramirez, R. A. Carletto, C. Tonetti, F. T. Giachet, A. Varesano and C. Vineis, Wool keratin film plasticized by citric acid for food packaging, *Food Packag. Shelf Life*, 2017, **12**, 100–106.

99. S. Bahram, M. Rezaei, M. Soltani, A. Kamali, S. M. Ojagh and M. Abdollahi, Whey protein concentrate edible film activated with cinnamon essential oil,*J. Food Process. Preserv.*, 2014, **38**(3), 1251–1258.
100. S. J. Calva-Estrada, M. Jiménez-Fernández and E. Lugo-Cervantes, Protein-based films: Advances in the development of biomaterials aplicable to food packaging, *Food Eng. Rev.*, 2019, **11**(2), 78–92.
101. C. Chang and M. T. Nickerson, Effect of plasticizer-type and genipin on the mechanical, optical, and water vapor barrier properties of canola protein isolate-based edible films, *Eur. Food Res. Technol.*, 2014, **238**(1), 35–46.
102. B. G. Erdem and S. Kaya, Edible film fabrication modified by freeze drying from whey protein isolate and sunflower oil: Functional property evaluation, *Food Packag. Shelf Life*, 2022, **33**, 100887.
103. E. J. Hopkins, C. Chang, R. S. H. Lam and M. T. Nickerson, Effects of flaxseed oil concentration on the performance of a soy protein isolate-based emulsion-type film, *Food Res. Int.*, 2015, **67**, 418–425.
104. S. Galus and J. Kadzińska, Whey protein edible films modified with almond and walnut oils, *Food Hydrocolloids*, 2016, **52**, 78–86.
105. S. P. L. F. Oliveira, L. C. Bertan, C. M. V. B.D. Rensis, A. P. Bilck and P. C. B. Vianna, Whey protein-based films incorporated with oregano essential oil, *Polímeros*, 2017, **27**(2), 158–164.
106. N. Sanla-Ead, A. Jangchud, V. Chonhenchob and P. Suppakul, Antimicrobial activity of cinnamaldehyde and eugenol and their activity after incorporation into cellulose-based packaging films, *Packag. Technol. Sci.*, 2012, **25**(1), 7–17.
107. A. N. Adilah, B. Jamilah, M. A. Noranizan and Z. A. N. Hanani, Utilization of mango peel extracts on the biodegradable films for active packaging, *Food Packag. Shelf Life*, 2018, **16**, 1–7.
108. M. C. Condés, M. C. Añón, A. N. Mauri and A. Dufresne, Amaranth protein films reinforced with maize starch nanocrystals, *Food Hydrocolloids*, 2015, **47**, 146–157.
109. N. Bandara, A. Akbari, Y. Esparza and J. Wu, Canola Protein: A Promising Protein Source for Delivery, Adhesive, and Material Applications, *J. Am. Oil Chem. Soc.*, 2018, **95**(8), 1075–1090.
110. N. Bandara and J. Wu, Chemically Modified Canola Protein–Nanomaterial Hybrid Adhesive Shows Improved Adhesion and Water Resistance, *ACS Sustainable Chem. Eng.*, 2018, **6**(1), 1152–1161.
111. N. P. Bandara, in *Nanoengineered and Biomimetic Protein-derived Adhesives with Improved Adhesion Strength and Water Resistance*, ERA, 2017.
112. N. Bandara and J. Wu, Randomly Oriented Strand Board Composites from Nanoengineered Protein-Based Wood Adhesive, *ACS Sustainable Chem. Eng.*, 2018, **6**(1), 457–466.
113. N. Bandara, Y. Esparza and J. Wu, Graphite Oxide Improves Adhesion and Water Resistance of Canola Protein-Graphite Oxide Hybrid Adhesive, *Sci. Rep.*, 2017, **7**(1), 11538.

114. N. Bandara, Y. Esparza and J. Wu, Exfoliating nanomaterials in canola protein derived adhesive improves strength and water resistance, *RSC Adv.*, 2017, **7**(11), 6743–6752.
115. N. Bandara, H. Zeng and J. Wu, Marine mussel adhesion: biochemistry, mechanisms, and biomimetics, *J. Adhes. Sci. Technol.*, 2013, **27**(18–19), 2139–2162.
116. N. Bandara, L. Chen and J. Wu, Adhesive properties of modified triticale distillers grain proteins, *Int. J. Adhes. Adhes.*, 2013, **44**, 122–129.
117. M. N. Islam, F. Rahman, A. K. Das and S. Hiziroglu, An overview of different types and potential of bio-based adhesives used for wood products, *Int. J. Adhes. Adhes.*, 2022, **112**, 102992.
118. N. Bandara, L. Chen and J. Wu, Protein Extraction from Triticale Distillers Grains, *Cereal Chem.*, 2011, **88**(6), 553–559.
119. D. W. R. Balkenende, S. M. Winkler and P. B. Messersmith, Marine-inspired polymers in medical adhesion, *Eur. Polym. J.*, 2019, **116**, 134–143.
120. S. Khosravi, F. Khabbaz, P. Nordqvist and M. Johansson, Protein-based adhesives for particleboards, *Ind. Crops Prod.*, 2010, **32**(3), 275–283.
121. W. Y. Choi, C. M. Lee and H. J. Park, Development of biodegradable hot-melt adhesive based on poly-ε-caprolactone and soy protein isolate for food packaging system, *LWT--Food Sci. Technol.*, 2006, **39**(6), 591–597.
122. P. Katyal, F. Mahmoudinobar and J. K. Montclare, Recent trends in peptide and protein-based hydrogels, *Curr. Opin. Struct. Biol.*, 2020, **63**, 97–105.
123. S. Kapoor and S. C. Kundu, Silk protein-based hydrogels: Promising advanced materials for biomedical applications, *Acta Biomater.*, 2016, **31**, 17–32.
124. T. A. Debele and W.-P. Su, Polysaccharide and protein-based functional wound dressing materials and applications, *Int. J. Polym. Mater. Polym. Biomater.*, 2022, **71**(2), 87–108.
125. S. P. Ndlovu, K. Ngece, S. Alven and B. A. Aderibigbe, Gelatin-Based Hybrid Scaffolds: Promising Wound Dressings, *Polymers*, 2021, **13**(17), 2959.
126. V. Đorđević, B. Balanč, A. Belščak-Cvitanović, S. Lević, K. Trifković, A. Kalušević, I. Kostić, D. Komes, B. Bugarski and V. Nedović, Trends in Encapsulation Technologies for Delivery of Food Bioactive Compounds, *Food Eng. Rev.*, 2015, **7**(4), 452–490.
127. S. Hong, D. W. Choi, H. N. Kim, C. G. Park, W. Lee and H. H. Park, Protein-based nanoparticles as drug delivery systems, *Pharmaceutics*, 2020, **12**(7), 604.
128. B. Hu, X. Liu, C. Zhang and X. Zeng, Food macromolecule based nanodelivery systems for enhancing the bioavailability of polyphenols, *J. Food Drug Anal.*, 2017, **25**(1), 3–15.

CHAPTER 2

Recent Advances in the Biosynthesis and Metabolic Engineering of Storage Lipids and Proteins in Seeds

KALLUM MCDONALD[a], STACY D. SINGER[b] AND GUANQUN CHEN*[a]

[a]Department of Agricultural, Food and Nutritional Science, Faculty of Life, Environmental, and Nutritional Science, University of Alberta, Edmonton, Alberta, T6G 2P5, Canada; [b]Agriculture & Agri-Food Canada, Lethbridge Research and Development Centre, Lethbridge, Alberta, T1J 4B1, Canada
*E-mail: gc24@ualberta.ca

2.1 Introduction

During the life cycle of terrestrial plants, a set of developmental processes occur in the embryo, and storage molecules accumulate to nourish the seedling during germination and into the early stages of growth. The biosynthesis of storage molecules, such as triacylglycerol (TAGs), which accrue in oil bodies (OBs), and seed storage proteins (SSPs), which accumulate in protein storage vacuoles, is a fundamental process in numerous plants and consequently has been the subject of intense study for several decades (for a review, see ref. 1–4). Beyond their importance in sustaining seedling growth, storage lipids and proteins have valuable and diverse

Smart Materials No. 41
Functional Materials from Lipids and Proteins
Edited by Nandika Bandara and Aman Ullah

Published by the Royal Society of Chemistry, www.rsc.org

applications in human and animal food products, as well as biodiesel and chemical industries.[5–8] The consumption of vegetable oils has driven the steady growth in the cultivation of oilseed crops.[9] Plants can also produce unusual fatty acids (FAs) with important nutritional and industrial applications.[8,10,11,12] The plant-based protein industry has also undergone rapid growth in recent years and is projected to increase further in the coming decades, with substantial economic potential through applications in human and animal diets, pharmaceuticals, industrial uses, and many more.[2,13,14] To fulfill these growing needs, more research is needed to understand the pathways and regulatory mechanisms involved in storage protein and lipid biosynthesis and to develop novel strategies for increasing the accumulation of these valuable storage compounds in seeds.

This chapter summarizes the biosynthesis of storage lipids (primarily TAGs) and SSPs in seeds and outlines the recent advances in the characterization of these pathways, as well as metabolic engineering strategies to increase their accumulation. The unusual FA production in plants is not discussed in this chapter; readers interested in this topic may choose to consult the papers in ref. 11,15. This information will facilitate the study of TAG and SSP biosynthesis, the development of lipid- and protein-enhanced crops, and the generation of lipid- and protein-based functional materials to fulfill the aforementioned societal, environmental, and economic demands.

2.2 *De Novo* Fatty Acid Biosynthesis and Acyl Editing

The major biosynthetic processes involved in the accumulation of seed storage oil are FA synthesis and TAG assembly (depicted in Figure 2.1). Seed FAs are first synthesized in plastids and then transported to the endoplasmic reticulum (ER) for acyl editing and TAG synthesis.[16,17] Additionally, regulatory factors often play diverse and important roles in mediating storage lipid biosynthesis, especially transcription factors (TFs).[18]

The *de novo* synthesis of FAs involves the consumption of the carbon substrate acetyl-CoA, which is the building block of acyl chains. To supply the plastid with carbon substrates for FA synthesis, pyruvate transporters, such as bile acid:sodium symporter family protein 2 (BASS2), facilitate the entry of pyruvate into the plastid.[19] Once imported, FA synthesis begins through the conversion of pyruvate to acetyl-CoA by the pyruvate dehydrogenase complex (PDC).[20] Acetyl-CoA then acts as the substrate for acetyl-CoA carboxylase (ACCase), which catalyzes the conversion of acetyl-CoA into malonyl-CoA.[21,22] This is considered the first step in FA synthesis because this reaction commits acetyl-CoA to acyl chain assembly.[17]

Subsequently, acyl-CoA molecules are joined to acyl carrier proteins (ACPs) for further extension.[23] From this point, the acyl chain is assembled through a series of reactions catalyzed by ketoacyl synthases (KAS family members) in a process known as condensation, which elongates

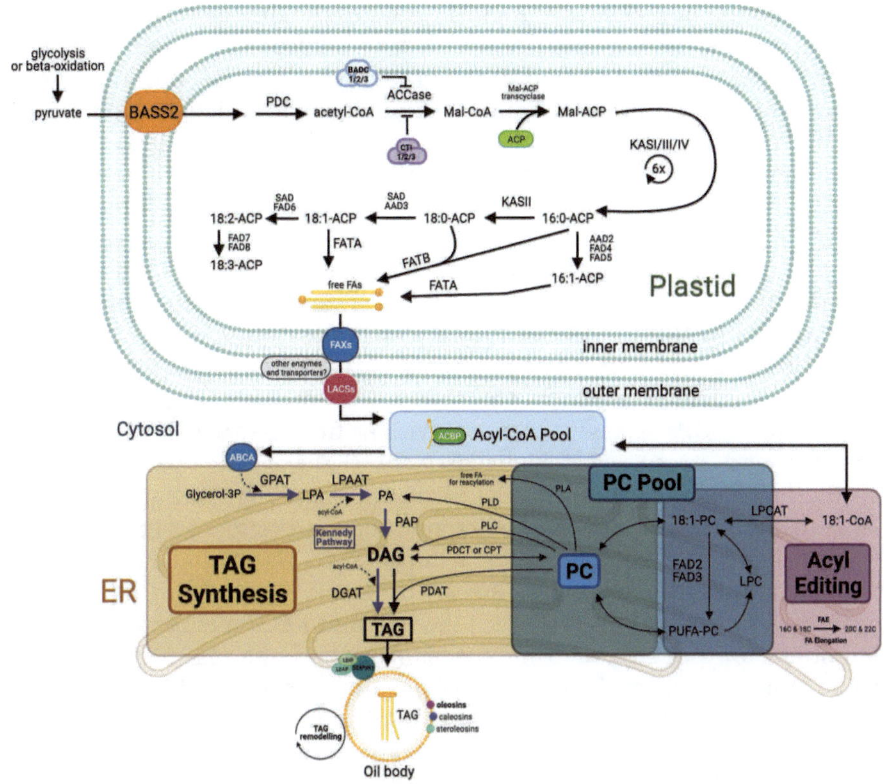

Figure 2.1 Overview of FA and TAG biosynthesis pathways in plant seeds. The fundamental proteins and chemicals involved in seed lipid biosynthesis, as well as recent discoveries and other key members with the potential for seed oil engineering, are included here. This figure is not a comprehensive rendering of all lipid biosynthesis enzymes and pathways. This figure is made using Biorender.com. BASS2, bile acid:sodium symporter family protein 2; PDC, pyruvate dehydrogenase complex; BADC (1/2/3), biotin/lipoyl attachment domain containing (1/2/3); CTI (1/2/3), carboxyltransferase interactor (1/2/3); ACCase, acetyl-CoA carboxylase; Mal, malonyl; ACP, acyl carrier protein; KAS, ketoacyl synthase; AAD (2/3), acyl-acyl carrier protein desaturase (2/3); FAD (4/5/6/7/8), fatty acid desaturase (4/5/6/7/8); FAT (A/B), fatty acid thioesterase (A/B); SAD, stearoyl-acyl carrier protein desaturase; FA, fatty acid; FAX, fatty acid exporter; LACS, long-chain acyl-CoA synthetase; ACBP, acyl-CoA-binding protein; ABCA, ATP-binding cassette A; GPAT, glycerol-3-phosphate acyltransferase; LPA, lysophosphatidic acid; LPAAT, lysophosphatidic acid acyltransferase; PA, phosphatidic acid; PAP, phosphatidic acid phosphatase; DAG, diacylglycerol; DGAT, diacylglycerol acyltransferase; TAG, triacylglycerol; LDAP, lipid droplet-associated protein; LDIP, LDAP-interacting protein; PDAT, phospholipid:diacylglycerol acyltransferase; PDCT, phosphatidylcholine:diacylglycerol cholinephosphotransferase; CPT, CDP-choline:diacylglycerol cholinephosphotransferase; PL (A/B/C), phospholipase (A/B/C); PC, phosphatidylcholine; LPCAT, lysophosphatidylcholine acyltransferase; LPC, lysophosphatidylcholine; PUFA, polyunsaturated fatty acid; FAE, fatty acid elongase.

the ACP-bound acyl chain.[24,25] KAS III catalyzes the first elongation reaction between malonyl-ACP and acetyl-CoA to produce a 4-carbon acyl-ACP, which is extended by 2-carbon subunits *via* KAS I catalysis until a 16:0-ACP is assembled.[23,26] A cycle of condensation, reduction, dehydration, and a second reduction takes place to add 2-carbon increments. This 16:0-ACP can either be lysed by acyl-ACP thioesterase B (fatty acid thioesterase B [FATB]), liberating 16:0 from the ACP, or be catalyzed by KAS II to produce 18:0-ACP.[27] The 18:0-ACP can be cleaved by FATB to produce free 18:0 or the acyl chain can be converted into 18:1-ACP, which is catalyzed by stearoyl-CoA-desaturase (SAD).[27,28] Alternatively, 16:0- and 18:0-ACP can undergo desaturation to 16:1- or 18:1-ACP by acyl-ACP desaturases (AADs).[28–30] Acyl-ACP thioesterase A (fatty acid thioesterase A [FATA]) can then hydrolyze the 18:1-ACP to yield free 18:1.[27] The thioesterase-mediated cleavage of acyl chains from ACPs is essential for providing free FAs, which can then be shuttled and processed for downstream pathways, in particular TAG synthesis. Fatty acid exporter (FAX) members, including FAX1 and FAX2, are FA transporters that provide an important contribution to the export of lipids from the plastid and positively impact seed oil accumulation.[31–34] The free FAs are then metabolically activated through linkage to CoA by long-chain acyl-CoA synthetases (LACSs).[35,36] FAX and LACS members likely play important roles in acyl shuttling mechanisms through plastid membranes and enable the accumulation of acyl-CoAs in the cytosol, which is a necessary precondition for acyl editing and TAG synthesis.[34] Multiple FAX homologs have been reported in the model plant *Arabidopsis*, and their precise functions require further exploration. Studies of LACS homologs from a diversity of plant species have revealed that these enzymes are essential for acyl chain activation and acyl trafficking between organelles (*e.g.* from plastid to ER) and also play a role in seed oil accumulation and FA composition.[37–41] Higher plants generally possess several LACS homologs exhibiting certain levels of functional redundancy, as well as a diversity of effects on lipid metabolism, and their precise functions in seed oil accumulation and FA composition remain to be investigated further.[39,42,43]

The synthesized acyl-CoAs persist in the cytosol and can be used for the biosynthesis of acyl lipids in the ER. Acyl-CoA-binding proteins (ACBPs) bind to acyl-CoAs and enable the maintenance of an acyl-CoA pool, thus playing an intermediary role in lipid biosynthesis.[44,45] The transportation of acyl-CoA to the ER for acyl editing and TAG assembly may be facilitated by acyl transporters.[45] ABCA9 has been identified as one such transporter, and its activity contributes substantially to seed oil accumulation, although other unknown ER-localized acyl-CoA importers may also contribute to the transport process.[46,47]

Acyl chains can further be modified (also known as acyl editing) to increase their diversity, and subsequently the FA profile of the seed oil, with desaturation and elongation being the two main modification reactions.[48] Desaturation occurs on phosphatidylcholine (PC), and the synthesis of

18:2-PC and 18:3-PC from 18:1-PC catalyzed by fatty acid desaturase 2 and 3 (FAD2 and FAD3), respectively, are two common reactions in plant seeds.[49,50] The Lands cycle comprises de-acylation and re-acylation between lysophosphatidylcholine (LPC) and PC, catalyzed by lysophosphatidyl acyltransferase (LPCAT) and phospholipase A and acyl-CoA synthase, and plays an important role in the exchange of acyl-chains between the PC pool and acyl-CoA pool. Elongation catalyzed by fatty acid elongase (FAE) in the ER is the key step in producing very-long-chain fatty acids (20C or more), which includes a sequence of condensation, reduction, dehydration, and a second reduction, much as with fatty acid synthase in the plastid.[51] However, the fatty acid elongation reaction is different from the *de novo* fatty acid synthesis in several ways, such as the reaction positions (ER *vs.* plastid) and the carrier of the carbon substrate (malonyl-CoA *vs.* malonyl-ACP). While desaturation and elongation are the main derivations of fatty acids, other modification reactions can also take place. These further enzymatic reactions may give rise to the synthesis of unusual fatty acids, such as ricinoleic acid through the hydroxylation of 18:1-PC.[52–54] It should be noted that although the desaturation and elongation of acyl chains in the ER are pivotal for seed oil accumulation, desaturation also occurs in the plastid, catalyzed by FAD4/5/6/7/8, and the activity of these enzymes is important in lipid biosynthesis.[29,55]

2.3 TAG Biosynthesis and Storage

Acyl-CoA can serve as an acyl donor for three acyltransferase reactions producing TAG in the Kennedy pathway.[56,57] Glycerol-3-phosphate acyltransferase (GPAT) catalyzes the first attachment of an acyl chain to the glycerol backbone, thereby converting glycerol-3-phosphate (G3P) into lysophosphatidic acid (LPA) and ultimately promoting oil accumulation.[58–61] The glycerol backbone in LPA is then acylated again by lysophosphatidic acid acyltransferase (LPAAT), yielding phosphatidic acid (PA).[62] The phosphate group is then cleaved from PA through the catalytic activity of phosphatidic acid phosphatase (PAP), which generates diacylglycerol (DAG).[63] Finally, the third and final acylation is catalyzed by diacylglycerol acyltransferase (DGAT), resulting in the production of TAG. As the enzyme catalyzing the last committed step in acyl-CoA-dependent TAG biosynthesis, DGAT has been studied extensively for decades.[64] An alternative TAG biosynthetic pathway involves the enzyme phospholipid:diacylglycerol acyltransferase (PDAT), which transfers an acyl group from PC to DAG to yield TAG.[64–66] The relative contribution of DGAT and PDAT to TAG formation may differ depending on the plant species.[67] The LPCAT-catalyzed regeneration of PC from LPC is important in supplying PDAT with the necessary PC substrates for TAG synthesis.[67] Moreover, the lipid biosynthesis activity of both DGAT and LPCAT is also important for proper plant development, as demonstrated by sterile pollen in *dgat1/lpcat2* double-knockout mutants.[68]

Other reactions also contribute to the Kennedy pathway for TAG biosynthesis. As PC is a major substrate for acyl editing, the entry and exit of fatty acids into PC and crosstalk between the PC pool and DAG/TAG are important in determining FA composition. CDP-choline:DAG cholinephosphotransferase (CPT) and PC:DAG cholinephosphotransferase (PDCT) facilitate the interconversion between DAG and PC.[69–71] This interconversion allows DAG to be derived from the PC pool through the activity of PAP in combination with either phospholipase C (PLC) or phospholipase D (PLD).[72,73] It is well established that these dynamic reactions are important in the biosynthesis of seed oils with various FA profiles in different plant species.

TAGs accumulate in OBs in the cytosol and serve as a major energy source for seedling growth.[74–76] OBs are storage organelles for TAG, consisting of a single-layered phospholipid membrane, and are synthesized at the ER.[74,75,77,78] Several oleosin proteins are found in abundance in OB membranes, contributing to their stability and facilitating the separation of OBs from the ER.[75,79–81] Oleosins are important for proper oil accumulation, and the expression of *OLEOSIN* genes is promoted by several major transcription factors.[82,83] Other OB membrane proteins, such as caleosin (CLO1) and steroleosin (HSD1), also contribute to OB stability and are important for long-term seed oil stability.[75,80] Recent studies have revealed that OB biogenesis also involves other enzymes, such as OB-associated protein (lipid droplet-associated protein [LDAP]), LDAP-interacting protein (LDIP), and SEIPIN proteins in *Arabidopsis*, and the mechanisms underlying the OB formation may involve different sets of OB-packaging proteins in different species.[84–87] However, numerous *OLEOSIN* genes remain uncharacterized, and hence, further research is warranted.[88]

2.4 Regulation of Lipid Biosynthesis

During seed development and the accumulation of seed oil, the expression of lipid biosynthesis genes is subjected to regulation by a variety of mechanisms (Figure 2.2). Because of its involvement in this crucial step in FA synthesis, ACCase is an important target for regulation. The biotin carboxyl carrier protein (BCCP) subunit of ACCase is a known target for biotin/lipoyl attachment domain-containing (BADC) proteins, which are negative regulators of ACCase activity.[89,90] ACCase activity can also be regulated by carboxyltransferase interactor (CTI)1/2/3, which are plastid envelope proteins that dock ACCase at the plastid membrane, reducing its activity.[91] The flow of carbon into FA synthesis is thereby reduced through the inhibition of ACCase mediated by CTIs. However, the knockout of *CTI* genes in *Arabidopsis* did not affect the quantity of seed oil, which may imply that other unidentified regulatory factors are involved and warrants further research.[91]

Regulating the expression of lipid-related genes is necessary to coordinate the development of the seed as well as the synthesis and sequestration

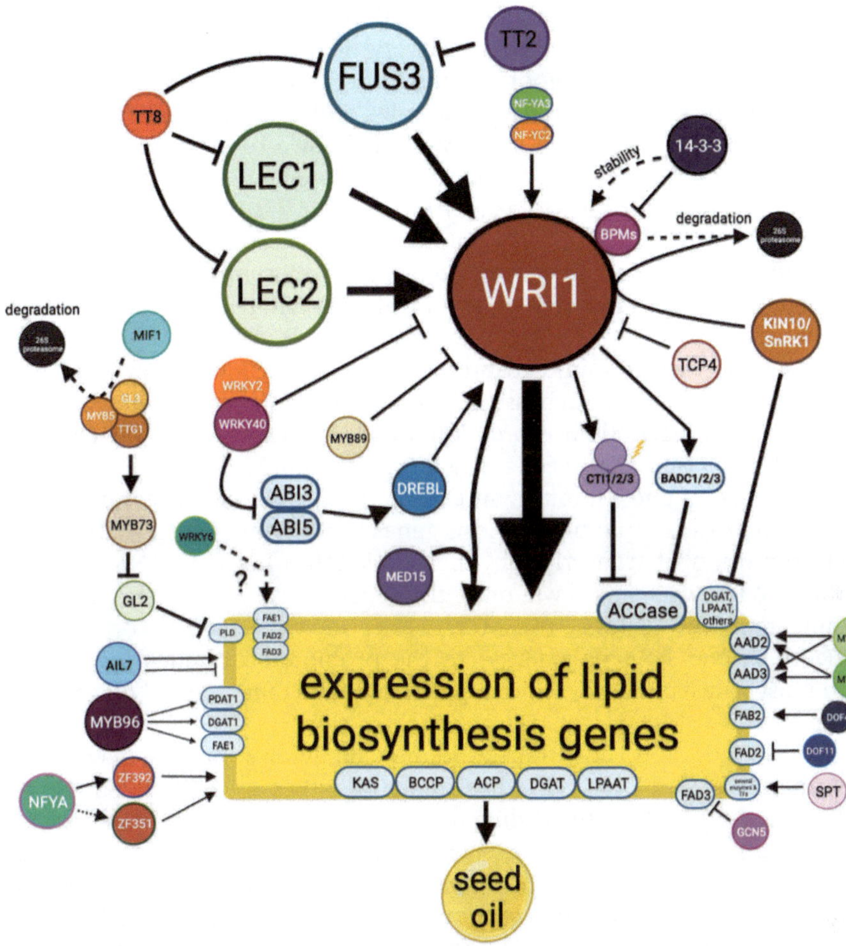

Figure 2.2 Overview of lipid biosynthesis regulation in plant seeds. The fundamental regulatory and transcription factors involved in seed lipid biosynthesis, as well as recent discoveries and other key members with the potential for seed oil engineering are included here. This figure is not a comprehensive rendering of all regulators and transcription factors in lipid biosynthesis. This figure is made using Biorender.com. TT (2/8), transparent testa (2/8); LEC (1/2), leafy cotyledon (1/2); FUS3, FUSCA3; NF-Y (A/A3/C2), nuclear factor of the Y box (A/A3/C2); BPM, BTB/POZMATH 1; SnRK1, sucrose non-fermenting-1-related kinase 1; TCP4, TCP family transcription factor 4; BADC (1/2/3), biotin/lipoyl attachment domain containing (1/2/3); CTI (1/2/3), carboxyltransferase interactor (1/2/3); DREBL, dehydration-responsive element-binding protein L; MED15, MEDIATOR 15; MYB (5/73/89/96/115/118), MYB-family domain protein (5/73/89/96/115/118); WRKY (2/6/40), WRKY-family transcription factor; ABI, abscisic acid insensitive; MIF1, mini zinc finger 1; TTG1, transparent testa glabra 1; GL (2/3), glabra (2/3); AIL7, aintegumenta-like 7; DOF (4/11), DNA binding with one finger (4/11); SPT, SPATULA; GCN5, general control nonderepressible 5; PLD, phospholipase D; FAE1, fatty acid elongase 1; FAD (2/3), fatty acid

of TAGs for storage. TFs have manifold influential roles in mediating lipid biosynthesis in seeds. Since the discovery of the *WRINKLED1* (*WRI1*) TF, which is one of the most important regulatory factors in lipid biosynthesis, there has been extensive study of the roles and interactions of TFs in the regulation of lipid biosynthesis.[18,92] WRI1 is known to promote the expression of more than twenty lipid biosynthesis-related genes, including *BCCP2* (encoding a subunit of ACCase), *ACP1*, and *KASI*.[89,93] Loss-of-function *wri1* mutants consequently accumulate up to 80% less seed oil, illustrating the substantial importance of WRI in storage lipid biosynthesis.[92] The diverse interactions and regulatory targets involving WRI1 have been recently reviewed.[94]

WRI1 was recently found to promote the expression of negative regulators of FA synthesis, such as *BADC1/2/3*. WRI1 binds directly to *BADC* promoters and drives their expression, providing a fine-tuning mechanism of FA synthesis that allows cells to achieve lipid homeostasis by promoting *BADC* expression and inhibiting ACCase activity.[89] While the recruitment of RNA polymerase to lipid biosynthesis genes is necessary for the WRI1-mediated promotion of transcription, the relationship between RNA polymerase and WRI1 was unknown until the recent discovery of the mediator complex subunit 15 (MED15). MED15 was shown to act in concert with WRI1 to deliver RNA polymerase to the promoters of genes involved in lipid biosynthesis pathways and to positively contribute to seed oil accumulation.[95]

Several pre- and post-transcriptional regulatory factors have been identified that influence WRI1 itself, including LEC1, LEC2, FUS3, 14-3-3, BMPs, and SnRK1 (KIN10), leading to downstream effects on the expression of oil biosynthesis genes.[18] TFs, such as MYB89 and TCP4, have also been shown to directly repress *WRI1* expression, leading to reduced expression of oil biosynthesis genes and consequently seed oil accumulation.[96,97] There are also post-translational regulatory mechanisms for WRI1. For example, 14-3-3 proteins are regulatory proteins capable of binding to and stabilizing phosphorylated WRI1, thus reducing WRI1 degradation, promoting the expression of lipid biosynthesis genes, and enhancing seed oil synthesis.[98,99] The TF DRBEL, from the DREB family, was also recently identified as a direct positive regulator of major lipid TFs, such as *WRI1, ABI3, and ABI5*, illustrating the complexity of interactions between transcriptional regulators that modulate lipid biosynthesis.[100]

Several other TFs contribute to the regulation of lipid biosynthesis gene expression and seed oil accumulation. For example, numerous MYB-class

Figure 2.2 *(continued)* desaturase (2/3); PDAT, phospholipid;diacylglyerol acyltransferase; DGAT1, diacylglycerol acyltransferase 1; KAS, ketoacyl synthase; BCCP, biotin carboxyl carrier protein; LPAAT, lysophosphatidic acid acyltransferase; ACCase, acetyl-CoA carboxylase; AAD (2/3), acyl-acyl carrier protein desaturase (2/3); FAB2, fatty acid biosynthesis 2.

TFs have also been found to play roles in lipid biosynthesis; in particular, MYB96 strongly impacts seed oil accumulation by driving the expression of *DGAT1* and *PDAT1*, which are the main enzymes involved in *de novo* and derived TAG synthesis, respectively.[101] On the other hand, MYB5 functions as part of a transcription activation complex that stimulates the expression of *GL2*, which encodes a known repressor of seed oil synthesis.[102] MYB5 is negatively regulated by MYB Interacting Factor 1 (MIF1), which binds to and targets MYB5 for degradation, thereby reducing the GL2-mediated repression of oil synthesis.[102] Similarly, MYB73 promotes oil biosynthesis by decreasing the expression of oil-reducing *GL2*.[103] Some MYB TFs also have specific impacts on FA composition, such as MYB115 and MYB118, which stimulate the synthesis of unsaturated FAs in the seed by promoting the transcription of two acyl desaturation genes, *AAD2* and *AAD3*.[104]

Several TFs from the TZF family have been shown to strongly promote the expression of lipid-related genes and promote oil synthesis. ZF392 forms a regulatory mechanism with two other TFs, ZF351 and NF-YA, to promote oil accumulation by driving the expression of key genes involved in FA synthesis (*BCCP2*, *KASIII*), TAG synthesis (*DGAT1*), TAG storage (*OLEO1*), and transcriptional regulation of lipid genes (*WRI1*).[105] However, further research is needed to characterize the roles of other TZF family members in modulating oil biosynthesis.

Aintegumenta-like 7 (AIL7), a TF in the AIL/PLETHORA (PLT) subfamily, appears to have a broad influence on the seed oil content and FA composition through the transcriptional regulation of a multitude of genes involved in lipid biosynthesis, including FA synthesis (*KASI*, *ACCase*, and *ACP*), acyl-CoA activation (*LACS9*), TAG synthesis (*LPAAT* and *PDAT*), acyl editing (*FAD3*, *FAE1*, and *KCS*), and OB formation (*OLEO1*).[106] Further characterization is needed to discern precisely how AIL7 modulates the expression of these downstream lipid biosynthesis genes. In addition, other AIL subfamily genes may also contribute to the regulation of lipid biosynthesis, as the upregulation of *AIL5* and *AIL6* has been observed in *Arabidopsis* seedlings following the induced overexpression of *FUS3*, which encodes a TAG-enhancing TF.[107]

Furthermore, several TFs have roles in repressing the expression of genes in oil biosynthesis pathways. For example, numerous TFs from the WRKY family, including WRKY2, WRKY6, and WRKY40, are known negative regulators of lipid biosynthesis that inhibit the transcription of lipid-promoting genes, such as *WRI1* and *ABI5*.[108,109] Similarly, the KIN10 kinase (otherwise known as SnRK1) negatively regulates WRI1 and DGAT1 through phosphorylation, resulting in reduced expression of lipid biosynthesis genes and diminished TAG synthesis.[110,111] The regulation of lipid biosynthesis can also occur at the epigenetic level. One such regulatory factor, GCN5, deactivates several lipid-related genes (in particular *FAD3*) through histone acetylation, thereby reducing desaturation of 18:2 to 18:3 and modifying the FA composition of the seed.[112]

The research described here illustrates the substantial progress that has been made in terms of characterizing the major TFs and regulatory mechanisms involved in seed lipid biosynthesis. However, the regulation of lipid biosynthesis by TFs is complex and requires a substantial amount of further exploration. For example, a recent transcriptomic study in canola has identified a vast network of TFs that are putatively involved in seed oil accumulation and that have yet to be properly characterized and situated within transcriptional regulation networks of lipid biosynthesis.[113]

2.5 Highlights in Seed Oil Engineering

Much of the research on seed oil engineering is focused on increasing seed oil accumulation or altering the properties of the seed oil by changing the FA composition. As numerous studies have investigated the genetic engineering of seed oil biosynthesis, we summarize them in Table 2.1 but only discuss some representative examples in this section. For reader's convenience, the storage oil and FA engineering studies described in Table 2.1 are organized alphabetically by species name and cover the model organism *Arabidopsis* and a diversity of crop species. The genetic engineering of the seed oil content and FA composition specifically through CRISPR/Cas9 has recently been reviewed elsewhere.[156]

As *de novo* FA synthesis and acyl editing are essential components of the seed oil content and fatty acid profiles, numerous genes involved in these processes have been engineered in recent studies. For example, the transport of carbon substrates into the plastid in preparation for FA synthesis is an essential process but has lacked research attention. The plastid-localized transporter BASS2, which facilitates the import of pyruvate to supply the carbon substrate for FA synthesis, has been found to increase oil accumulation when overexpressed in *Arabidopsis* seeds.[19] Several proteins involved in the FA synthesis pathway have also been identified as promising targets for seed oil engineering. The overexpression of genes encoding ACCase subunits leads to an increase in the seed oil content, as does the downregulation and/or knockout of several *BADC* genes (encoding ACCase repressors).[90,148,157] The engineering of FA synthesis can also be used to modify the types of FA that are synthesized. The overexpression of certain *FATB* members has been demonstrated to increase the proportions of short- and medium-chain FAs, while the overexpression of *LACS9* increased the abundance of long-chain FAs at the expense of medium-chain FAs.[38,117,136] The overexpression of *FAD2* or *SAD1* has been shown to increase levels of unsaturated FAs.[115,118]

Lipid transporters have also been the subject of genetic engineering. For example, overexpression of *ABCA9* resulted in a substantial increase in the seed TAG content.[46] Notably, overexpression of either *FAX1* or *FAX2* greatly increased the seed oil content while simultaneously increasing the seed protein content, with no impact on the FA composition.[33,34] The

Table 2.1 Recent genetic manipulations of oil content and FA composition in plant seeds.

Species	Gene	Expression	Target pathway or mechanism	Transformation method	Phenotypic outcome	Ref.
A. thaliana and *C. sativa*	*ZmLEC1*	Seed-specific OE	Transcription	*A. tum*	↑Oil	114
A. thaliana and *Z. mays*	*ZmSAD1*	Seed-specific OE	FA desaturation	*A. tum*, particle bombardment	↑PUFAs	115
Arabidopsis thaliana	*AAD2*, *AAD3*, and *MYB115*	Seed-specific OE	FA desaturation	*A. tum*	↑Omega-7 FAs	116
Arabidopsis thaliana	*CnFATB3*	Seed-specific OE	FA desaturation	*A. tum*	↑MCFAs and LCFAs	117
Arabidopsis thaliana	*VfFAD2* and *VfDGAT2*	OE	FA desaturation and Kennedy	*A. tum*	↑Oil, 18:3, 20:3	118
Arabidopsis thaliana	*MCAMT*	Seed-specific OE	FA synthesis	*A. tum*	↑Oil	119
Arabidopsis thaliana	*SsDGAT2*	OE	Kennedy	*A. tum*	↑18:1 and ↓18:3	120
Arabidopsis thaliana	*AtDGAT1*, *RcFAH12*, and *RcDGAT2*	OE and mutation	Kennedy	*A. tum*	↑Unusual FAs	121
Arabidopsis thaliana	*Seipin1* and *RcFAH12*	Seed-specific OE	OBs and FA hydroxylation	*A. tum*	↑ HFAs	122
Arabidopsis thaliana	*GmPLDγ*	OE	PC exchange	*A. tum*	↑Oil and ↑LCFA	123
Arabidopsis thaliana	*AtPDAT*	OE	PC exchange	*A. tum*	↑Oil	65
Arabidopsis thaliana	*BASS2*	Seed-specific OE	Pyruvate transport	*A. tum*	↑Oil	19
Arabidopsis thaliana	*AIL7*	Seed-specific OE	Transcription	*A. tum*	↓Oil, ↓LCFA, and ↑UFA	106
Arabidopsis thaliana	*SPATULA*	OE	Transcription	*A. tum*	↑Oil and ↓SSP	124
Arabidopsis thaliana	*DGAT2*	Seed-specific OE	Kennedy	*A. tum*	↑PUFA and ↓18:0	125
Arabidopsis thaliana	*GmLEC2a*	Seed-specific OE	Transcription	*A. tum*	↑Oil and ↑SSP	83
Arabidopsis thaliana	*WRI1*, *DGAT1*, and *SDP1*	Seed-specific OE	Transcription, Kennedy, and lipid turnover	*A. tum*	↑Oil and ↓seed number	126
Arabidopsis thaliana	*AtFAX1*	Seed-specific OE	Transport	*A. tum*	↑Oil, ↑SSP, and ↑yield	33
Arabidopsis thaliana	*OvFAD2* and *OvFAE1*	Seed-specific OE	FA desaturation and elongation	*A. tum*	Dihydroxy PUFA synthesized	12
Arachis hypogea	*AtLEC1*	Seed-specific OE	Transcription	*A. tum*	↑Oil and modified FA profile	127
B. napus and *N. benthamiana*	*BnGPDH*, *BnGPAT*, *DGAT*, *ScGPDH*, and *ScLPAAT*	Seed-specific OE	Kennedy	*A. tum*	↑Oil	128

(*continued*)

Table 2.1 (*continued*)

Species	Gene	Expression	Target pathway or mechanism	Transformation method	Phenotypic outcome	Ref.
Brassica juncea	*AtDGAT1*	Seed-specific OE	Kennedy	*A. tum*	↑Oil and ↓18:1 and 18:2	129
Brassica napus	*BnFAD2*	Gene-edited allele	FA desaturation	CRISPR/Cas9 and *Ag. tum*	↑18:1 and ↓18:2	130
Brassica napus	*PgFADX* and *PgFAD2*	Seed-specific OE	FA desaturation	*A. tum*	Punicic acid synthesized	131
Brassica napus	*BnFATB2* and *BnSADs*	Seed-specific OE and miRNA	FA desaturation and FA synthesis	*A. tum*	↑16:0 and 18:0	132
Brassica napus	*SsDGAT1*	OE	Kennedy	*A. tum*	↑Oil	133
Brassica napus	*BnLACS2*	OE	Kennedy	*A. tum*	↑Oil and ↓protein	40
Brassica napus	*GmDOF4* and *GmDOF11*	OE	Transcription and FA desaturation	*A. tum*	↑18:1 and ↓18:2 and 18:3	134
Camelina sativa	*pPLAIIId*	Seed-specific OE	Acyl editing	*A. tum*	↑Oil and ↓seed cellulose	135
Camelina sativa	*CpuFATB3*, *CvFATB1*, and *CnLPAAT*	Seed-specific OE	FA desaturation and Kennedy	*A. tum*	↑10:0, 12:0, 14:0, and 16:0	136
Camelina sativa	*LaKCS*, *AtKCR*, *AtHCD* and *AtECR*	Seed-specific OE	FA synthesis	*A. tum*	↑24:1	137
Camelina Sativa	*AtDGAT1* and *ScGPD1*	Seed-specific OE	Kennedy	*A. tum*	↑Oil and seed weight	138
Camelina sativa	*DGAT1A*, *DGAT1B*, and *DGAT1C*	Seed-specific OE	Kennedy	*A. tum*	↑Oil and number of cells in seeds	139
Camelina sativa	*AtWRI1*	Seed-specific OE	Transcription	*A. tum*	↑Oil and seed weight	140
Camelina sativa	*FAX1/ABCA9*	OE	Transport	*A. tum*	↑Oil and seed weight	47
Camelina sativa	*MYB96A* and *DGAT1C*	Seed-specific OE	Transcription and Kennedy	*A. tum*	↑Oil	141
Carthamus tinctorius	*FAD2.2* and *FATB*	Seed-specific RNAi	FA desaturation and synthesis	*A. tum*	↑18:1	142
Crambe abyssinica	*CaFAD2* and *CaFAE1*	Seed-specific RNAi	FA desaturation and elongation	*A. tum*	↑18:1 and ↓22:1	143
Glycine max	*SiDGAT1*	OE	Kennedy	*A. tum*	↑Oil, ↓protein, ↓sugars, and changed FA profile	144

(*continued*)

Table 2.1 (*continued*)

Species	Gene	Expression	Target pathway or mechanism	Transformation method	Phenotypic outcome	Ref.
Glycine max	*GmDGAT1 variants*	Seed-specific OE	Kennedy	Recombination-mediated cassette exchange	↑Oil, ↑protein, and ↓carbohydrates	145
Glycine max	*GmDGAT2A*	Seed-specific OE	Kennedy	*A. tum*	↑Oil and 18:2	146
Glycine max	*DGAT1A* and *DGAT1B*	Seed-specific OE	Kennedy	*A. tum*	↑Oil and ↓ protein	147
Glycine max	*ZF392*	OE	Transcription	*A. tum*	↑ Oil	105
Gossypium hirsutum	*GhACCase (BCCP1, BC1, CTa2*, and *CTb)*	Seed-specific OE	FA synthesis	*A. tum*	↑Oil	148
Gossypium hirsutum	*GhDGAT1*	Seed-specific OE	Kennedy	*A. tum*	↑Oil, ↓protein, and ↑16:0, 18:0, and 18:1	149
Gossypium hirsutum	*GhDOF1*	OE	Transcription	*A. tum*	↑ Oil and ↓seed protein	150
Jatropha curcas	*DGAT1* and *DGAT2*	OE	Kennedy	*A. tum*	↑Oil and 18:2 and ↓protein, 16:0, 18:0, and 18:1	151
Lepidium campestre	*AtWRI1, AtHb2*, and *BvHb2*	Seed-specific OE	Transcription and hemoglobin	*A. tum*	↑Oil	152
Nicotiana tabacum	*SiFAD7*	OE	FA desaturation	*A. tum*	↑18:3 and ↓18:2	153
Oryza sativa	*ACBP2*	OE	Acyl pool	*A. tum*	↑Oil and LCFAs	154
Vernonia galamensis	*DGAT1A*	Seed-specific OE	Kennedy	Particle bombardment	↑Oil and ↓carbohydrate	155

simultaneous overexpression of *FAX1*/*ABCA9* was shown to substantially increase the seed oil content, likely by enhancing the shuttling of lipids out of the plastid (FAX1) and into the ER (ABCA9) for TAG synthesis.[47]

Much research has investigated the enzymes in the Kennedy pathway for their potential in seed lipid engineering. For instance, overexpression of LPAATs results in increased seed oil content and increased unsaturated FA levels, likely by elevating several lipid biosynthesis pathways.[128,158] Furthermore, as the main enzyme responsible for synthesizing TAGs in the final step in the Kennedy pathway, DGAT is an obvious and well-established candidate for the genetic engineering of seed oil.[126,133,144,159,160] Co-expression of DGAT with other genes has also been used to increase the seed oil content. For example, the combination of seed-specific overexpression of yeast *GPD1* to increase the availability of the G3P backbone substrate with overexpression of *DGAT1* to increase TAG synthesis led to increases in both the seed oil content and seed yield.[138] Some DGATs may also be promising for modifying the seed FA composition; for example, the overexpression of *SbDGAT2* increased the proportion of 18:1 at the expense of 18:3.[120] DGAT continues to be a key target for increasing seed oil through augmenting TAG synthesis.[64]

Studies have also aimed to improve the seed oil content by targeting the proteins involved in TAG storage organelles. OLEO1 is the major protein in the membranes of oil bodies, and its overexpression in seeds increased the expression of several lipid biosynthesis-related genes, including *DGAT1*, leading to increased seed oil accumulation, despite reduced OB size.[88,161] Overexpression of several oleosins from *Brassica napus* L. (*OLE1*, *OLE2*, and *OLE4*) increased the seed oil content, 18:2 content, and OB size.[78] Altering the expression of oleosins may increase the rate of budding of OBs from the ER, resulting in smaller OBs, which may prove to be a limiting factor in using oleosins to engineer a higher oil content.[162] This occurrence may be related to the *SEIPIN* family, which localizes to ER–OB junctions and is involved in determining the OB size.[163] Overexpression of *SEIPIN1* increases OB size and seed oil accumulation, while *SEIPIN1* downregulation reduces the OB size and oil content.[163] Thus, there may be potential in combining overexpression of both seipins and oleosins to determine whether the number and size of OBs can be increased simultaneously to achieve an additive effect on seed oil accumulation.

TFs also hold an abundance of potential for seed oil engineering. WRI1 is a key target for genetic engineering because of its central regulatory role in oil biosynthesis.[152,164] Overexpression of *WRI1* has been shown to increase the expression of *PDC* and *ACCase*, which are responsible for the initial channeling of carbon into FA synthesis, resulting in greater seed oil accumulation.[140] Protecting post-translational WRI1 from degradation may also be a viable strategy to engineer seed oil, as demonstrated by the increased oil accumulation achieved by overexpression of a 14-3-3 protein.[99]

Genes in other pathways may also be targeted for the metabolic engineering of lipid biosynthesis. For example, chalcone synthase (CHS) was recently found to influence seed oil pathways despite having no obvious role in seed storage oil biosynthesis. Overexpression of *CHS* was followed by upregulation of several genes involved in seed maturation and FA synthesis, resulting in FA composition shifts that increased the abundance of unsaturated FAs, especially 18:3, in the seed oil.[165] This reveals a previously unknown effect of CHS on lipid biosynthesis, although further exploration is needed to clarify the mechanisms involved and determine whether *CHS* holds potential in seed oil engineering.

2.6 Seed Storage Protein Biosynthesis

SSPs allow for the long-term storage of nitrogen to supply seedling growth and accumulate during the late stages of seed development.[166,167] Upon germination, they are broken down into amino acids (AAs) that supply protein biosynthesis for seedling growth.[3,4] SSPs also contribute to seed longevity by shielding the seed from oxidative damage, thereby slowing the pace of seed deterioration, and are important for proper germination.[168,169] The stability of SSPs themselves can be enhanced through sumoylation by SIZ1, which post-translationally modifies SSPs to improve their longevity.[169]

SSP biosynthesis, as distinct from AA biosynthesis, differs substantially from lipid biosynthesis in that it does not consist of sequential enzyme-catalyzed reactions with chemical intermediates. Rather, SSP genes that have been transcribed into mRNAs are then translated by ER-localized ribosomes into proteins, followed by folding, processing, and transport.[167,170] After the transport and processing of SSPs, they accumulate within protein storage vacuoles for storage.[171] As with most cellular functions, there is extensive transcriptional regulation of genes encoding SSPs, AA transporters, and other associated proteins.[3] This chapter does not cover the pathways of AA biosynthesis and instead focuses on factors that affect the accumulation of SSPs (especially TFs).

The two main storage protein classes in *Arabidopsis* and *B. napus* are 12S globulins (cruciferin [CRU]) and 2S albumins (napins [NAP]).[3] Numerous other major SSPs exist and are often named after the species in which they are abundant, such as glycinin in *Glycine max* (soybean) and zeins in *Zea mays* (corn).[3,172,173] SSPs are typically multimeric assemblies of identical or related proteins; in *Arabidopsis*, napins are encoded by a five-member gene family (*At2S1–At2S5*), while cruciferins are encoded by four genes (*AtCRA1/CRU1*, *AtCRA2*, *AtCRB/CRU2*, and *AtCRC/CRU3*).[3] In legumes, such as faba bean (*Vicia faba* L.) and pea (*Pisum sativum* L.), the main SSPs are vicilins and legumins.[4,6]

Several studies have demonstrated that SSP composition has remarkable plasticity because knocking out one or more SSP-encoding genes typically results in negligible or small impacts on the overall SSP content.[174,175]

Instead of reducing the overall SSP content, the knockout or downregulation of SSP genes, including even the most important SSPs, leads to large compensatory shifts in the abundance of other SSPs.[176–179] The redistributions in SSP composition reported in these studies are typically accompanied by changes in the AA and FA profiles as well. These studies demonstrate the impressive plasticity of maturing seeds with knockdowns or knockouts of major SSPs to adjust their proteomes and provide insights into engineering seed proteins.

2.7 Regulation of Seed Storage Protein Biosynthesis

The expression of SSP genes is regulated by numerous TFs (Figure 2.3), but a select set of TFs play an outsized role in seed maturation, development, and accumulation of storage compounds.[4,180,181] These include master TFs, such as ABI3, ABI5, FUS3, LEC1, and LEC2, which are essential for SSP accumulation.[3,180,182] The expression of numerous SSP genes is directly promoted by these master TFs.[4,183–188] There is also regulatory interaction between these master TFs, with LEC1 able to directly promote the expression of *FUS3* and *ABI3*, which has a positive effect on the expression of SSP genes as well as the overall seed protein content.[189] The transcription-promoting action of ABI3 can be enhanced through synergistic interaction with the regulatory protein RGL3, thereby augmenting SSP gene expression and seed protein accumulation.[190] FUS3 can contribute to the seed protein content through indirect means, such as by repressing the expression of *TTG1*, a repressor of genes encoding napin, thereby supporting seed protein accumulation.[191] There is partial functional redundancy between these key TFs, making it challenging to precisely define the extent of their roles and targets.

In addition to the key TFs discussed in the above paragraph, numerous other TFs are also known to affect the SSP gene expression and seed protein content.[192] MYC2, MYC3, and MYC4 are negative regulators of cruciferin synthesis but positively regulate napin synthesis, with a net negative effect on the seed protein content.[193] Another negative regulator, GZF1 directly represses expression of the SSP gene *GluB-1* and attenuates the seed protein content.[194] A pair of TFs, RISBZ1 and RPBF, has been shown to function in tandem to synergistically promote SSP accumulation, specifically globulins and prolamins.[195,196] NF-YC12 directly promotes the expression of *GS1;3*, which is involved in AA synthesis and subsequently increases the seed protein content.[173] Some TFs have broad, negative effects on SSP gene expression, such as the MYB-family TF ORDORANT1 and the NAC-family member TuSPR, both of which attenuate seed protein levels. Other NAC TFs, such as NAC019, NAC20, NAC26, NAC128, and NAC130, have been shown to foster SSP accumulation by promoting the expression of several SSP-encoding genes.[197–201]

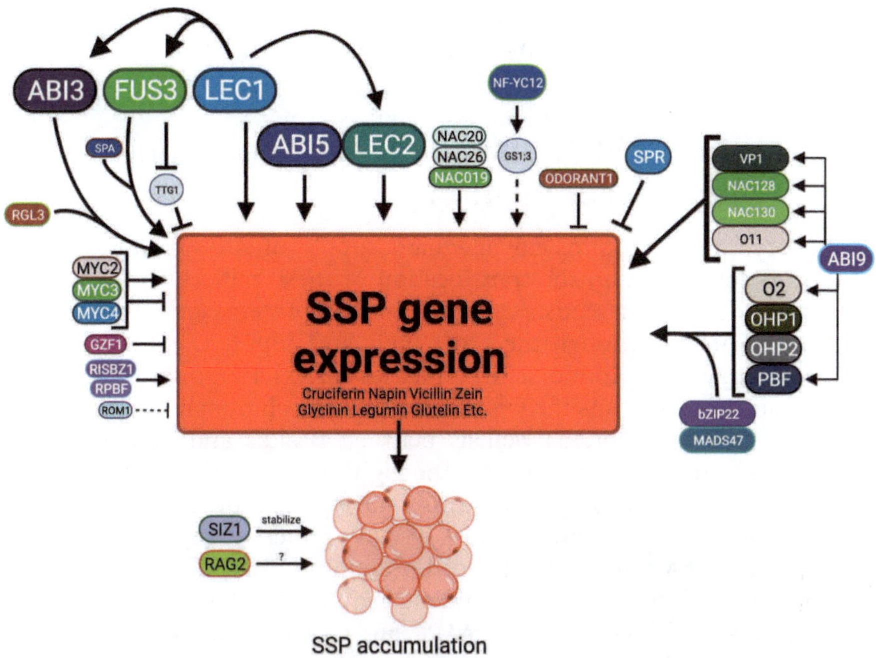

Figure 2.3 Overview of storage protein biosynthesis regulation in plant seeds. The fundamental regulatory and transcription factors involved in seed protein biosynthesis, as well as recent discoveries and other key members with the potential for seed protein engineering, are included here. This figure is not a comprehensive rendering of all regulators and transcription factors in seed protein biosynthesis. This figure is made using Biorender.com. ABI (3/5/9), abscisic acid insensitive (3/5/9); FUS3, FUSCA3; LEC (1/2), leafy cotyledon (1/2); NAC (19/20/26/128/130), NAC-family transcription factor (19/20/26/128/130); NY-FC12, nuclear factor of the Y box C12; GS1;3, glutamine synthase 1;3; SPR, storage protein repressor; VP1, viviparous 1; O (2/11), opaque (2/11); OHP (1/2), O2 heterodimerizing protein (1/2); bZIP22, basic leucine zipper transcription factor 22; PBF, prolamin-box-binding factor; MADS47, MADS-box transcription factor 47; MYC (2/3/4), MYC-family basic helix-loop-helix transcription factor (2/3/4); GZF1, GDNF-inducible zinc finger protein 1; RISBZ1, rice seed b-Zipper 1; RPBF, rice prolamin-box-binding factor; ROM1, regulator of MAT1; SIZ1, E3 SUMO ligase SIZ1; RAG2, rice albumin gene 2; RGL3, RGA-like protein 3; SPA, storage protein activator; TTG1, transparent testa glabra 1.

The expression of zein-encoding genes is regulated by several major TFs, including O2, O7, PBF, OHP1, and OHP2, all of which positively influence the zein content in seeds.[25,178,202] Both bZIP22 and MADS47 can interact with other major zein-promoting TFs, such as PBF1, OHP1, and OHP2, to stimulate SSP gene expression and biosynthesis.[172,203] ABI9 was found to be a positive regulator of seed protein accumulation by promoting the expression of several SSP-related TFs, including O2, PBF1, NAC128,

NAC130, O11, and VP1.[179,204] The transcriptional regulation of zeins in corn has recently been described in a detailed review.[205]

2.8 Amino Acid Transporters and Seed Storage Protein Accumulation

SSP biosynthesis requires an adequate supply of AA substrates and is therefore dependent on the translocation of these molecules through the phloem to developing seeds.[3,206,207] AA transport proteins are involved in the movement of AAs through the plant body and supply them to the seeds for SSP synthesis.[207] Amino acid permeases (AAPs) are important long-distance transporters of AAs throughout the plant, including root-to-shoot and source-to-sink transport as well as between tissues and between organelles.[207] *AAP1* and *AAP8* are known to be highly expressed in seeds, and their function is essential for providing AAs during seed development for SSP accumulation.[192,208–211] However, loss of AAP8 function does not eliminate source-to-sink AA transport, and seed protein does not decrease, implying compensatory function by yet-unidentified transporters.[207,212] Several AA transporters, such as OMT, AAP2, AAP6, and CAT6, are all known to provide important contributions in AA transport to developing seeds and increasing seed protein content.[192,207,211,213–216] Another class of AA transporters called, Usually Multiple Acids Move In and Out Transporters (UmamiTs), are known to play important roles in supplying AAs to developing seeds.[217] Several of these transporters, such as UmamiT18, have been shown to be important in supplying AAs to developing embryos, and single-knockout mutants for these genes produce substantially smaller seeds with reduced AA content.[217,218] However, the loss of function of some transporters (*e.g.*, UmamiT24 and UmamiT25) does not reduce the AA content or seed protein content upon maturity, suggesting some level of redundancy within the UmamiT transporter family.[219] Although the importance of these UmamiT members in AA transport and seed development has been partially explored, the potential for using them in engineering increased SSP content has not yet been explored.

2.9 Highlights in Seed Storage Protein Engineering

Much of the research on SSP biosynthesis pathways is focused on the regulation of SSP gene expression. Numerous recent studies involving the genetic engineering of SSPs are summarized in Table 2.2. The SSP engineering studies described in Table 2.2 are organized alphabetically by species name and cover the model organism *Arabidopsis* and a diversity of crop species.

Transcriptional regulators are obvious targets for seed protein engineering; for example, overexpression of genes encoding major TFs, such as

Table 2.2 Recent genetic manipulations of the SSP content in plant seeds.

Species	Gene	Expression	Target pathway or mechanism	Transformation method	Phenotypic outcome	Ref.
Arabidopsis thaliana	*OsASN1*	ZmUBI promoter OE	AA biosynthesis	*A. tum*	↑SSPs	220
Arabidopsis thaliana	*SAT1*	OE (bundle sheath cell-specific)	AA biosynthesis	*A. tum*	↑SSPs	221
Arabidopsis thaliana	*IbEXP1*	OE	Cell wall development	*A. tum*	↑SSPs	222
Arabidopsis thaliana	*FAX1*	Seed-specific OE	FA transport	*A. tum*	↑SSP and ↑oil	33
Arabidopsis thaliana	*FAX2*	Seed-specific OE	FA transport	*A. tum* (CRISPR)	↑SSP and ↑oil	34
Arabidopsis thaliana	*GmLEC2a*	OE	Transcription	*A. tum*	↑SSPs and ↑oil	83
Arabidopsis thaliana	*TTG1*	Knockout mutant	Transcription	T-DNA insertion	↑SSPs	191
Arabidopsis thaliana	*RGL3*	OE	Transcription	*A. tum*	↑SSPs	190
Glycine max	*AAP6a*	OE (35S)	AA transport	*A. tum*	↑SSPs and ↑NUE	216
Oryza sativa	*AAP6a*	OE (35S)	AA transport	*A. tum*	↑SSPs	215
Oryza sativa	*OsRAG2*	OE (ZmUBI1 promoter)	SSP biosynthesis	*A. tum*	↑SSPs	223
Oryza sativa	*GZF1*	RNAi (seed-specific)	Transcription	*A. tum*	↑SSPs	194
Oryza sativa	*NF-CY12*	OE (ZmUBI1 promoter)	Transcription	*A. tum*	↑SSPs	173
Pisum sativum	*PsAAP1(3 a)*	OE	AA transport	*A. tum*	↑SSPs	211
Pisum sativum	*PsAAP1*	Phloem-specific OE	AA transport	*A. tum*	↑SSPs	224
Pium sativum	*PsAAP1*	Phloem-specific OE	AA transport	*A. tum*	↑SSPs and ↑NUE	225
Triticum aestivum	*TaAAP13*	Endosperm-specific OE	AA transport	Particle bombardment	↑SSPs	226
Triticum aestivum	*TuSPR*	RNAi	Transcription	Particle bombardment	↑SSPs	198
Triticum aestivum	*ODORANT1*	RNAi	Transcription	Particle bombardment	↑SSPs	197

LEC2, *ABI19*, and *FUS3*, have been shown to increase SSP gene expression and seed protein content, while knockdown of *ORDORANT1* and *SPR* (SSP-repressing TFs) enhanced seed protein accumulation.[83,188,197,198,204] The transport and synthesis of AAs have also been explored for increasing the seed protein content through genetic interventions. For example, the overexpression of the AA-synthesizing genes (*e.g.*, *ASN1*) or AA transporters (*e.g.*, *AAP1*, *AAP6*, and *AAT*) improved the nitrogen use efficiency and led to concomitant increases in the seed protein content.[215,216,220,225,227] These studies demonstrate that the synthesis and movement of AAs are promising areas for enhancing seed protein levels. A unique approach to enhancing the SSP content is through overexpression of *RAG2*, an amylase/trypsin inhibitor, although the mechanisms involved are unclear.[223]

As mentioned previously, the overexpression of the lipid transporters *FAX1* and *FAX2* greatly increases the seed oil content while simultaneously increasing the seed protein content.[33,34] This is an unusual outcome, given that seed oil and protein accumulation typically has an inverse relationship due to the competition of their biosynthesis pathways for carbon resources, indicating that FAX transporters may be promising targets for simultaneously engineering increased storage oil and protein contents in seeds.[228]

2.10 Conclusion

There has been impressive progress in the characterization of the pathways and regulation of seed storage lipid and protein biosynthesis in recent years. This chapter reviewed the fundamentals of seed storage lipid and protein biosynthesis, situated recent advancements within the existing pathways, and highlighted promising avenues for the use of genetic engineering to enhance lipid and protein contents in seeds. However, knowledge gaps persist, especially in areas such as acyl interconversion between CoA and PC, intracellular lipid transport, OB biogenesis, AA transport, and transcriptional regulation networks. As such, further research to characterize the interactions and networks of genes and proteins involved in lipid and protein biosynthesis will be essential for developing strategies to engineer plants with increased levels of seed storage substances.

Abbreviations

A. tum	*Agrobacterium tumefaciens*
AA	Amino acid
AAD	Acyl-acyl carrier protein desaturase
ABCA	ATP-binding cassette A
ABI	Abscisic acid insensitive
ACBP	Acyl-CoA-binding protein

ACCase	Acetyl-CoA carboxylase
ACP	Acyl carrier protein
AIL	Aintegumenta-like
BADC	Biotin/lipoyl attachment domain containing
BASS	Bile acid sodium symporter family protein
BCCP	Biotin carboxyl carrier protein
BPM	BTB/POZMATH
bZIP	Basic leucine zipper transcription factor
CAT	Cationic amino acid transporter
CHS	Chalcone synthase
CLO	Caleosin
CoA	Co-enzyme A
CPT	CDP-choline:diacylglycerol cholinephosphotransferase
CRU/CRA/CRB/ CRC	Cruciferins
CTI	Carboxyltransferase interactor
DAG	Diacylglycerol
DGAT	Diacylglycerol acyltransferase
DOF	DNA-binding with one finger
DREBL	Dehydration-responsive element-binding protein L
ER	Endoplasmic reticulum
FA	Fatty acid
FAB2	Fatty acid biosynthesis 2
FAD	Fatty acid desaturase
FAE	Fatty acid elongase
FAT	Fatty acid thioesterase
FAX	Fatty acid exporter
FUS3	FUSCA3
G3P	Glycerol-3-phosphate
GCN	General control nonderepressible
GL	Glabra
GluB	Glutelin B
GPAT	Glycerol-3-phosphate acyltransferase
GPD	Glycerol-3-phosphate dehydrogenase
GS	Glutamine synthase
GZF	GDNF-inducible zinc finger protein
HSD	Steroleosin
KAS	Ketoacyl synthase
LACS	Long-chain acyl-CoA synthetase
LCFA	Long-chain fatty acid
LDAP	Lipid droplet-associated protein
LEC	Leafy cotyledon
LPA	Lysophosphatidic acid
LPAAT	Lysophosphatidic acid acyltransferase

LPC	Lysophosphatidylcholine
LPCAT	Lysophosphatidylcholine acyltransferase
MADS	MADS-box transcription factor
Mal	Malonyl
MED	Mediator
MIF	Mini zinc finger
MUFA	Monounsaturated fatty acid
MYB	MYB-family domain protein
MYC	MYC-family basic helix-loop-helix transcription factor
NAC	NAC-family transcription factor
NAP	Napin
NF-Y	Nuclear factor of the Y box
NUE	Nitrogen use efficiency
O	Opaque
OB	Oil body
OE	Overexpression/overexpressing
OHP	O2 heterodimerizing protein
OLEO/OLE	Oleosin
OMT	Oxoglutarate/malate translocator
PA	Phosphatidic acid
PAP	Phosphatidic acid phosphatase
PBF	Prolamin-box-binding factor
PC	Phosphatidylcholine
PDAT	Phospholipid diacylglyerol acyltransferase
PDC	Pyruvate dehydrogenase complex
PDCT	Phosphatidylcholine diacylglycerol cholinephosphotransferase
PLA/B/C/D	Phospholipase A/B/C/D
PUFA	Polyunsaturated fatty acid
RAG	Rice albumin gene
RGL	Repressor of GA-like protein
RISBZ	Rice seed b-Zipper
RNAi	RNA interference
ROM	Regulator of MAT1
RPBF	Rice prolamin-box-binding factor
SAD	Stearoyl-acyl carrier protein desaturase
SIZ1	E3 SUMO ligase SIZ1
SnRK	Sucrose non-fermenting-1-related kinase
SPA	Storage protein activator
SPR	Storage protein repressor
SPT	Spatula
SSP	Seed storage protein
TAG	Triacylglycerol
TCP	TCP family transcription factor

TF	Transcription factor
TTG	Transparent testa glabra
TZF	Tandem CCCH zinc finger
UmamiT	Usually multiple acids move in and out transporter
VP	Viviparous
WRI	Wrinkled
WRKY	WRKY-family transcription factor
TT	Transparent testa

Acknowledgements

We acknowledge the support from Alberta Innovates [2020F055R, 2020F053R, and 2019F111R], Alberta Canola Producers Commission [ACPC-137-AR20/2020F055R], Results Driven Agriculture Research [RDAR, 2019F111R], Canadian Poultry Research Council [AMN122], SaskCanola and Western Grains Research Foundation [CARP2022.17], the Canada Foundation for Innovation (CFI), the Canada Research Chairs Program, and the Natural Sciences and Engineering Research Council of Canada [RGPIN-2016–05926].

References

1. P. D. Bates, *Biochim. Biophys. Acta*, 2016, **1861**, 1214.
2. T. Vanhercke, C. C. Wood, S. Stymne, S. P. Singh and A. G. Green, *Plant Biotechnol. J.*, 2013, **11**, 197.
3. T. Fujiwara, E. Nambara, K. Yamagishi, D. B. Goto and S. Naito, *Arabidopsis Book*, 2002, **1**.
4. R. D. Thompson, in *eLS*, John Wiley & Sons, Ltd, 2018, https://doi.org/10.1002/9780470015902.a0023693.
5. M. Rahman and M. Michalak, in *In Breeding Oilseed Crops for Sustainable Production*, ed. Surinder Gupta, Academic Press, Washington MA, 1st edn, 2016, vol. **15**, p. 361.
6. A. O. Warsame, D. M. O'Sullivan and P. Tosi, *J. Agric. Food Chem.*, 2018, **66**, 12617.
7. T. A. McKeon, D. G. Hayes, D. F. Hildebrand and R. J. Weselake, in *Industrial Oil Crops*, Academic Press, Washington MA, 1st edn, 2016.
8. S. D. Singer and R. J. Weselake, in *Plant Bioproducts*, ed. Chen Guanqun, Springer Publishing, New York NY, 2018, vol. **5**, p. 59.
9. M. El-Hamidi and F. A. Zaher, *Bull. Natl. Res. Cent.*, 2018, **42**, DOI: 10.1186/s42269-018-0019-0.
10. J.Ohlrogge, *Plant Physiol.*, 1994, **104**, 821.
11. E. B. Cahoon and Y. Li-Beisson, *Curr. Opin. Plant Biol.*, 2020, **55**, 66.

12. X. Li, A. M. Teitgen, A. Shirani, J. Ling, L. Busta, R. E. Cahoon, W. Zhang, Z. Li, K. D. Chapman, D. Berman, C. Zhang, R. E. Minto and E. B. Cahoon, *Nat. Plants*, 2018, **4**, 711–720.
13. M. Henchion, M. Hayes, A. Mullen, M. Fenelon and B. Tiwari, *Foods*, 2017, **6**, DOI: 10.3390/foods6070053.
14. V. R. Dirisala, R. R. Nair, K. Srirama, P. N. Reddy, K. R. S. S. Rao, N. Satya Sampath Kumar and G. Parvatam, *Acta Physiol. Plant*, 2017, **39**, DOI: 10.1007/s11738-016-2315-3.
15. K. R. Lee, G. Q. Chen and H. U. Kim, *Plant Cell Rep.*, 2015, **34**, 603.
16. J. B. Ohlrogge, D. N. Kuhn and P. K. Stumpf, *Proc. Natl. Acad. Sci.*, 1979, **76**, 1194.
17. J. J. Thelen and J. B. Ohlrogge, *Metab. Eng.*, 2002, **4**, 12.
18. Q. Kong, Y. Yang, L. Guo, L. Yuan and W. Ma, *Front. Plant. Sci.*, 2020, **11**, DOI: 10.3389/fpls.2020.00024.
19. E.-J. Lee, M. Oh, J.-U. Hwang, Y. Li-Beisson, I. Nishida and Y. Lee, *Front. Plant. Sci.*, 2017, **8**, DOI: 10.3389/fpls.2017.00194.
20. M. Williams and D. D. Randall, *Plant Physiol.*, 1979, **64**, 1099.
21. T. Konishi, K. Shinohara, K. Yamada and Y. Sasaki, *Plant Cell Physiol.*, 1996, **37**, 117.
22. J. Ke, R. H. Behal, S. L. Back, B. J. Nikolau, E. S. Wurtele and D. J. Oliver, *Plant Physiol.*, 2000, **123**, 497.
23. S. Baud, B. Dubreucq, M. Miquel, C. Rochat and L. Lepiniec, *Arabidopsis Book*, 2008, **6**, DOI: 10.1199/tab.0113.
24. A. P. Brown, V. Affleck, T. Fawcett and A. R. Slabas, *J. Exp. Bot.*, 2006, **57**, 1563.
25. J. Yang, C. Ji and Y. Wu, *Genetics*, 2016, **204**, 581.
26. K. Dehesh, H. Tai, P. Edwards, J. Byrne and J. G. Jaworski, *Plant Physiol.*, 2001, **125**, 1103.
27. J. J. Salas and J. B. Ohlrogge, *Arch. Biochem. Biophys.*, 2002, **403**, 25.
28. J. Shanklin and C. Somerville, *Proc. Natl. Acad. Sci.*, 1991, **88**, 2510.
29. J. Browse and C. Somerville, *Annu. Rev. Plant Physiol. Plant Mol. Biol.*, 1991, **42**, 467.
30. F. M. Bryant, O. Munoz-Azcarate, A. A. Kelly, F. Beaudoin, S. Kurup and P. J. Eastmond, *Plant Physiol.*, 2016, **172**, 154.
31. N. Li, I. L. Gügel, P. Giavalisco, V. Zeisler, L. Schreiber, J. Soll and K. Philippar, *PLoS Biol.*, 2015, **13**, DOI: 10.1371/journal.pbio.1002053.
32. N. Li, H. Meng, S. Li, Z. Zhang, X. Zhao, S. Wang, A. Liu, Q. Li, Q. Song, X. Li, L. Guo, H. Li, J. Zuo and K. Luo, *Plant Physiol.*, 2020, **182**, 1910.
33. Y. Tian, X. Lv, G. Xie, J. Zhang, Y. Xuand F. Chen, *Biochem. Biophys. Res. Commun.*, 2018, **500**, 370.
34. Y. Tian, X. Lv, G. Xie, L. Wang, T. Dai, X. Qin, F. Chenand Y. Xu, *Plant Cell Physiol.*, 2019, **60**, 2231.
35. J. A. Aznar-Moreno, R. Sánchez, S. K. Gidda, E. Martínez-Force, A. J. Moreno-Pérez, M. Venegas Calerón, R. Garcés, R. T. Mullen and J. J. Salas, *Front. Plant Sci.*, 2018, **9**, DOI: 10.3389/fpls.2018.01496.

36. J. M. Shockey, M. S. Fulda and J. A. Browse, *Plant Physiol.*, 2002, **129**, 1710.
37. Y.-S. Zheng, H. Chen, Y. Yuan, Y. Wang, L. Chen, X. Liu and D.-D. Li, *Ind. Crops Prod.*, 2018, **122**, 252.
38. Y. Wang, J. Zhao, H. Chen, Q. Zhang, Y. Zhengand D. Li, *Sci. Hortic*, 2021, **276**, DOI: 10.1016/j.scienta.2020.109751.
39. D. Jessen, C. Roth, M. Wiermer and M. Fulda, *Plant Physiol.*, 2015, **167**, 351.
40. L.-N. Ding, S.-L. Gu, F.-G. Zhu, Z.-Y. Ma, J. Li, M. Li, Z. Wang and X.-L. Tan, *BMC Plant Biol.*, 2020, **20**, DOI: 10.1186/s12870-020-2240-x.
41. Y. Xu, R. Holic, D. Li, X. Pan, E. Mietkiewska, G. Chen, J. Ozga and R. J. Weselake, *Biochem. J.*, 2018, **475**, 1473.
42. L. Zhao, T. M. Haslam, A. Sonntag, I. Molina and L. Kunst, *Plant Cell Physiol.*, 2019, **60**, 1041.
43. H. Zhao, D. K. Kosma and S. Lü, *Front. Plant Sci.*, 2021, **12**, DOI: 10.3389/fpls.2021.640996.
44. S. Xiao and M.-L. Chye, *Prog. Lipid Res.*, 2011, **50**, 141.
45. Z. Guo, Z. Ye, R. P. Haslam, L. V. Michaelson, J. A. Napier and M. Chye, *Plant Direct*, 2019, **3**.
46. S. Kim, Y. Yamaoka, H. Ono, H. Kim, D. Shim, M. Maeshima, E. Martinoia, E. B. Cahoon, I. Nishida and Y. Lee, *Proc. Natl. Acad. Sci.*, 2013, **110**, 773.
47. G. Cai, G. Wang, S.-C. Kim, J. Li, Y. Zhou and X. Wang, *Biotechnol. Biofuels*, 2021, **14**, DOI: 10.1186/s13068-021-01899-w.
48. S. Stymne and A. K. Stobart, *Biochem. J.*, 1984, **223**, 305.
49. J. Okuley, J. Lightner, K. Feldmann and J. Browse, *Plant Cell*, 1994, **6**, 147.
50. V. Arondel, B. Lemieux, I. Hwang, S. Gibson, H. M. Goodman and C. R. Somerville, *Science*, 1992, **258**, 1353.
51. M. Rossak, M. Smith and L. Kunst, *Plant Mol. Biol.*, 2001, **46**, 717.
52. R. A. Moreau and P. K. Stumpf, *Plant Physiol.*, 1981, **67**, 672.
53. F. J. van de Loo, P. Broun, S. Turner and C. Somerville, *Proc. Natl. Acad. Sci.*, 1995, **92**, 6743.
54. P. D. Bates and J. Browse, *Front. Plant Sci.*, 2012, **3**, DOI: 10.3389/fpls.2012.00147.
55. M. I. Gurr, J. L. Harwood, K. N. Frayn, D. J. Murphy and R. H. Michell, in *In Lipids: biochemistry, biotechnology and health*, Wiley-Blackwell, Oxford, 6th edn, 2016.
56. E. P. Kennedy and S. B. Weiss, *J. Biol. Chem.*, 1956, **222**, 193.
57. S. B. Weiss, E. P. Kennedy and J. Y. Kiyasu, *J. Biol. Chem.*, 1960, **235**, 40.
58. R. K. Jain, M. Coffey, K. Lai, A. Kumar and S. L. MacKenzie, *Biochem. Soc. Trans.*, 2000, **28**, 958.
59. S. D. Singer, G. Chen, E. Mietkiewska, P. Tomasi, K. Jayawardhane, J. M. Dyer and R. J. Weselake, *J. Exp. Bot.*, 2016, **67**, 4627.

60. Y. Bai, Y. Shen, Z. Zhang, Q. Jia, M. Xu, T. Zhang, H. Fang, X. Yu, L. Li, D. Liu, X. Qi, Z. Chen, S. Wu, Q. Zhang and C. Liang, *Int. J. Mol. Sci.*, 2021, **22**, DOI: 10.3390/ijms22020785.
61. K. N. Jayawardhane, S. D. Singer, R. J. Weselake and G. Chen, *Lipids*, 2018, **53**, 469.
62. M. W. Lassner, C. K. Levering, H. M. Davies and D. S. Knutzon, *Plant Physiol.*, 1995, **109**, 1389.
63. Y. Nakamura, R. Koizumi, G. Shui, M. Shimojima, M. R. Wenk, T. Ito and H. Ohta, *Proc. Natl. Acad. Sci.*, 2009, **106**, 20978.
64. Y. Xu, K. M. P. Caldo, D. Pal-Nath, J. Ozga, M. J. Lemieux, R. J. Weselake and G. Chen, *Lipids*, 2018, **53**, 663.
65. W. Banas, A. S. Carlsson and A. Banas, *J. Agric. Sci.*, 2014, **6**, 65.
66. U. Ståhl, A. S. Carlsson, M. Lenman, A. Dahlqvist, B. Huang, W. Banas, A. Banas and S. Stymne, *Plant Physiol.*, 2004, **135**, 1324.
67. M. Zhang, J. Fan, D. C. Taylor and J. B. Ohlrogge, *Plant Cell*, 2009, **21**, 1885.
68. L. Wang, Q. Li, Q. Xia, W. Shen, G. Selvaraj and J. Zou, *Lipids*, 2020, **55**, 457–467.
69. C. R. Slack, L. C. Campbell, J. A. Browse and P. G. Roughan, *Biochim. Biophys. Acta*, 1983, **754**, 10.
70. C. R. Slack, P. G. Roughan, J. A. Browse and S. E. Gardiner, *Biochim. Biophys. Acta*, 1985, **833**, 438.
71. C. Lu, Z. Xin, Z. Ren, M. Miquel and J. Browse, *Proc. Natl. Acad. Sci.*, 2009, **106**, 18837.
72. Z. Novotna, O. Valentova, J. Martinec, T. Feltl and K. Nokhrina, *Biochem. Soc. Trans.*, 2000, **28**, 817.
73. J. Lee, R. Welti, W. T. Schapaugh and H. N. Trick, *Plant Biotechnol. J.*, 2011, **9**, 359.
74. L. Y. Yatsu and T. J. Jacks, *Plant Physiol.*, 1972, **49**, 937.
75. J. T. C. Tzen, *ISRN Bot.*, 2012, **2012**, DOI: 10.5402/2012/173954.
76. J. Gu, H. Chao, H. Wang, Y. Li, D. Li, J. Xiang, J. Gan, G. Lu, X. Zhang, Y. Long and M. Li, *Front. Plant Sci.*, 2017, **7**, DOI: 10.3389/fpls.2016.01989.
77. A. Frey-Wyssling, E. Grieshaber and K. Mühlethaler, *J. Ultrastruct. Res.*, 1963, **8**, 506.
78. K. Chen, Y. Yin, S. Liu, Z. Guo, K. Zhang, Y. Liang, L. Zhang, W. Zhao, H. Chao and M. Li, *BMC Plant Biol.*, 2019, **19**, DOI: 10.1186/s12870-019-1891-y.
79. A. H. C. Huang, *Annu. Rev. Plant Physiol. Plant Mol. Biol.*, 1992, **43**, 177.
80. P. Jolivet, E. Roux, S. d'Andrea, M. Davanture, L. Negroni, M. Zivy and T. Chardot, *Plant Physiol. Biochem.*, 2004, **42**, 501.
81. C.-Y. Huang and A. H. C. Huang, *Plant Physiol.*, 2017, **174**, 2248.
82. H. U. Kim, S.-J. Jung, K.-R. Lee, E. H. Kim, S.-M. Lee, K. H. Roh and J.-B. Kim, *FEBS Open Bio*, 2014, **4**, 25.

83. S. Manan, M. Z. Ahmad, G. Zhang, B. Chen, B. U. Haq, J. Yang and J. Zhao, *Front. Plant Sci.*, 2017, **8**, DOI: 10.3389/fpls.2017.01604.
84. M. Pyc, S. K. Gidda, D. Seay, N. Esnay, F. K. Kretzschmar, Y. Cai, N. M. Doner, M. S. Greer, J. J. Hull, D. Coulon, C. Bréhélin, O. Yurchenko, J. de Vries, O. Valerius, G. H. Braus, T. Ischebeck, K. D. Chapman, J. M. Dyer and R. T. Mullen, *Plant Cell*, 2021, **33**, 3076.
85. D. Coulon, L. Brocard, K. Tuphile and C. Bréhélin, *Biochimie*, 2020, **169**, 29.
86. M. S. Greer, Y. Cai, S. K. Gidda, N. Esnay, F. K. Kretzschmar, D. Seay, E. McClinchie, T. Ischebeck, R. T. Mullen, J. M. Dyer and K. D. Chapman, *Plant Cell*, 2020, **32**, 2932.
87. K. D. Chapman, M. Aziz, J. M. Dyer and R. T. Mullen, *Biochem. J.*, 2019, **476**, 1929.
88. D. Zhang, H. Zhang, Z. Hu, S. Chu, K. Yu, L. Lv, Y. Yang, X. Zhang, X. Chen, G. Kan, Y. Tang, Y.-Q. C. An and D. Yu, *PLoS Genet.*, 2019, **15**, DOI: 10.1371/journal.pgen.1008267.
89. H. Liu, Z. Zhai, K. Kuczynski, J. Keereetaweep, J. Schwenderand J. Shanklin, *Plant Physiol.*, 2019, **181**, 55.
90. M. J. Salie, N. Zhang, V. Lancikova, D. Xu and J. J. Thelen, *Plant Cell*, 2016, **28**, 2312.
91. Y. Ye, K. Nikovics, A. To, L. Lepiniec, E. T. Fedosejevs, S. R. Van Doren, S. Baud andand J. J. Thelen, *Nat. Commun.*, 2020, **11**, DOI:10.1038/s41467-020-20014-5.
92. N. Focks and C. Benning, *Plant Physiol.*, 1998, **118**, 91.
93. K. Maeo, T. Tokuda, A. Ayame, N. Mitsui, T. Kawai, H. Tsukagoshi, S. Ishiguro and K. Nakamura, *Plant J.*, 2009, **60**, 476.
94. Q. Kong, L. Yuan and W. Ma, *Plants*, 2019, **8**, DOI: 10.3390/plants8070238.
95. M. J. Kim, I.-C. Jang and N.-H. Chua, *Plant Physiol.*, 2016, **171**, 1951.
96. Q. Kong, Y. Yang, P. M. Low, L. Guo, L. Yuan and W. Ma, *Plant Signaling Behav.*, 2020, **15**, DOI: 10.1080/15592324.2020.1812878.
97. D. Li, C. Jin, S. Duan, Y. Zhu, S. Qi, K. Liu, C. Gao, H. Ma, M. Zhang, Y. Liao and M. Chen, *Plant Physiol.*, 2017, **173**, 1211.
98. Q. Kong and W. Ma, *Plant Signaling Behav.*, 2018, **13**, DOI: 10.1080/15592324.2018.1482176.
99. W. Ma, Q. Kong, J. J. Mantyla, Y. Yang, J. B. Ohlrogge and C. Benning, *Plant J.*, 2016, **88**, 228.
100. Y.-Q. Zhang, X. Lu, F.-Y. Zhao, Q.-T. Li, S.-L. Niu, W. Wei, W.-K. Zhang, B. Ma, S.-Y. Chenand J.-S. Zhang, *Sci. Rep.*, 2016, **6**, DOI: 10.1038/srep34307.
101. H. G. Lee, B.-Y. Park, H. U. Kim and P. J. Seo, *Plant Biotechnol. Rep.*, 2015, **9**, 161.
102. T. Cheng, P. Zhao, Y. Ren, J. Zou and M. X. Sun, *New Phytol.*, 2021, **229**, 2152.

103. Y.-F. Liu, Q.-T. Li, X. Lu, Q.-X. Song, S.-M. Lam, W.-K. Zhang, B. Ma, Q. Lin, W.-Q. Man, W.-G. Du, G.-H. Shui, S.-Y. Chen and J.-S. Zhang, *BMC Plant Biol.*, 2014, **14**, DOI: 10.1186/1471-2229-14-73.
104. M. A. Troncoso-Ponce, G. Barthole, G. Tremblais, A. To, M. Miquel, L. Lepiniec and S. Baud, *Plant Cell*, 2016, **28**, 2666.
105. L. Lu, W. Wei, Q. Li, X. Bian, X. Lu, Y. Hu, T. Cheng, Z. Wang, M. Jin, J. Tao, C. Yin, S. He, W. Man, W. Li, Y. Lai, W. Zhang, S. Chen and J. Zhang, *New Phytol.*, 2021, **231**, 661.
106. S. D. Singer, K. N. Jayawardhane, C. Jiao, R. J. Weselake and G. Chen, *Plant Cell Rep.*, 2021, **40**, 1647.
107. M. Zhang, X. Cao, Q. Jia and J. Ohlrogge, *Plant J.*, 2016, **88**, 95.
108. G. Song, X. Li, R. Munir, A. R. Khan, W. Azhar, M. U. Yasin, Q. Jiang, I. Bancroft and Y. Gan, *Physiol. Plant.*, 2020, **169**, 612.
109. W.-C. Yeap, F.-C. Lee, D. K. Shabari Shan, H. Musa, D. R. Appleton and H. Kulaveerasingam, *Plant J.*, 2017, **91**, 97.
110. Z. Zhai, H. Liu and J. Shanklin, *Plant Cell*, 2017, **29**, 871.
111. K. M. P. Caldo, W. Shen, Y. Xu, L. Hanley-Bowdoin, G. Chen, R. J. Weselake and M. J. Lemieux, *Plant J.*, 2018, **96**, 287.
112. T. Wang, J. Xing, X. Liu, Z. Liu, Y. Yao, Z. Hu, H. Peng, M. Xin, D.-X. Zhou, Y. Zhang and Z. Ni, *Plant J.*, 2016, **88**, 794.
113. A. Rajavel, S. Klees, J.-S. Schlüter, H. Bertram, K. Lu, A. O. Schmitt and M. Gültas, *Int. J. Mol. Sci.*, 2021, **22**, DOI: 10.3390/ijms22031033.
114. Y. Zhu, L. Xie, G. Q. Chen, M. Y. Lee, D. Loque and H. V. Scheller, *Biotechnol. Biofuels*, 2018, **11**, DOI: 10.1186/s13068-018-1049-4.
115. H. Du, M. Huang, J. Huand J. Li, *BMC Plant Biol.*, 2016, **16**, DOI: 10.1186/s12870-016-0827-z.
116. H. Ettaki, M. A. Troncoso-Ponce, A. To, G. Barthole, L. Lepiniec and S. Baud, *PLoS One*, 2018, **13**, DOI: 10.1371/journal.pone.0192156.
117. Y. Yuan,, L. Gao, R. Sun, T. Yu, Y. Liang,, D. Liand Y. Zheng, *Sci. Hortic.*, 2017, **223**, 5.
118. Y. Chen, Q. Cui, Y. Xu, S. Yang, M. Gaoand Y. Wang, *Mol. Gen. Genet.*, 2015, **290**, 1605.
119. S. H. Jung, R. J. Kim, K. J. Kim, D. H. Lee and M. C. Suh, *Plant Cell Physiol.*, 2019, **60**, 1239.
120. Y. Wang, D. Peng, L. Zhang, X. Tan, D. Yuan, X. Liu and B. Zhou, *Plant Mol. Biol. Rep.*, 2016, **34**, 638.
121. H. van Erp, J. Shockey, M. Zhang, N. D. Adhikari and J. Browse, *Plant Physiol.*, 2015, **168**, 36.
122. D. Lunn, J. G. Wallis and J. Browse, *Plant Cell Physiol.*, 2018, **59**, 205.
123. Y. Bai, G. Jing, J. Zhou, S. Li, R. Bi, J. Zhao, Q. Jia, Q. Zhang and W. Zhang, *Plant Sci.*, 2020, **290**, DOI:10.1016/j.plantsci.2019.110298.
124. B. Liu, C. Hua, G. Song, M. Wu, R. Cui, A. Zhang, Y. Liu, L. Huang, A. Yan, I. Ali, A. R. Khan and Y. Gan, *Plant Growth Regul.*, 2017, **82**, 111.
125. Y. Jin, Y. Yuan, L. Gao, R. Sun, L. Chen, D. Li and Y. Zheng, *Front. Plant Sci.*, 2017, **8**, DOI: 10.3389/fpls.2017.01791.

126. H. van Erp, A. A. Kelly, G. Menard and P. J. Eastmond, *Plant Physiol.*, 2014, **165**, 30.
127. G. Tang, P. Xu, W. Ma, F. Wang, Z. Liu, S. Wan and L. Shan, *Front. Plant Sci.*, 2018, **9**, DOI: 10.3389/fpls.2018.00260.
128. F. Liu, Y. Xia, L. Wu, D. Fu, A. Hayward, J. Luo, X. Yan, X. Xiong, P. Fu, G. Wu and C. Lu, *Gene*, 2015, **557**, 163.
129. S. Savadi, V. Naresh, V. Kumar and S. R. Bhat, *Botany*, 2016, **94**, 177.
130. A. Okuzaki, T. Ogawa, C. Koizuka, K. Kaneko, M. Inaba, J. Imamura and N. Koizuka, *Plant Physiol. Biochem.*, 2018, **131**, 63.
131. Y. Xu, E. Mietkiewska, S. Shah, R. J. Weselake and G. Chen, *Metab. Eng.*, 2020, **62**, 20.
132. J.-Y. Sun, J. Hammerlindl, L. Forseille, H. Zhang and M. A. Smith, *Plant Biotechnol. J.*, 2014, **12**, 624.
133. D. Peng, L. Zhang, X. Tan, D. Yuan, X. Liu and B. Zhou, *Mol. Breeding.*, 2016, **36**, DOI: 10.1007/s11032-016-0543-2.
134. Q. Sun, J. Xue, L. Lin, D. Liu, J. Wu, J. Jiang and Y. Wang, *Agronomy*, 2018, **8**, DOI: 10.3390/agronomy8100222.
135. M. Li, F. Wei, A. Tawfall, M. Tang, A. Saettele and X. Wang, *Plant Biotechnol. J.*, 2015, **13**, 766.
136. H. J. Kim, J. E. Silva, H. S. Vu, K. Mockaitis, J.-W. Nam and E. B. Cahoon, *J. Exp. Bot.*, 2015, **66**, 4251.
137. D. Huai, Y. Zhang, C. Zhang, E. B. Cahoon and Y. Zhou, *PLoS One*, 2015, **10**, DOI: 10.1371/journal.pone.0131755.
138. S. Chhikara, H. M. Abdullah, P. Akbari, D. Schnell and O. P. Dhankher, *Plant Biotechnol. J.*, 2018, **16**, 1034.
139. H. Kim, J. H. Park, D. J. Kim, A. Y. Kim and M. C. Suh, *Plant Biotechnol. Rep.*, 2016, **10**, 141.
140. D. An and M. C. Suh, *Plant Biotechnol. Rep.*, 2015, **9**, 137.
141. R. J. Kim, H. U. Kim and M. C. Suh, *Ind. Crops Prod.*, 2019, **138**, 111475.
142. C. C. Wood, S. Okada, M. C. Taylor, A. Menon, A. Mathew, D. Cullerne, S. J. Stephen, R. S. Allen, X.-R. Zhou, Q. Liu, J. G. Oakeshott, S. P. Singh and A. G. Green, *Plant Biotechnol. J.*, 2018, **16**, 1788.
143. X. Li, D. Mei, Q. Liu, J. Fan, S. Singh, A. Green, X.-R. Zhou and L.-H. Zhu, *Plant Biotechnol. J.*, 2016, **14**, 323.
144. Z. Wang, M. Yang, Y. Sun, Q. Yang, L. Wei, Y. Shao, G. Bao and W. Li, *Mol. Breed.*, 2019, **39**, DOI: 10.1007/s11032-019-1016-1.
145. K. Roesler, B. Shen, E. Bermudez, C. Li, J. Hunt, H. G. Damude, K. G. Ripp, J. D. Everard, J. R. Booth, L. Castaneda, L. Feng and K. Meyer, *Plant Physiol.*, 2016, DOI: 10.1104/pp.16.00315.
146. G. Jing, D. Tang, Y. Yao, Y. Su, Y. Shen, Y. Bai, W. Jing, Q. Zhang, F. Lin, D. Guo and W. Zhang, *Biochem. Biophys. Res. Commun.*, 2021, **568**, 143.
147. J. Zhao, R. Bi, S. Li, D. Zhou, Y. Bai, G. Jing, K. Zhang and W. Zhang, *J. Plant Physiol.*, 2019, **242**, DOI: 10.1016/j.jplph.2019.153019.

148. Y. Cui, Z. Liu, Y. Zhao, Y. Wang, Y. Huang, L. Li, H. Wu, S. Xu and J. Hua, *Plant Mol. Biol*, 2017, **35**, 287.
149. P. Wu, X. Xu, J. Li, J. Zhang, S. Chang, X. Yang and X. Guo, *J. Crop Sci.*, 2021, **9**, 487.
150. Y. Su, W. Liang, Z. Liu, Y. Wang, Y. Zhao, B. Ijaz and J. Hua, *J. Plant Physiol.*, 2017, **218**, 222.
151. T.-T. Zhang, H. He, C.-J. Xu, Q. Fu, Y.-B. Tao, R. Xu and Z.-F. Xu, *Plants*, 2021, **10**, DOI: 10.3390/plants10040699.
152. E. Ivarson, N. Leiva-Eriksson, A. Ahlman, S. Kanagarajan, L. Bülow and L.-H. Zhu, *Front. Plant Sci.*, 2017, **7**, DOI: 10.3389/fpls.2016.02032.
153. R. K. Bhunia, A. Chakraborty, R. Kaur, M. K. Maiti and S. K. Sen, *Plant Cell Rep.*, 2016, **35**, 213.
154. Z. Guo, R. P. Haslam, L. V. Michaelson, E. C. Yeung, S. Lung, J. A. Napier and M. Chye, *Plant J.*, 2019, **100**, 1132.
155. M. AL-Amery, B. Downie, S. DeBolt, M. Crocker, K. Urschel, B. Goff, N. Teets, J. Gollihue and D. Hildebrand, *Biocatal. Agric. Biotechnol.*, 2019, **21**, DOI: 10.1016/j.bcab.2019.101303.
156. U. Subedi, K. N. Jayawardhane, X. Pan, J. Ozga, G. Chen, N. A. Foroud and S. D. Singer, *Lipids*, 2020, **55**, 495.
157. J. Keereetaweep, H. Liu, Z. Zhai and J. Shanklin, *Plant Physiol.*, 2018, **177**, 208.
158. S. Chen, Y. Lei, X. Xu, J. Huang, H. Jiang, J. Wang, Z. Cheng, J. Zhang, Y. Song, B. Liao and Y. Li, *PLoS One*, 2015, **10**, DOI: 10.1371/journal.pone.0136170.
159. C. Jako, A. Kumar, Y. Wei, J. Zou, D. L. Barton, E. M. Giblin, P. S. Covello and D. C. Taylor, *Plant Physiol.*, 2001, **126**, 861.
160. D. C. Taylor, Y. Zhang, A. Kumar, T. Francis, E. M. Giblin, D. L. Barton, J. R. Ferrie, A. Laroche, S. Shah, W. Zhu, C. L. Snyder, L. Hall, G. Rakow, J. L. Harwood and R. J. Weselake, *Botany*, 2009, **87**, 533.
161. Y. Lu, M. Chi, L. Li, H. Li, M. Noman, Y. Yang, K. Ji, X. Lan, W. Qiang, L. Du, H. Li and J. Yang, *Front. Plant Sci.*, 2018, **9**, DOI: 10.3389/fpls.2018.01393.
162. Y. Song, X.-D. Wang and R. J. Rose, *Plant Cell Rep.*, 2017, **36**, 1519.
163. Y. Cai, J. M. Goodman, M. Pyc, R. T. Mullen, J. M. Dyer and K. D. Chapman, *Plant Cell*, 2015, **27**, 2616.
164. J. Ye, C. Wang, Y. Sun, J. Qu, H. Mao and N.-H. Chua, *Front. Plant Sci.*, 2018, **9**, DOI: 10.3389/fpls.2018.01479.
165. L. Peng, Q. Hao, S. Men, X. Wang, W. Huang, N. Tong, M. Chen, Z. Liu, X. Ma and Q. Shu, *Physiol. Plant.*, 2021, **172**, 64.
166. R. B. Goldberg, G. de Paiva and R. Yadegari, *Science*, 1994, **266**, 605.
167. E. M. Herman and B. A. Larkins, *Plant Cell*, 1999, **11**, 601.
168. T.-P. Nguyen, G. Cueff, D. D. Hegedus, L. Rajjou and L. Bentsink, *J. Exp. Bot.*, 2015, **66**, 6399.
169. J. S. Kwak, S.-I. Kim, S. W. Park, J. T. Song and H. S. Seo, *Biochem. Biophys. Res. Commun.*, 2019, **519**, 761.

170. R. S. Boston, P. V. Viitanen and E. Vierling, *Plant. Mol. Biol.*, 1996, **32**, 191.
171. M. J. Chrispeels, *Annu. Rev. Plant Physiol. Plant Mol. Biol.*, 1991, **42**, 21.
172. C. Li, Y. Yue, H. Chen, W. Qi and R. Song, *Plant Cell*, 2018, **30**, 2402.
173. Y. Xiong, Y. Ren, W. Li, F. Wu, W. Yang, X. Huang and J. Yao, *J. Exp. Bot.*, 2019, **70**, 3765.
174. W. J. Lyzenga, M. Harrington, D. Bekkaoui, M. Wigness, D. D. Hegedus and K. L. Rozwadowski, *BMC Plant Biol.*, 2019, **19**, DOI: 10.1186/s12870-019-1873-0.
175. H. Rolletschek, J. Schwender, C. König, K. D. Chapman, T. Romsdahl, C. Lorenz, H.-P. Braun, P. Denolf, K. Van Audenhove, E. Munz, N. Heinzel, S. Ortleb, T. Rutten, S. McCorkle, T. Borysyuk, A. Guendel, H. Shi, M. Vander Auwermeulen, S. Bourot and L. Borisjuk, *Plant Cell*, 2020, **32**, 2383.
176. M. A. Schmidt, W. B. Barbazuk, M. Sandford, G. May, Z. Song, W. Zhou, B. J. Nikolau and E. M. Herman, *Plant Physiol.*, 2011, **156**, 330.
177. T. Kawakatsu, S. Hirose, H. Yasuda and F. Takaiwa, *Plant Physiol.*, 2010, **154**, 1842.
178. Z. Zhang, J. Yang and Y. Wu, *Plant Cell*, 2015, **27**, 1162.
179. X. Zheng, Q. Li, C. Li,, D. An, Q. Xiao, W. Wangand Y. Wu, *Plant Cell*, 2019, **31**, 2613.
180. H. Bäumlein, S. Miséra, H. Luerßen, K. Kölle, C. Horstmann, U. Wobus and A. J. Müller, *Plant J.*, 1994, **6**, 379.
181. A. Fatihi, C. Boulard, D. Bouyer, S. Baud, B. Dubreucq and L. Lepiniec, *Plant. Sci.*, 2016, **250**, 198.
182. W. Reidt, T. Wohlfarth, M. Ellerström, A. Czihal, A. Tewes, I. Ezcurra, L. Rask and H. Bäumlein, *Plant J.*, 2000, **21**, 401.
183. I. Ezcurra, M. Ellerström, P. Wycliffe, K. Stålberg and L. Rask, *Plant Mol. Biol.*, 1999, **40**, 699.
184. T. Kroj, G. Savino, C. Valon, J. Giraudat and F. Parcy, *Development*, 2003, **130**, 6065.
185. P. Lara, L. Oñate-Sánchez, Z. Abraham, C. Ferrándiz, I. Díaz, P. Carbonero and J. Vicente-Carbajosa, *J. Biol. Chem.*, 2003, **278**, 21003.
186. G. Mönke, L. Altschmied, A. Tewes, W. Reidt, H.-P. Mock, H. Bäumlein and U. Conrad, *Planta*, 2004, **219**, 158.
187. S. A. Braybrook, S. L. Stone, S. Park, A. Q. Bui, B. H. Le, R. L. Fischer, R. B. Goldberg and J. J. Harada, *Proc. Natl. Acad. Sci.*, 2006, **103**, 3468.
188. F. Sun, X. Liu, Q. Wei, J. Liu, T. Yang, L. Jia, Y. Wang, G. Yang and G. He, *Front. Plant Sci.*, 2017, **8**, DOI: 10.3389/fpls.2017.01133.
189. Y. Kagaya, R. Toyoshima, R. Okuda, H. Usui, A. Yamamoto and T. Hattori, *Plant Cell Physiol.*, 2005, **46**, 399.
190. Y. Hu, L. Zhou, Y. Yang, W. Zhang, Z. Chen, X. Li, Q. Qian, F. Kong, Y. Li, X. Liu and X. Hou, *Plant Physiol.*, 2021, **185**, 1697.
191. M. Chen, B. Zhang, C. Li, H. Kulaveerasingam, F. T. Chew and H. Yu, *Plant Physiol.*, 2015, **169**, 391.

192. K. Gacek, I. Bartkowiak-Broda and J. Batley, *Front. Plant Sci.*, 2018, **9**, DOI: 10.3389/fpls.2018.00890.
193. C. Gao, S. Qi, K. Liu, D. Li, C. Jin, Z. Li, G. Huang, J. Hai, M. Zhang and M. Chen, *Plant Physiol. Biochem.*, 2016, **108**, 63.
194. Y. Chen, A. Sun, M. Wang, Z. Zhu and P. B. F. Ouwerkerk, *Plant Mol. Biol.*, 2014, **84**, 621.
195. M. P. Yamamoto, Y. Onodera, S. M. Touno and F. Takaiwa, *Plant Physiol.*, 2006, **141**, 1694.
196. T. Kawakatsu, M. P. Yamamoto, S. M. Touno, H. Yasuda and F. Takaiwa, *Plant J.*, 2009, **59**, 908.
197. G. Luo, L. Shen, Y. Song, K. Yu, J. Ji, C. Zhang, W. Yang, X. Li, J. Sun, K. Zhan, D. Cui, Y. Wang, C. Gao, D. Liu and A. Zhang, *Plant Biotechnol. J.*, 2021, **19**, 1863.
198. L. Shen, G. Luo, Y. Song, J. Xu, J. Ji, C. Zhang, E. Gregová, W. Yang, X. Li, J. Sun, K. Zhan, D. Cui, D. Liu and A. Zhang, *Plant Biotechnol. J.*, 2021, **19**, 992.
199. J. Wang, Z. Chen, Q. Zhang, S. Meng and C. Wei, *Plant Physiol.*, 2020, **184**, 1775.
200. Z. Zhang, J. Dong, C. Ji, Y. Wu and J. Messing, *Proc. Natl. Acad. Sci.*, 2019, **116**, 11223.
201. Y. Gao, K. An, W. Guo, Y. Chen, R. Zhang, X. Zhang, S. Chang, V. Rossi, F. Jin, X. Cao, M. Xin, H. Peng, Z. Hu, W. Guo, J. Du, Z. Ni, Q. Sun and Y. Yao, *Plant Cell*, 2021, **33**, 603.
202. G. Wang, X. Sun, G. Wang, F. Wang, Q. Gao, X. Sun, Y. Tang, C. Chang, J. Lai, L. Zhu, Z. Xu and R. Song, *Genetics*, 2011, **189**, 1281.
203. Z. Qiao, W. Qi, Q. Wang, Y. Feng, Q. Yang, N. Zhang, S. Wang, Y. Tang and R. Song, *PLoS Genet.*, 2016, **12**, DOI: 10.1371/journal.pgen.1005991.
204. T. Yang, L. Guo, C. Ji, H. Wang, J. Wang, X. Zheng, Q. Xiao and Y. Wu, *Plant Cell*, 2021, **33**, 104.
205. C. Li and R. Song, *Theor. Appl. Genet.*, 2020, **133**, 1443.
206. W. B. Frommer, S. Hummel and J. W. Riesmeier, *Proc. Natl. Acad. Sci.*, 1993, **90**, 5944.
207. G. Yang, Q. Wei, H. Huang and J. Xia, *Plants*, 2020, **9**, DOI: 10.3390/plants9080967.
208. M. Miranda, L. Borisjuk, A. Tewes, U. Heim, N. Sauer, U. Wobus and H. Weber, *Plant J.*, 2001, **28**, 61.
209. A. Sanders, R. Collier, A. Trethewy, G. Gould, R. Sieker and M. Tegeder, *Plant J.*, 2009, **59**, 540.
210. K. Weigelt, H. Küster, R. Radchuk, M. Müller, H. Weichert, A. Fait, A. R. Fernie, I. Saalbach and H. Weber, *Plant J.*, 2008, **55**, 909.
211. J. E. Grant, A. Ninan, N. Cripps-Guazzone, M. Shaw, J. Song, I. Pet ĺk, O. E. Novák, M. Tegeder and P. E. Jameson, *Funct. Plant Biol.*, 2021, **48**, DOI: 10.1071/FP21011.
212. J. P. Santiago and M. Tegeder, *Plant Physiol.*, 2016, **171**, 508.

213. E. Riebeseel, R. E. Häusler, R. Radchuk, T. Meitzel, M.-R. Hajirezaei, R. J. N. Emery, H. Küster, A. Nunes-Nesi, A. R. Fernie, W. Weschke and H. Weber, *Plant J.*, 2009, **61**, 350.
214. R. Schmidt, H. Stransky and W. Koch, *Planta*, 2007, **226**, 805.
215. B. Peng, H. Kong, Y. Li, L. Wang, M. Zhong, L. Sun, G. Gao, Q. Zhang, L. Luo, G. Wang, W. Xie, J. Chen, W. Yao, Y. Peng, L. Lei, X. Lian, J. Xiao, C. Xu, X. Li and Y. He, *Nat. Commun.*, 2014, **5**, DOI: 10.1038/ncomms5847.
216. S. Liu, D. Wang, Y. Mei, T. Xia, W. Xu, Y. Zhang, X. You, X. Zhang, L. Li and N. N. Wang, *Plant Biotechnol. J.*, 2020, **18**, 1749.
217. B. Müller, A. Fastner, J. Karmann, V. Mansch, T. Hoffmann, W. Schwab, M. Suter-Grotemeyer, D. Rentsch, E. Truernit, F. Ladwig, A. Bleckmann, T. Dresselhaus and U. Z. Hammes, *Curr. Biol.*, 2015, **25**, 3126.
218. F. Ladwig, M. Stahl, U. Ludewig, A. A. Hirner, U. Z. Hammes, R. Stadler, K. Harter and W. Koch, *Plant Physiol.*, 2012, **158**, 1643.
219. J. Besnard, C. Zhao, J.-C. Avice, S. Vitha, A. Hyodo, G. Pilot and S. Okumoto, *J. Exp. Bot.*, 2018, **69**, 5221.
220. S. Lee, J. Park, J. Lee, D. Shin, A. Marmagne, P. O. Lim, C. Masclaux-Daubresse, G. An and H. G. Nam, *Plant Cell Physiol.*, 2020, **61**, 1309.
221. X. Xiang, Y. Wu, J. Planta, J. Messing and T. Leustek, *Plant Biotechnol. J.*, 2018, **16**, 1057.
222. J. M. Bae, M. S. Kwak, S. A. Noh, M.-J. Oh, Y.-S. Kim and J. S. Shin, *Transgenic Res.*, 2014, **23**, 657.
223. W. Zhou, X. Wang, D. Zhou, Y. Ouyang and J. Yao, *Plant Biotechnol. J.*, 2017, **15**, 568.
224. L. Zhang, M. G. Garneau, R. Majumdar, J. Grant and M. Tegeder, *Plant J.*, 2015, **81**, 134.
225. M. Perchlik and M. Tegeder, *Plant Physiol.*, 2017, **175**, 235.
226. Y. Wan, Y. Wang, Z. Shi, D. Rentsch, J. L. Ward, K. Hassall, C. A. Sparks, A. K. Huttly, P. Buchner, S. Powers, P. R. Shewry and M. J. Hawkesford, *PLoS One*, 2021, **16**, DOI: 10.1371/journal.pone.0246763.
227. Y. Zhou, H. Cai, J. Xiao, X. Li, Q. Zhang and X. Lian, *Theor. Appl. Genet.*, 2009, **118**, 1381.
228. M. Kanai, S. Mano, M. Kondo, M. Hayashi and M. Nishimura, *Plant Biotechnol. J.*, 2016, **14**, 1241.

CHAPTER 3

Protein-based Materials for Functional Packaging

SARAH JANKHANI, BOON PENG CHANG AND TIZAZU H. MEKONNEN*

Department of Chemical Engineering, Institute of Polymer Research, Waterloo Institute of Nanotechnology, University of Waterloo, Waterloo, Ontario, Canada
*E-mail: tmekonnen@uwaterloo.ca

3.1 Introduction

Functional packaging plays a crucial role in the food industry by keeping the quality of food products during transportation, storage, and end-use. Moreover, it eases the marketing of the products by providing the possibility to communicate with customers through texts, graphical pictures, and logos to support them make informed purchases.[1] In addition, packages protect food products from quick quality deterioration and contamination by restricting their contact with the external environment.[2] Several factors can cause the spoilage of edible products, including microbial contamination, insect overrun, endogenous degradation, and chemical or physical changes.[1] Packaging mainly protects edible products from microbial spoilage, contamination, and oxidation.

Several materials are used in the packaging industry, including glasses, metals, paperboards, and polymeric materials. Synthetic polymer films and coating barriers are commonly used in the food packaging industries due to their particular mechanical properties and characteristics, such as gas and

Smart Materials No. 41
Functional Materials from Lipids and Proteins
Edited by Nandika Bandara and Aman Ullah

Published by the Royal Society of Chemistry, www.rsc.org

liquid impermeability, hydrophobicity, antimicrobial protection, low cost, and stability.[3] During the past few decades, petroleum-based containers and films have been widely used in food packages owing to their excellent mechanical properties and low production cost.[1] However, the consumption of petroleum-based packages increases the consumption of petroleum sources, which can increase the price of packages in the future due to the non-renewable nature of these resources. Moreover, petroleum-based packages are non-degradable, and their incineration disposal results in the emission of carbon monoxide, hydrochloric acid, chlorine, dioxin, nitrides, and many other dangerous materials, which can cause serious environmental problems and human safety concerns.[4,5]

By taking all this into account, the development of bio-sourced and degradable materials to replace petroleum-based ones in food packaging is necessary. Unlike petroleum-based materials, biodegradable materials are sustainable, and their post-consumption disposal is environmentally friendly. Sometimes, they can be decomposed by microorganisms, releasing no hazardous emissions.[6] Due to biodegradability, processability, availability, and low cost and being harmless to edible products and the environment, biopolymers can be used in food packaging as promising replacements for traditional food packaging materials.[7] Biopolymers are not only biodegradable but also produced from renewable resources; therefore, it can be concluded that biopolymers originate from nature and will return to nature after disposal. Figure 3.1 shows the closed-loop life cycle of biopolymer packaging materials.

Biopolymers can be obtained from diverse resources, including waste and byproducts.[9] Figure 3.2 presents the classification of biopolymers into three main groups based on their extraction and production methods. The first group contains biopolymers that can be obtained directly from biomass. These biopolymers are classified into subsets based on their primary feedstock, including polysaccharide-, protein-, and lipid-based. The second group includes biopolymers that can be synthesized from bio-derived monomers. Polyesters, such as polylactic acid (PLA), are included in this group. The last group is biopolymers produced directly by microorganisms, such as polyhydroxyalkanoates (PHA), bacterial cellulose, and xanthan–pullulan.[8]

Biopolymer materials can be utilized as films and coating layers in the food packaging industry. The film is a thin layer developed by drying the molded material in specific shapes, which can be used as food wraps. On the other hand, a coating is a thin layer of material that covers the product, acts as a barrier, and restricts the interaction between the food product and the environment during transportation and storage. Commonly, a coating is applied directly over the product surface by sparing or dipping the product in the coating material. Moreover, a coating can be used in combination with other types of packaging, such as paperboards, to improve its efficiency.[2,6]

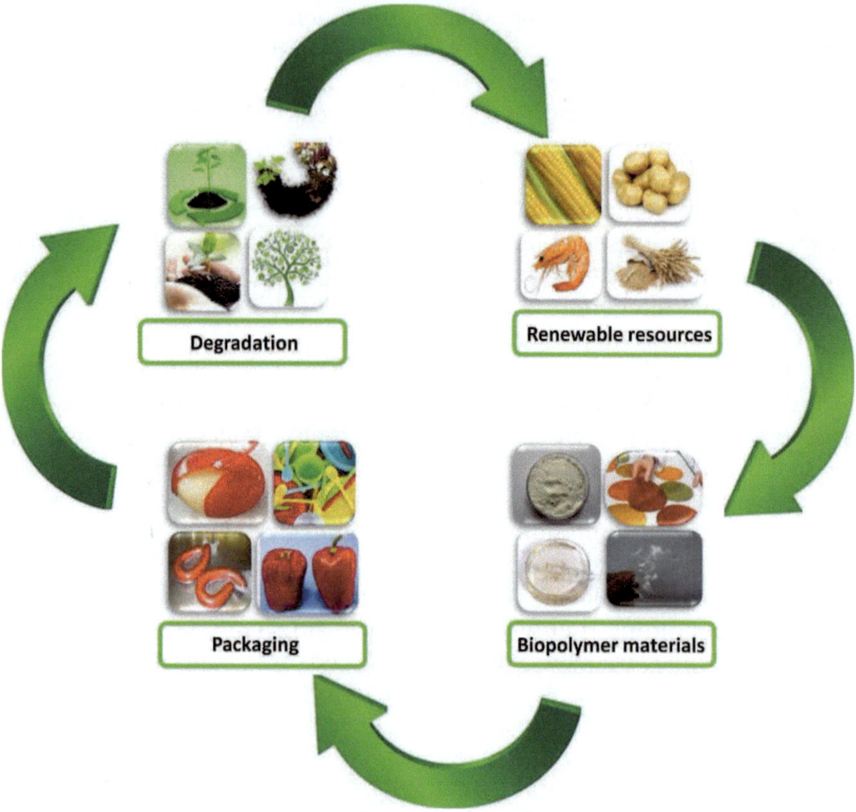

Figure 3.1 Life cycle of food packages from renewable sources. Reproduced from ref. 8 with permission from Elsevier, Copyright 2018.

As mentioned earlier, biopolymers can be naturally produced from polysaccharides. Polysaccharide-based biopolymers are inexpensive and readily available. Starch and its derivatives have a great film-forming ability, exhibiting notable mechanical properties, such as strength, elasticity, and transparency.[10] However, due to high water solubility and water vapor permeability (WVP), they, as packaging materials, allow microorganisms to grow, hence causing weight shrinkage and accelerating microbial spoilage. Studies show that varying the type of starch does not change the mechanical properties of starch-based films; however, the WVP and wettability decrease in films containing more amylose.[8] On the other hand, lipids exhibit excellent barrier properties and hydrophobicity. However, due to their weak mechanical properties, such as inelasticity, lipid films are mostly used as coating layers. Moreover, to compensate for this weakness, lipids can be used to develop composite films and emulsified films and coatings, consisting of both hydrophobic lipid and hydrophilic layers.[8]

Among all types of biopolymers, the protein-based group is getting more attention for food packaging in recent years. Unlike polysaccharides

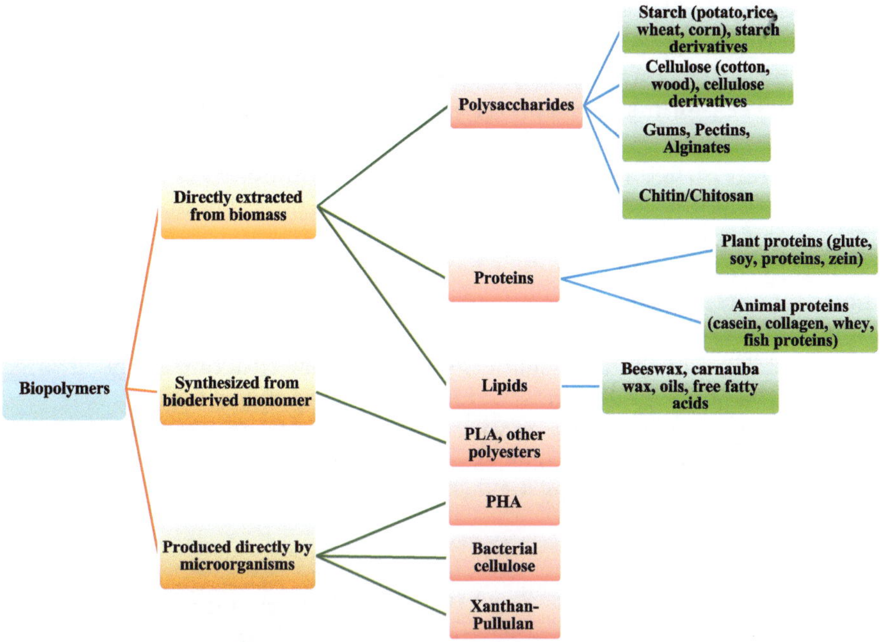

Figure 3.2 Classification of biopolymers based on their origin. Reproduced from ref. 8 with permission from Elsevier, Copyright 2018.

and lipids, proteins have a high intermolecular bonding potential, which contributes to the greater mechanical properties of protein-based films compared to lipid- and polysaccharide-based films.[11,12] Moreover, protein films are naturally hydrophilic; therefore, they can adhere to polar surfaces easily while impermeable to nonpolar gases, such as oxygen and carbon dioxide. However, protein films are not water resistant, and they indicate poor functionality as a water vapor barrier. Despite that, high-molecular-weight proteins are insoluble or weakly soluble in water, and because of their gas barrier properties, they still sufficiently fulfill food packaging material requirements.[11,13]

Food packaging materials require an excellent film-forming property because usually protein-based materials are utilized in combination with different substrates, such as plastics and papers, as films and barriers to make full use of them. For example, whey protein has been applied on paper, low-density polyethylene (LDPE), and polypropylene (PP) as coating layers and successfully decreased the oil permeability and WVP.[11] Polyethylene (PE), PP, and polyethylene terephthalate (PET) films containing gelatin-based coatings showed a remarkable improvement in barrier properties against oxygen and ultraviolet (UV) radiation.[11]

Intrinsically, proteins are one of the most complicated groups of biopolymers naturally synthesized from the limitless sequential arrangements of 20 different types of amino acids. In comparison to other biopolymers and polymers, proteins, due to their sequence-specific chemistry, have complicated structures, which enable them to fold into hierarchical structures. There are four different degrees of structures for proteins, *i.e.*, primary, secondary, tertiary, and quaternary (see Figure 3.3). The primary structure is a simple linear sequence of amino acids. The primary structure starts with N-terminal amino acids and ends with C-terminal amino acids held by covalent or peptide bonds (Figure 3.3A). The secondary structures, such as α-helix and β-sheet, result from the formation of hydrogen bonding between the –CO and –NH groups (backbone interactions) (Figure 3.3B). The tertiary structure is the arrangement of the secondary structure in the space due to side-chain interactions (Figure 3.3C). On the other hand, the quaternary structure is the arrangement of the polypeptide chains to form a larger complex (Figure 3.3D).

Moreover, proteins have a variety of functions and structures. Some proteins have structural roles, such as collagen and keratin. Globular proteins are mostly soluble in water and have an enzymatic role, while

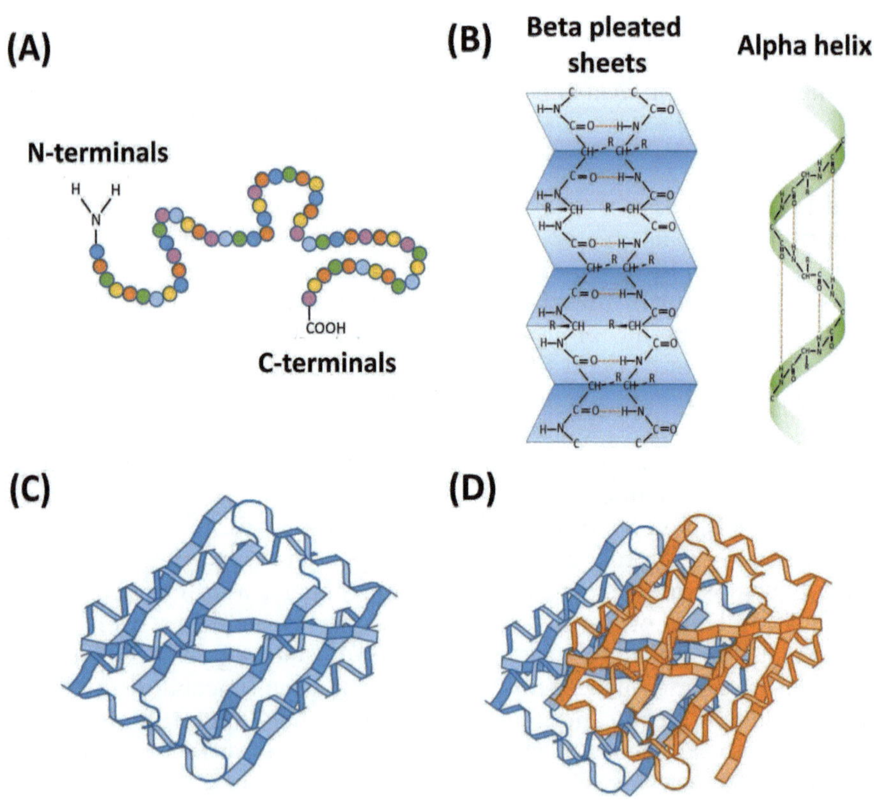

Figure 3.3 Four main structures of proteins and their degrees of arrangement: (A) Primary structure, (B) secondary structure, (C) tertiary structure, and (D) quaternary structure.[4–6]

membrane proteins usually act as receptors. The structure of proteins is not rigid and is able to switch between different structures to accomplish different tasks and functions.[4] Several types of proteins with film-forming ability have been reported to be applicable as basic materials for edible films and coating barriers in the food packaging industry. Among these, globular proteins are the most investigated and utilized, especially due to their amphiphilic behavior, making them suspendable in water and interactive with polar plasticizers. Table 3.1 shows an overview of some studied proteins obtained from different resources for food packaging, films, and coatings.[11]

The capabilities of protein-based biopolymers, including excellent gas barrier properties, biodegradability, and acceptable mechanical properties, make them one of the most appealing materials for developing and enhancing food packaging. Meanwhile, the poor resistance of protein films against water vapor and the brittle nature of protein-based biopolymers caused by the strong, cohesive energy density have yet to be discussed. It has been reported that the addition of some polysaccharides, such as carboxymethyl cellulose (CMC), pectin, and agar, can enhance the mechanical properties and increase the WVP of films.[35] Moreover, adding natural plasticizers to protein-based biopolymers can enhance the mechanical properties by increasing the viscoelasticity and extensibility.[4]

A plasticizer needs to be added to the raw material during the preparation process to modify the 3D structure of the proteins. As a result of adding plasticizers, the attractive intermolecular force in the polymeric material decreases, and the free volume increases consequently. Moreover, plasticizers can join polymers by chemical or physical reactions, increasing the elasticity and flexibility of polymers while reducing the cohesion and rigidity in the structure. Therefore, the protein-based films will not present fragility and brittleness during thermo-mechanical or casting fabrication and end-use.

Plasticizers are low-molecular-weight, non-volatile compounds that can enhance the flexibility and processability of polymers by reducing the tension of deformation, the glass transition temperature (T_g), density, and viscosity while increasing the flexibility of polymer chains.[36] Therefore, plasticizers make polymers tougher and hence increase their practical utility. Plasticizers are classified into two main groups: internal plasticizers and external plasticizers. External plasticizers interact with the polymer chain without forming any chemical bond. On the contrary, internal plasticizers are either co-polymerized into the polymer structure or reacted with the original polymer chemically and become a part of the polymer.[36] Generally, internal plasticizers have bulky structures, preventing polymers from getting close together, leading to an increase in the T_g and softening the polymer.

In the sector of biopolymer films, plasticizers can be categorized into water-soluble plasticizers (hydrophilic) and water-insoluble plasticizers

Table 3.1 Overview of commonly used proteins in the food packaging industry.

Source	Tertiary structure	Protein	Ref.
Plant protein	Globular protein	Wheat gluten	14
		Rice (bran) protein	15
		Corn zein	16
		Soy protein	17
		Pea protein	18
		Phaseolin	19
		Peanut protein	20
		Winged bean protein	21
		Pistachio protein	22
		Lupin protein	23
		Sunflower protein	24
		Cottonseed protein	25
		Amaranth protein	26
		Sorghum kafiri	27
		Potato protein	28
		Sesame	29
Animal protein	Fibrillar protein	Casein	7
		Gelatin/collagen	30
		Keratin	31
		Fish myofibrillar protein	12
		Silk protein	32
	Globular protein	Whey protein	7
		Egg albumin	33
Fungi	Globular protein	Hydrophobins	34

(hydrophobic). Hydrophilic plasticizers are typically water-soluble, and their use at high concentrations causes an increased level of water diffusion through the polymer films. This leads to low functionality in food packaging purposes due to low resistance to water vapors. On the contrary, hydrophobic plasticizers need to be emulsified in the aqueous phase of the protein–polymer dispersion during the plasticization. They may prevent microvoid formation in the film, which decreases the water permeability; nevertheless, hydrophobic plasticizers can result in phase separation, which reduces the flexibility of polymer films and can cause the formation of discontinuous zones during film drying.[36–38]

The outcome of using a single plasticizer varies from one protein to another. For instance, the molecular weight and the position of the hydroxyl groups of polyol plasticizers are two key factors that highly affect the plasticizing ability when added to protein materials.[3] Due to the hygroscopic nature of biopolymers and plasticizers, water can be considered the strongest plasticizer. In addition to water, polyol monosaccharides, disaccharides, and oligosaccharides are commonly used as plasticizers.[3] Table 3.2 shows some of the common plasticizers utilized in protein-based films and coatings.

Table 3.2 Overview of some natural plasticizers used in protein-based films and coatings.

System of application	Plasticizer	Ref.
Zein	Oleic and linoleic acids	39
Whey protein	Glycine (GLY) and sorbitol	40
Wheat gluten	Saturated fatty acids	41
	Glycerin	42
Caseinate–pullulan	Water and sorbitol	43
Whey protein/beeswax emulsion	GLY	44
Gelatin	GLY and sorbitol	45
	Sucrose, oleic acid, citric acid, tartaric acid, malic acid, PEG of different molecular weights (300, 400, 600, 800, 1500, 4000, 10 000, and 20 000 g mol^{-1}), sorbitol, mannitol, EG, DEG, TEG, EA, diethanolamine (DEA), and TEA	46
Pigskin gelatin	GLY	47
	Sorbitol	48
Bovine gelatin	Fatty acids	49
	Sorbitol	48
Fish myofibrillar protein	Glycerin and water	50
Fish muscle proteins	GLY, PG, DEG, and EG	51
Water-soluble fish proteins	GLY and polyethylene glycol (PEG)	52
Fish skin protein	Fatty acids (FA) and sucrose esters (FASE)	53
Feather keratin	GLY	31
Peanut protein	Glycerin, sorbitol, PEG, and PG	20
Sunflower protein	Saturated fatty acids (FA)	41

3.2 Physical and Barrier Properties of Protein-based Packaging Films

Several types of physical, chemical, and biochemical modifications have been used to improve the physical characteristics of protein-based films and coatings in order to enhance their functionality as food packaging materials or in other applications (Figure 3.4). Mechanical and barrier properties of protein-based films and coatings determine the intensity and the way these modifications affect them. In Figure 3.4, mechanical properties, such as the tensile strength (TS), the elongation at break or fracture strain, and the elastic modulus or Young's modulus, and the barrier properties, such as the WVP, relative humidity (RH), and the oxygen permeability (OP), are considered for different films in several physical modifications.[54]

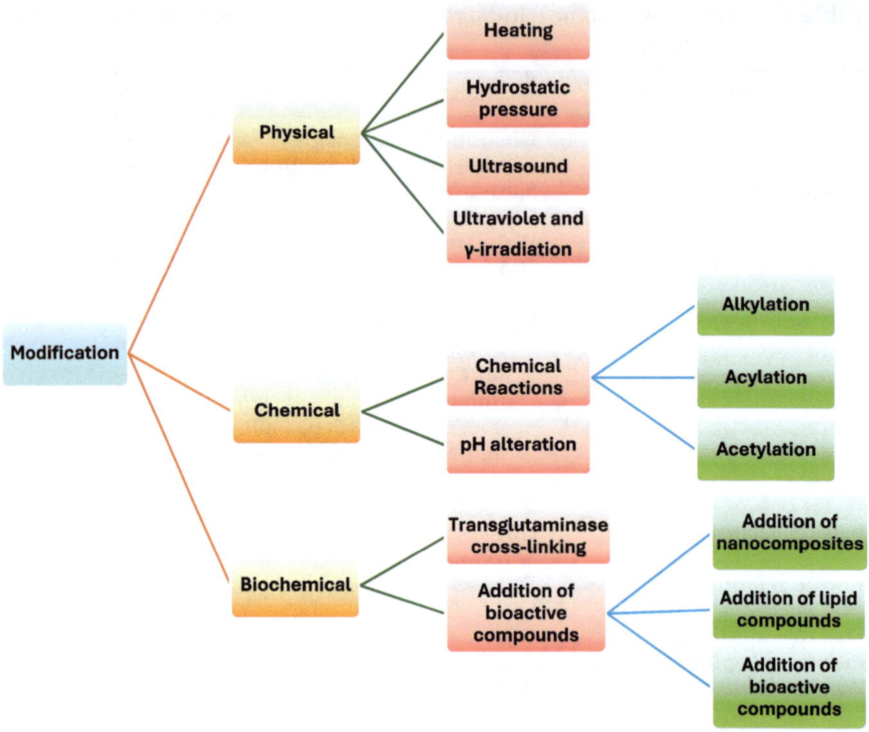

Figure 3.4 Classification of methods for modifying physical, chemical, and biochemical properties of protein-based films.

3.2.1 Physical Modifications

3.2.1.1 Heating

In high-temperature environments, proteins have a natural tendency to aggregate irreversibly to form gel networks. Increasing the temperature will increase the mobility of peptide chains, resulting in changing the structure and aggregation state. Moreover, increasing the temperature of proteins results in various inter- and intramolecular hydrophobic and electrostatic interactions along with hydrogen and disulfide bonds because of the existence of cysteine residues. Therefore, it can be concluded that the properties of these proteins highly depend on the temperature as well as the ionic strength of the environment they are in and the presence of other molecules. Heat treatment or heat curing destroys the structure of proteins and causes rearrangement. The initiation of thermal denaturation for some commonly used proteins in food packaging is investigated in previous studies. For instance, the thermal denaturation for wheat gluten (WG) and soy protein is reported to occur at 90 °C,[55–57] while this value is around 70 °C for whey protein.[58] However, these values can be strongly affected by the protein concentration and the type of the solvent.

Moreover, the TS of protein-based films, such as whey-, soy-, and gluten-based films, increases if the protein films undergo a high-temperature process, mainly due to intra- and intermolecular disulfide crosslinking, hydrogen bonding, and hydrophobic bonding.[55,59] In addition, it has been reported that some proteins (*e.g.*, soy protein) form smoother and more transparent films as a result of heat treatment, resulting in lower WVP and higher elasticity.[54,60] Table 3.3 summarizes the effect of thermal treatment on the mechanical properties of various protein-derived films.

3.2.1.2 *Hydrostatic Pressure*

The effect of pressure on proteins is usually compensated by the volume change.[63] The intermolecular hydrophobic and electrostatic interactions in proteins could be disrupted under pressure, resulting in volume reduction. Moreover, the hydrostatic pressure can stabilize protein hydrogen bonds without affecting the covalent bonds and reduce their volume. It can be concluded that increasing the pressure can result in changing the quaternary, tertiary, and secondary structures of proteins while leaving the

Table 3.3 Summary of previous studies on the effect of thermal treatment on the tensile strength and elongation of protein-based materials.

Modification	Tensile strength (MPa)	Elongation (%)	Ref.
WPI: 70 °C: 5 min	3.4 ± 0.4	7 ± 1	61
WPI: 70 °C: 10 min	3.3 ± 0.1	8 ± 2	61
WPI: 70 °C: 15 min	4.5 ± 0.2	9 ± 0.4	61
WPI: 70 °C: 20 min	4.9 ± 0.2	17 ± 5	61
WPI: 80 °C: 5 min	3.8 ± 0.3	7 ± 1	61
WPI: 80 °C: 10 min	7 ± 2	18 ± 3	61
WPI: 80 °C: 15 min	12 ± 2	17 ± 4	61
WPI: 80 °C: 20 min	14 ± 2	18 ± 4	61
WPI: 90 °C: 5 min	8 ± 2	3 ± 1	62
WPI: 90 °C: 10 min	12 ± 2	14 ± 3	61
WPI: 90 °C: 15 min	12 ± 2	16 ± 3	61
WPI: 90 °C: 20 min	13 ± 2	16 ± 5	61
WPI: 100 °C: 5 min	8 ± 3	14 ± 3	61
WPI: 100 °C: 10 min	10 ± 3	14 ± 4	61
WPI: 100 °C: 15 min	12 ± 2	15 ± 5	61
WPI: 100 °C: 20 min	9 ± 2	18 ± 3	61
SPI control	11.2 ± 2.0	10.2 ± 5.5	60
SPI: 60 °C: 24 h	11	180	55
SPI: 70 °C: 24 h	9	170	55
SPI: 80 °C: 24 h	13	160	55
WG control	1.7 ± 0.3	501 ± 46	59
WG: 80 °C: 15 min	2.4 ± 0.4	391 ± 58	59
WG: 95 °C: 15 min	2.5 ± 0.3	386 ± 75	59
WG: 110 °C: 15 min	3.1 ± 0.4	327 ± 58	59
WG: 125 °C: 15 min	6.3 ± 0.5	275 ± 12	59
WG: 140 °C: 15 min	7.3 ± 1.2	170 ± 26	59
WG: 140 °C: 15 min	4.2 ± 1.1	326 ± 49	59

primary structure unchanged. However, it should be considered that the reaction of proteins with the increase in pressure cannot be predicted without testing them individually.[28,54,63,64]

It has been reported that large aggregates of whey proteins can be isolated under high hydrostatic pressures. Therefore, hydrophobic groups will be available for intermolecular sulfhydryl bond formation.[65] Comparing it to heat treatment, protein gels obtained from inducing hydrostatic pressure are less rigid but more soluble in water and more water permeable.[66] Therefore, pressure-processed proteins may not be appealing to food packaging material development. However, previous studies show that by increasing the pressure, harder gels can be obtained with lower solubility and higher breaking stress.[67]

3.2.1.3 Ultrasound

Ultrasound is an acoustic wave with a frequency greater than 20 kHz.[54,68] Low-energy ultrasound has a frequency between 100 kHz and 1 MHz, with intensities lower than 1 W cm^{-2}, and high-energy ultrasound has a frequency between 16 and 100 kHz, with intensities above 1 W cm^{-2}.[54] As a result of cyclic compression and decompression under high-intensity ultrasound, cavitation takes place.[54] Moreover, shearing forces caused by the intense collapse of gas bubbles are strong and can break covalent bonds of proteins in aqueous solutions. In addition, in some spots in the solution, the pressure and temperature can increase tremendously. Along with these shear forces, the thermal decomposition of proteins occurs under this situation, resulting in the formation of free hydroxyl and hydrogen radicals that can denature the structure of proteins.[54]

An investigation on whey protein films demonstrated that ultrasonic processing can significantly increase the strength, elasticity, and hydrophobicity of the films.[54,69] However, the WVP of films remained unchanged. The main observation in this study was the enhancement of the homogeneous distribution of the film by sonication. Additionally, the sonication process results in the formation of smaller particles in the solution, and the input energy increases the molecular interactions; therefore, the molecular order increases and results in the formation of films with higher strength.[54,69–71] Investigations on the effect of ultrasonic processing on the whey protein coating of frozen fish show that sonicated coatings exhibit lower lipid oxidation compared to untreated coatings. Furthermore, it was reported that soy protein-containing films exhibit better vapor and oxygen barrier properties after treatment with low-frequency ultrasound.[54,72]

3.2.1.4 UV and γ-Irradiation

Double bonds and aromatic rings of proteins convert the absorbed radiations into chemical energy, which can induce photoisomerization. Radiations, such as UV, result in the formation of intermolecular covalent bonds in proteins.[33,73] Additionally, some radiations, such as gamma and high-frequency UV, cause the oxidation of amino acids, break covalent bonds, form protein free radicals, and produce free oxygen radicals by radiolysis of water. Therefore, it can be deduced that irradiation has both a direct and an indirect influence on proteins.[73,74]

Investigations on the impact of UV irradiation on the whey protein films show that the tensile strength of films increases significantly, while barrier properties and elongation at breaking remain unchanged. Moreover, it has been reported that the tensile strength of cast soy milk protein increases linearly by increasing the UV irradiation intensity.[75,76] It was also shown that the β-strand and β-sheet contents of proteins increase by exposure of whey protein isolate (WPI) solutions to γ-irradiation. Therefore, the fracture strength of whey protein gels and films increases.[77] In another study, investigations show an increase in the fracture strength of both soy protein isolate (SPI) and SPI/WPI blend films as a result of γ-irradiation (128 kGy) treatment prior to casting.[78] However, the WVP of SPI film was reduced after the treatment, while the WVP of blend film remained unchanged.[79] In addition, an increase in puncture strength and puncture deformation of SPI and SPI/WPI blend films is observed after γ-irradiation treatment (32 kGy), and the effect of irradiation on the SPI film was greater than that on the blend film. Furthermore, it was noted that γ-treated gluten films (10 kGy) displayed an increase in TS and WVP but a decrease in elongation.[59] However, increasing the radiation doses up to 40 kGy can have an inverse effect, which could result in a reduction in gluten solubility because of the depolymerization and fracture of covalent linkages.[80,81]

3.2.2 Chemical Modifications

The chemical modification of protein-based films and coatings takes place *via* two main methods: first, reaction with chemical agents, and second, pH alteration. The protein side chains play a key role in determining the modification methods. Additionally, the pH and the position of amino acids in the proteins can affect the reaction rates.[54,82,83] Table 3.4 presents the classification of amino acids based on their reactive functional group and the commonly used chemical modification platforms.[54,83,84]

3.2.2.1 Modification of Functional Groups of Proteins

Alkylation, one of the common protein chemical modification platforms, is a chemical process involving the addition or substitution of an alkyl group into an organic substrate. In protein alkylation, the existence of lysine is essential because protein alkylation occurs at the amino group of the protein side chain. Generally, reductive alkylation is not commonly used for protein modification. However, slight enhancements have been reported in some previous studies.[83,85] For food proteins with nucleophilic amino acid residues, an acylation reaction can be used for

Table 3.4 Chemical modification methods based on the side chain of protein amino acid residues.[a]

Side chain	Amino acids	Chemical modification methods	Whey protein (mol%)	Soy protein (mol%)	Wheat protein (mol%)
Amino	Lysine	Alkylation, acylation, and TG	10.5	6.1	N/A
	Arginine		2.5	8.3	1.8
Carboxyl	Glutamine	Amidation, esterification, and TG	11.2	N/A	N/A
	Glutamine acid		6.2	24.5	45.8
	Asparagine		3.1	12	2.9
	Aspartic acid		6.9	N/A	N/A
Disulfide	Cysteine	Reduction and oxidation	2.8	0.03	N/A
Imidazole	Histidine	Alkylation and oxidation	1.5	2.8	1.6
Indole	Tryptophan	Alkylation and oxidation	2	N/A	N/A
Phenolic	Tyrosine	Acylation	3.6	3.5	3.5
	Tryptophan		2	N/A	N/A
	Histidine		1.5	2.8	1.6
	Phenylalanine		3.2	5.4	5.2
Sulfhydryl	Cysteine	Alkylation and oxidation	2.8	0.03	N/A
Thioether	Methionine	Alkylation and oxidation	N/A	N/A	N/A

modification. These nucleophilic amino acids react with the carbonyl group of acylating agents, resulting in the improvement of functional properties of proteins[83] (Figure 3.5).

In addition, acetylation, which is a specific form of acylation, can be used to modify the physical properties of proteins containing amino, hydroxyl, or thiol groups in the presence of acid or base catalysts, such as zinc chloride and sulfuric acid. Acetylation of proteins with acetic anhydrides results in the formation of covalent bonds between the acetyl groups and the amino groups of the proteins (Figure 3.6). Therefore, the number of amino groups decreases and causes the partial unfolding of the backbone.[83]

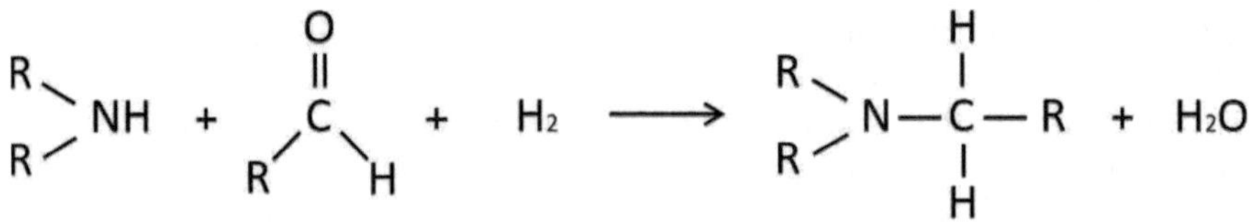

Figure 3.5 Reductive alkylation.

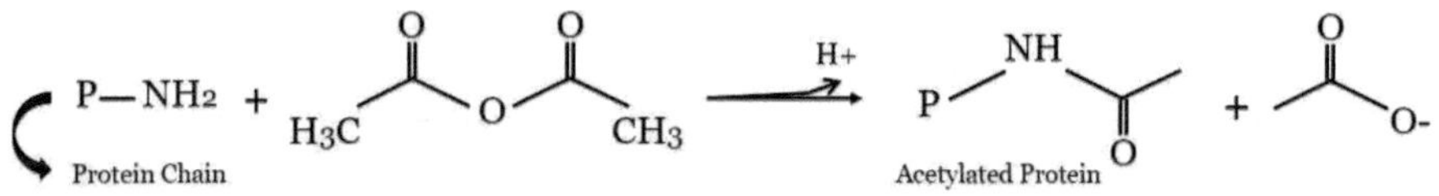

Figure 3.6 Acetylation of protein chains.

As a result of this, it has been reported that the gel strength of SPI and its water-binding capacity have decreased. However, the acetylation process is not successful enough in modifying proteins for food packaging applications because it can only slightly modify the tensile strength, WVP, and OP of the native protein films.[86]

In conclusion, limited studies have been conducted on the chemical modification of protein-based films and coatings, such as alkylation, acylation, and acetylation. Most of these studies reported slight improvements in the physical properties, such as rigidity, WVP, and OP, which affect the application of protein-based films and coatings in the food packaging industry.

3.2.2.2 *pH Alteration*

The net charge of proteins highly depends on their environmental pH. At low pH, proteins generally have a net positive charge, whereas at high pH values, they generally have a net negative charge. Therefore, the functional properties of proteins can be affected by the pH value.[87] For example, the intermolecular interactions of SPI and WG films are highly affected by pH. SPI film exhibited tremendously good film formation properties at pH ranges 1–3 and 6–12, while it can hardly form films at pH 4–5.[54] Moreover, WG films showed the best film formation ability at pH ranges 2–4 and 9–13, while they show inferior film formation at pH 5–8.[54] Furthermore, previous studies show that SPI films formed at 6–11 pH values compared to those formed at acidic pH exhibit higher tensile strength and elongation at fracture while displaying lower WVP.[88] Moreover, the WG films formed under 9–13 pH values have considerably higher tensile strength compared to those formed under acidic pH.[54,87,88]

Based on several studies summarized in Table 3.5, changing the pH of the film-forming solution affects both the mechanical and barrier properties of protein-based films. Moreover, the isoelectric point of the proteins has a crucial role as the WVP and the OP of the films are maximum at this point.[88] Additionally, both WPI and SPI films showed an improvement in elongation at fracture break by increasing the pH.[54]

3.2.2.3 *Biochemical Modifications of Proteins*

In addition to physical and chemical modifications, biochemical modifications can be used to enhance the physical and barrier properties of protein-based films and coatings. In the food packaging industry, the

Table 3.5 Tensile strength and elongation at break values for modified WPI and SPI under pH alteration.

Modification pH alteration	Tensile strength (MPa)	Elongation at break (%)	Ref.
WPI: pH 7	N/A	40 ± 4	89
WPI: pH 8	N/A	54 ± 5	89
WPI: pH 9	N/A	66 ± 7	89
SPI: pH 6	3.5 ± 0.2	72 ± 13.2	88
SPI: pH 8	3.6 ± 0.4	139.5 ± 19.5	88
SPI: pH 10	3.6 ± 0.1	169.3 ± 9.3	88
SPI: pH 12	1.3 ± 0.5	66.5 ± 31.6	88

WVP, tensile strength, film-forming properties, and OP are the key features that need to be improved. To fulfill the improvement of these features, crosslinking by transglutaminase (TG) enzyme, addition of bioactive compounds, and fabrication of composite films can be used.[54]

Crosslinking is a process of forming bonds between polymer chains and a crosslinking agent. Crosslinking can be classified into three main categories. If the polymer chains are covalently bonded, regardless of the incorporation of a crosslinker, it is called chemical crosslinking (Figure 3.7A and B). When polymer chains are non-covalently bonded, it is called physical crosslinking (Figure 3.7C). Additionally, if the crosslinking occurs by the incorporation of an enzyme, it is called enzymatic crosslinking (Figure 3.7D).

Transglutaminase is an enzyme that catalyzes the acyl transfer reaction, which takes place between the glutamine side chain and the lysine side chain of proteins. In this reaction, a bond is formed between the γ-carboxyl group of glutamine and the ε-amino group of lysine, resulting in crosslinking of the protein chains. Owing to its high number of glutamine side chains, WG is practically used for crosslinking *via* the transglutaminase enzyme. Moreover, it was observed that heating protein solutions, which increases the number of accessible glutamine and lysine residues, can increase the degree of crosslinking.[54,91,92]

Generally, transglutaminase crosslinking increases the cohesive energy of proteins within the matrix, hence increasing their tensile strength.[54] However, a considerable increase in the cohesive energy of proteins may result in the rigidity of the films, which makes them unsuitable for many food packaging applications. Moreover, in some cases, transglutaminase crosslinking may significantly increase the WVP and make the films unsuitable for packaging products where moisture levels need to be controlled. However, high levels of WVP are sometimes required for certain fruits and vegetables that require higher levels of humidity, and in such instances, transglutaminase-crosslinked protein films can be suitable.[54,93]

By taking all these into account, some studies on transglutaminase crosslinking have successfully modified protein-based films for food packaging applications. For example, TG-crosslinked WPI and casein films

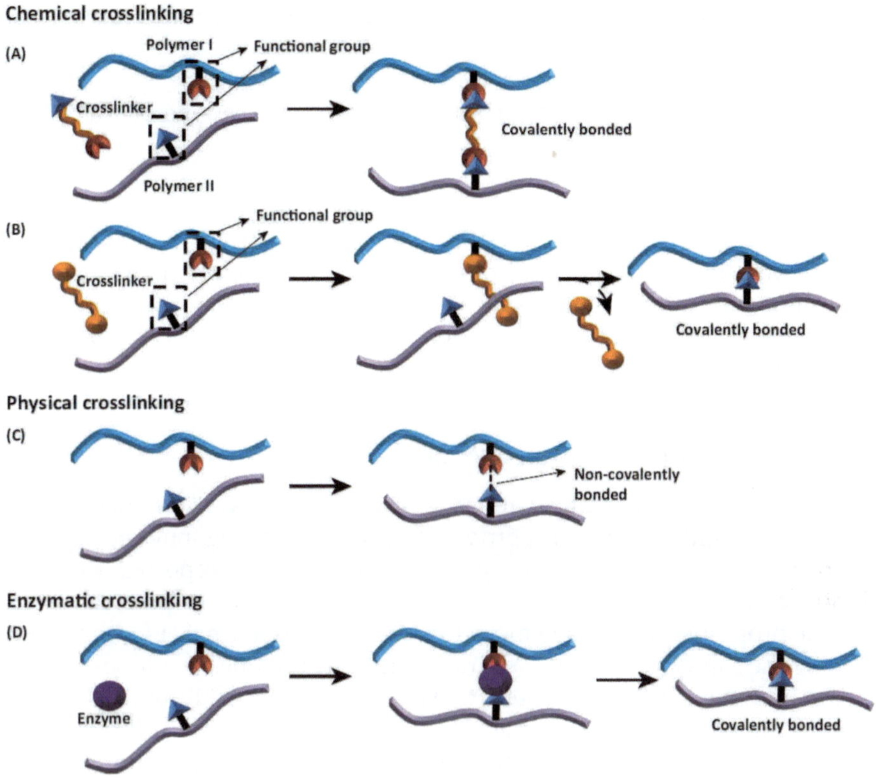

Figure 3.7 Schematic of the crosslinking categories. (A) Chemical crosslinking with the crosslinker incorporated into the bond. (B) Chemical crosslinking with the crosslinker not incorporated into the bond. (C) Physical crosslinking. (D) Enzymatic crosslinking. Adapted from ref. 90 with permission from Elsevier, Copyright 2015.

showed slightly lower WVP.[94] Moreover, crosslinked chitosan–whey protein edible films formed by using microbial transglutaminase as a catalyst were found to have high mechanical resistance, lower deformability, modified oxygen and carbon dioxide barrier properties, and lower permeability to water vapors.[95] Several studies on transglutaminase-crosslinked whey proteins showed improvements in gas barrier properties and WVP.[54,96–99]

In summary, crosslinking by using transglutaminase results in increased tensile strength in both WPI and SPI films. In WPI films, the tensile strength has increased from ~6 to 13 MPa, while only a small increase can be observed in SPI films.[93,100] The main reason for this can be the variation in the available side chains between SPI and WPI as well as their interaction with the enzyme. However, as mentioned earlier, in many of the reported studies, the WVP of WPI and SPI remained unchanged with a few exceptions.[93,100]

3.3 Composite Films and Addition of Bioactive Compounds

In recent years, novel approaches for further modifications of the barrier and mechanical properties of proteins have been investigated. Generally, these approaches can be physical, chemical, and biochemical in nature and mostly focus on modifying the protein structure and/or interactions between protein molecules. Among these approaches, methods that include the fabrication of multiphase films, including blend and composite films, are much investigated.[101] In this section, some of these methods, including the addition of nanoparticles, antimicrobial materials, lipid materials, and bioactive compounds, are reviewed.

3.3.1 Addition of Nanomaterials

In the case of WPI, many studies reported significant enhancement in the tensile and barrier properties of films by adding nanoparticles to fabricate nanocomposites. For instance, it has been reported that WPI films, containing less than 1 wt.% TiO_2 nanoparticles, provide higher tensile properties and elastic modulus compared to normal WPI films while having the same WVP.[102] In a separate study, WPI/nanoclay composite films show improved WVP compared to WPI films.[103] In addition, WPI-based films blended with zein nanoparticles (ZNP) demonstrated improved water vapor barrier and mechanical properties, making them appealing for food packaging applications.[104] Furthermore, montmorillonite nanoclay (MMT), due to its unique platelet structure, caused substantial TS and elongation at break enhancement of SPI while also reducing the WVP compared to the native SPI films.[105,106] SPI nanocomposite films filled with starch nanocrystals (SNCs) are reported to be transparent and homogeneous, with lower water association. As expected, these SPI films became less extendable by increasing the amount of SNC.[107] A previous study on soy proteins also shows that as a result of adding peanut protein nanoparticles, the tensile strength, WVP, and thermal stability of the obtained films have improved considerably.[108]

In addition to whey- and soy protein-based films, WG films have also been modified by adding nanoparticles. Studies show that WG/MMT films have lower WVP and slightly higher tensile strength compared to neat WG films while maintaining almost similar oxygen and carbon dioxide permeability.[14,109] Therefore, WG/MMT films have a high potential for functional food packaging.

3.3.2 Addition of Antimicrobial Materials

The development of edible films containing antimicrobial materials is a promising method for enhancing food packaging technology.

Antimicrobial food packaging can prevent the growth of bacteria, hence increasing the shelf-life of food products.[110] However, the incorporation of antimicrobial materials could negatively affect the mechanical and optical properties of films. Therefore, various tuning processes and optimizations are needed during the film fabrication.[17] One successful example of adding antimicrobial materials to protein-based films is the addition of an antifungal agent and an antibacterial agent, for example, natamycin and thymol, respectively, to the bilayer films of SPI and PLA.[107] The SPI/PLA films have high transparency, strong adhesion between layers, and acceptable mechanical properties compared to regular SPI films, and the antibacterial additives are able to inhibit the growth of mold, yeast, and two strains of bacteria. Another successful example is adding rutin and epicatechin to SPI films as crosslinking agents and as a natural way to improve the specific properties of SPI films. As a result of the addition of rutin and epicatechin, the SPI films showed significantly improved tensile strength than normal SPI films. However, the epicatechin was found to increase the WVP.[17,111]

3.3.3 Addition of Lipid Material

As mentioned earlier, a significant weakness of protein-based food packaging materials is their water permeability or hydrophilicity. Therefore, the addition of hydrophobic materials, such as lipids and waxes, can compensate for this weakness to significant levels. In the case of whey protein films, the incorporation of viscoelastic milk fat and beeswax showed better improvement in WVP when compared to candelia or carnauba wax.[112] Moreover, it has been reported that the tensile strength and elongation at break could significantly increase as the beeswax particle size in whey protein films shrinks.[113] However, lipids can negatively affect the functionality of protein-based films. For instance, mechanical properties, such as tensile strength and elongation, had decreased at high lipid concentrations.[113] Furthermore, the addition of lipids can sometimes result in brittleness and affect the transparency of films, which is usually undesirable for food packaging applications.[112,114]

3.4 Barrier Properties

Both mechanical properties and barrier properties loom large in the efficiency of protein-based and other food packaging materials. As mentioned earlier, protein-based edible films have better physical properties than polysaccharide- and lipid-based films. However, protein-based materials have high WVP, which restricts their application in the food packaging industry.[115] To tackle the challenge, it is essential to enhance the barrier properties of protein-based films, such as oxygen permeability and WVP.

Since oxygen is responsible for microbial spoilage, chemical spoilage, vitamin loss, aging, and many other degradation processes, functional food packages need to provide an adequate barrier to oxygen in order to increase the quality and shelf-life of food products.[116] The degree of protection of edible films and coatings relies on their oxygen barrier properties. Many factors, such as temperature, polymer chain interaction, hydrophilic/hydrophobic properties of materials, and the ambiance humidity, can affect the oxygen permeability of edible films and coatings. For example, gas permeability increases at higher temperatures, and films containing hydrocolloids may have higher resistance to oxygen compared to synthetic polymers due to the abundance of bonds between chains, inhibiting oxygen transmission.[117,118] Moreover, water molecules in humid environments can act as plasticizers by interacting with the film-forming polymers and increase oxygen permeability.[116] Owing to their ordered, hydrogen-bonded network, proteins are inherently good oxygen barriers.[116,119] The oxygen permeability of some commonly used protein-based films is summarized in Table 3.6.

3.4.1 Impact of Antioxidants

To reduce the impact of permeated oxygen on protein film-packaged food products, antioxidants could be incorporated into the formulations. This is because the antioxidants can protect the packaged food from oxidation.[125] Based on their mechanisms of action, antioxidants can be categorized into

Table 3.6 Oxygen permeability of some commonly used protein-based films.

Protein film	Oxygen permeability ($mL\ \mu m\ m^{-2}\ d^{-1}\ Pa^{-1}$)	Condition	Ref.
Caseinate	1.44×10^5	23 °C and 55 ± 2% RH	120
Whey protein	1.85×10^4	23 °C and 55 ± 2% RH	120
Zein	1.8	23 °C and 55 ± 2% RH	121
Bovine gelatin	4.80×10^{-3}	23 °C and 55 ± 2% RH	30
Zein : PEG + glycerol (2.6 : 1)	3.9×10^{-2} to 9×10^{-2}	25 °C and 0% RH	122
Gluten : glycerol (2.5 : 1)	6.1×10^{-3}	25 °C and 0% RH	
SPI : glycerol (2.4 : 1)	6.1×10^{-3}	25 °C and 0% RH	
WPI : glycerol (2.3 : 1)	7.61×10^{-1}	23 °C and 50% RH	
WPI : sorbitol (2.3 : 1)	4.3×10^{-3}	23 °C and 50% RH	
Collagen	$<4 \times 10^{-5}$ to 5×10^{-4}	RT °C and 0% RH	
	2.3×10^{-2}	RT °C and 63% RH	
	8.9×10^{-1}	RT °C and 93% RH	
WPI : Gly (60 : 40)	1.27×10^{-1}	NA	123
WPI : Gly (60 : 40) + antioxidant	1.31×10^{-1}	NA	
WPI : Gly (50 : 50)	8.7×10^{-2}	37 °C and 21% RH	124
WPI : Gly (50 : 50)	5.2×10^{-2}	29 °C and 21% RH	

primary and secondary antioxidants. Primary antioxidants refer to free radical acceptors that prevent oxidation by interrupting or inhibiting the initiation or propagation step of the oxidation process. On the contrary, secondary antioxidants decrease the rate of oxidation without converting free radicals into more stable compounds. Moreover, based on their source and origin, antioxidants can be classified into natural antioxidants and synthetic antioxidants.[125]

Plant extracts from green tea, ginkgo leaves, ginger, and grape seeds are common sources of natural antioxidants due to their high polyphenol, flavone, glycoside, tannin, and flavonoid contents.[126] Moreover, oleoresins and essential oils are rich sources of natural antioxidants.[127] Polyphenols in borage seeds and leaves also have high antioxidant capacity that improves the barrier properties of films and coatings.[127] In addition to natural antioxidants, synthetic antioxidants, such as

Table 3.7 Summary of natural and synthetic antioxidants in packaging films and coatings.[a]

Antioxidant	Protein-based film and the antioxidant concentration (%)	Application	Ref.
Butylated hydroxyanisol (BHA)	0.01 in corn zein film	Cooked turkey	16
Fenugreek	0.2 in egg albumin coating	Diced raw and cooked poultry breast meat	128
Oregano	1 in whey protein film	Beef muscle	129
Pimento	1 in whey protein	Beef muscle	129
Rosemary extract	2–2.5 in gelatin film	Cold-smoked sardine	130
Borage extract	4 in gelatin film	Food simulants	131
Butylated hydroxytoluene (BHT)	0.01 in PLA film	Food simulants	132
Ginseng extract	5 in alginate film	Food simulants	133
Cinnamon essential oil	2 in chitosan coating	Rainbow trout	134
Chitosan	Chitosan coating	Salmon fillet	135
Ferulic acid	4 in sodium caseinate	Food simulants	127
Mustard extract	14 in xanthan film	Smoked salmon	136
Green tea extract	0.3 in gelatin film	Food simulants	137
Ascorbic acid	1 in alginate coating	Kent mangoes	118
Thyme essential oil	5 in soy protein film	Ground beef patties	138
Olive leaf extract	10 in PLA film	Film simulants	139
Citric acid	1.5 in LLDPE/starch film	Beef	140
Clove essential oil	2 in chitosan coating	Cooked pork sausages	141
Pomegranate seed extract	5 in CMC film	Food simulants	142
Tocopherol	5 in CMC film	Food simulants	143
Aloe vera gel	50 in gelatin coating	Fresh cut oranges	144
Black tea	10 in gelatin film	Fresh cut oranges	144

[a]Adapted from ref. 117 with permission from Elsevier, Copyright 2019.

tert-butylhydroquinone (TBHQ), butylated hydroxyanisol (BHA), butylated hydroxytoluene (BHT), and propyl gallate (PG), are widely used to improve the functionality of films and coatings. Studies on some natural and synthetic antioxidants used in packaging films and coatings are summarized in Table 3.7.

The poor water vapor resistance of protein films in comparison with other polymers is one of the most severe limitations that prohibit their extensive industrial-scale use in food packaging. Therefore, it is essential to increase the moisture resistance of protein-based films, especially for applications in highly humid environments. As mentioned earlier, the WVP of protein-based films can sometimes decrease, while some physical and chemical modifications increase their mechanical strength. Thus, the modification methods should be optimized to keep the WVP of the films at an optimal level while enhancing the film-forming properties.

On the other hand, some modifications can simultaneously improve mechanical properties and barrier properties. For example, the WVP of films obtained from mixed calcium caseinate (CAS) and WPI demonstrated a significant reduction after γ-irradiation treatment. The highest reduction was observed for CAS : WPI (25 : 75), with a final permeability of 0.69 g mm m^{-2} d^{-1} mm Hg^{-1}.[145] Moreover, it has been reported that the WVP of zein-based films decreases under γ-irradiation treatment compared to non-irradiated zein film. In addition, as shown in Figure 3.8, the WVP values decrease by increasing the radiation dose.[146] A study on sesame protein isolate-based edible films showed that exposure to UV irradiation results in the formation of covalent bonds in the structure of films, resulting in chain mobility restriction, and an increase in the surface hydrophobicity that subsequently reduces the water permeability.[29]

Additionally, an investigation on the effect of γ-irradiation on gluten-based films exhibited that γ-irradiation acts as a cross-linking agent, resulting in the aggregation of cleaved polypeptide chains and the reduction in the water vapor diffusion rate through the film. Consequently, irradiated gluten-based films showed a 29% decrease in the WVP value compared to untreated films. As presented in Figure 3.9, increasing the irradiation dose increases the treatment effects in reducing the WVP of the gluten-based films.[147]

In addition to γ-irradiation, UV treatments can enhance the WVP of protein-based films. An investigation on WG, corn zein, egg albumin, and sodium caseinate films showed enhancement in the WVP of the films as a result of UV treatment. The study showed a significant decrease in the WVP of egg albumin films after UV treatment, while only a slight change has been observed in the WVP of corn zein, WG, and sodium caseinate films.[33]

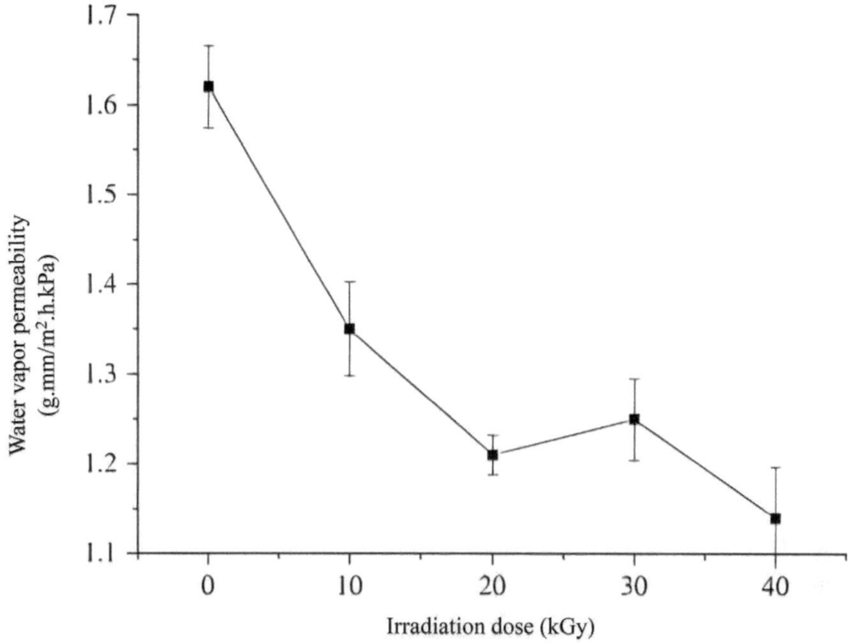

Figure 3.8 Effect of increasing the γ-irradiation dose on WVP of zein-based films. Adapted from ref. 146 with permission from Elsevier, Copyright 2009.

3.4.2 Impact of Plasticizers on Barrier Properties

Some studies showed that, besides physical treatment, the use of appropriate plasticizers can alter the WVP of protein films. In recent years, many researchers have focused on the use of low-volatility compounds as plasticizers, such as glycerol, sorbitol, some monosaccharides, and several fatty acids, for improving the WVP of edible films.[148] A study on gelatin-based films showed that the addition of two types of polyol plasticizers, such as glycerol and sorbitol, to gelatin films resulted in lower WVP values compared to the addition of a single polyol plasticizer.[148] Moreover, the study showed that the films that were plasticized by polyols demonstrated lower WVP compared to those plasticized by monosaccharide plasticizers. In another study, films plasticized by polyethylene glycol (PEG) and tributyl citrate have shown low WVP probably due to the hydrophobic characteristics of such plasticizers that restrict water vapor diffusion through the film.[148] Due to the hydrophilic nature of proteins, films containing a high portion of gelatin exhibited higher WVP; therefore, the existence of polysaccharides in the films improves the crosslinking of gelatin and lowers the WVP by decreasing the free volume of the polymeric matrix.[149] The formation of flexible protein structures as a result of their interactions with plasticizers could further improve the barrier properties of protein-based

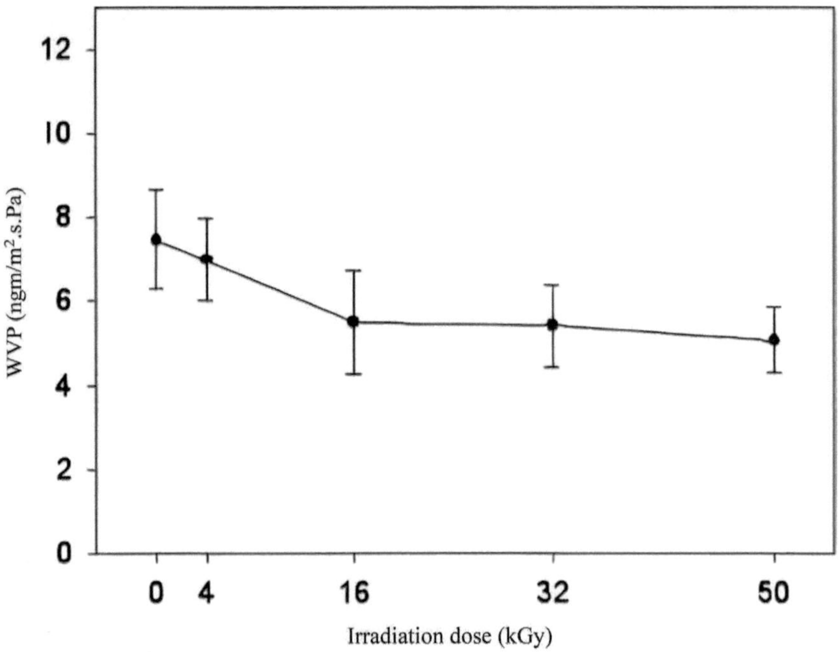

Figure 3.9 Effect of different doses of γ-irradiation on the WVP of gluten-based films. Adapted from ref. 147 with permission from Elsevier, Copyright 2005.

films.[150] Figure 3.10 shows the formation of network structures and bonding of quil keratin protein structures plasticized with different plasticizers.

A study on zein protein films showed that the unplasticized films have lower WVP values compared to those films plasticized with glucose and galactose. However, plasticizing with fructose showed a decrease in the WVP.[152] The reason can be the lower hydrophobicity of fructose compared to glucose and galactose. Moreover, research on milk protein-based films demonstrated that at the same plasticizer content, sorbitol had the best functionality in decreasing the WVP compared to sodium lactate and glycerin. However, plasticized films with all three components (sorbitol, lactate, and glycerin) exhibited lower WVP than control films.[153] A summary of other recent studies on the influence of different plasticizers on protein-based films is shown in Table 3.8.

3.5 Conclusions

Due to biodegradability, flexibility, transparency, and good barrier and film-forming properties, protein-based biopolymers are convenient replacements for petroleum-based food packages. Many studies have

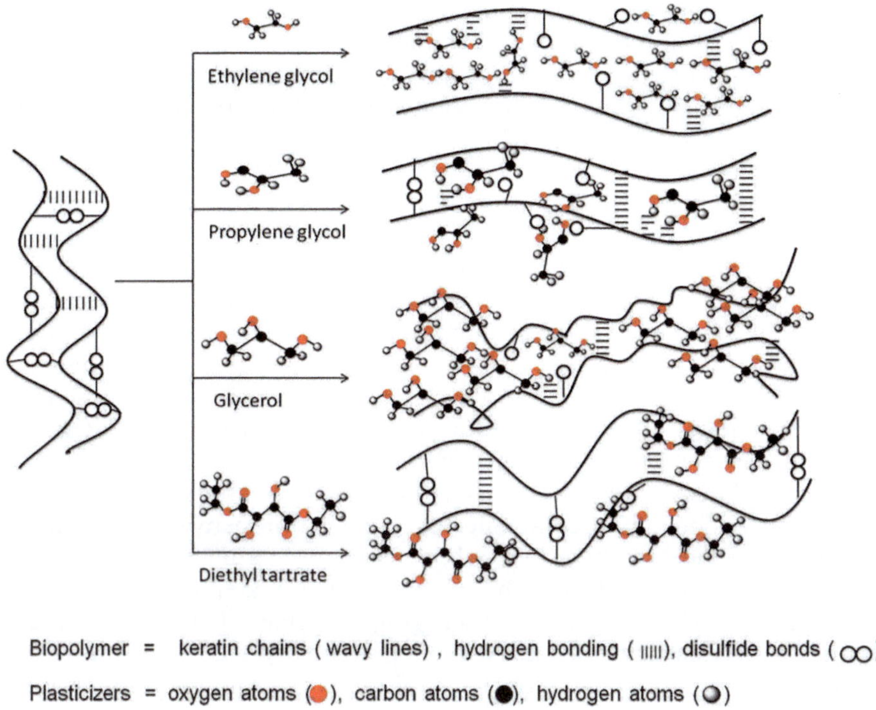

Figure 3.10 Plasticization of quil keratin with various plasticizers, *i.e.*, ethylene glycol, propylene glycol, glycerol, and diethyl tartrate. Adapted from ref. 151 with permission from the American Chemical Society, Copyright 2011.

Table 3.8 Water vapor permeability values of protein-based films plasticized with different plasticizers.

Protein/plasticizer (wt./wt.)	WVP (g mm m^{-2} d^{-1} kPa^{-1})	Ref.
WPI : Gly = 1 : 1–2 : 1	116–144	154
WPI : Xyl = 1 : 1–2 : 1	84–89	154
WPI : Sor = 1 : 1–2 : 1	84–112	154
WPI : Gly = 1 : 1–1.6 : 1	119.8–154.56	40
WPI : Sor = 1 : 1–1.6 : 1	61.92–84.72	40
WPI : PEG 200 = 1 : 1	134.64	40
WPI : PEG 400 = 1 : 1	129.6	40
WG : Gly : Suc = 15 : 6 : 0	121.55	42
WG : Gly : Suc = 15 : 4 : 2	99.14	42
WG : Gly : Suc = 15 : 3 : 3	81.90	42
WG : Gly : Suc = 15 : 0 : 6	15.52	42
WG : Gly : Sor = 15 : 3 : 3	105.17	42
SC : Gly = 0.89 : 1–1.67 : 1	5.4–11.4	155
SC : PEG 400 = 0.81 : 1–1.32 : 1	14.8–22.6	155
Zein : Gly = 2.3 : 1	35.52	153
Zein : Gly = 5.67 : 1	24.24	153

investigated the functionality and characterization of different types of protein-based films and coatings containing whey protein, wheat gluten, soy protein, corn zein, casein, gelatin, and keratin during the past few years. Studies highlighted in this chapter revealed that the majority of them confirmed the functionality of protein-based films as a good packaging material alternative for various food products. However, physical and chemical modifications are necessary to enhance the characteristics of protein-based films and coatings to ensure that they can be an alternative to conventional petroleum-based packaging. Several methods have been investigated to improve the WVP and film-forming properties of protein-based films. Some of the most common physical, chemical, and biochemical modification methods are discussed, especially those that can effectively influence WPI-, SPI-, WG-, zein-, and casein-based films. However, the efficiency of these techniques depends on their intensities as well as the concentration and presence of other components in the preparation methods. Moreover, plasticizers could be introduced as an efficient modification method to enhance the water vapor barrier property of protein-based films due to their inherent hydrophilic nature. According to the results of some recent studies, protein-based biopolymer materials can meet the requirements of a range of food packaging materials. However, the modification of barrier properties is usually required. There is still a substantial interest in protein-based biopolymer films and room for improvement mainly related to their mechanical and barrier properties as well as functional attributes for packaging applications.

Abbreviations

CMC	Carboxymethyl cellulose
DEA	Diethanolamides
GLY	Glycine
LDPE	Low-density polyethylene
PE	Polyethylene
PEG	Polyethylene glycol
PET	Polyethylene terephthalate
PHA	Polyhydroxyalkanoates
PHB	Polyhydroxybutyrate
PHBV	Poly(3-hydroxybutyrate-co-3-hydroxyvalerate)
PLA	Polylactic acid
PP	Polypropylene
SPI	Soy protein isolate
WG	Wheat gluten
WPI	Whey protein isolate
WVP	Water vapor permeability
Xyl	Xylitol

References

1. V. García Ibarra, R. Sendón and A. Rodríguez-Bernaldo De Quirós, in *Antimicrobial Food Packaging Based on Biodegradable Materials*, Elsevier Inc, 2016, https://doi.org/10.1016/B978-0-12-800723-5.00029-2.
2. P. Tyagi, K. S. Salem, M. A. Hubbe and L. Pal, Advances in barrier coatings and film technologies for achieving sustainable packaging of food products – A review, *Trends Food Sci. Technol.*, 2021, **115**, 461–485.
3. H. Chen, J. Wang, Y. Cheng, C. Wang, H. Liu, H. Bian, Y. Pan, J. Sun and W. Han, Application of Protein-Based Films and Coatings for Food Packaging: A Review, *Polymers*, n.d., DOI: 10.3390/polym11122039.
4. T. H. Mekonnen, M. Misra and A. K. Mohanty, Processing, Performance, and Applications of Plant and Animal Protein-Based Blends and Their Biocomposites, in *Biocomposites*, ed. M. Misra, J. K. Pandey and A. K. Mohanty, Woodhead Publishing, 2015, pp. 201–235, https://doi.org/10.1016/B978-1-78242-373-7.00017-2.
5. T. H. Mekonnen, P. G. Mussone, P. Choi and D. C. Bressler, Development of proteinaceous plywood adhesive and optimization of its lap shear strength, *Macromol. Mater. Eng.*, 2015, **300**, 198–209.
6. T. H. Mekonnen, *Valorization of waste protein biomass for bio-based plastics, composites, and adhesives development*, PhD Thesis, University of Alberta, 2014, https://doi.org/10.7939/R3WP9TC41.
7. H. Chen, J. Wang, Y. Cheng, C. Wang, H. Liu, H. Bian, Y. Pan, J. Sun and W. Han, Application of protein-based films and coatings for food packaging: A review, *Polymers*, 2019, **11**, DOI: 10.3390/polym11122039.
8. S. Popović, V. Lazić, N. Hromiš, D. Šuput and S. Bulut, Biopolymer Packaging Materials for Food Shelf-Life Prolongation, in *Biopolymers for Food Design*, 2018, pp. 223–277, http://dx.doi.org/10.1016/B978-0-12-811449-0.00008-6.
9. T. H. Mekonnen, P. G. Mussone, P. Choi and D. C. Bressler, Adhesives from waste protein biomass for oriented strand board composites: development and performance, *Macromol. Mater. Eng.*, 2014, **299**, 1003–1012.
10. E. Ojogbo, E. O. Ogunsona and T. H. Mekonnen, Chemical and physical modifications of starch for renewable polymeric materials, *Mater. Today Sustain.*, 2020, **7–8**, 100028.
11. M. B. Coltelli, F. Wild, E. Bugnicourt, P. Cinelli, M. Lindner, M. Schmid, V. Weckel, K. Müller, P. Rodriguez, A. Staebler, L. Rodríguez-Turienzo and A. Lazzeri, State of the art in the development and properties of protein-based films and coatings and their applicability to cellulose based products: An extensive review, *Coatings*, 2016, **6**, 1–59.
12. B. Cuq, C. Aymard, J.-L. Cuq and S. Guilbert, Edible Packaging Films Based on Fish Myofibrillar Proteins: Formulation and Functional Properties, *J. Food Sci.*, 1995, **60**, 1369–1374.

13. *Edible Films and Coatings for Food Applications*, ed. K. Huber and M. Embuscado, Springer-Verlag, New York, 2009, http://doi.org/10.1007/978-0-387-92824-1.
14. M. R. Guilherme, L. H. C. Mattoso, N. Gontard, S. Guilbert and E. Gastaldi, Composites: Part A Synthesis of nanocomposite films from wheat gluten matrix and MMT intercalated with different quaternary ammonium salts by way of hydroalcoholic solvent casting, *Composites, Part A*, 2010, **41**, 375–382.
15. Y. J. Shin, S. A. Jang and K. Bin Song, Preparation and mechanical properties of rice bran protein composite films containing gelatin or red algae, *Food Sci. Biotechnol.*, 2011, **20**, 703–707.
16. T. J. Herald, K. A. Hachmeister, S. Huang and J. R. Bowers, Corn zein packaging materials for cooked turkey, *J. Food Sci.*, 1996, **61**, 415–418.
17. A. González and C. I. A. Igarzabal, Soy protein – Poly (lactic acid) bilayer films as biodegradable material for active food packaging, *Food Hydrocolloids*, 2013, **33**, 289–296.
18. V. M. Azevedo, A. C. S. de Oliveira, S. V. Borges, J. C. Raguzzoni, M. V. Dias and A. L. R. Costa, Pea protein isolate nanocomposite films for packaging applications: Effect of starch nanocrystals on the structural, *Emirates J. Food Agric.*, 2020, **32**, 495–504.
19. C. V. L. Giosafatto, P. Di Pierro, P. Gunning, A. Mackie, R. Porta and L. Mariniello, Characterization of Citrus pectin edible films containing transglutaminase-modified phaseolin, *Carbohydr. Polym.*, 2014, **106**, 200–208.
20. A. Jangchud and M. S. Chinnan, Properties of peanut protein film: Sorption isotherm and plasticizer effect, *LWT--Food Sci. Technol.*, 1999, **32**, 89–94.
21. M. U. Makeri, F. Abdulmannan, M. A. Ilowefah, C. Chiemela, S. M. Bala and K. Muhammad, Comparative physico-chemical, functional and structural characteristics of winged bean [Psophocarpus tetragonolobus DC] and soybean [Glycine max.] Protein isolates, *J. Food Meas. Charact.*, 2017, **11**, 835–846.
22. M. Javanmard, Shelf life of whey protein-coated pistachio kernel (Pistacia vera L.), *J. Food Process Eng.*, 2008, **31**, 247–259.
23. E. B. Ceresino, E. Johansson, H. H. Sato, T. S. Plivelic, S. A. Hall, J. Bez and R. Kuktaite, Lupin protein isolate structure diversity in frozen-cast foams: Effects of transglutaminases and edible fats, *Molecules*, 2021, **26**, DOI: 10.3390/molecules26061717.
24. F. Ayhllon-Meixueiro, C. Vaca-Garcia and F. Silvestre, Biodegradable films from isolate of sunflower (Helianthus annuus) proteins, *J. Agric. Food Chem.*, 2000, **48**(7), 3032–3036.
25. J. G. de Oliveira Filho, J. M. Rodrigues, A. C. F. Valadares, A. B. de Almeida, T. M. de Lima, K. P. Takeuchi, C. C. F. Alves, H. A. de F.

Sousa, E. R. da Silva, F. H. Dyszy and M. B. Egea, Active food packaging: Alginate films with cottonseed protein hydrolysates, *Food Hydrocolloids*, 2019, **92**, 267–275.

26. A. López-Rubio, A. Blanco-Padilla, K. Oksman and S. Mendoza, Strategies to improve the properties of amaranth protein isolate-based thin films for food packaging applications: Nano-layering through spin-coating and incorporation of cellulose nanocrystals, *Nanomaterials*, 2020, **10**, 1–12.
27. C. Gao, J. Taylor, N. Wellner, Y. B. Byaruhanga, M. L. Parker, E. N. C. Mills and P. S. Belton, Effect of preparation conditions on protein secondary structure and biofilm formation of kafirin, *J. Agric. Food Chem.*, 2005, **53**, 306–312.
28. A. K. Baier and D. Knorr, Influence of high isostatic pressure on structural and functional characteristics of potato protein, *Food Res. Int.*, 2015, **77**, 753–761.
29. N. Fathi, H. Almasi and M. K. Pirouzifard, Effect of ultraviolet radiation on morphological and physicochemical properties of sesame protein isolate based edible films, *Food Hydrocolloids*, 2018, **85**, 136–143.
30. R. J. Avena-Bustillos, B. Chiou, C. W. Olsen, P. J. Bechtel, D. A. Olson and T. H. McHugh, Gelation, Oxygen Permeability, and Mechanical Properties of Mammalian and Fish Gelatin Films, *J. Food Sci.*, 2011, **76**, 519–524.
31. G. Rocha Plácido Moore, S. Maria Martelli, C. Gandolfo, P. José do Amaral Sobral and J. Borges Laurindo, Influence of the glycerol concentration on some physical properties of feather keratin films, *Food Hydrocolloids*, 2006, **20**, 975–982.
32. B. Marelli, M. A. Brenckle, D. L. Kaplan and F. G. Omenetto, Silk Fibroin as Edible Coating for Perishable Food Preservation, *Sci. Rep.*, 2016, **6**, 1–11.
33. J. W. Rhim, A. Gennadios, D. Fu, C. L. Weller and M. A. Hanna, Properties of ultraviolet irradiated protein films, *LWT--Food Sci. Technol.*, 1999, **32**, 129–133.
34. V. C. Lo, Q. Ren, C. Pham, V. Morris, A. Kwan and M. Sunde, Fungal Hydrophobin Proteins Produce Self-Assembling Protein Films with Diverse Structure and Chemical Stability, *Nanomaterials*, 2014, **4**, 827–843.
35. M. Letendre, G. D'Aprano, M. Lacroix, S. Salmieri and D. St-Gelais, Physicochemical properties and bacterial resistance of biodegradable milk protein films containing agar and pectin, *J. Agric. Food Chem.*, 2002, **50**, 6017–6022.
36. M. G. A. Vieira, M. A. Da Silva, L. O. Dos Santos and M. M. Beppu, Natural-based plasticizers and biopolymer films: A review, *Eur. Polym. J.*, 2011, **47**, 254–263.

37. R. Bodmeier and O. Paeratakul, Plasticizer uptake by aqueous colloidal polymer dispersions used for the coating of solid dosage forms, *Int. J. Pharm.*, 1997, **152**, 17–26.
38. J. Siepmann, O. Paeratakul and R. Bodmeier, Modeling plasticizer uptake in aqueous polymer dispersions, *Int. J. Pharm.*, 1998, **165**, 191–200.
39. F. X. B. Santosa and G. W. Padua, Tensile properties and water absorption of zein sheets plasticized with oleic and linoleic acids, *J. Agric. Food Chem.*, 1999, **47**, 2070–2074.
40. T. H. McHugh, J.-F. Aujard and J. M. Krochta, Plasticized Whey Protein Edible Films: Water Vapor Permeability Properties, *J. Food Sci.*, 1994, **59**, 416–419.
41. M. Pommet, A. Redl, M. H. Morel and S. Guilbert, Study of wheat gluten plasticization with fatty acids, *Polymer*, 2002, **44**, 115–122.
42. G. Cherian, C. W. Gennadios and P. Chinachoti, Thermomechanical behavior of wheat gluten films: effect of sucrose, glycerin, and sorbitol, *Cereal Chem.*, 1995, **72**, 1–6.
43. E. Kristo and C. G. Biliaderis, Water Sorption and Thermo-Mechanical Properties of Water/Sorbitol-Plasticized Composite Biopolymer Films: Caseinate–Pullulan Bilayers and Blends, *Food Hydrocolloids*, 2006, **20**(7), 1057–1071.
44. G. Galietta, L. Di Gioia, S. Guilbert and B. Cuq, Mechanical and Thermomechanical Properties of Films Based on Whey Proteins as Affected by Plasticizer and Crosslinking Agents, *J. Dairy Sci.*, 1998, **81**, 3123–3130.
45. M. Thomazine, R. A. Carvalho and P. J. A. Sobral, Physical properties of gelatin films plasticized by blends of glycerol and sorbitol, *J. Food Sci.*, 2005, **70**, 172–176.
46. N. Cao, X. Yang and Y. Fu, Effects of various plasticizers on mechanical and water vapor barrier properties of gelatin films, *Food Hydrocolloids*, 2009, **23**, 729–735.
47. P. Bergo and P. J. A. Sobral, Effects of plasticizer on physical properties of pigskin gelatin films, *Food Hydrocolloids*, 2007, **21**, 1285–1289.
48. P. J. A. Sobral, F. C. Menegalli, M. D. Hubinger and M. A. Roques, Mechanical, water vapor barrier and thermal properties of gelatin based edible films, *Food Hydrocolloids*, 2001, **15**, 423–432.
49. L. C. Bertan, P. S. Tanada-Palmu, A. C. Siani and C. R. F. Grosso, Effect of fatty acids and "Brazilian elemi" on composite films based on gelatin, *Food Hydrocolloids*, 2005, **19**, 73–82.
50. P. J. A. Sobral, E. S. Monterrey-Q and A. M. Q. B. Habitante, Glass transition study of Nile Tilapia myofibrillar protein films plasticized by glycerin and water, *J. Therm. Anal. Calorim.*, 2002, **67**, 499–504.
51. F. M. Vanin, P. J. A. Sobral, F. C. Menegalli, R. A. Carvalho and A. M. Q. B. Habitante, Effects of plasticizers and their concentrations on

thermal and functional properties of gelatin-based films, *Food Hydrocolloids*, 2005, **19**, 899–907.
52. N. E. Suyatma, L. Tighzert, A. Copinet and V. Coma, Effects of hydrophilic plasticizers on mechanical, thermal, and surface properties of chitosan films, *J. Agric. Food Chem.*, 2005, **53**, 3950–3957.
53. A. Jongjareonrak, S. Benjakul, W. Visessanguan and M. Tanaka, Fatty acids and their sucrose esters affect the properties of fish skin gelatin-based film, *Eur. Food Res. Technol.*, 2006, **222**, 650–657.
54. J. Zink, T. Wyrobnik, T. Prinz and M. Schmid, Physical, chemical and biochemical modifications of protein-based films and coatings: An extensive review, *Int. J. Mol. Sci.*, 2016, DOI: 10.3390/ijms17091376.
55. K. M. Kim, C. L. Weller, M. A. Hanna and A. Gennadios, Heat Curing of Soy Protein Films at Atmospheric and Sub-atmospheric Conditions, *J. Food Sci.*, 2006, DOI: 10.1111/j.1365-2621.2002.tb10663.x.
56. H. Singh and F. MacRitchie, Changes in proteins induced by heating gluten dispersions at high temperature, *J. Cereal Sci.*, 2004, **39**, 297–301.
57. C. M. Lakemond, H. H. J. de Jongh, M. Hessing, H. Gruppen and A. G. J. Voragen, Heat Denaturation of Soy Glycinin: Influence of pH and Ionic Strength on Molecular Structure, *J. Agric. Food Chem.*, 2000, 1991–1995.
58. N. Parris, J. M. Purcell and S. M. Ptashkin, Thermal Denaturation of Whey Proteins in Skim Milk, *J. Agric. Food Chem.*, 1991, 2167–2170.
59. V. Micard, R. Belamri, M. H. Morel and S. Guilbert, Properties of chemically and physically treated wheat gluten films, *J. Agric. Food Chem.*, 2000, **48**, 2948–2953.
60. Y. M. Stuchell and J. M. Krochta, Enzymatic Treatments and Thermal Effects on Edible Soy Protein Films, *J. Food Sci.*, 1994, **59**, 1332–1337.
61. M. Pérez-Gago and J. M. Krochta, Denaturation Time and Temperature Effects on Solubility, Tensile Properties, and Oxygen Permeability of Whey Protein Edible Films, *J. Food Sci.*, 2001, **66**, 705–710.
62. C. Vachon, H. Yu, R. Yefsah, R. Alain and M. Lacroix, Mechanical and Structural Properties of Milk Protein Edible Films Cross-Linked by Heating and γ -Irradiation, *Polymers*, 2000, 3202–3209.
63. V. V. Mozhaev, K. Heremans, J. Frank, P. Masson and C. Balny, High pressure effects on protein structure and function, *Proteins Struct. Funct. Genet.*, 1996, **24**, 81–91.
64. K. Heremans, High pressure effects on proteins and other biomolecules, *Annu. Rev. Biophys. Bioeng.*, 1982, **11**, 1–21.
65. V. Ya and T. Haertlé, Reducer driven baric denaturation and oligomerisation of whey proteins, *J. Biotechnol.*, 2000, **79**, 205–209.
66. C. J. Schaschke, Induced structural change to β -lactoglobulin by combined pressure and temperature, *Biochem. Eng. J.*, 2000, **5**, 73–76.
67. C. Kanno, T. Mu, T. Hagiwara, M. Ametani and N. Azuma, Gel Formation from Industrial Milk Whey Proteins under Hydrostatic Pressure:

Effect of Hydrostatic Pressure and Protein Concentration, *J. Agric. Food Chem.*, 1998, **8561**, 417–424.
68. J. Keränen, J. Hakala and V. Hongisto, The Sound Insulation of Façades at Frequencies 5–5000 Hz, *Build. Environ.*, 2019, **156**, 12–20.
69. D. M. Kadam, M. Thunga, S. Wang, M. R. Kessler, D. Grewell, B. Lamsal and C. Yu, Preparation and characterization of whey protein isolate films reinforced with porous silica coated titania nanoparticles, *J. Food Eng.*, 2013, **117**, 133–140.
70. R. Banerjee, H. Chen and J. Wu, Milk protein-based edible film mechanical strength changes due to ultrasound process, *J. Food Sci.*, 1996, **61**, 824–828.
71. S. Guilbert, N. Gontard and B. Cuq, Technology and applications of edible protective films, *Packag. Technol. Sci.*, 1995, **8**, 339–346.
72. Z. Wang, X. X. Sun, Z. X. Lian, X. X. Wang, J. Zhou and Z. S. Ma, The effects of ultrasonic/microwave assisted treatment on the properties of soy protein isolate/microcrystalline wheat-bran cellulose film, *J. Food Eng.*, 2013, **114**, 183–191.
73. A. Gennadios, J. W. Rhim, A. Handa, C. L. Weller and M. A. Hanna, Ultraviolet radiation affects physical and molecular properties of soy protein films, *J. Food Sci.*, 1998, **63**, 225–228.
74. W. W. Nawar, Reaction mechanisms in the radiolysis of fats: A review, *J. Agric. Food Chem.*, 1978, **26**, 21–25.
75. B. R. A. Hurley, A. Ouzts, J. Fischer and T. Gomes, PAPER PRESENTED AT IAPRI WORLD CONFERENCE 2012 Effects of Private and Public Label Packaging on Consumer Purchase Patterns, *Packag. Technol. Sci.*, 2013, **29**, 399–412.
76. Z. Ustunol and B. Mert, Water solubility, mechanical, barrier, and thermal properties of cross-linked whey protein isolate-based films, *J. Food Sci.*, 2004, **69**, DOI: 10.1111/j.1365-2621.2004.tb13365.x.
77. K. Cieśla, S. Salmieri and M. Lacroix, γ-irradiation influence on the structure and properties of calcium caseinate-whey protein isolate based films. Part 2. Influence of polysaccharide addition and radiation treatment on the structure and functional properties of the films, *J. Agric. Food Chem.*, 2006, **54**, 8899–8908.
78. M. Lacroix, T. C. Le, B. Ouattara, H. Yu, M. Letendre, S. F. Sabato, M. A. Mateescu and G. Patterson, Use of γ-irradiation to produce films from whey, casein and soya proteins: Structure and functionals characteristics, *Radiat. Phys. Chem.*, 2002, **63**, 827–832.
79. S. F. Sabato, B. Ouattara, H. Yu, G. D'Aprano, C. Le Tien, M. A. Mateescu and M. Lacroix, Mechanical and barrier-properties of cross-linked soy and whey protein based films, *J. Agric. Food Chem.*, 2001, **49**, 1397–1403.
80. D. Brault, G. D'Aprano and M. Lacroix, Formation of Free-Standing Sterilized Edible Films from Casemates, *J. Agric. Food Chem.*, 1997, **45**, 2964–2969.

81. H. Köksel, H. D. Sapirstein, S. Çelik and W. Bushuk, Effects of gamma-irradiation of wheat on gluten proteins, *J. Cereal Sci.*, 1998, **28**, 243–250.
82. D. G. Smyth, Chemical modifications of proteins, *FEBS Lett.*, 1977, **74**, 148–149.
83. J. J. Kester and T. Richardson, Modification of Whey Proteins to Improve Functionality, *J. Dairy Sci.*, 1984, **67**, 2757–2774.
84. F. Hammann and M. Schmid, Determination and quantification of molecular interactions in protein films: A review, *Materials*, 2014, **7**, 7975–7996.
85. F. F. Shih, Interaction of soy isolate with polysaccharide and its effect on film properties, *J. Am. Oil Chem. Soc.*, 1994, **71**, 1281–1285.
86. V. M. Ghorpade, H. Li, A. Gennadios and M. A. Hanna, Chemically modified soy protein films, *Trans. Am. Soc. Agric. Eng.*, 1995, **38**, 1805–1808.
87. M. Wihodo and C. I. Moraru, Physical and chemical methods used to enhance the structure and mechanical properties of protein films: A review, *J. Food Eng.*, 2013, **114**, 292–302.
88. A. Gennadios, A. H. Brandenburg, C. L. Weller and R. F. Testin, Effect of pH on Properties of Wheat Gluten and Soy Protein Isolate Films, *J. Agric. Food Chem.*, 1993, **41**, 1835–1839.
89. M. B. Pérez-Gago and J. M. Krochta, Water vapor permeability of whey protein emulsion films as affected by pH, *J. Food Sci.*, 1999, **64**, 695–698.
90. N. Reddy, R. Reddy and Q. Jiang, Crosslinking biopolymers for biomedical applications, *Trends Biotechnol.*, 2015, **33**, 362–369.
91. J. S. Wang, M. M. Zhao, X. Q. Yang, Y. M. Jiang and C. Chun, Gelation behavior of wheat gluten by heat treatment followed by transglutaminase cross-linking reaction, *Food Hydrocolloids*, 2007, **21**, 174–179.
92. E.-T. Nayef, T. H. Mekonnen, P. Mussone, D. Bressler and P. Choi, Nonisothermal DSC Study of Epoxy Resins Cured with Hydrolyzed Specified Risk Material, *Ind. Eng. Chem. Res.*, 2013, **52**, 8189–8199.
93. M. Yildirim and N. S. Hettiarachchy, Properties of films produced by cross-linking whey proteins and 11s globulin using transglutaminase, *J. Food Sci.*, 1998, **63**, 248–252.
94. J. H. Oh, B. Wang, P. D. Field and H. A. Aglan, Characteristics of edible films made from dairy proteins and zein hydrolysate cross-linked with transglutaminase, *Int. J. Food Sci. Technol.*, 2004, **39**, 287–294.
95. P. Di Pierro, B. Chico, R. Villalonga, L. Mariniello, A. E. Damiao, P. Masi and R. Porta, Chitosan-whey protein edible films produced in the absence or presence of transglutaminase: Analysis of their mechanical and barrier properties, *Biomacromolecules*, 2006, **7**, 744–749.
96. Y. Zhu, A. Rinzema, J. Tramper and J. Bol, Original paper Microbial transglutaminase – a review of its production and application in food processing, *Appl. Microbiol. Biotechnol.*, 1995, **44**, 277–282.

97. G. A. H. DeJong and S. J. Koppelman, Transglutaminase Catalyzed Reactions, *J. Food Sci.*, 2002, **67**, 2798–2806.
98. V. Den Truong, D. A. Clare, G. L. Catignani and H. E. Swaisgood, Cross-Linking and Rheological Changes of Whey Proteins Treated with Microbial Transglutaminase, *J. Agric. Food Chem.*, 2004, **52**, 1170–1176.
99. A. S. Eissa, C. Puhl, J. F. Kadla and S. A. Khan, Enzymatic cross-linking of β-lactoglobulin: Conformational properties using FTIR spectroscopy, *Biomacromolecules*, 2006, **7**, 1707–1713.
100. C. H. Tang, Y. Jiang, Q. B. Wen and X. Q. Yang, Effect of transglutaminase treatment on the properties of cast films of soy protein isolates, *J. Biotechnol.*, 2005, **120**, 296–307.
101. V. Falguera, J. P. Quintero, A. Jiménez, J. A. Muñoz and A. Ibarz, Edible films and coatings: Structures, active functions and trends in their use, *Trends Food Sci. Technol.*, 2011, **22**, 292–303.
102. J. J. Zhou, S. Y. Wang and S. Gunasekaran, Preparation and characterization of whey protein film incorporated with TiO_2 nanoparticles, *J. Food Sci.*, 2009, **74**, DOI: 10.1111/j.1750-3841.2009.01270.x.
103. R. Sothornvit, J. W. Rhim and S. I. Hong, Effect of nano-clay type on the physical and antimicrobial properties of whey protein isolate/clay composite films, *J. Food Eng.*, 2009, **91**, 468–473.
104. P. Oymaci and S. A. Altinkaya, Improvement of barrier and mechanical properties of whey protein isolate based food packaging films by incorporation of zein nanoparticles as a novel bionanocomposite, *Food Hydrocolloids*, 2016, **54**, 1–9.
105. P. Kumar, K. P. Sandeep, S. Alavi, V. D. Truong and R. E. Gorga, Preparation and characterization of bio-nanocomposite films based on soy protein isolate and montmorillonite using melt extrusion, *J. Food Eng.*, 2010, **100**, 480–489.
106. P. Kumar, K. P. Sandeep, S. Alavi, V. D. Truong and R. E. Gorga, Effect of Type and Content of Modified Montmorillonite on the Structure and Properties of Bio-Nanocomposite Films Based on Soy Protein Isolate and Montmorillonite, *J. Food Sci.*, 2010, **75**, 46–56.
107. A. González and C. I. Alvarez Igarzabal, Nanocrystal-reinforced soy protein films and their application as active packaging, *Food Hydrocolloids*, 2015, **43**, 777–784.
108. X. Li, N. Ji, C. Qiu, M. Xia, L. Xiong and Q. Sun, The effect of peanut protein nanoparticles on characteristics of protein- and starch-based nanocomposite films: A comparative study, *Ind. Crops Prod.*, 2015, **77**, 565–574.
109. S. Tunc, H. Angellier, Y. Cahyana, P. Chalier, N. Gontard and E. Gastaldi, Functional properties of wheat gluten / montmorillonite nanocomposite films processed by casting, *J. Membr. Sci.*, 2007, **289**, 159–168.

110. E. O. Ogunsona, R. Muthuraj, E. Ojogbo, O. Valerio and T. H. Mekonnen, Engineered nanomaterials for antimicrobial applications: A review, *Appl. Mater. Today.*, 2020, **18**, 100473.
111. K. Friesen, C. Chang and M. Nickerson, Incorporation of phenolic compounds, rutin and epicatechin, into soy protein isolate films: Mechanical, barrier and cross-linking properties, *Food Chem.*, 2015, **172**, 18–23.
112. T. H. Shellhammer and J. M. Krochta, Whey Protein Emulsion Film Performance as Affected by Lipid Type and Amount, *J. Food Sci.*, 1997, **62**, 390–394.
113. M. B. Pérez-Gago and J. M. Krochta, Lipid Particle Size Effect on Water Vapor Permeability and Mechanical Properties of Whey Protein / Beeswax Emulsion Films, *J. Agric. Food Chem.*, 2001, **49**, 996–1002.
114. A. Jiménez, M. J. Fabra, P. Talens and A. Chiralt, Effect of lipid self-association on the microstructure and physical properties of hydroxypropyl-methylcellulose edible films containing fatty acids, *Carbohydr. Polym.*, 2010, **82**, 585–593.
115. T. Bourtoom, Edible protein films: Properties enhancement, *Int. Food Res. J.*, 2009, **16**, 1–9.
116. J. Bonilla, L. Atarés, M. Vargas and A. Chiralt, Edible films and coatings to prevent the detrimental effect of oxygen on food quality: Possibilities and limitations, *J. Food Eng.*, 2012, **110**, 208–213.
117. S. Sahraee, J. M. Milani, J. M. Regenstein and H. S. Kafil, Protection of foods against oxidative deterioration using edible films and coatings: A review, *Food Biosci.*, 2019, **32**, DOI: 10.1016/j.fbio.2019.100451.
118. R. M. Robles-Sánchez, M. A. Rojas-Graü, I. Odriozola-Serrano, G. González-Aguilar and O. Martin-Belloso, Influence of alginate-based edible coating as carrier of antibrowning agents on bioactive compounds and antioxidant activity in fresh-cut Kent mangoes, *LWT--Food Sci. Technol.*, 2013, **50**, 240–246.
119. L. Yang and A. T. Paulson, Effects of lipids on mechanical and moisture barrier properties of edible gellan film, *Food Res. Int.*, 2000, **33**, 571–578.
120. P. M. Tomasula, W. C. Yee and N. Parris, Oxygen permeability of films made from CO_2-precipitated casein and modified casein, *J. Agric. Food Chem.*, 2003, **51**, 634–639.
121. B. Ghanbarzadeh, A. R. Oromiehie, M. Musavi, P. M. Falcone, Z. E. D-Jomeh and E. R. Rad, Study of mechanical properties, oxygen permeability and AFM topography of zein films plasticized by polyols, *Packag. Technol. Sci.*, 2007, **20**, 155–163.
122. K. S. Miller and J. M. Krochta, Oxygen and aroma barrier properties of edible films: A review, *Trends Food Sci. Technol.*, 1997, **8**, 228–237.
123. J. H. Han and J. M. Krochta, Physical properties of whey protein coating solutions and films containing antioxidants, *J. Food Sci.*, 2007, **72**, DOI: 10.1111/j.1750-3841.2007.00358.x.

124. J. I. Maté and J. M. Krochta, Oxygen uptake model for uncoated and coated peanuts, *J. Food Eng.*, 1998, **35**, 299–312.
125. K. S. Eça, T. Sartori and F. C. Menegalli, Films and Edible Coatings Containing Antioxidants - A Review, *Braz. J. Food Technol.*, 2014, **17**(2), DOI: 10.1590/bjft.2014.017.
126. J. Li, Z. Wu, C. Huang, H. Liu, R. Huang and L. Li, Mechanical properties of cyanate ester/epoxy nanocomposites modified with plasma functionalized MWCNTs, *Compos. Sci. Technol.*, 2014, **90**, 166–173.
127. M. J. Fabra, A. Hambleton, P. Talens, F. Debeaufort and A. Chiralt, Effect of ferulic acid and α-tocopherol antioxidants on properties of sodium caseinate edible films, *Food Hydrocolloids*, 2011, **25**, 1441–1447.
128. D. B. Armitage, N. S. Hettiarachchy and M. A. Monsoor, Natural antioxidants as a component of an egg albumen film in the reduction of lipid oxidation in cooked and uncooked poultry, *J. Food Sci.*, 2002, **67**, 631–634.
129. M. Oussalah, S. Caillet, S. Salmiéri, L. Saucier and M. Lacroix, Antimicrobial and antioxidant effects of milk protein-based film containing essential oils for the preservation of whole beef muscle, *J. Agric. Food Chem.*, 2004, **52**, 5598–5605.
130. J. Gómez-Estaca, L. Bravo, M. C. Gómez-Guillén, A. Alemán and P. Montero, Antioxidant properties of tuna-skin and bovine-hide gelatin films induced by the addition of oregano and rosemary extracts, *Food Chem.*, 2009, **112**, 18–25.
131. J. Gómez-Estaca, B. Giménez, P. Montero and M. C. Gómez-Guillén, Incorporation of antioxidant borage extract into edible films based on sole skin gelatin or a commercial fish gelatin, *J. Food Eng.*, 2009, **92**, 78–85.
132. Y. Byun, Y. T. Kim and S. Whiteside, Characterization of an antioxidant polylactic acid (PLA) film prepared with α-tocopherol, BHT and polyethylene glycol using film cast extruder, *J. Food Eng.*, 2010, **100**, 239–244.
133. K. Norajit, K. M. Kim and G. H. Ryu, Comparative studies on the characterization and antioxidant properties of biodegradable alginate films containing ginseng extract, *J. Food Eng.*, 2010, **98**, 377–384.
134. S. M. Ojagh, M. Rezaei, S. H. Razavi and S. M. H. Hosseini, Effect of chitosan coatings enriched with cinnamon oil on the quality of refrigerated rainbow trout, *Food Chem.*, 2010, **120**, 193–198.
135. B. W. S. Souza, M. A. Cerqueira, H. A. Ruiz, J. T. Martins, A. Casariego, J. A. Teixeira and A. A. Vicente, Effect of Chitosan-based coatings on the shelf life of Salmon (Salmo salar), *J. Agric. Food Chem.*, 2010, **58**, 11456–11462.

136. I. H. Kim, H. J. Yang, B. S. Noh, S. J. Chung and S. C. Min, Development of a defatted mustard meal-based composite film and its application to smoked salmon to retard lipid oxidation, *Food Chem.*, 2012, **133**, 1501–1509.
137. J. Wu, S. Chen, S. Ge, J. Miao, J. Li and Q. Zhang, Preparation, properties and antioxidant activity of an active film from silver carp (Hypophthalmichthys molitrix) skin gelatin incorporated with green tea extract, *Food Hydrocolloids*, 2013, **32**, 42–51.
138. B. Kodal Coşkun, E. Çalikoğlu, Z. Karagöz Emiroğlu and K. Candoğan, Antioxidant active packaging with soy edible films and oregano or thyme essential oils for oxidative stability of ground beef patties, *J. Food Qual.*, 2014, **37**, 203–212.
139. B. Marcos, C. Sárraga, M. Castellari, F. Kappen, G. Schennink and J. Arnau, Development of biodegradable films with antioxidant properties based on polyesters containing α-tocopherol and olive leaf extract for food packaging applications, *Food Packag. Shelf Life*, 2014, **1**, 140–150.
140. A. V. Júnior, N. Fronza, F. B. Foralosso, D. Dezen, E. Huber, J. H. Z. dos Santos, R. A. F. Machado and M. G. N. Quadri, Biodegradable Duo-functional Active Film: Antioxidant and Antimicrobial Actions for the Conservation of Beef, *Food Bioprocess Technol.*, 2015, **8**, 75–87.
141. S. Lekjing, A chitosan-based coating with or without clove oil extends the shelf life of cooked pork sausages in refrigerated storage, *Meat Sci.*, 2016, **111**, 192–197.
142. M. Nemazifard, G. Kavoosi, Z. Marzban and N. Ezedi, Physical, mechanical, water binding, and antioxidant properties of cellulose dispersions and cellulose film incorporated with pomegranate seed extract, *Int. J. Food Prop.*, 2017, DOI: 10.1080/10942912.2016.1219369.
143. S. M. Martelli, C. Motta, T. Caon, J. Alberton, I. C. Bellettini, A. C. P. do Prado, P. L. M. Barreto and V. Soldi, Edible carboxymethyl cellulose films containing natural antioxidant and surfactants: α-tocopherol stability, in vitro release and film properties, *LWT--Food Sci. Technol.*, 2017, **77**, 21–29.
144. M. Radi, E. Firouzi, H. Akhavan and S. Amiri, Effect of gelatin-based edible coatings incorporated with Aloe vera and black and green tea extracts on the shelf life of fresh-cut oranges, *J. Food Qual.*, 2017, **2017**, DOI: 10.1155/2017/9764650.
145. B. Ouattara, L. T. Canh, C. Vachon, M. A. Mateescu and M. Lacroix, Use of γ-irradiation cross-linking to improve the water vapor permeability and the chemical stability of milk protein films, *Radiat. Phys. Chem.*, 2002, **63**, 821–825.
146. E. A. Soliman and M. Furuta, Influence of γ-irradiation on mechanical and water barrier properties of corn protein-based films, *Radiat. Phys. Chem.*, 2009, **78**, 651–654.

147. S. L. Lee, M. S. Lee and K. B. Song, Effect of gamma-irradiation on the physicochemical properties of gluten films, *Food Chem.*, 2005, **92**, 621–625.
148. N. Suderman, M. I. N. Isa and N. M. Sarbon, The effect of plasticizers on the functional properties of biodegradable gelatin-based film: A review, *Food Biosci.*, 2018, **24**, 111–119.
149. S. Fakhreddin Hosseini, M. Rezaei, M. Zandi and F. F. Ghavi, Preparation and functional properties of fish gelatin-chitosan blend edible films, *Food Chem.*, 2013, **136**, 1490–1495.
150. S. J. Calva-Estrada, M. Jiménez-Fernández and E. Lugo-Cervantes, Protein-Based Films: Advances in the Development of Biomaterials Applicable to Food Packaging, *Food Eng. Rev.*, 2019, 78–92.
151. A. Ullah, T. Vasanthan, D. Bressler, A. L. Elias and J. Wu, Bioplastics from feather quill, *Biomacromolecules*, 2011, **12**, 3826–3832.
152. N. Parris and D. R. Coffin, Composition Factors Affecting the Water Vapor Permeability and Tensile Properties of Hydrophilic Zein Films, *J. Agric. Food Chem.*, 1997, **45**, 1596–1599.
153. N. Parris and D. R. Coffin, Composition Factors Affecting the Water Vapor Permeability and Tensile Properties of Hydrophilic Films, *J. Agric. Food Chem.*, 1995, **43**, 1432–1435.
154. N. B. Shaw, F. J. Monahan, E. D. O'Riordan and M. O'Sullivan, Physical properties of WPI films plasticized with glycerol, xylitol, or sorbitol, *J. Food Sci.*, 2002, **67**, 164–167.
155. D. C. W. Siew, C. Heilmann, A. J. Easteal and R. P. Cooney, Solution and film properties of sodium caseinate/glycerol and sodium caseinate/polyethylene glycol edible coating systems, *J. Agric. Food Chem.*, 1999, **47**, 3432–3440.

CHAPTER 4

Lipid-based Materials in Edible and Non-edible Food Packaging

DEBARSHI NATH[a], PRIYANKA CHAKRABORTY[b], SUBROTA HATI[c], BIRENDRA KUMAR MISHRA[b] AND PREETAM SARKAR*[a]

[a]Department of Food Process Engineering, National Institute of Technology Rourkela, India; [b]Department of Rural Development and Agricultural Production, North-Eastern Hill University, Tura Campus, India; [c]Department of Dairy Microbiology, SMC College of Dairy Science, Kamdhenu University, Anand, India
*E-mail: sarkarpreetam@nitrkl.ac.in

4.1 Introduction

Food packaging is used to preserve the packaged products from external factors, such as light, temperature, and humidity. All these external factors are responsible for the rapid degradation of food products and reduced shelf life. Moreover, packaging protects the food materials from other external factors, such as contaminants, dust, odor, and external shocks.[1] Therefore, proper packaging of food products is of paramount importance, and as such, numerous research studies have been undertaken to develop non-toxic packaging materials for food products. Conventional packaging materials are generally produced from petroleum-based plastics, which are non-biodegradable and pose a great threat to the environment. Plastics are the most popular and easily available resources for food packaging due to their superior tensile properties, low price, and easy manufacturing.

Smart Materials No. 41
Functional Materials from Lipids and Proteins
Edited by Nandika Bandara and Aman Ullah

Published by the Royal Society of Chemistry, www.rsc.org

These plastics, when discarded, create litter and pose an acute waste management problem.[2] A report from the United States Environmental Protection Agency stated that the total municipal waste produced in 2014 was 260 million tonnes, of which 14.32 million tonnes came from petroleum-based packaging wastes.[3] These synthetic materials have numerous adverse effects on humans.[4,5] Furthermore, due to several technical and economic challenges, the recycling of synthetic packaging wastes is limited; in fact, less than 3% of waste plastic is recycled globally.[6] Therefore, people focus on biodegradable packaging films and coatings to reduce the wastage generated by petroleum-based packaging materials. A significant interest in and innovative research on edible and non-edible packaging have been observed in the food packaging industries as a result of increasing consumer demand for safe, high-quality, nutritious food products with extended shelf lives along with environmental awareness of scarce natural resources and the environmental repercussions of packaging waste.[7,8]

With this in view, different biomaterials, such as polysaccharides, proteins, and lipids, are gaining worldwide attention for their applications in packaging systems to formulate edible and non-edible films and coatings capable of replacing conventional plastic materials.[6] Polysaccharide- and protein-based films and coatings have high mechanical properties and an excellent barrier to oxygen at low and intermediate relative humidity, but their barrier to water vapor is poor due to their hydrophilic nature. On the other hand, films and coatings made using lipid compounds have excellent moisture and gas barrier attributes but are typically opaque and inflexible.[9] Some lipids have favorable properties concerning film formation. These properties include compatibility with other film-forming ingredients, excellent moisture and gas barrier properties, and the ability to act as a thickening agent in film-forming solutions, making them attractive candidates for greater use in edible and non-edible films and coatings.[9] Different lipids used in the film-forming process include essential oils, waxes, resins, and fatty acids. These compounds enhance the functionality of the films and coatings by imparting antimicrobial and antioxidant properties and improving their hydrophobicity.[9,10] Essential oils are the most common lipid compounds used in preparing edible and non-edible films/coatings. This is because of their robust antimicrobial and antioxidant activities, ultimately leading to the improved shelf life of food products.[1] Therefore, composite lipid-based films and coatings are effective in preserving the quality of food products. However, their activity is largely dependent on the chemical structure of the lipid particles and their arrangement and interaction with other film components (*e.g.* polysaccharides, proteins, and bioactive compounds).[9]

This chapter discusses the different types of lipids used for synthesizing edible and non-edible films/coatings. The effects of lipids on the properties of films, along with some study findings that showcase the ability of lipids to enhance the shelf life of food products, have also been highlighted in this chapter.

4.2 Different Types of Lipids Used in Films and Coatings

4.2.1 Essential Oils

Essential oils are chemical compounds that are derived from plants. In addition to antimicrobial and antioxidant activities, essential oils are also well recognized for other health benefits, such as anti-cancer, anti-diabetic, anti-anxiety, and anti-inflammatory properties.[11] The type of soil utilized, the local climate, and the stage of development when the plant is harvested all affect the quality and attributes of essential oils. However, they also depend on the extraction method and the solvent used.[12,13]

The use of essential oil in food packaging and as a food additive is currently the subject of extensive investigation because of its potential to replace synthetic chemicals when incorporated with polymers owing to the detrimental effects of synthetic chemicals. They are loaded into the packaging system, so the compound is gradually released into the atmosphere around the food product. Numerous plants, including *Melaleuca alternifolia* (tea), *Salvia rosmarinus* (rosemary), *Thymus vulgaris* (thyme), *Zingiber officinale* (ginger), *Cinnamomum verum* (cinnamon), *Syzygium aromaticum* (clove), and *Lavandula angustifolia* (lavender), produce different kinds of essential oils.[14] The FDA, however, vigorously contests the product's safety and potential health benefits and restricts its sale in the market because the FDA has not yet approved the claim of diagnosis, treatment, and illness prevention.

4.2.2 Waxes

Waxes are a heterogeneous class of lipids consisting of non-polar substances and hydrocarbons. Chemically, waxes are an ester of long-chain aliphatic acid with a long-chain aliphatic alcohol. Waxes are widely used for packaging baked goods, confectionaries, sweets, and dairy products because of their semi-transparency and hydrophobic nature. The waxes can be categorized into natural, petroleum-based, and synthetic waxes.[15] Petroleum waxes include montan wax and paraffin wax, while natural waxes include candelilla wax, soy wax, beeswax, and carnauba wax.[16] Almost 70% of the waxes available in the market are petroleum-based waxes. However, dwindling oil reserves, fluctuations in crude oil prices, and concerns about the long-term viability of petroleum-based products have prompted a search for alternatives.[17] Due to their abundance and biodegradability, bio-derived waxes are a promising source for developing renewable and eco-friendly wax alternatives. Partial or full hydrogenation is the most prevalent process for synthesizing bio-based waxes. Because waxes have high moisture barrier properties, they are used in food packaging materials. As a result, they can protect food products from moisture present in the environment or prevent moisture loss from the food.[16] This property of waxes is attributed to their lack of polar constituents or significant hydrophilic portions that could react with water molecules, resulting in the obstruction of moisture passage through them.[9] The application of wax coatings to fresh produce-type products that require cold storage improves respiratory functions and prevents perspiration, thereby extending shelf life.[18]

4.2.3 Fatty Acids and Triglycerides

Fatty acids are the main components of triacylglycerols (also known as triglycerides), phospholipids, and other complex lipids. These fatty acids and lipids are widely present in nature. Therefore, triacylglycerols, and thus fatty acids, are the primary contributors to the dietary fat[19] in humans. Fatty acids are polar lipids mainly used as dispersing agents and emulsifiers. Besides these applications, fatty acids with a carbon atom range of 14–18 are used to develop films and coatings.[6] Most fatty acids generated from vegetable oils are generally recognized as safe (GRAS) and are widely used in producing edible and non-edible films/coatings.[9,20] The most important characteristics of fatty acids and lipids derived from them depend on their degree of saturation, chain length, and physical state, which significantly affect the properties of films and coatings. Melting points of fatty acids generally increase with the chain length and decrease with the number of double bonds. Additionally, the water vapor permeability (WVP) of fatty acid-incorporated films and coatings is largely affected by the chain length and the degree of saturation of the fatty acids.[9] Some examples of fatty acids used in food packaging systems are linolenic acid, linoleic acid, oleic acid, and stearic acid, among many others.

Triglycerides are esters of fatty acids derived from glycerol and three fatty acids. They exhibit enhanced polarity compared to waxes that are insoluble in water; however, they form a stable monolayer by spreading at the interface. The hydrophobic nature of triglycerides primarily depends on their structure, where long-chain triglycerides are insoluble, while short-chain triglycerides are partly soluble in water.[7,9] This has been attributed to the increased mobility of hydrocarbon chains and reduced efficiency of lateral packing of acyl chains caused by the decreased chain interactions in van der Waals attraction.[7] Table 4.1 presents different types of lipid materials used in films and coatings.

Table 4.1 Commonly used lipids in food packaging and their sources.

Name of the lipid compound	Derived from	Ref.
Basil essential oil	*Ocimum basilicum*	21
Oregano essential oil	*Origanum vulgare*	22
Cinnamon essential oil	*Cinnamomum cassia*	22
Rosemary essential oil	*Salvia rosmarinus*	22
Citronella essential oil	*Cymbopogon nardus*	23
Peppermint essential oil	*Mentha × piperita* L.	24
Bergamot essential oil	*Citrus bergamia*	25
Lemon essential oil	*Citrus limon*	25
Beeswax	*Apis mellifera*	26
Candelilla wax	*Euphorbia antisyphilitica*	26
Sunflower wax	*Helianthus annuus*	26
Carnauba wax	*Copernicia cerifera*	26

4.3 Applications of Lipid-based Edible and Non-edible Films and Coatings

4.3.1 Essential Oil-incorporated Films and Coatings

The application of essential oils as a preservative and natural food additive has been the subject of numerous studies due to their potential to replace synthetic food additives and preservatives, as these synthetic chemicals adversely affect the nutritional content of food products and can harm consumers in the long run. These factors have immensely contributed to the popularity of essential oils.[1] Essential oils can be impregnated into food packaging films and coatings for their release into the food matrix over time. An edible film or wrap is any substance used to wrap meals to enhance their shelf life, which can be eaten along with the food with or without further removal of the coating.[27] Furthermore, they can also be combined with metallic and organic nanomaterials to further improve the functionality of the packaging system through a synergistic effect.

Hashemi and Mousavi Khaneghah,[28] in their research, prepared basil seed gum-based edible film incorporated with different concentrations of oregano essential oil (1–6% v/v). Upon loading of essential oil, the WVP of the films decreased significantly (4.33 to 3.82 × 10^{-11} g s^{-1} m^{-2} Pa^{-1}). This significant reduction in WVP values was due to the decreased binding ability and the increased tortuosity. This indicated that the mean path length the water molecule covers to penetrate the film matrix would be greater in the case of films with 6% oil concentration. The transparency of the films also increased from 2.11 to 4.77. This indicated that the film became more transparent with increased oil loading due to the transparent and clear nature of the oregano essential oil. The incorporation of essential oils also enhanced the active properties of the biopolymer films. Essential oils inhibited foodborne pathogens, such as *Salmonella typhimurium*, *Bacillus cereus*, *Escherichia coli*, *Staphylococcus aureus,* and

Pseudomonas aeruginosa. The highest zone of inhibition (ZOI), measuring 42.3 mm, was reported for 6% oil loaded with *B. cereus*. Furthermore, the antioxidant activity of the films also increased with an increase in oil concentration, as evident from the 2,2′-azinobis-(3-ethylbenzothiazoline-6-sulfonate) assay (ABTS), 2,2′-diphenyl-1-picrylhydrazyl radical assay (DPPH), and ferric-reducing ability tests.

In a study by Agarwal *et al.*,[29] xyloglucan-rich tamarind kernel powder was incorporated with geraniol at different concentrations (0.5–3.0% w/w) to prepare antimicrobial food packaging films. The authors reported that mechanical attributes, such as tensile strength (TS) and Young's modulus (YM), reduced after the incorporation of geraniol. It was concluded that this decrease in tensile properties was due to the disruption of the biopolymer network upon geraniol impregnation, which replaces the strong biopolymer–biopolymer network with a weak biopolymer–lipid network. The water vapor permeability rate (WVTR) of films reduced with the increase in oil concentration in the polymer matrix. The reason for the reduction is due to the increase in tortuosity, preventing the penetration of moisture through the film surface. Furthermore, the active films demonstrated profound antimicrobial activity against *B. cereus* and *E. coli* as geraniol is a monoterpene alcohol potent against Gram-positive and Gram-negative bacteria.

Similarly, Giteru *et al.*[30] synthesized kafirin-based films by incorporating citral for food packaging applications. The TS of the films reduced, while the elongation at break (EAB) improved upon oil loading. Likewise, the barrier properties of the citral-loaded films were also enhanced, as evident from the decreasing trend in WVTR and oxygen permeability (OP) values. The citral-loaded active films also showed profound antimicrobial activity against *Campylobacter jejuni*, *Pseudomonas fluorescens*, and *L. monocytogenes*. Similar results were reported for other essential oil-based biopolymer films, such as carvacrol, citral,[31,32] cinnamon essential oil,[33] *Zataria multiflora* essential oil,[34] and rosemary essential oil.[35]

Nanoparticles are also incorporated into the polymer matrix with essential oil for enhanced functionality of the biopolymer films. Some nanoparticles in the film matrix also facilitate the sustained release of essential oils to the surrounding headspace and preserve the structure of the active agents even at high temperatures.[36] Along with preserving active agents, they reinforce the film properties and enhance the tensile, barrier, and optical properties of the films and coatings.

Ejaz *et al.* prepared bovine skin gelatin (BSG) films by loading ZnO (2%) and clove essential oil (25% and 50% w/w) for shrimp packaging.[37] They reported that the TS of the films decreased, while the EAB increased upon oil addition. This was attributed to the plasticization effect of clove essential oil. The addition of essential oil also negatively affected the OP of the films. However, the light transmission through the BSG/ZnO/clove essential oil films decreased significantly with an increased oil concentration. This was because of the scattering of light rays by the ZnO and oil droplets. The transparency of the films also decreased due to the intense color imparted by the clove essential oil. The BSG/ZnO/clove essential oil films exhibited the highest log reduction in the growth of *L. monocytogenes* and *S. typhimurium*. The hydrophobic oil and Zn^{2+} ion could successfully penetrate the bacterial cell membrane and alter the structure of the cell wall. This would create more pores in the cell structure from which the intracellular materials diffuse, leading to cell inhibition. This robust antimicrobial activity of oil and ZnO contributed to the increased shelf life of shrimp during refrigerated storage.

Likewise, Souza *et al.*,[38] in their study, prepared chitosan films loaded with montmorillonite (MMT) and ginger essential oil. The authors found that the TS and EAB of chitosan/MMT films improved, while YM decreased compared to the control film. However, after the incorporation of ginger essential oil, the TS, EAB, and YM reduced significantly, which was attributed to the formation of a discontinuous matrix and the replacement of strong polar bonds of chitosan with weaker bonds formed between chitosan and ginger essential oil. The oil-loaded chitosan/MMT films recorded lower %transmittance values than pristine chitosan and chitosan/MMT films. The oil droplets and MMT platelets in the film matrix scattered the light rays, increasing the opacity of the films. Interestingly, the barrier properties of active films (both OP and WVP) deteriorated with an increase in oil concentration. The decrease in WVP was due to the possible interaction between the biopolymer and ginger essential oil, which facilitates the transport of moisture, while the decrease in OP was because of the plasticization effect exerted by the essential oil, which increases the intermolecular space among the polymer chains and promotes the transport of gas.

Similarly, Carvalho *et al.*[39] used the whey protein isolate and cellulose nanofibers to prepare biocomposite films. Thyme essential oil was loaded at different concentrations (20–40% w/w), and its effect was studied. The incorporation of essential oils increased the crystallinity index and glass transition temperature but decreased the WVP. In the case of mechanical properties, with an increase in oil concentration, the TS, EAB, YM, and puncture strength decreased uniformly. A similar trend was reported for films loaded with cinnamon essential oil/TiO_2,[40] rosemary essential oil/ZnO,[41] and nettle essential oil/cellulose nanofiber.[42]

The essential oil-based coating systems are also used to preserve the quality of food products. These are also incorporated with nanomaterials for controlled delivery of active agents and directly coated over the food product. Coatings based on pectin, sodium alginate, chitosan, and gums have been useful in combining different active agents to prepare antimicrobial coatings and prevent water loss.[27] Zhang *et al.* formulated chitosan-based coatings loaded with nano-encapsulated *Paulownia tomentosa* essential oil to retain the quality and increase the shelf life of pork chops at 4 °C for 16 days.[43] The authors reported that the antimicrobial coatings successfully inhibited the growth of microorganisms [characterized by total viable count (TVC), *Pseudomonas* spp. counts, and lactic acid bacteria count] and prevented the lipid oxidation in pork chops, as indicated by the thiobarbituric acid reactive substance (TBARS) value. Moreover, the encapsulation of essential oil facilitated the sustained release of active ingredients, leading to superior antimicrobial and antioxidant activities. The coated samples also maintained a good sensory score throughout the storage period.

Similarly, in another study, chitosan–carvacrol formulation coating was incorporated with caprylic acid to extend the shelf life of Pacific white shrimps (*Litopenaeus vannamei*) for 10 days of iced storage.[44] The chitosan and carvacrol coating delayed the bacterial growth and sustained the pH and total volatile basic nitrogen (TVBN) values of the coated shrimps compared to the uncoated ones. Furthermore, this coating protected the color of samples and retained the sensorial and textural attributes of shrimps. The incorporation of caprylic acid had a synergistic effect on the preservative actions of the chitosan–carvacrol coatings, which was evidenced by the increase in the efficacy of delay in TVBN and total plate count. The caprylic acid-incorporated coatings also retained the textural properties to a greater extent. Furthermore, the chitosan–carvacrol and chitosan–carvacrol–caprylic acid coatings showed a significant reduction in water loss due to the hydrophobic nature of the active ingredients. The shelf life of the coated samples was acceptable until the 6th day of storage, establishing its application for preserving marine products. Hosseini *et al.* synthesized fish skin gelatin films incorporated with oregano essential oil.[45] This antimicrobial coating was applied on rainbow trout (*Oncorhynchus mykiss*) fillets stored at 4 °C. The essential oil-based active coating lowered the TVBN values compared to control samples. The microbial growth was also inhibited in the active coated samples as the samples exhibited lower total viable and psychrotrophic count values. The coating also delayed the lipid oxidation of samples, as evident from low peroxide and TBARS values. Therefore, adding oregano essential oil to the coating system significantly improved the quality parameters of rainbow trout fillets and can be used as a preservative coating for other food products.

Similarly, in another study, Khaledian *et al.*[46] prepared cellulose nanofiber-based edible coatings impregnated with ginger essential oil and citric acid to preserve the shelf life of barbecue chicken. The optimum formulation for the coating was cellulose nanofiber (CNF)/2% ginger essential oil/citric acid 1%, which enhanced the shelf life of samples for 6 days compared to the control sample. The two active ingredients showed a synergistic effect on the quality parameters of chicken samples. The active coated samples had lesser microbial growth on the chicken surface and lower TBARS and peroxide values than the control sample. Moreover, incorporating ginger essential oil improved the odor, texture, taste, and overall acceptability of coated samples. Similar results were reported for

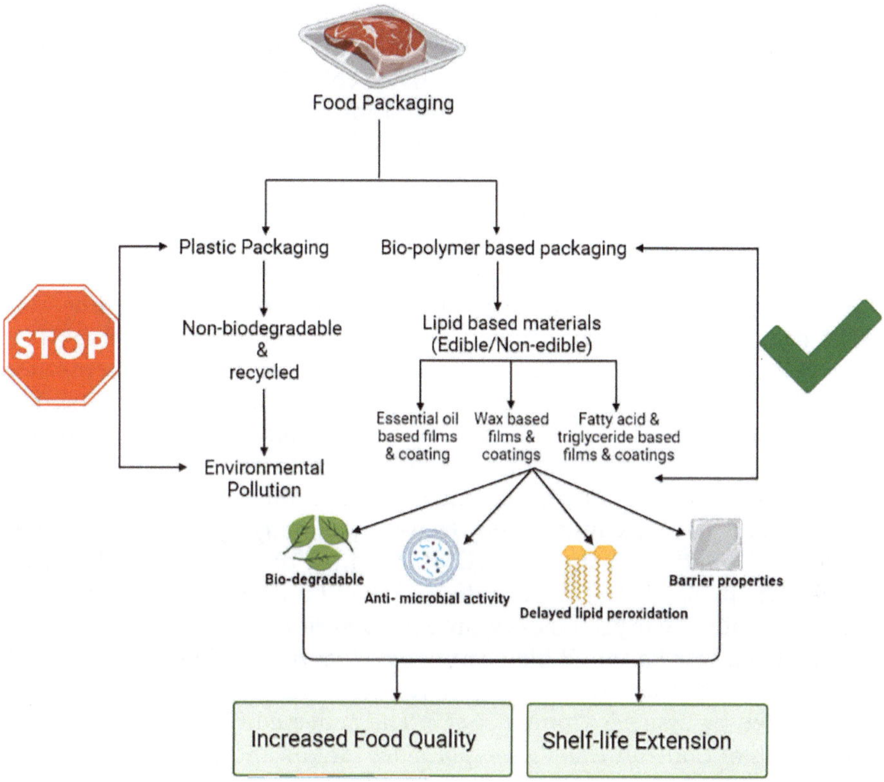

Figure 4.1 An overview of plastic packaging *versus* biopolymer-based packaging.

chitosan/*Thymus capitatus* essential oil,[47] sodium alginate/galbanum gum/*Ziziphora persica* essential oil,[48] and chitosan/*Citrus sinensis* essential oil.[49] Figure 4.1 illustrates the different types of lipid compounds used in food packaging and their beneficial effects on the quality parameters and shelf life of food products.

4.3.2 Wax-based Films and Coatings

Despite the excellent characteristics of wax as a hydrophobic substance, its use as an ingredient in food packaging materials has somewhat been poorly documented compared to other materials. In a study by Haruna *et al.*,[50] konjac glucomannan film-based emulsified composite films were prepared by incorporating carnauba wax at different concentrations (4–20% w/w). The addition of carnauba wax significantly enhanced the mechanical, barrier, and hydrophobic properties of composite films. The wax-incorporated films reduced the solubility by 5 times and increased the contact angle and strength by 2 times relative to the pristine konjac glucomannan films.

Moreover, the WVTR of the composite films increased by 48%, while the EAB increased by 61% as compared to the control film. A higher concentration of wax also improved the homogeneity, cohesiveness, and smoothness of the film. A similar result was reported for cassava starch/cashew gum/carnauba wax[51] and fish protein/carnauba wax.[52]

Similarly, palm wax was incorporated into fish gelatin films at different concentrations (15–60%) by Syahida *et al.*[53] The palm wax-incorporated film was more flexible, thicker, and opaque than the pristine gelatin films. The TS of the films increased significantly to 15% wax concentration, following which it decreased. Moreover, the film solubility, swelling, and WVP decreased upon adding palm wax. The authors reported that 15% w/w palm wax was the optimum concentration as higher-concentration films exhibited reduced TS, YM, and WVP. This has been attributed to the development of discontinuity in the polymer matrix owing to the increase in wax concentration, which created pinholes and cracks in the film surface. The strong polymer–polymer interaction was replaced by weak polymer–wax interaction. However, a study by Klangmuang and Sothornvit[54] reported that beeswax, when added to hydroxypropyl methylcellulose/nanoclay films, slightly affected the film properties and improved the WVP, while the oxygen barrier property deteriorated.

Candelilla wax-based edible coating was incorporated with ellagic acid, and the effect of the active coating on the shelf life of avocados for 6 weeks was studied by Saucedo-Pompa *et al.*[55] The active edible coating inhibited the growth of *Colletotrichum gloeosporioides* fungus and delayed the change in weight loss and appearance of the coated samples. In another study, Ag/TiO_2 nanoparticles and paraffin wax were incorporated into low-density polyethylene films.[56] Due to the addition of paraffin, the mechanical properties improved significantly; however, the optical properties were not affected at all. Without the presence of paraffin, the mechanical properties degraded, which was attributed to the formation of aggregates due to the presence of nanoparticles, while the presence of paraffin improved the dispersion and wetting of the film matrix. Moreover, the wax-based nanocomposite coating prevented the growth of *E. coli* and *S. albidoflavus* by 95% and 85%, respectively. The antimicrobial activity of the coating was because of the presence of Ag/TiO_2 nanoparticles, and paraffin further improved the antimicrobial activity by dispersing the nanoparticles properly in the polymer matrix.

In a recent study by Bucio *et al.*,[15] the mechanical and physical properties of paraffin wax, candelilla wax, and beeswax were studied. Cheese samples were coated with the waxes and stored at 30 °C for 5 weeks. Beeswax was the best option for cheese coating as it did not show any cracks after storage, while the samples coated with paraffin wax and candelilla wax exhibited microcracks with reduced moisture, weight, and size. The beeswax-coated cheese was ductile, while the other samples were brittle with fractures.

4.3.3 Fatty Acid- and Triglyceride-based Films and Coatings

The application of fatty acids in forming edible films has increasingly become popular due to their hydrophobic properties. The films would exhibit low moisture penetration and high moisture resistance, which would tremendously help preserve the quality of food products. In a study by Ma *et al.*,[57] tara gum-based edible films were incorporated with oleic acid at different concentrations (5–20% w/w) to prepare novel food packaging materials. The authors reported no chemical interaction between oleic acid and tara gum; however, the addition of oleic acid results in the formation of micropores in the film matrix. With an increasing concentration of oleic acid, the TS of the films decreased (57.4 to 26.8 MPa), while the contact angle increased from 76.75° to 107.28°. In contrast, the thermal stability of the edible films was also enhanced upon adding oleic acid. As expected, the WVP of the films decreased significantly from 0.694 to 0.526 $g\ s^{-1}\ m^{-1}\ Pa^{-1} \times 10^{-10}$ with an increase in the concentration of oleic acid. Similarly, chitosan- and oleic acid-based emulsified films were prepared by Aguirre-Loredo *et al.*[58] The WVP of the films decreased from 8.07 to 6.16 $g\ m\ s^{-1}\ m^{-2}\ Pa^{-1} \times 10^{-10}$ with an increase in oleic acid concentration, likely due to the increased viscosity of the emulsion phase, ensuring the stabilization of emulsion during the film-forming process. However, the oxygen permeability of the films increased, which was attributed to the increase in the plasticization effect of the biopolymer backbone by water molecules. A similar result was reported in another study on chitosan/oleic acid film.[59]

Fatty acids are also combined with wax-based films to increase the functionality of biopolymer films. Fabra *et al.*,[60] in their study, used oleic acid and beeswax to prepare sodium caseinate composite films. The authors reported that the tensile properties, such as flexibility, elasticity, and stretchability of the films, increased due to the plasticization effect of the additives, while TS and YM of the films decreased with an increase in oleic acid concentration compared to the control. Moreover, the WVP of the composite films decreased due to the hydrophobic nature of oleic acid and beeswax. In another study, Fabra *et al.*[61] reported that sodium caseinate incorporated with a mixture of oleic acid and beeswax decreased the transparency and gloss value of the films compared to the control film. This has been attributed to the formation of lipid droplets on the surface and internal parts of the film, which prevented the penetration of the light rays. A similar result was reported for cassava starch, carnauba wax, and stearic acid.[62]

Konjac glucomannan films were coated with stearic acid (5–60% wt) and shellac (12 g) to prepare moisture-resistant coating by Wei *et al.*[63] The WVP of the films decreased from 2.63 to 0.37 $\times 10^{-11}\ g\ m^{-1}\ s^{-1}\ Pa^{-1}$ due to the interfacial adhesion between konjac glucomannan substrate and coating layer along with the homogeneous mixing of stearic acid and shellac in the coating system. Moreover, 20% wt stearic acid composite films exhibited the highest TS (41 MPa) and YM (2069 MPa). Moreover, the contact angle also increased from 68° to 101.2° as the stearic acid content increased to 40% wt. The enhancement in the film properties was attributed to the moisture resistance ability and homogeneity of shellac–stearic acid coating. Similarly, in another study, rice starch and iota-carrageenan-based edible coatings were developed by incorporating stearic acid at different concentrations (0.3–0.9% w/w).[64] The TS, EAB, and transparency decreased with an increased stearic acid concentration. Interestingly, the WVP of the films increased after the incorporation of stearic acid. It was attributed to the hydrophobic–hydrophilic ratio in the polymer matrix. The formulation with 0.3% stearic acid was considered the optimum concentration for developing films with good barrier, mechanical, and physical properties. The stearic acid in the polymer matrix imparts hydrophobic characteristics, leading to structural changes in the film surface. A similar result was reported for sodium caseinate and stearic acid films.[65]

Fatty acids, when incorporated into biopolymer systems, can also control the release of active ingredients from film matrices. In a study, the effect of oleic acid on the release of tetrahydrocurcumin from chitosan–carboxymethyl cellulose film matrix was investigated.[66] It was reported that oleic acid negatively affected the release of tetrahydrocurcumin because of the stability of tetrahydrocurcumin encapsulation in the oleic acid droplet. In Table 4.2, different studies that investigate the effect of lipid-based films and coatings in enhancing the shelf life and preserving the quality of food products are presented.

4.4 Conclusion

Due to environmental issues and consumer concerns, bio-based materials are being investigated as alternatives to petrochemicals for packaging applications. To preserve the food quality and enhance the safety standards of food products, lipid-based edible and non-edible films and coatings are used as they possess no environmental or health risks. Lipids, such as essential oils, are used widely due to their hydrophobic, antimicrobial, and antioxidant properties. Films and coatings, when loaded with active ingredients, successfully enhance the shelf life of food products considerably due to their ability to inhibit microbial growth and delay lipid peroxidation. The ability of biopolymer-based food packaging materials to carry and control the release of active ingredients has been thoroughly investigated in recent studies. Other lipids, such as natural waxes and fatty acids, are also used in film formation. They impart superior moisture and light barrier properties to the films and coatings and could be used in formulating moisture-resistant packaging or coating systems. Other film properties, such as mechanical, physical, and thermal properties, vary depending on the concentration of lipids, the biopolymer used, and film-forming methods. Therefore, lipid-based films and coating systems

Table 4.2 Application of lipid-based antimicrobial coatings on real food systems.

Sl no.	Biopolymer	Lipids used	Nanofiller	Food product	Target microorganism	Ref.
1.	Cassava starch	Cinnamon essential oil	Bentonite	Pork meatballs	*Escherichia coli* *Salmonella typhimurium* *Staphylococcus aureus*	67
2	Chitosan/carboxymethyl cellulose	*Ziziphora clinopodioides* essential oil	Montmorillonite	Minced camel's meat	*Listeria monocytogenes* *Escherichia coli*	68
3	Chitosan	Ginger essential oil	Montmorillonite	Poultry meat	*Bacillus cereus* *Salmonella enterica*	69
4	Quince seed mucilage	Oregano and thyme essential oils		Rainbow trout fillets	*Pseudomonas* spp. *Enterobacteriaceae* H_2S-producing bacteria	70
5	Farsi gum	Clove and thyme essential oils		Rainbow trout fillets	Psychrotrophic bacteria counts	71
6	Chitosan	Thyme essential oil		Karish cheese	Psychrotrophic bacteria counts Yeast and mold count	72
7	Gelatin	Palm wax/lemongrass essential oil		Ground beef	Psychrotrophic bacterial counts Mesophilic bacterial counts	73
8		Paraffin/cinnamon and oregano essential oil		Cherry Tomatoes	*Alternaria alternata*	74
9	Alginate	Oleic acid		Strawberries and raspberries	Human noroviruses (NoVs) Hepatitis A virus	75
10	Chitosan	*Mentha spicata* essential oil	ZnO	Rainbow trout fillets	Psychrotrophic count, *Pseudomonas* spp. *Enterobacteriaceae*	76

(*continued*)

Table 4.2 *(continued)*

Sl no.	Biopolymer	Lipids used	Nanofiller	Food product	Target microorganism	Ref.
11	Chitosan	Essential oil combination (cinnamon, citrus extract, Asian formulation, and Mediterranean formulation)	Ag	Strawberries	*Escherichia coli* *Listeria monocytogenes*, *Salmonella typhimurium*, *Aspergillus niger*	77
12	Fish protein isolate/fish skin gelatin	Basil leaf essential oil	ZnO	Sea bass slices	*Pseudomonas* spp., H_2S-producing bacteria *Enterobacteriaceae* Psychrophilic bacteria	78
13	Whey protein isolate	Rosemary essential oil	Cellulose nanofiber/ TiO_2 nanoparticle	Lamb meat	*Pseudomonas* spp. count, *Enterobacteriaceae* count *Staphylococcus aureus* *Listeria monocytogenes* *Escherichia coli*	79
14	Gelatin/chitosan	*Ferulago angulate* essential oil		Turkey meat	Coliform counts *Staphylococcus aureus*	80

hold great promise in replacing traditional packaging materials, which will assist in the mitigation of non-biodegradable packaging wastes.

References

1. R. Ribeiro-Santos, M. Andrade, N. R. de Melo and A. Sanches-Silva, *Trends Food Sci. Technol.*, 2017, **61**, 132–140.
2. C. R. Álvarez-Chávez, S. Edwards, R. Moure-Eraso and K. Geiser, *J. Cleaner Prod.*, 2012, **23**, 47–56.
3. U. EPA, *US Environ. Prot. Agency.*, 2014, 1–13.
4. K.-O. Honikel, *Meat Sci.*, 2008, **78**, 68–76.
5. M. L. T. Polônio and F. Peres, *Cad. Saude Publica*, 2009, **25**, 1653–1666.
6. S. A. Mohamed, M. El-Sakhawy and M. A.-M. El-Sakhawy, *Carbohydr. Polym.*, 2020, **238**, 116178.
7. T. Janjarasskul and J. M. Krochta, *Annu. Rev. Food Sci. Technol.*, 2010, **1**, 415–448.
8. C. V. Dhumal, K. Pal and P. Sarkar, *Polym.-Plast. Technol. Mater.*, 2019, **58**, 255–269.
9. J. W. Rhim and T. H. Shellhammer, in *Innovations in food packaging*, Elsevier, 2005, pp. 362–383.
10. U. Amin, M. U. Khan, Y. Majeed, M. Rebezov, M. Khayrullin, E. Bobkova, M. A. Shariati, I. M. Chung and M. Thiruvengadam, *Int. J. Biol. Macromol.*, 2021, **183**, 2184–2198.
11. R. A. Anderson, Z. Zhan, R. Luo, X. Guo, Q. Guo, J. Zhou, J. Kong, P. A. Davis and B. J. Stoecker, *J. Tradit. Complementary Med.*, 2016, **6**, 332–336.
12. A. I. Hussain, F. Anwar, S. T. H. Sherazi and R. Przybylski, *Food Chem.*, 2008, **108**, 986–995.
13. M. Khajeh, Y. Yamini, N. Bahramifar, F. Sefidkon and M. R. Pirmoradei, *Food Chem.*, 2005, **91**, 639–644.
14. S. Sharma, S. Barkauskaite, A. K. Jaiswal and S. Jaiswal, *Food Chem.*, 2020, 128403.
15. A. Bucio, R. Moreno-Tovar, L. Bucio, J. Espinosa-Dávila and F. Anguebes-Franceschi, *Coatings*, 2021, **11**, 261.
16. P. Tyagi, K. S. Salem, M. A. Hubbe and L. Pal, *Trends Food Sci. Technol.*, 2021, **115**, 461–485.
17. T. Fei and T. Wang, *Curr. Opin. Food Sci.*, 2017, **16**, 7–14.
18. S. Despond, E. Espuche, N. Cartier and A. Domard, *J. Appl. Polym. Sci.*, 2005, **98**, 704–710.
19. P. C. Calder, *J. Parenter. Enteral Nutr.*, 2015, **39**, 18S–32S.
20. E. Baldwin, M. Nisperos, R. Hagenmaier and R. Baker, *Food Technol.*, 1997, **51**, 56–62.
21. M. A. Ch, S. B. Naz, A. Sharif, M. Akram and M. A. Saeed, *Br. J. Pharm. Res.*, 2015, **7**, 330–339.
22. S. Burt, *Int. J. Food Microbiol.*, 2004, **94**, 223–253.

23. L. S. Wei and W. Wee, *Iran. J. Microbiol.*, 2013, **5**, 147.
24. G. Mahendran and L. U. Rahman, *Phytother. Res.*, 2020, **34**, 2088–2139.
25. S. d. Moufida and B. Marzouk, *Phytochemistry*, 2003, **62**, 1283–1289.
26. Y. Lan, in *Encyclopedia of Food Chemistry*, ed. L. Melton, F. Shahidi and P. Varelis, Academic Press, Oxford, 2019, pp. 312–316, https://doi.org/10.1016/B978-0-08-100596-5.22344-5.
27. M. E. Martiñon, R. G. Moreira, M. E. Castell-Perez and C. Gomes, *LWT--Food Sci. Technol.*, 2014, **56**, 341–350.
28. S. M. B. Hashemi and A. Mousavi Khaneghah, *Prog. Org. Coat.*, 2017, **110**, 35–41.
29. S. Agarwal, M. Hoque, N. Bandara, K. Pal and P. Sarkar, *Food Packag. Shelf Life*, 2020, **26**, 100562.
30. S. G. Giteru, R. Coorey, D. Bertolatti, E. Watkin, S. Johnson and Z. Fang, *Food Chem.*, 2015, **168**, 341–347.
31. C. Vilas Dhumal, K. Pal and P. Sarkar, *J. Mater. Sci.: Mater. Med.*, 2019, **30**, 117.
32. C. V. Dhumal, J. Ahmed, N. Bandara and P. Sarkar, *Food Packag. Shelf Life*, 2019, **21**, 100380.
33. Y. Han, M. Yu and L. Wang, *Food Packag. Shelf Life*, 2018, **15**, 35–42.
34. H. Hashemi Gahruie, E. Ziaee, M. H. Eskandari and S. M. H. Hosseini, *Carbohydr. Polym.*, 2017, **166**, 93–103.
35. M. Abdollahi, M. Rezaei and G. Farzi, *Int. J. Food Sci. Tech.*, 2012, **47**, 847–853.
36. G. Biddeci, G. Cavallaro, F. Di Blasi, G. Lazzara, M. Massaro, S. Milioto, F. Parisi, S. Riela and G. Spinelli, *Carbohydr. Polym.*, 2016, **152**, 548–557.
37. M. Ejaz, Y. A. Arfat, M. Mulla and J. Ahmed, *Food Packag. Shelf Life*, 2018, **15**, 113–121.
38. V. G. L. Souza, J. R. A. Pires, C. Rodrigues, P. F. Rodrigues, A. Lopes, R. J. Silva, J. Caldeira, M. P. Duarte, F. B. Fernandes and I. M. Coelhoso, *Coatings*, 2019, **9**, 700.
39. R. A. Carvalho, A. C. S. de Oliveira, T. A. Santos, M. V. Dias, M. I. Yoshida and S. V. Borges, *J. Polym. Environ.*, 2020, **28**, 231–241.
40. E. Arezoo, E. Mohammadreza, M. Maryam and M. N. Abdorreza, *Int. J. Biol. Macromol.*, 2020, **157**, 743–751.
41. M. Alizadeh-Sani, E. M. Kia, Z. Ghasempour and A. Ehsani, *J. Polym. Environ.*, 2021, **29**, 588–598.
42. S. E. M. Kalajahi, A. Alizadeh, H. Hamishehkar, H. Almasi and N. Asefi, *J. Polym. Environ.*, 2021, 1–12.
43. H. Zhang, X. Li and H. Kang, *Lwt*, 2019, **116**, 108580.
44. Q. Wang, J. Lei, J. Ma, G. Yuan and H. Sun, *Int. J. Biol. Macromol.*, 2018, **106**, 123–129.
45. S. F. Hosseini, M. Rezaei, M. Zandi and F. F. Ghavi, *J. Aquat. Food Prod. Technol.*, 2016, **25**, 835–842.
46. Y. Khaledian, M. Pajohi-Alamoti and B. Bazargani-Gilani, *J. Food Process. Preserv.*, 2019, **43**, e14114.

47. K. Martínez, M. Ortiz, A. Albis, C. Gilma Gutiérrez Castañeda, M. E. Valencia and C. D. Grande Tovar, *Biomolecules*, 2018, **8**, 155.
48. H. Hamedi, M. Kargozari, P. M. Shotorbani, N. B. Mogadam and M. Fahimdanesh, *Food Hydrocolloids*, 2017, **72**, 35–46.
49. M. Sheikh, S. Mehnaz and M. B. Sadiq, *J. Sci. Food Agric.*, 2021, **101**, 6248–6257.
50. M. H. Haruna, Y. Wang and J. Pang, *J. Food Sci. Technol.*, 2019, **56**, 5138–5145.
51. D. C. Rodrigues, C. A. Caceres, H. L. Ribeiro, R. F. de Abreu, A. P. Cunha and H. M. Azeredo, *Food Hydrocolloids*, 2014, **38**, 147–151.
52. V. P. Romani, B. Olsen, M. P. Collares, J. R. M. Oliveira, C. Prentice and V. G. Martins, *Food Hydrocolloids*, 2020, **108**, 106087.
53. S. N. Syahida, M. R. Ismail-Fitry,Z. M. A. a. Ainun and Z. A. N. Hanani, *Food Packag. Shelf Life*, 2020, **23**, 100437.
54. P. Klangmuang and R. Sothornvit, *LWT--Food Sci. Technol.*, 2016, **65**, 222–227.
55. S. Saucedo-Pompa, R. Rojas-Molina, A. F. Aguilera-Carbó, A. Saenz-Galindo, H. de La Garza, D. Jasso-Cantú and C. N. Aguilar, *Food Res. Int.*, 2009, **42**, 511–515.
56. N. Hosseini Nasab, M. M. Jalili and S. Farrokhpay, *J. Appl. Polym. Sci.*, 2018, **135**, 45913.
57. Q. Ma, D. Hu, H. Wang and L. Wang, *Food Hydrocolloids*, 2016, **56**, 127–133.
58. R. Y. Aguirre-Loredo, A. I. Rodríguez-Hernández and N. Chavarría-Hernández, *CyTA--J. Food*, 2014, **12**, 305–312.
59. M. Vargas, A. Albors, A. Chiralt and C. González-Martínez, *Postharvest Biol. Technol.*, 2006, **41**, 164–171.
60. M. J. Fabra, P. Talens and A. Chiralt, *J. Food Eng.*, 2008, **85**, 393–400.
61. M. J. Fabra, P. Talens and A. Chiralt, *Food Hydrocolloids*, 2009, **23**, 676–683.
62. M. Chiumarelli and M. D. Hubinger, *Food Hydrocolloids*, 2014, **38**, 20–27.
63. X. Wei, J. Pang, C. Zhang, C. Yu, H. Chen and B. Xie, *Carbohydr. Polym.*, 2015, **118**, 119–125.
64. R. Thakur, B. Saberi, P. Pristijono, J. Golding, C. Stathopoulos, C. Scarlett, M. Bowyer and Q. Vuong, *Int. J. Biol. Macromol.*, 2016, **89**, 952–960.
65. E. Rezvani, G. Schleining, G. Sümen and A. R. Taherian, *J. Food Eng.*, 2013, **116**, 598–605.
66. E. Le Goué, C. Gardrat, M. Romain, M. Rollini, C. Moresoli and V. Coma, *Food Hydrocolloids*, 2022, **124**, 107202.
67. B. Iamareerat, M. Singh, M. B. Sadiq and A. K. Anal, *J. Food Sci. Technol.*, 2018, **55**, 1953–1959.
68. A. Khezrian and Y. Shahbazi, *Int. J. Biol. Macromol.*, 2018, **106**, 1146–1158.

69. V. G. Souza, J. R. Pires, É. T. Vieira, I. M. Coelhoso, M. P. Duarte and A. L. Fernando, *Coatings*, 2018, **8**, 177.
70. M. Jouki, F. T. Yazdi, S. A. Mortazavi, A. Koocheki and N. Khazaei, *Int. J. Food Microbiol.*, 2014, **174**, 88–97.
71. P. Dehghani, S. M. H. Hosseini, M.-T. Golmakani, M. Majdinasab and S. Esteghlal, *Food Hydrocolloids*, 2018, **77**, 677–688.
72. M. Al-Moghazy,H. S. El-sayed, H. H. Salama and A. A. Nada, *Food Biosci.*, 2021, 101230.
73. S. Nurul Syahida, M. R. Ismail-Fitry, Z. M. A. Ainun and Z. A. Nur Hanani, *Food Packag. Shelf Life*, 2021, **28**, 100640.
74. A. Rodriguez-Lafuente, C. Nerin and R. Batlle, *J. Agric. Food Chem.*, 2010, **58**, 6780–6786.
75. I. Falcó, P. L. Flores-Meraz, W. Randazzo, G. Sánchez, A. López-Rubio and M. J. Fabra, *Food Hydrocolloids*, 2019, **87**, 611–618.
76. Y. Shahbazi and N. Shavisi, *J. Aquat. Food Prod. Technol.*, 2018, **27**, 986–997.
77. S. Shankar, D. Khodaei and M. Lacroix, *Food Hydrocolloids*, 2021, **117**, 106750.
78. Y. A. Arfat, S. Benjakul, K. Vongkamjan, P. Sumpavapol and S. Yarnpakdee,*J. Food Sci. Technol.*, 2015, **52**, 6182–6193.
79. M. A. Sani, A. Ehsani and M. Hashemi, *Int. J. Food Microbiol.*, 2017, **251**, 8–14.
80. H. R. Naseri, F. Beigmohammadi, R. Mohammadi and E. Sadeghi, *J. Food Process. Preserv.*, 2020, **44**, e14558.

CHAPTER 5

Role of Lipids and Proteins as Additives in Sensory and Functionality Enhancers in Food Systems

M. WEERAWARNA N. R. P.*

Massey University, New Zealand
*E-mail: M.Weerawarna@massey.ac.nz

5.1 Advances in Protein-based Additives

5.1.1 Plant Protein-based Meat Substitutes and Analogues

The recent trends in consumer food choice behaviors have created a demand for plant-based, sustainable, or ethically sourced alternative protein ingredients.[1,2] However, omnivores tend to refrain from consuming meat analogues purported to have compromised taste.[3] The combined approach of using alternative protein ingredients and processing methods produces novel products that mimic the sensory properties of meat in specific meal contexts, catering to omnivores or flexitarian consumers.[4]

Enhancing the appearance of meat analogues using optimized cooking methods was suggested to mimic visual cues of actual meat products, creating motivating expectations on meat analogues before consumption.[5,6] Visual perceptions of meat analogues primarily depend on alternative

Smart Materials No. 41
Functional Materials from Lipids and Proteins
Edited by Nandika Bandara and Aman Ullah

Published by the Royal Society of Chemistry, www.rsc.org

protein ingredients. Sharima-Abdullah *et al.*[7] explored the ratio of chickpea flour to texturized vegetable protein to increase consumer acceptability of meatless nuggets. The increase in chickpea flour has increased the scores for the color and appearance of the meatless nuggets. Therefore, choosing alternative protein ingredients in meatless product formulations is important.

The substitution of meat with plant proteins negatively impacts the aroma and flavor profiles of meatless alternative products.[8] Plant proteins have created beany, grassy, or pulse-related aromas and flavors that were primarily not preferred by omnivores. These off-flavors of plant proteins were masked by adding flavoring agents, such as herbs, spices, and flavor enhancers, to the formulations of meat analogues. For example, extruded peanut (peanut flour and peanut protein concentrate) *vs.* soy-based mince alternatives were studied using a seasoned puff pastry.[9] Applying beef flavor, soy sauce, tomato powder, and red pepper to peanut protein-based mince puff scored higher for acceptable meat flavor and lower dislike rates. Formulating meat analogues with flavorings relevant to meal context is beneficial to increase consumer acceptability.

Recreating textural properties (texture, mouthfeel, and juiciness) of conventional meat or meat products in meat analogues remains challenging.[8,10] Therefore, careful selection of plant proteins was considered to achieve the physicochemical properties in meat analogues to resemble conventional meat. These physicochemical properties were mainly driven by fat- and water-holding capacity, gelling properties, and emulsifying properties of plant proteins. Currently, blending various plant proteins and ingredients to enhance the textural properties of meat analogues is practiced.[11] For example, Argel *et al.*[12] used lentils, chickpeas, and bean flour as meat substitutes to partially replace pork in low-fat burger patties. This pulse flour mixture has increased oil and water retention during cooking, increased cooking yield, and reduced hardness in low-fat burger patties, making them sensorially acceptable. Pulse flour primarily contains starch, fiber, and proteins. Starch molecules create heat-induced gel matrices with water, increasing water-holding capacity in food systems. Similarly, proteins bind with water and increase water-holding potential. These three-dimensional gel matrices primarily incorporate water, starch, animal proteins, plant proteins, and fiber into the food system *via* hydrogen, van der Waals, electrostatic, and hydrophobic interactions[13] and compensate for lost textural properties in low-fat and combined animal–plant protein foods, such as burger patties. Wheat proteins (gluten) were also used to achieve viscoelastic properties in meat-free or meat-reduced chicken sausage production.[14]

5.1.2 Insect Proteins

Edible insects are another novel sustainable source of protein with less environmental impact than conventional animal meat production.[15] Insects or insect-based ingredients have been a part of food culture in some Eastern African, Latin American, and Asian populations.[16,17] In comparison, insects are not familiar as a source of protein for some consumers.[17–19] Therefore, consumer acceptance of insect proteins widely depends on food neophobia status,[20] safety or allergen concerns,[21,22] nutritional quality, and sensory aspects.[15] Insects are identified as rich in proteins, fibers, and minerals[23]; however, they are perceived to be dangerous and disgusted by some consumers.[24,25]

Recent advances in the processing of insects have increased their application as an ingredient in other familiar products (for example, cricket flour in pasta[26] and brownies,[27] protein fortification in bread with mealworm,[28] and insect burger patties), rather than the consumption of traditionally cooked whole insect (for example cricket), and hence increased consumer acceptability.[21,29–33] However, even a 5% replacement of meat with insects has compromised the structural and physical properties of sausage, resulting in a crumbly texture that was sensorially undesirable.[34] High amounts of fats and/or sodium are added to these combinatorial products to compensate for their lost sensory qualities.[35,36]

5.1.3 Non-dairy Gels

Plant protein gels are also increasingly demanding as non-dairy yogurt alternatives. Brückner-Gühmann *et al.*[37] characterized the structural and sensory properties of lactic acid-fermented oat protein concentrate gels. Oat contains about 15% high-quality proteins.[38] Oat protein concentrate contains proteins and starch. Gelatinized starch in oat protein concentrates formed a gel matrix, which incorporated oat proteins. This has prevented the aggregation of oat proteins during acidification and improved the textural properties, *i.e.* creamy, moist, soft, and smooth, and consumer liking for oat protein gels.[37] Therefore, oat protein concentrate shows potential as a relatively novel functional protein-based additive in developing plant-based acidified gels. Pea proteins, containing eight essential amino acids and being less allergenic than soy, were considered in non-dairy yogurt processing.[39] However, unpleasant flavors, such as grassy notes and hay-like flavors, have reduced the sensory acceptability of pea protein yogurts or other products.[40–42] Different levels of denaturation and aggregation of pea proteins during pea milk processing were utilized to optimize the gel firmness and textural properties of pea protein yogurts.[39] A comparative study on sensory profiles of soy-, coconut-, cashew-, almond-, and hemp-based yogurts *vs.* a dairy yogurt revealed that some plant-based yogurts were similarly appreciated to dairy-based products in the presence of specific hydrocolloids, sweeteners, and flavors in formulations.[43]

Therefore, consumer acceptability of non-dairy gels can be increased by optimizing the sensory profiles with relevant additives and processing technologies.[44]

5.1.4 Edible Coatings and Films

Protein-based edible coatings and films are an emerging category of biodegradable food packaging.[45,46] Spraying, brushing, and electrospraying methods coat these proteins onto different foods.[47] These coatings are animal- or vegetable-origin proteins, mainly sourced from the by-products used in the production of biodegradable coatings.[48] These proteins are biopolymers and vary in composition and structure.[49] Therefore, proteins were processed using plasticization, cross-linking, nanotechnology, and composite films[50] to enhance the functional properties of the films to closely represent the synthetic flexible films used in food packaging.[45] For example, the applicability of protein coatings in reducing fat uptake and moisture loss of food during deep fat frying has been explored.[51,52] A comparative study on whey protein and soy protein isolate showed whey protein as a better coating for deep-fried cassava in reducing fat absorption.[53] Myofibrillar protein recovered from by-products of fish filleting and poultry processing was studied as coatings to reduce fat absorption in animal muscle-based foods due to product compatibility.[51] Denaturation of water-soluble fish proteins to assist gelation and thus coating[54] and cross-linking to enhance their moisture barrier properties[55] were reported. Additionally, these animal by-product-based myofibrillar proteins could be used as a sustainable nutritional additive in combinatorial meat analogues. Furthermore, applying the protein coatings and the consequent reduction in fat uptake and moisture loss during deep fat frying will influence the sensory profiles of the product and, hence, are worth investigating.

Protein-based films have shown better gas and aroma barrier properties and mechanical properties and act as potential carriers for active compounds compared to lipid or polysaccharide films.[56] Active compounds, such as antimicrobial agents,[57–59] antioxidants,[59–61] aroma, and flavor compounds,[62,63] were incorporated into protein films. Castro *et al.*,[64] for example, investigated if whey protein concentrate films with green tea extract could prevent the lipid oxidation of fresh salmon and optimize storage stability. The results indicated the successful prevention of lipid oxidation of fresh salmon up to 14 days of storage. Lipid oxidation is a primary cause of undesirable sensory properties such as off-flavors in food.[65] The use of active whey protein concentrate films can extend the sensory acceptability of fresh salmon during storage. The potential of protein coatings as carriers (retention and release) for aroma and flavor compounds, such as *n*-hexanal in soy protein films[62]; garlic and pepper powder in whey protein films; thymol, cinnamaldehyde, and vanillin in soy protein isolate films,[66] has been explored. Therefore, it is evident that protein-based films are a potential sustainable source of food packaging and sensory quality enhancers. However, abundant raw materials, mechanical properties, safety, and cost of producing these protein films will need further investigation.

Regardless of the research on protein films as carriers for aroma and flavor compounds, the impact of proteins in the film on the sensory properties of packaged food is inconclusive. García Arteaga *et al.*[67] studied different enzymatic hydrolysis parameters on the sensory properties of pea protein isolates. The researchers showed that the bitterness of pea protein hydrolysates was correlated with the degree of hydrolysis. Therefore, the findings from this study will assist in developing more sensory-appealing pea protein isolates, which are demanding ingredients due to their high protein content. Pea protein isolates with better sensory quality will benefit the manufacturing of protein films as sensory enhancers. Furthermore, soy proteins and whey proteins or their isolates are widely studied in protein-based films. However, soy and whey proteins produce beany or grassy flavors, bitterness, astringency, and other undesirable sensory properties. The impact of the sensory profile of the proteins used in films or coating formation on the packaged food will need careful consideration.

5.2 Advances in Lipid-based Additives as Sensory Enhancers

Lipids contribute to various functionalities in foods but are not limited to enhancing sensory perception. Incorporating lipids as delivery systems and hence the food structure, for example, emulsions, double emulsions, nanocarriers, and polymers, is a recent research interest in optimizing sensory profiles of novel functional food systems.

5.2.1 Micro- or Nanocarriers and Emulsions

The addition of bioactive compounds is reported to cause undesirable sensory properties, such as bitterness and astringency, in fortified products. These bioactives are unstable during product processing, storage, and digestion upon exposure to temperature, pressure, oxygen, pH, and/or light.[68–71] The instability of these bioactives further compromises sensory acceptability by creating off colors, tastes, and flavors in the fortified products.[72,73] Therefore, researchers and the food industry focus on developing emulsions and double emulsions with lipids as micro- or nanodelivery systems for bioactive compounds.[74–76] Phenolic compounds, for example, are a key interest in functional foods and pharmaceutical industries. However, phenolic compounds have low bioavailability and create undesirable flavors. Faridi Esfanjani *et al.*[77] and Martínez-Álvarez *et al.*[78] studied lipid-based nanoencapsulation *vs.* microencapsulation to overcome these limitations on incorporating phenolic compounds into food systems.

Healthy fats were added to reformulate processed meat products to reduce animal fat (trans and unsaturated fats).[79,80] Animal fat in processed meats contributes to rheological and sensory properties (texture, juiciness, palatability, color, flavor, and aroma).[81–83] The potential of bio-based healthy fats containing monounsaturated and polyunsaturated fatty acids of vegetable or marine origin in reformulated meat products was investigated.[84–86] However, unsaturated fatty acids remain in a liquid state at room temperature and compromise the structure of processed meat products. Furthermore, these healthy fats are prone to lipid oxidation, producing off-flavors in these meat products. Structuring healthy liquid oils *via* encapsulation into oleogels and emulsion hydrogels was explored to achieve properties of saturated fats[87] and enhance the texture of reformulated meat products.[88] Furthermore, reformulating meat products with encapsulated oils was advantageous to avoid undesirable flavors associated with most healthy plant- or marine-based oils.[89] However, encapsulated oils did not achieve the appearance and consistency of visible animal fat, which consumers prefer in some meat products.[85] Espert *et al.*[90] studied sunflower oil–water emulsions in cocoa cream to reduce fat digestibility while maintaining consistency and consumer liking. These delivery systems and food structures are targeted to enhance the sensory acceptability of foods with low digestible calories.

5.2.2 Microbial Polyunsaturated Fatty Acids

Microbial oils are suggested as an alternative source of polyunsaturated fatty acids (PUFAs).[91] Fatty fish or fish oil is the primary source of PUFAs in food additives or supplements. However, the use of fish oil-based PUFAs is becoming debatable due to the contamination of fatty fish with dioxins and heavy metals.[92] While the supply of quality PUFAs from fish is becoming limited, the demand for PUFAs increases, widening the gap in the available food-grade PUFAs. Hence, oil-producing (oleaginous) microorganisms are a novel prospective source of PUFA production.[92,93]

Yeast,[94,95] fungi,[96,97] and microalgae[98,99] were studied for their potential for PUFA production under batch culture and limited nutrients (nitrogen). However, most of these microbial oils were produced as biofuels. The use of genetically engineered oleaginous microorganisms[100,101] and the source of substrates[102–104] pose limitations for consumer acceptability but are cost-effective for the production of edible microbial lipids. Therefore, wider research will be required to achieve a less costly but efficient production process of edible microbial lipids with higher consumer willingness to purchase.

References

1. R. Tso, A. J. Lim and C. G. Forde, A Critical Appraisal of the Evidence Supporting Consumer Motivations for Alternative Proteins, *Foods*, 2021, **10**(1), 24.
2. M. Kaczorowski, *A year of innovation in meat substitutes and eggs, 2020*, 2020.
3. M. F. S. Reipurth, L. Hørby, C. G. Gregersen, A. Bonke and F. J. A. Perez Cueto, Barriers and facilitators towards adopting a more plant-based diet in a sample of Danish consumers, *Food Qual. Prefer.*, 2019, **73**, 288–292.
4. M. Fiorentini, A. J. Kinchla and A. A. Nolden, Role of Sensory Evaluation in Consumer Acceptance of Plant-Based Meat Analogs and Meat Extenders: A Scoping Review, *Foods*, 2020, **9**(9), 1334.
5. J. Delwiche, The impact of perceptual interactions on perceived flavor, *Food Qual. Prefer.*, 2004, **15**(2), 137–146.
6. I. Gómez, F. C. Ibañez and M. J. Beriain, Physicochemical and sensory properties of sous vide meat and meat analog products marinated and cooked at different temperature-time combinations, *Int. J. Food Prop.*, 2019, **22**(1), 1693–1708.
7. N. Sharima-Abdullah, C. Hassan, N. Arifin and N. Huda-Faujan, Physicochemical properties and consumer preference of imitation chicken nuggets produced from chickpea flour and textured vegetable protein, *Int. Food Res. J.*, 2018, **25**(3), 1016–1025.
8. M. A. Asgar, A. Fazilah, N. Huda, R. Bhat and A. A. Karim, Nonmeat Protein Alternatives as Meat Extenders and Meat Analogs, *Compr. Rev. Food. Sci. Food. Saf.*, 2010, **9**(5), 513–529.
9. D. Rehrah, M. Ahmedna, I. Goktepe and J. Yu, Extrusion parameters and consumer acceptability of a peanut-based meat analogue, *Int. J. Food Sci. Technol.*, 2009, **44**(10), 2075–2084.
10. A. Bakhsh, S.-J. Lee, E.-Y. Lee, N. Sabikun, Y.-H. Hwang and S.-T. Joo, A Novel Approach for Tuning the Physicochemical, Textural, and Sensory Characteristics of Plant-Based Meat Analogs with Different Levels of Methylcellulose Concentration, *Foods*, 2021, **10**(3), 560.
11. K. Kyriakopoulou, J. K. Keppler and A. J. van der Goot, Functionality of Ingredients and Additives in Plant-Based Meat Analogues, *Foods*, 2021, **10**(3), 600.
12. N. S. Argel, N. Ranalli, A. N. Califano and S. C. Andrés, Influence of partial pork meat replacement by pulse flour on physicochemical and sensory characteristics of low-fat burgers, *J. Sci. Food Agric.*, 2020, **100**(10), 3932–3941.
13. Z. E. Sikorski, J. Pokorny and S. Damodaran, Physical and Chemical Interactions of Components in Food Systems, in *Fennema's Food Chemistry*, ed. S. Damodaran, K. L. Parkin and O. R. Fennema, CRC Press, Boca Raton, FL, 4th edn, 2007.

14. M. H. Kamani, M. S. Meera, N. Bhaskar and V. K. Modi, Partial and total replacement of meat by plant-based proteins in chicken sausage: evaluation of mechanical, physico-chemical and sensory characteristics, *J. Food Sci. Technol.*, 2019, **56**(5), 2660–2669.
15. M. Mishyna, J. Chen and O. Benjamin, Sensory attributes of edible insects and insect-based foods – Future outlooks for enhancing consumer appeal, *Trends Food Sci. Technol.*, 2020, **95**, 141–148.
16. J. Tao and Y. O. Li, Edible insects as a means to address global malnutrition and food insecurity issues, *Food Qual. Saf.*, 2018, **2**(1), 17–26.
17. L. H. Fasolin, R. N. Pereira, A. C. Pinheiro, J. T. Martins, C. C. P. Andrade and O. L. Ramos, *et al.*, Emergent food proteins – Towards sustainability, health and innovation, *Food Res. Int.*, 2019, **125**, 108586.
18. A. C. Hoek, P. A. Luning, P. Weijzen, W. Engels, F. J. Kok and C. de Graaf, Replacement of meat by meat substitutes. A survey on person- and product-related factors in consumer acceptance, *Appetite*, 2011, **56**(3), 662–673.
19. F. Vanhonacker, E. J. Van Loo, X. Gellynck and W. Verbeke, Flemish consumer attitudes towards more sustainable food choices, *Appetite*, 2013, **62**, 7–16.
20. E. P. Köster, Diversity in the determinants of food choice: A psychological perspective, *Food Qual. Prefer.*, 2009, **20**(2), 70–82.
21. C. Mutungi, F. G. Irungu, J. Nduko, F. Mutua, H. Affognon and D. Nakimbugwe, *et al.*, Postharvest processes of edible insects in Africa: A review of processing methods, and the implications for nutrition, safety and new products development, *Crit. Rev. Food Sci. Nutr.*, 2019, **59**(2), 276–298.
22. M. Testa, M. Stillo, G. Maffei, V. Andriolo, P. Gardois and C. M. Zotti, Ugly but tasty: A systematic review of possible human and animal health risks related to entomophagy, *Crit. Rev. Food Sci. Nutr.*, 2017, **57**(17), 3747–3759.
23. R. Musundire, C. J. Zvidzai, C. Chidewe, B. K. Samende and F. A. Manditsera, Nutrient and anti-nutrient composition of Henicus whellani (Orthoptera: Stenopelmatidae), an edible ground cricket, in south-eastern Zimbabwe, *Int. J. Trop. Insect Sci.*, 2014, **34**(4), 223–231.
24. G. Sogari, D. Bogueva and D. Marinova, Australian Consumers' Response to Insects as Food, *Agriculture*, 2019, **9**(5), 108.
25. P. M. Poortvliet, L. Van der Pas, B. C. Mulder and V. Fogliano, Healthy, but Disgusting: An Investigation Into Consumers' Willingness to Try Insect Meat, *J. Econ. Entomol.*, 2019, **112**(3), 1005–1010.
26. A. Duda, J. Adamczak, P. Chełmińska, J. Juszkiewicz and P. Kowalczewski, Quality and Nutritional/Textural Properties of Durum Wheat Pasta Enriched with Cricket Powder, *Foods*, 2019, **8**(2), 46.
27. C. E. Gurdian, D. D. Torrico, B. Li, G. Tuuri and W. Prinyawiwatkul, Effect of Informed Conditions on Sensory Expectations and Actual

Perceptions: A Case of Chocolate Brownies Containing Edible-Cricket Protein, *Foods*, 2021, **10**(7), 1480.
28. A. Roncolini, V. Milanović, F. Cardinali, A. Osimani, C. Garofalo and R. Sabbatini, *et al.*, Protein fortification with mealworm (Tenebrio molitor L.) powder: Effect on textural, microbiological, nutritional and sensory features of bread, *PLoS One*, 2019, **14**(2), e0211747.
29. A. Orkusz, W. Wolańska, J. Harasym, A. Piwowar and M. Kapelko, Consumers' Attitudes Facing Entomophagy: Polish Case Perspectives, *Int. J. Environ. Res. Public Health*, 2020, **17**(7), 2427.
30. C. Hartmann and M. Siegrist, Becoming an insectivore: Results of an experiment, *Food Qual. Prefer.*, 2016, **51**, 118–122.
31. S. Piha, T. Pohjanheimo, A. Lähteenmäki-Uutela, Z. Křečková and T. Otterbring, The effects of consumer knowledge on the willingness to buy insect food: An exploratory cross-regional study in Northern and Central Europe, *Food Qual. Prefer.*, 2018, **70**, 1–10.
32. M. Castro and E. Chambers, Consumer Avoidance of Insect Containing Foods: Primary Emotions, Perceptions and Sensory Characteristics Driving Consumers Considerations, *Foods*, 2019, **8**(8), 351.
33. R. Caparros Megido, C. Gierts, C. Blecker, Y. Brostaux, É. Haubruge and T. Alabi, *et al.*, Consumer acceptance of insect-based alternative meat products in Western countries, *Food Qual. Prefer.*, 2016, **52**, 237–243.
34. J. Scholliers, L. Steen and I. Fraeye, Partial replacement of meat by superworm (Zophobas morio larvae) in cooked sausages: Effect of heating temperature and insect:Meat ratio on structure and physical stability, *Innovative Food Sci. Emerging Technol.*, 2020, **66**, 102535.
35. Impossible-Foods, What Are The Nutrition Facts?, Available from: https://faq.impossiblefoods.com/hc/en-us/articles/360018.
36. Bugfoundation, Germany's First Insect Burger, Available from: https://www.bugfoundation.com/our-burger.html.
37. M. Brückner-Gühmann, M. Banovic and S. Drusch, Towards an increased plant protein intake: Rheological properties, sensory perception and consumer acceptability of lactic acid fermented, oat-based gels, *Food Hydrocolloids*, 2019, **96**, 201–208.
38. O. E. Mäkinen, N. Sozer, D. Ercili-Cura and K. Poutanen, in *Sustainable Protein Sources*, ed. S. R. Nadathur, J. P. D. Wanasundara and L. Scanlin, Academic Press, San Diego, 2017, pp. 105–119.
39. W. Ma, C. Zhang, X. Kong, X. Li, Y. Chen and Y. Hua, Effect of pea milk preparation on the quality of non-dairy yoghurts, *Food Biosci.*, 2021, **44**, 101416.
40. W. S. U. Roland, L. Pouvreau, J. Curran, F. van de Velde and P. M. T. de Kok, Flavor Aspects of Pulse Ingredients, *Cereal Chem.*, 2017, **94**(1), 58–65.

41. M. Trikusuma, L. Paravisini and D. G. Peterson, Identification of aroma compounds in pea protein UHT beverages, *Food Chem.*, 2020, **312**, 126082.
42. C. Zhang, Y. Hua, X. Li, X. Kong and Y. Chen, Key volatile off-flavor compounds in peas (Pisum sativum L.) and their relations with the endogenous precursors and enzymes using soybean (Glycine max) as a reference, *Food Chem.*, 2020, **333**, 127469.
43. N. Grasso, L. Alonso-Miravalles and J. A. O'Mahony, Composition, Physicochemical and Sensorial Properties of Commercial Plant-Based Yogurts, *Foods*, 2020, **9**(3), 252.
44. E. Pontonio and C. G. Rizzello, Milk Alternatives and Non-Dairy Fermented Products: Trends and Challenges, *Foods*, 2021, **10**(2), 222.
45. S. J. Calva-Estrada, M. Jiménez-Fernández and E. Lugo-Cervantes, Protein-Based Films: Advances in the Development of Biomaterials Applicable to Food Packaging, *Food Eng. Rev.*, 2019, **11**(2), 78–92.
46. H. Chen, J. Wang, Y. Cheng, C. Wang, H. Liu and H. Bian, *et al.*, Application of Protein-Based Films and Coatings for Food Packaging: A Review, *Polymers*, 2019, **11**(12), 2039.
47. B. Hassan, S. A. S. Chatha, A. I. Hussain, K. M. Zia and N. Akhtar, Recent advances on polysaccharides, lipids and protein based edible films and coatings: A review, *Int. J. Biol. Macromol.*, 2018, **109**, 1095–1107.
48. Z. A. M. Adilah and Z. A. N. Hanani, Active packaging of fish gelatin films with Morinda citrifolia oil, *Food Biosci.*, 2016, **16**, 66–71.
49. N. K. Dubey and R. Dubey, Edible films and coatings: An update on recent advances, in *Biopolymer-Based Formulations*, ed. K. Pal, I. Banerjee, P. Sarkar, D. Kim, W.-P. Deng and N. K. Dubey, *et al.*, Elsevier, 2020, pp. 675–695.
50. C. V. Dhumal and P. Sarkar, Composite edible films and coatings from food-grade biopolymers, *J. Food Sci. Technol.*, 2018, **55**(11), 4369–4383.
51. S. Albert and G. S. Mittal, Comparative evaluation of edible coatings to reduce fat uptake in a deep-fried cereal product, *Food Res. Int.*, 2002, **35**(5), 445–458.
52. D. Ananey-Obiri, L. Matthews, M. H. Azahrani, S. A. Ibrahim, C. M. Galanakis and R. Tahergorabi, Application of protein-based edible coatings for fat uptake reduction in deep-fat fried foods with an emphasis on muscle food proteins, *Trends Food Sci. Technol.*, 2018, **80**, 167–174.
53. D. De Grandi Castro Freitas, S. A. G. Berbari, P. Prati, F. M. Fakhouri, F. P. Collares Queiroz and E. Vicente, Reducing fat uptake in cassava product during deep-fat frying, *J. Food Eng.*, 2009, **94**(3), 390–394.
54. K. I. Iwata, S. H. Ishizaki, A. K. Handa and M. U. Tanaka, Preparation and characterization of edible films from fish water-soluble proteins, *Fisheries Sci.*, 2000, **66**(2), 372–378.

55. T. Bourtoom, Edible films and coatings: characteristics and properties, *Int. Food Res. J.*, 2008, **15**(3), 237–248.
56. M. Mellema, Mechanism and reduction of fat uptake in deep-fat fried foods, *Trends Food Sci. Technol.*, 2003, **14**(9), 364–373.
57. C. M. B. S. Pintado, M. A. S. S. Ferreira and I. Sousa, Control of pathogenic and spoilage microorganisms from cheese surface by whey protein films containing malic acid, nisin and natamycin, *Food Control.*, 2010, **21**(3), 240–246.
58. C. Ibarguren, G. Céliz, A. S. Díaz, M. A. Bertuzzi, M. Daz and M. C. Audisio, Gelatine based films added with bacteriocins and a flavonoid ester active against food-borne pathogens, *Innovative Food Sci. Emerging Technol.*, 2015, **28**, 66–72.
59. P. Kaewprachu, K. Osako, S. Benjakul, P. Suthiluk and S. Rawdkuen, Shelf life extension for Bluefin tuna slices (Thunnus thynnus) wrapped with myofibrillar protein film incorporated with catechin-Kradon extract, *Food Control.*, 2017, **79**, 333–343.
60. M. Gallego, M. Gordon, F. Segovia and P. M. Almajano, Gelatine-based antioxidant packaging containing Caesalpinia decapetala and Tara as a coating for ground beef patties, *Antioxidants*, 2016, **5**(2), 10.
61. F. Liu, R. J. Avena-Bustillos, B. S. Chiou, Y. Li, Y. Ma and T. G. Williams, *et al.*, Controlled-release of tea polyphenol from gelatin films incorporated with different ratios of free/nanoencapsulated tea polyphenols into fatty food simulants, *Food Hydrocolloids*, 2017, **62**, 212–221.
62. F. M. Monedero, A. Hambleton, P. Talens, F. Debeaufort, A. Chiralt and A. Voilley, Study of the retention and release of n-hexanal incorporated into soy protein isolate–lipid composite films, *J. Food Eng.*, 2010, **100**, 33–138.
63. A. Ket-On, P. Natkritta, A. Somwangthanaroj, T. Janjarasskul and K. Tananuwong, Properties and storage stability of whey protein edible film with spice powders, *J. Food Sci. Technol.*, 2016, **53**(7), 2933–2942.
64. F. V. R. Castro, M. A. Andrade, A. Sanches Silva, M. F. Vaz and F. Vilarinho, The Contribution of a Whey Protein Film Incorporated with Green Tea Extract to Minimize the Lipid Oxidation of Salmon (Salmo salar L.), *Foods*, 2019, **8**(8), 327.
65. D. J. McClements and E. A. Decker, Lipids, in *Fennema's Food Chemistry*, ed. S. Damodaran, K. L. Parkin and O. R. Fennema, CRC Press, FL, 4th edn, 2007.
66. B. C. Hu, M. Chen and Z. Wang, Release of thymol, cinnamaldehyde and vanillin from soy protein isolate films into olive oil, *Packag. Technol. Sci.*, 2012, **25**, 97–106.
67. V. García Arteaga, M. Apéstegui Guardia, I. Muranyi, P. Eisner and U. Schweiggert-Weisz, Effect of enzymatic hydrolysis on molecular

weight distribution, techno-functional properties and sensory perception of pea protein isolates, *Innovative Food Sci. Emerging Technol.*, 2020, **65**, 102449.
68. J. A. Rothwell, A. Medina-Remón, J. Pérez-Jiménez, V. Neveu, V. Knaze and N. Slimani, *et al.*, Effects of food processing on polyphenol contents: A systematic analysis using Phenol-Explorer data, *Mol. Nutr. Food Res.*, 2015, **59**(1), 160–170.
69. P. Bandyopadhyay, A. K. Ghosh and C. Ghosh, Recent developments on polyphenol–protein interactions: effects on tea and coffee taste, antioxidant properties and the digestive system, *Food Funct.*, 2012, **3**(6), 592–605.
70. M. Smuda and M. A. Glomb, Maillard Degradation Pathways of Vitamin C, *Angew. Chem., Int. Ed.*, 2013, **52**(18), 4887–4891.
71. M. Yousefi, A. Ehsani and S. M. Jafari, Lipid-based nano delivery of antimicrobials to control food-borne bacteria, *Adv. Colloid Interface Sci.*, 2019, **270**, 263–277.
72. Z. Zorić, S. Pedisić, D. B. Kovačević, D. Ježek and V. Dragović-Uzelac, Impact of packaging material and storage conditions on polyphenol stability, colour and sensory characteristics of freeze-dried sour cherry (prunus cerasus var. Marasca), *J. Food Sci. Technol.*, 2016, **53**(2), 1247–1258.
73. D. J. McClements, Encapsulation, protection, and release of hydrophilic active components: potential and limitations of colloidal delivery systems, *Adv. Colloid Interface Sci.*, 2015, **219**, 27–53.
74. N. P. Aditya, Y. G. Espinosa and I. T. Norton, Encapsulation systems for the delivery of hydrophilic nutraceuticals: Food application, *Biotechnol. Adv.*, 2017, **35**(4), 450–457.
75. N. P. Aditya and S. Ko, Solid lipid nanoparticles (SLNs): delivery vehicles for food bioactives, *RSC Adv.*, 2015, **5**(39), 30902–30911.
76. E. Dickinson, Advances in food emulsions and foams: reflections on research in the neo-Pickering era, *Curr. Opin. Food Sci.*, 2020, **33**, 52–60.
77. A. Faridi Esfanjani, E. Assadpour and S. M. Jafari, Improving the bioavailability of phenolic compounds by loading them within lipid-based nanocarriers, *Trends Food Sci. Technol.*, 2018, **76**, 56–66.
78. Ó. Martínez-Álvarez, M. M. Calvo and J. Gómez-Estaca, Recent Advances in Astaxanthin Micro/Nanoencapsulation to Improve Its Stability and Functionality as a Food Ingredient, *Mar. Drugs*, 2020, **18**(8), 406.
79. R. Domínguez, B. Bohrer, P. E. S. Munekata, M. Pateiro and J. M. Lorenzo, Recent Discoveries in the Field of Lipid Bio-Based Ingredients for Meat Processing, *Molecules*, 2021, **26**(1), 190.
80. J. D. Rios-Mera, E. Saldaña, I. Patinho, M. M. Selani and C. J. Contreras-Castillo, Advances and gaps in studies on healthy meat products and their relationship with regulations: The Brazilian scenario, *Trends Food Sci. Technol.*, 2021, **110**, 833–840.

81. S. Barbut, A. G. Marangoni, U. Thode and B. E. Tiensa, Using Canola Oil Organogels as Fat Replacement in Liver Pâté, *J. Food Sci.*, 2019, **84**(9), 2646–2651.
82. A. K. F. I. Câmara, M. V. Geraldi, P. K. Okuro, M. R. Maróstica, R. L. da Cunha and M. A. R. Pollonio, Satiety and in vitro digestibility of low saturated fat Bologna sausages added of chia mucilage powder and chia mucilage-based emulsion gel, *J. Funct. Foods*, 2020, **65**, 103753.
83. F. A. L. de Carvalho, P. E. S. Munekata, M. Pateiro, P. C. B. Campagnol, R. Domínguez and M. A. Trindade, *et al.*, Effect of replacing backfat with vegetable oils during the shelf-life of cooked lamb sausages, *LWT*, 2020, **122**, 109052.
84. R. Domínguez, R. Agregán, A. A. Gonçalves and J. M. Lorenzo, Effect of fat replacement by olive oil on the physico-chemical properties, fatty acids, cholesterol and tocopherol content of pâté, *Grasas Aceites*, 2016, **67**(2), e133.
85. J. C. Barros, P. E. Munekata, F. A. L. de Carvalho, R. Domínguez, M. A. Trindade and M. Pateiro, *et al.*, Healthy beef burgers: Effect of animal fat replacement by algal and wheat germ oil emulsions, *Meat Sci.*, 2021, **173**, 108396.
86. R. Domínguez, M. Pateiro, P. E. Sichetti Munekata, P. C. Bastianello Campagnol and J. M. Lorenzo, Influence of partial pork backfat replacement by fish oil on nutritional and technological properties of liver pâté, *Eur. J. Lipid Sci. Technol.*, 2017, **119**(5), 1600178.
87. C. de Souza Paglarini, G. de Figueiredo Furtado, A. R. Honório, L. Mokarzel, V. A. da Silva Vidal and A. P. B. Ribeiro, *et al.*, Functional emulsion gels as pork back fat replacers in Bologna sausage, *Food Struct.*, 2019, **20**, 100105.
88. R. T. Heck, J. M. Lorenzo, B. A. Dos Santos, A. J. Cichoski, C. R. de Menezes and P. C. B. Campagnol, Microencapsulation of healthier oils: an efficient strategy to improve the lipid profile of meat products, *Curr. Opin. Food Sci.*, 2021, **40**, 6–12.
89. Z. Bolger, N. P. Brunton and F. J. Monahan, Impact of inclusion of flaxseed oil (pre-emulsified or encapsulated) on the physical characteristics of chicken sausages, *J. Food Eng.*, 2018, **230**, 39–48.
90. M. Espert, A. Bresciani, T. Sanz and A. Salvador, Functionality of low digestibility emulsions in cocoa creams. Structural changes during in vitro digestion and sensory perception, *J. Funct. Foods*, 2019, **54**, 146–153.
91. G. Christophe, P. Fontanille and C. Larroche, Research and Production of Microbial Polyunsaturated Fatty Acids, in *Bioprocessing for Biomolecules Production*, ed. G. Molina, V. K. Gupta, B. N. Singh and N. Gathergood, 2019, pp. 167–194.
92. C. Ratledge and C. Lippmeier, in *Fatty Acids*, ed. M. U. Ahmad, AOCS Press, 2017, pp. 237–278.

93. B. Bharathiraja, S. Sridharan, V. Sowmya, D. Yuvaraj and R. Praveenkumar, Microbial oil - A plausible alternate resource for food and fuel application, *Bioresour. Technol.*, 2017, **233**, 423–432.
94. D. Murphy, The biogenesis and functions of lipid bodies in animals, plants and microorganisms, *Prog. Lipid. Res.*, 2001, **40**, 325–438.
95. K. Mlícková, E. Roux, K. Athenstaedt, S. d'Andrea, G. Daum and T. Chardot, *et al.*, Lipid accumulation, lipid body formation, and acyl coenzyme A oxidases of the yeast *Yarrowia lipolytica*, *Appl. Environ. Microbiol.*, 2004, **70**(7), 3918–3924.
96. S. Papanikolaou, M. Galiotou-Panayotou, S. Fakas, M. Komaitis and G. Aggelis, Lipid production by oleaginous Mucorales cultivated on renewable carbon sources, *Eur. J. Lipid Sci. Technol.*, 2007, **109**(11), 1060–1070.
97. X. Tang, H. Chen, Y. Q. Chen, W. Chen, V. Garre and Y. Song, *et al.*, Comparison of Biochemical Activities between High and Low Lipid-Producing Strains of Mucor circinelloides: An Explanation for the High Oleaginicity of Strain WJ11, *PLoS One*, 2015, **10**(6), e0128396.
98. L.-J. Ren, X.-M. Sun, X.-J. Ji, S.-L. Chen, D.-S. Guo and H. Huang, Enhancement of docosahexaenoic acid synthesis by manipulation of antioxidant capacity and prevention of oxidative damage in Schizochytrium sp., *Bioresour. Technol.*, 2017, **223**, 141–148.
99. C. Shene, Y. Chisti, D. Vergara, C. Burgos-Díaz, M. Rubilar and M. Bustamante, Production of eicosapentaenoic acid by Nannochloropsis oculata: Effects of carbon dioxide and glycerol, *J. Biotechnol.*, 2016, **239**, 47–56.
100. F. Qi, M. Zhang, Y. Chen, X. Jiang, J. Lin and X. Cao, *et al.*, A lignocellulosic hydrolysate-tolerant Aurantiochytrium sp. mutant strain for docosahexaenoic acid production, *Bioresour. Technol.*, 2017, **227**, 221–226.
101. T. Sakamoto, E. Sakuradani, T. Okuda, H. Kikukawa, A. Ando and S. Kishino, *et al.*, Metabolic engineering of oleaginous fungus Mortierella alpina for high production of oleic and linoleic acids, *Bioresour. Technol.*, 2017, **245**, 1610–1615.
102. M. Čertík, Z. Adamechová and L. Guothová, Simultaneous enrichment of cereals with polyunsaturated fatty acids and pigments by fungal solid state fermentations, *J. Biotechnol.*, 2013, **168**(2), 130–134.
103. M. Demir, I. Turhan, A. Kucukcetin and Z. Alpkent, Oil production by Mortierella isabellina from whey treated with lactase, *Bioresour. Technol.*, 2013, **128**, 365–369.
104. Y. Liang, N. Sarkany, Y. Cui, J. Yesuf, J. Trushenski and J. W. Blackburn, Use of sweet sorghum juice for lipid production by *Schizochytrium limacinum* SR21, *Bioresour. Technol.*, 2010, **101**(10), 3623–3627.

CHAPTER 6

Protein- and Lipid-based Bioactive Compounds

THARUSHI S. SHINALI[a], JIE YU[a,b], RUIJING QU[a,c] AND NAN SHANG*[a]

[a]College of Engineering, China Agricultural University, Beijing, China; [b]College of Food Science, Sichuan Agricultural University, Yaan, China; [c]School of Biotechnology and Food Science, Tianjin University of Commerce, Tianjin, China
*E-mail: nshang@cau.edu.cn

6.1 Introduction

Proteins and lipids can be considered indispensable macronutrients in the human diet, providing energy and significant nutritional benefits. Recently, the biological activity of these protein- and lipid-based compounds has also been considered in novel research studies due to their promising biofunctional properties in human health and disease prevention as natural compounds.

Accordingly, several natural protein-derived peptides and protein hydrolysates are known to have functional or biologically active components. However, most peptides are biologically inactive within the parent protein molecules. Bioactive peptides (BPs) are generally short sequences of amino acids with 2–20 residues, and they should be released from protein sources to exhibit their biofunctional properties.[1] Bioactive proteins possess diverse functional properties, including antioxidant, antimicrobial, and other therapeutic potentials against several chronic diseases, including

Smart Materials No. 41
Functional Materials from Lipids and Proteins
Edited by Nandika Bandara and Aman Ullah

Published by the Royal Society of Chemistry, www.rsc.org

cancers and cardiovascular diseases. The structure of the BPs plays a key role in determining their specific physiological activities. Therefore, significant structure–activity parameters should be considered in the production and processing of BPs for biotechnological applications.[2] The bioavailability of the peptides after oral administration is another key factor to be considered in incorporating them into functional food, nutraceutical applications, and drug delivery.[3]

On the other hand, fatty acids, phytosterols, phenolic lipids, and fat-soluble vitamins are the most important dietary bioactive lipids.[4] The bioactivity of lipids enables a wide range of benefits toward human health, including immune regulation, inflammation and maintaining homeostasis, inhibition of osteoblastic differentiation while promoting osteoclastic differentiation in bone health, and prevention/treatment of chronic diseases.[4–6] Different metabolic mechanisms and pathways of lipid components within the human body, their chemical composition, and specific structural features enable the role of lipids beyond basic nutrition, facilitating the bioactivity of lipids. However, most health benefits are related to the type of fatty acid contained. It was reported that the most prominent bioactive lipids are polyunsaturated fatty acids (PUFAs), and its *cis*-isomer configuration is known to have significant bioactivity.[7] These PUFAs are considered essential fatty acids as they cannot be synthesized in the body and must be obtained from food. Based on the chemical structure, there are two types of PUFAs: omega-3 (alpha-linolenic acid [ALA], eicosapentaenoic acid [EPA], and docosahexaenoic acid [DHA]) and omega-6. Similar to BPs, bioactive lipid components have also received much attention in various applications.

Besides the health benefits, the biocompatibility of bioactive proteins and lipids also facilitates various functional applications in different fields, including food preservation, packaging, and medical science. This chapter will discuss the sources of the bioactive compounds, novel production strategies, and significant properties of these bioactive protein and lipid materials toward human health and their functional applications.

6.2 Protein- and Lipid-based Bioactive Compounds for Human Health

6.2.1 Protein-based Bioactive Compounds

6.2.1.1 Sources of Bioactive Protein

Protein-rich animal, marine and aquatic, plant, and microbial origins are major BP and amino acid sources. BPs derived from animal origins of meat, fish, egg, and milk have distinct health benefits.[8] Especially, bovine milk, egg, and other dairy products from livestock production have been identified as important sources of BPs.[9,10] The hydrolysates of actin, myosin, and troponin proteins from muscle sources have been recognized to provide BPs with angiotensin I-converting enzyme (ACE)-inhibitory activity.[11] Slaughterhouse blood from the meat production food chain can also be considered an emerging source of potential BPs. It has been reported that BPs derived from bovine and porcine blood sources show ACE-inhibitory activity, antioxidant activity, antimicrobial properties, mineral-binding ability, and opioid activity.[12] Various fungal and bacterial sources, including *Agaricus bisporus*, *Tricholoma matsutake*, *Bacillus amyloliquefaciens*, and *Lactobacillus paracasei*, can be considered potent sources of peptides with antimicrobial, antioxidant, and antihypertensive properties.[13–16] Whey protein generated as a liquid byproduct during coagulation of the casein in the dairy industry is an ample source of different bioactive proteins and peptides with specific functional and nutraceutical properties, such as antimicrobial, antioxidant, and antihypertensive properties. These bioactive proteins and peptides include albutensin A, lactoferricin, serorphin, α-lactorphin, β-lactorphin, β-lactotensin, bovine serum albumin (BSA), and immunoglobulin.[17]

The focus on peptides from marine and aquatic animals and their by-products as a source of bioactive proteins has also been increased recently. Studies revealed that many BPs with various functional properties could be extracted from marine and aquatic sources, such as sponges, sea slugs, solitary tunicate (*Styela clava*), soft corals, and other fish varieties.[18–20] Enzymatic hydrolysis of marine algae, algae processing waste, and microalgae species (Spirulina) produces biopeptides that exhibit different biological functions, depending on the amino acid sequence.[21,22] These functionalities include anticancer, antiatherosclerotic, antioxidant, antibacterial, and anti-inflammatory effects.[21]

Proteins derived from various parts of plants, such as leaves, fruits, and seeds, and their processing wastes are now considered potential sources of proteins. As the current research trend is moving toward biomolecules, these plant proteins are also being explored for the presence of biologically active proteins and peptides. The protein hydrolysates from seeds, fruits, leaves, rice bran, soy, cumin, common bean, oat, and lupine have been reported the presence of antidiabetic peptides that act as alpha-glucosidase inhibitors, alpha-amylase inhibitors, dipeptidyl peptidase-IV inhibitors, inhibitors of the glucose transporter system, and insulin mimetics.[23] Another study reported the *in vivo* effectiveness of five potent peptides in yellow field pea protein hydrolysate, which have the sequences of LTFPG, IIPLEN, LSSGDVF, IFENLQN, and FEGTVFENG for significant antihypertensive effects.[24] Bean and cottonseed sources also have important antimicrobial and antioxidant peptides.[25]

Pulse crops, such as chickpeas, lentils, dry peas, and lupine, are vital sources of proteins, carbohydrates, dietary fibers, minerals, and vitamins. Their nutritional characteristics have been associated with providing many health benefits and managing some chronic diseases. Pulses also contain antinutritional protein or non-protein compounds. Although antinutritional compounds in pulses cause some deleterious effects on human and animal health when they are consumed in raw form, recent studies revealed that these compounds could be used as potent proteins and peptides after adequate processing procedures to provide beneficial health properties in the management of chronic diseases, such as cancers, obesity, and hypertension.[26] These antinutritional protein compounds include lectins, trypsin, chymotrypsin, antifungal peptides, ribosome-inactivating proteins, and ACE inhibitors.[27,28]

Fruit processing wastes are also a rich source of proteins, and hydrolysis of these proteins produces BPs having various nutraceutical and pharmacological properties. Passion fruit seeds have been reported to have a peptide of *Pe*-AFP1 (5.0 kDa), which is capable of showing antifungal properties against the filamentous fungi *Trichoderma harzianum*, *Fusarium oxysporum*, and *Aspergillus fumigatus*.[29] The presence of amino acids, such as histidine and phenylalanine, in different fruit wastes was reported to show antioxidant properties.[30] Antimicrobial and antioxidant peptides have been identified to be present in agricultural and industrial waste, such as coffee ground waste and baijiu vinasse.[25] Therefore, the BPs derived from food processing wastes can be utilized in developing functional ingredients or products as an industrial application, thereby securing their significant benefits for human health.

6.2.1.2 Production of Functional Materials from Proteins

As BPs have received increasing attention to using them in industrial applications due to their beneficial properties, the consideration for effective industrial production of these bioactive materials has paved the way. The important bioactive fractions of proteins that can be utilized in functional applications are peptides. Therefore, BPs can be produced using different technologies, including enzymatic synthesis, microbial fermentation, chemical synthesis, and recombinant DNA technology.[8] The production of bioactive compounds follows various separation approaches to purify them to be used in industrial applications. Once the target peptides are extracted, they can be fractionated using advanced methods, such as ultrafiltration, electrodialysis with ultrafiltration membranes, sequential chromatography, and reverse-phase liquid chromatography, and then the molecular structures can be characterized using mass spectrometric methods, such as liquid chromatography–mass spectrometry and tandem mass spectrometry.[1,31] However, the lack of commercially viable large-scale production technologies for these bioactive compounds still hinders industrial applications in some fields.

The interested BPs or hydrolysates can be released from food-derived proteins by *in vitro* enzymatic hydrolysis using different types of single or multiple specific or nonspecific protease enzymes. However, various factors, including the degree of hydrolysis, time taken for the hydrolysis procedure, enzyme–substrate ratios, and pretreatments on proteins, affect the production process of BPs using enzymatic synthesis.[2] The protease enzymes that produce BPs are usually obtained from animal, plant, and microbial sources. Alemán *et al.*[32] produced bioactive protein hydrolysates with different molecular weights from giant squid gelatin using commercial proteases. The novel protein hydrolysates could be used as functional ingredients in numerous applications due to their anticancer, antihypertensive, and antioxidant properties. The results revealed that the most bioactive hydrolysates with antioxidant activity and cytotoxic effects on cancer cells could be prepared using alcalase and esperase proteases. Enzyme-assisted bioactive protein extraction methods have been found to generate peptides with higher ACE-inhibitory and antioxidant activities from marine algae species, and this method is considered an eco-friendly, nontoxic, high-bioactivity, and cost-effective technology used in large-scale productions.[1] Kim and Wijesekara[33] also reported that biopeptides could be produced by enzymatic hydrolysis of food proteins using α-chymotrypsin, papain, neutrase, and trypsin under optimal temperature and pH conditions of the reaction media. The authors further noted that the enzymatic hydrolysis to produce BPs had obtained much attention over the other production methods due to the lack of residual organic solvents or toxic chemicals in the final products.

Additionally, *in situ* microbial fermentation of various protein sources can produce bioactive proteins although it is not widely used for enzymatic synthesis. Lactic acid bacteria (LAB) and bifidobacteria are commonly used for the production process.[1] Different bioactive proteins with significant health benefits are found to be released during microbial fermentation of dairy products using LAB by hydrolyzing milk proteins with cell-surface proteinases.[34] Fermentation of soybean using *Bacillus subtilis*, LAB, and fungi species results in the hydrolysis of soybean proteins, such as glycinin and β-conglycinin, releasing BPs with bioactive properties, such as antihypertensive, antimicrobial, antioxidant, antidiabetic, and anticancer activities, depending on the sequence and composition of amino acids.[35] Microbial proteases hydrolyze parent proteins into peptides with different molecular weights and free amino acids that can act as bioactive materials. In addition, microbes are capable of removing hyper-allergic or antinutritional factors, such as trypsin inhibitors, glycinin, β-conglycinin, phytate, oligosaccharides raffinose and stachyose, and saponins in soybeans during the fermentation process.[36] In contrast, inconsistent production of peptides due to the changes in the activity of microorganisms under different conditions in the reaction medium is one of the major disadvantages encountered in this process. Moreover, much work on fermented food-derived BPs should be carried out to understand their effects on humans and expand their commercial production in the food, nutraceutical, and pharmaceutical industries.

Chemical synthesis of proteins enables the production of target bioactive proteins using specific chemicals to mediate the peptide bond formation.

Chemical synthesis mainly involves either solid-phase synthesis or synthesis in solutions. Currently, rapid, effective, and automated solid-phase synthesis methods of proteins are commonly used. However, the production of peptides containing more than 40 amino acid residues is limited by the stepwise assembly of amino acids using solid-phase synthesis, while the convergent assembly of protected or partially protected peptide segments both in solution and on solid phase enables the synthesis of larger peptides with more than 40 residues.[37] Using the methods of fragmentation of the protein into peptides by chemical methods and the organic synthesis of the solid phase of certain peptide fragments, Ialenti *et al.*[38] synthesized novel BPs with anti-inflammatory properties from the amino acid sequence of an interesting pharmacological protein seminal vesicle-IV (SV-IV) synthesized by rats. Another recent study has highlighted the promising therapeutic roles, including antioxidant, low-density lipoprotein (LDL) cholesterol-lowering, and anti-atherosclerosis activities, exhibited by chemically synthesized bioactive proteins.[31]

Recombinant DNA technology involves the production of recombinant proteins using cloning and gene expression in host cells. Microbial cells, including bacteria and yeast, marine algae, insect, mammalian, and plant cells are used as expression systems. The expression systems should be carefully selected to produce proteins with higher bioactivity, lower production costs, and higher production and purification yields.[39] A novel study has reported utilizing recombinant protein technology for the stable expression of the full-length human factor H, a promising biopharmaceutical associated with several human diseases.[40] Therefore, as a promising technology to produce bioactive proteins using genetic engineering approaches, its focus has emerged in the field.

Besides the production of BPs, as a novel concept, bioactive polyphenolic compounds can be engineered with functional proteins based on the ability to form protein–polyphenol interactions through covalent interactions, non-covalent interactions, and the self-assembling ability of protein–polyphenol conjugates into nanoparticles.[41] These protein–polyphenol aggregates may act as bioactive conjugates in functional food applications capable of enhancing different biological functions, including chemical stability, gastrointestinal stability, target delivery, cell uptakes, and other biological functionalities, such as antioxidant, anticancer, and anti-inflammatory activities.

6.2.2 Lipid-based Bioactive Compounds

6.2.2.1 Sources of Bioactive Lipids

Lipid-based bioactive compounds are derived from plant, animal, marine, or microbial sources. Olive oil is a rich source of oleic acid, a monounsaturated long-chain fatty acid. Linoleic acid (LA) is the parent compound of the omega-6 group. LA can be found in canola, safflower, and corn oils, while seeds, grains, nuts, including flaxseed and flaxseed oil, walnuts and walnut oil, and canola oil contain ALA.[4,42] Animal-origin food sources, such as pork, lard, turkey fat, and butter, are also competent sources of omega-6 fatty acids.[43] Conjugated linoleic acid (CLA) is an essential fatty acid composed of various positional and geometrical linoleic acid isomers. Its bioactivity is due to the conjugated form of double bonds. CLA can be mostly found in ruminant-origin (cow and sheep) meat and milk as rumen microorganisms convert LA into CLA, which is then absorbed by animal tissues.[44] Verardo *et al.*[45] reported pomegranate seed oil as a rich source of bioactive lipids containing more CLA, including punicic acid and several other critical fatty acids, such as oleic, linoleic, and palmitic acid.

The results of the study conducted by Soares *et al.*[46] showed that seaweed species could be considered a potential source of bioactive lipids, including notably higher amounts of poly- and monounsaturated fatty acids. The focus on the large-scale production of important PUFAs from microalgae species has emerged, as microalgae are rich sources of these fatty acids and more cost-effective than biofuels from microalgae.[47,48] EPA can be abundantly found in different classes of microalgae, including *Bacillariophyceae*, *Chlorophyceae*, *Cryptophyceae*, *Chrysophyceae*, *Eustigmatophyceae*, and *Prasinophyceae*, while DHA is predominant in dinoflagellate (*Crypthecodinium cohnii*) and species of *Schizochytrium*.[48] DHA is also found in marine fish and fish oils.[4]

Milk from various mammalian species, including human and bovine milk, has been found to have significant bioactive lipids, such as triacylglycerides, which are metabolized into mono- and diacylglycerides, saturated and polyunsaturated fatty acids, *trans*-fatty acids, and phospholipids including sphingomyelin.[49] Buttermilk and butter serum resulted as low-value products from the dairy industry and have also emerged as a source for functional food applications. It has been determined that they are potent sources of important bioactive lipids, such as phospholipids, sphingomyelin, choline, and long-chain PUFAs, such as DHA.[50,51] Sphingomyelin in butter serum has been found to have a proportion closer to that in human breast milk.[51] As sphingomyelin is one of the most significant sphingolipids in human breast milk, playing a role in the growth and development of breastfed infants, butter serum can be sourced to match the composition of human milk in infant milk formula production. Other than in milk, sphingomyelins have also been reported to be provided through the diet from animal origins, such as fish, meat, and eggs, and show potent health benefits in promoting health and preventing diseases.[4]

Phytosterols, which are plant-derived plant sterols and stanols, are also classified as lipid-based bioactive compounds that play a role in lowering cholesterol levels. Phytosterols are widely found in plant sources, including vegetable oils, cereals, and nuts.[52] The most abundant phytosterols (98%

of all phytosterols in plant sources) include β-sitosterol, campesterol, and stigmasterol.[4]

Phenolic lipids are mainly synthesized by plants as secondary metabolites and by animals, bacteria, and fungi. They are produced during regular development and respond to stress conditions caused by infection and ultraviolet or wound injury in plants, animals, and microorganisms.[53] These compounds are derivatives of mono- and dihydroxyphenols, where the chemical structure facilitates these phenolic lipids to have strong amphiphilic properties and thereby become biologically active lipids.[4] Phenolic lipids act as antioxidant, antigenotoxic, cytostatic, and antimicrobial agents in various applications.

6.2.2.2 Production of Lipid-based Functional Materials

Lipid-based biomolecules can be extracted using numerous conventional and non-conventional extraction techniques. Before the extraction process, various mechanical, physical, chemical, and enzymatic pretreatments are applied to disintegrate the parent cells and tissues, facilitating higher lipid extraction yields. A suitable solvent should be carefully selected for the extraction of lipids by solvent extraction methods, and the choice of solvents mainly depends on their nontoxicity, volatility, polarity, and mutagenicity. Generally, triglycerides can be efficiently extracted using nonpolar solvents, while phospholipids can be extracted using polar solvents.[54] Recently, the use of green solvents, including supercritical CO_2, plant-derived terpenes, such as D-limonene, *p*-cymene, and α-pinene, and ionic liquids has also emerged toward the extraction of lipid compounds from microalgae and yeast biomass.[55] Guderjan *et al.*[56] revealed that the recovery of phytosterols from maize germ oil was increased by 32.4% with the pulsed-electric field pretreatments during the oil extraction process. Health-beneficial fatty acids are commonly obtained by solvent extraction methods, such as solvent-accelerated extraction, ultrasound-assisted extraction, and microwave-assisted extraction. However, some methods are ineffective in obtaining high-quality final products due to the losses of bioactivity by thermal degradation and oxidation during extraction. Moreover, the risks related to human health and safety are other concerns associated with using organic solvents in traditional extraction methods. Therefore, as an alternative, supercritical CO_2 extraction provides a promising large-scale commercialized platform to extract lipid compounds, such as PUFAs and phytosterols, from different vegetal matrices with preserved bioactivities.[57]

In addition, structured lipids are produced by modifying or restructuring the naturally occurring lipids using chemical or enzymatic reactions. Chemical methods include interesterification approaches, such as acidolysis, alcoholysis, *trans*-esterification, and direct esterification, whereas enzymatic modification involves various lipase enzymes. These structured lipids can act as functional lipids in functional food, nutraceutical, and pharmaceutical applications. Senanayake and Shahidi[58] have reported ample evidence of these enzymatically and chemically modified lipids containing gamma-linolenic acid and *n*-3 PUFAs, which can provide specific health benefits. The structured lipid enriched with medium- and long-chain triacylglycerols was observed to be synthesized from the enzymatic interesterification of catfish oil and coconut oil using the lipase NS 40086 enzyme.[59] This novel functional lipid facilitates its applicability as a human milk fat substitute in infant formulas. Ji *et al.*[60] have produced a new dietary structured lipid with functional properties to resist obesity caused by soybean and coconut oil. Other than that, many recent studies have been directed at the synthesis of significantly structured lipids, such as medium- and long-chain triacylglycerols, modified phospholipids, low-calorie lipids, lipids rich in beneficial fatty acids, and *trans*-free or low-plastic fats with improved nutritional or functional properties.[61–64]

6.2.3 Significant Properties of Bioactive Compounds Toward Human Health

6.2.3.1 Protein-based Bioactive Compounds in Human Health

Bioactive proteins, peptides, and amino acids can be used in biotechnological applications with potential health benefits for humans based on various functional properties, such as antimicrobial activity, anticancer activity, antioxidant activity, antihypertensive properties, anti-inflammatory properties, and immunomodulatory properties. Biological activities indicated by proteins are closely related to the structure and size of bioactive molecules.[65] Especially, the amino acid sequence within the primary structure and secondary structure of food protein-derived peptides facilitates to exhibit important properties, including antimicrobial, antioxidant, mineral-binding, immunomodulatory, antithrombotic, hypocholesterolemic, anticancer, and antihypertensive properties. *In vivo* studies have shown that ACE-inhibitory peptides, released by enzymatic hydrolysis of intact proteins during food processing and gastrointestinal digestion, are capable of lowering blood pressure by controlling the vasoconstrictor effects of angiotensin II and potentiating the vasodilatory effects of bradykinin.[66] ACE-inhibitory peptides are generally considered to have alkaline or acidic amino acids in the N-terminal and larger amounts of hydrophobic and positively charged amino acids in the C-terminal of peptides. Furthermore, gut endogenous proteins generated by gastrointestinal digestion of dietary proteins, such as mucins, serum albumin, digestive enzymes, microbial proteins, and other secreted proteins, are also identified as potential sources of bioactive proteins.[67] These endogenous gut proteins contain peptide sequences that can exercise potent bioactivities, including ACE-inhibitory peptides.

Recently, there has been more interest in BPs derived from marine sources due to their health-promoting and therapeutic potential. A study conducted by Fan *et al.*[1] reported that marine algae-derived biopeptides have the potential to exhibit cytotoxic effects on human cancer cells through various mechanisms of action, such as apoptosis induction, tubulin–microtubule imbalance, and angiogenesis inhibition. In addition, bioactive protein hydrolysates or peptides obtained from marine sources indicate free radical-scavenging ability, resistance to gastrointestinal enzymes, ability to decrease systolic blood pressure, promising therapeutic effects on prevention of atherosclerosis, effective antimicrobial activities against disease-causing microorganisms, immunomodulatory properties, anti-ultraviolet radiation, and anti-osteoporosis activities.[1,68] Moreover, because of these significant bioactive properties, peptides have been used in some commercial applications as functional foods, including baby products, low-antigen food, mobile food, functional products for sports nutrition and blood pressure regulation, and products to promote mineral absorption.[65] Casein- and whey protein-derived ACE-inhibitory peptides and hydrolysates could be functional ingredients in developing food products and nutraceuticals to provide health benefits in managing several diseases.[69]

Additionally, bioavailability is a key factor in considering the target biofunctionalities exerted by these bioactive compounds, as their bioactivity needs to be retained during the gastrointestinal digestion process until reaching the specific action target. The bioavailability of bioactive proteins can be enhanced through modifications using chemical, physical, or combined methods in protein engineering, including enzyme inhibitors, carriers, or absorption enhancers, to protect peptides from enzymatic digestion in the gastrointestinal tract, resulting in more stable synthetic analogs, colloidal systems, nanoparticles, liposomes, emulsions, and microencapsulation.[3]

6.2.3.2 Lipid-based Bioactive Compounds in Human Health

Lipids play a major role in the occurrence and prevention of chronic diseases, such as cardiovascular diseases, cancer, osteoporosis, obesity, and diabetes in humans. In this context, the concern for the biological activity of lipids has been steeply increased. Different fatty acids benefit human health, especially long-chain PUFAs. Fatty acids are generally classified based on the length of the carbon chain, the degree of unsaturation, and the *cis*/*trans*-conformation. Some short-chain fatty acids provide significant protection to the human intestine tract. Butyrate, one of the commonly found short-chain fatty acids in milk, shows anti-colon cancer properties and promotes colonic health.[4] Tan *et al.*[70] have broadly discussed the anti-inflammatory, antitumorigenic, and antimicrobial roles and mechanisms of short-chain fatty acids, including acetate, propionate, and butyrate. Medium-chain fatty acids prevent obesity by reducing adipose tissues and providing therapeutic effects on type 2 diabetes, cancer, cardiovascular disease, and traumatic brain injury.[4]

Bioactive PUFAs and their metabolites also indicate antimicrobial effects. Especially, arachidonic acid shows effective antimicrobial activity against gram-positive and gram-negative bacteria, fungi, and enveloped viruses, including influenza and HIV.[71] Arachidonic acid, EPA, DHA, and their anti-inflammatory metabolites (lipoxin, resolvins, protectins, and maresins) can also be considered endogenous antiviral molecules, which are capable of inactivating enveloped viruses, including severe acute respiratory syndrome coronavirus 2 (SARS-Cov-2), severe acute respiratory syndrome (SARS), and Middle East respiratory syndrome (MERS).[71]

Phosphatidic acid and lysophosphatidic acid are lipid species involved in cellular processes, including the regulation of intracellular membrane transport, and are key intermediates in lipid biosynthesis.[72] The biological activity of these lipids in cells is mainly due to their ability to form intra- and intermolecular hydrogen bonds based on their chemical structure and ionization behavior in cell membranes.[73] Moreover, the formation of these hydrogen bonds facilitates lipid–protein interactions.

Due to the strong amphiphilic nature, phenolic lipids, including cardols, cardanols, hydroxybenzoic acid, and their derivatives, exert effective neuroprotective actions, especially against Alzheimer's disease, through the modulation of cell membranes to enhance membrane stability and fluidity while influencing the activity of receptors of signaling molecules.[53] Neuroprotective mechanisms of these bioactive compounds, particularly against Alzheimer's disease, include the inhibition of wrongful incision; antioxidant activity through their alkyl side chains attached to the phenol ring; inhibition of acetylcholine esterase *via* radical-scavenging activity, butyrylcholinesterase inhibitors, and selective metal-chelating agents; and free radical-scavenging properties.[53]

Phytosterols, including β-sitosterol, campesterol, and stigmasterol, can reduce the risk of cardiovascular diseases, cancers (colon, prostate, and breast), tumors, and rheumatoid arthritis; reduce plasma cholesterol and pro-inflammatory cytokine levels; and provide anti-inflammatory and antioxidant effects.[4] Numerous clinical studies provide ample evidence to implement phytosterols into medical practices as an effective therapeutic solution in lowering LDL cholesterol levels. However, more researches need to be performed to discuss the ability of long-term usage of phytosterols in treating cardiovascular diseases.[74] Due to a highly similar chemical structure to cholesterol, phytosterols exhibit significant effects in managing hypercholesterolemia *via* inhibiting cholesterol absorption in the gastrointestinal tract.

In addition, the Food and Drug Administration defines *trans*-fatty acids as "all unsaturated fatty acids that contain one or more isolated (*e.g.*, non-conjugated) double bonds in a *trans* configuration". However, conjugated linoleic acid is excluded from the definition of *trans*-fatty acid.[75] It is reported that the physiological effects of *trans*-fatty acids derived from ruminant origin and other processed edible fats are dependent on their isomeric profile.[76] High consumption of *trans*-fatty acids is linked with increased health risks in cardiovascular diseases, breast cancer, colon cancer, obesity, and diabetes.[77]

6.3 Potential Applications of Bioactive Protein and Lipid Compounds

6.3.1 Bioactive Proteins and Lipids as Functional Ingredients

In recent years, bioactive proteins and lipids have drawn much attention to be used as functional ingredients in functional foods, nutraceuticals, and pharmaceutical formulations due to the above-discussed physiological health benefits and potential to manage/prevent certain diseases.

6.3.1.1 Bioactive Protein-based Functional Ingredients

BPs can be added to or enriched in food or nutraceutical products by modifying the usual manufacturing process, and it was reported that several companies globally produce commercial food and nutraceutical products with BPs or hydrolysates as functional ingredients.[78] Besides health-promoting effects, bioactive proteins from fish and egg white can be effectively utilized in food and beverage applications as emulsifiers, binding agents, gelling agents, clarifying agents, and food plasticizers.[34] Proteins derived from different animal, plant, marine, and microbial sources release BPs with different physiological health benefits, such as antihypertensive, antimicrobial, antioxidant, anticancer, antidiabetic, antiobesogenic, anti-Alzheimer, and osteoprotective properties. Therefore, nutraceuticals and food products could be tailored with these BPs as a food-based solution in managing various chronic diseases. However, the immunogenicity and toxicity of peptides need to be carefully estimated prior to the consideration of BPs in food and therapeutic products, as their consumption may lead to several complications, such as intestinal wall disruption, erythrocyte and lymphocyte toxicity, free radical production, enzymopathic and immunopathic tissue damage, and cytotoxicity.[79] Furthermore, the biological activities of peptides can be affected by operational conditions applied to peptide production (*e.g.*, pH, temperature, time, and pretreatments on the protein source), food processing and storage conditions, food matrix, and the conditions in the gastrointestinal tract, including the presence of digestive enzymes and pH.[80,81] The BP-based delivery systems could be developed using encapsulation technologies, including edible coatings, emulsions, microcapsules, nanoliposomes, nanoparticles, and nanofibers, to improve the stability and bioavailability of peptides in functional foods, nutraceuticals, and drug delivery systems and also during gastrointestinal digestion. Various peptide-based drug delivery systems have emerged as effective therapeutic agents in biomedical applications. Self-assembling peptides, peptide–drug conjugates (composed of a BP, linker, and therapeutic agent), target peptides, and cell-penetrating peptides can act as carriers in peptide-based drug delivery systems, which are generally applied to the treatment of certain diseases, including cancer, neurodegenerative diseases, and cardiovascular diseases.[82]

6.3.1.1.1 Polyphenol–Protein Interactions in Bioactivity. Polyphenols are significant plant bioactive compounds with potential health benefits due to their antioxidant activity, anti-inflammatory activity, anti-aging properties, and anti-cancer activity. However, as reported in previous studies, these benefits are limited due to the lack of bioavailability, bioaccessibility, and stability of polyphenols.[41,83,84] Especially, these polyphenols are prone to degradation during gastrointestinal digestion.[84] Therefore, the concept of engineering polyphenols with proteins is recognized as a potent bioactive delivery system of protein–polyphenol aggregates to enhance their biological functions and stability in the gastrointestinal tract. Thermodynamic parameters determine the characteristics of protein–polyphenol interactions.[85]

An analysis of complexes of tea catechin with serum proteins, human serum albumin (HSA), and BSA using Fourier-transform infrared spectroscopy, UV–visible spectroscopy, and molecular docking indicates that more stable protein–polyphenol adducts are formed with larger catechins, and the stability of catechin–HSA is higher than catechin–BSA.[85] Another *in vitro* gastrointestinal model showed that the recovery index and antioxidant and anti-inflammatory bioactivity of blueberry and grape polyphenols conjugated with rice–pea protein are higher than those of unmodified polyphenols.[86] Interacted protein carriers facilitate the effective delivery of anthocyanin compounds to the lower gastrointestinal parts and make anthocyanin available for gut microbial catabolism, which involves the release of bioactive metabolites into circulation.[86] Developing soy protein–pectin–tea polyphenol, ternary nanoparticles have been used *via* photocatalysis under mild conditions, minimizing potential health and environmental concerns. The results depict that the ternary nanoparticles possess higher antioxidant activity due to the improved hydrogen-donating ability obtained through photocatalysis.[87] Therefore, significant biological functions of polyphenols can be enhanced by conjugating polyphenols with proteins as functional ingredients, ensuring their benefits to human health.

6.3.1.2 Bioactive Lipid-based Functional Ingredients

Similarly, as mentioned in the previous sections, bioactive lipid components can be enriched in different products as functional ingredients promoting health benefits.

Omega-3 fatty acids, including EPA and DHA, can act as nutraceuticals, and enriched diets have positive effects in preventing certain diseases, such as cardiovascular diseases and myocardial infarction, psoriasis, bowel diseases, mental illnesses, cancers, and bronchial asthma.[88,89] Meanwhile, PUFAs are highly susceptible to oxidation, making enriched foods have off-flavors and reduced bioactivity regarding desired physiological properties. Ruiz *et al.*[90] have discussed encapsulation technology for vegetable oils containing omega-3 polyunsaturated fatty acids to facilitate their effective use in functional food applications. Encapsulated vegetable oils can act as an excellent source of PUFAs with higher stability and bioavailability, allowing longer shelf life of enriched foods while maintaining their significant fatty acid composition. For the effective delivery of oils, they are encapsulated either in beads or capsules. Generally, encapsulation technologies involved in entrapping oils in beads are spray-drying, freeze-drying, emulsification, emulsion extrusion, solvent evaporation, and ionic gelation. In contrast, the coacervation method, extrusion through concentric nozzles, condensation and interfacial polymerization, and microfluidic devices are used in encapsulating oils in capsules.[90] It was reported that omega-3 microcapsules have been manufactured commercially as a functional ingredient for the food industry using spray-drying techniques under the brands of Dry n-3 (BASF, Germany), Supercoat omega-3 (The Wright Group, USA), and Omega-3 powder (Arjuna, India).[91] The same study also reported that different food products, such as infant foods, animal feeds, dairy products, meat products, cereals, and bakery products fortified with PUFAs, including DHA and EPA, are commercially available around the globe. Phytosterols are also promising functional food ingredients that can be added to different food products, such as salad dressing, milk and yogurt drinks, bakery products, chocolates, and cereal bars, and it was evident that dietary phytosterols provide significant effects in lowering serum cholesterol and reducing colon, prostate, and breast cancers.[92] Some regulatory frameworks have also promoted the utilization of phytosterols in functional foods or supplements. Accordingly, phytosterol-enriched formulations are classified as "food" in the European Union, and thereby, these formulations are freely available for purchase without any health professional advice.[93]

Lipid-based delivery systems, such as self-emulsifying and self-micro-emulsifying drug delivery systems, have also gained much attention in drug delivery and other functional applications in the food industry. As the efficacy of lipid-based delivery systems upon oral administration is determined by their fate in the gastrointestinal tract, a comprehensive understanding of the effects of gastrointestinal digestion and *in vitro* tests to predict their fate in the human body needed to be considered in the formulation of lipid-based delivery systems, especially for drug delivery.[94] Corzo-Martinez *et al.*[95] have developed a novel bioactive lipid-based delivery system from ratfish liver oil by enzymatic glycerolysis, with potential to incorporate into highly bioavailable and efficient nutritional supplements and functional foods. Ratfish oil contains more bioactive alkylglycerols, which possess special health benefits, including anticarcinogenic and immunomodulatory properties.

Nanostructured lipid carriers (NLCs) are also emerging as an effective delivery system with higher bioaccessibility and stability for lipophilic bioactive compounds in drug delivery. They are composed of both liquid and solid lipids as the core matrix and contain liquid oils, resulting in imperfect crystals, which allows NLCs to have higher bioactive loading capacity, protection from degradation, and minimized expulsion of the bioactive drugs to the outer medium.[96] Another study has highlighted the applications of NLCs through dermal, oral, pulmonary, and ocular administration routes in effectively delivering related drugs to treat certain diseases, including cancers and dermal diseases.[97]

6.3.2 Protein-based Delivery Systems

Nutrient and bioactive delivery vehicle development has become a burgeoning field in response to the worldwide growing demand for healthy foods and supplements.[98] However, common nutrients or bioactives were chemically unstable and susceptible to loss and degradation by environmental conditions (*e.g.*, temperature, pH, and light) during processing or storage, as well as to the low bioavailability during gastrointestinal digestion.[99] Therefore, developing a wide range of food delivery systems to improve the stability of nutrients or bioactives was imperative. Food-grade protein colloidal systems have attracted significant attention for their great potential in fabricating nutrient delivery vehicles due to biocompatibility, biodegradability, and structural and functional tunability.[100]

6.3.2.1 Protein-based Hydrogels

Hydrogels are three-dimensional solid networks fabricated by physically or chemically cross-linked hydrophilic polymeric structures that can entangle a lot of water or other biological liquids inside their network.[101] The advantages of hydrogels as nutrient and bioactive delivery vehicles included relatively compact structures for gastric digestion resistance, tunable porosity, and diffusive transport property for controlled release. Protein-based hydrogels have been demonstrated to possess high loading capacity and controlled release of bioactives.[102] The soy, zein, whey, and β-lactoglobulin protein-based hydrogels were among the most used protein hydrogels for nutrient and bioactive delivery applications.[103] The protein-based hydrogels could be classified into three classes: physical (*e.g.*, heat and ionic), chemical (*e.g.*, glutaraldehyde and genipin), and biochemical (*e.g.*, enzymatic) cross-linking hydrogels.[100] Cross-linking conditions, type, and protein concentration regulated the network structure of hydrogels. It was found that the network structure of hydrogels, such as morpho-structural properties, pore size, surface area, network formation morphology, and degree of cross-linking, was closely related to the release behavior of nutrients and bioactives in hydrogels.[104] Maltais *et al.* developed filamentous and particulate soy protein hydrogels using different concentrations of calcium chloride and found that the release behavior of riboflavin in filamentous hydrogels differed from that of particulate hydrogels.[105] Zhang *et al.* showed that the soy protein isolate–κ-carrageenan (SPI-KC) composite hydrogels exhibited a more uniform and compact network structure with sustained-release properties.[106] Alavi *et al.* developed whey protein and KC composite hydrogels and found that the composite hydrogels showed high capacity and physical stability of curcumin.[107] Ding *et al.* developed soy protein/soy polysaccharide hydrogels for retention and controlled release of folic acid and found that hydrogels could protect folic acid, and the photodegradation rate was only 17%.[108] Protein-based hydrogels have been proven to improve the stability of nutrients and bioactives against chemical, biological, or physical degradation.

6.3.2.2 Protein-based Particles

Proteins can self-assemble into nano-/microparticles and be used to protect and deliver nutrients and bioactives.[109] Protein particles can be prepared using various methods, such as heat and mechanical, which became more suitable for constructing nutrient and bioactive delivery systems than the native states of the protein.[110] Protein particles can be generally divided into water-soluble protein particles, water-insoluble protein particles, and composite protein particles. Whey protein, egg white protein, soy protein, and gelatin were typical water-soluble proteins.[111,112] Zein, gliadin, and kafirin were promising materials for forming water-insoluble protein particles. Due to the insoluble nature of these proteins, the anti-solvent method has become a popular technique for preparing insoluble protein-based particles.[100,113] The formation of protein–polysaccharide complexes is a practical method to generate well-dispersed nanoparticles, whose properties were controlled by several factors, including the type and concentrations of proteins and polysaccharides, ionic strength, and pH.[114] Hydrophilic nutrients (*e.g.*, vitamin C and anthocyanin) and lipophilic nutrients (*e.g.*, curcumin and flavonoids) could be efficiently entrapped in protein particle vehicles and protected from degradation, possessing, and increased thermal and light stability.[115,116] Diarrassouba *et al.*[117] studied self-assembled β-lactoglobulin and egg white lysozyme complex microspheres as potential carriers for vitamin D3 with up to 90% encapsulation efficiency. Chang *et al.* showed that the chemical stability and antioxidant activity of curcumin significantly increased after encapsulation in egg white protein particles.[118]

6.3.2.3 Protein-based Emulsion and Emulsion Gel

The emulsion is assembled by combining two immiscible fluids to form a suspension of one fluid dispersed as small droplets throughout the other fluid,[119] including nanoemulsion, Pickering emulsion, high internal-phase emulsion, multilayer emulsion, and multiple emulsion.[120] The emulsion gel is a complex colloidal system consisting of emulsion and gel networks. These emulsion-based systems could be used to encapsulate nutrients and bioactives.[121] Protein has been widely utilized as an emulsion-based material due to its high functional properties, such as emulsification and gelation capabilities.[122] Encapsulation and controlled release of lipophilic nutrients (*e.g.*, polyunsaturated fatty acids, lutein, and curcumin) was the major application of protein-based emulsion and emulsion gel.[110] Proteins could be hydrolyzed and degraded by natural enzymes in the gastrointestinal tract, and the encapsulated lipophilic nutrient and bioactive were released in the micellar form.[123] The release and bioavailability of the encapsulated lipophilic nutrient and bioactive were controlled by many factors, including interfacial properties, droplet size, cross-linking network, and surface charge.[124] Ji *et al.*[125] showed that pea protein isolate–chitosan emulsion exhibited a high retention rate (93%) of eicosapentaenoic acid during storage. Zhang *et al.*[122] prepared soy protein–whey protein isolate (WPI) composite emulsion gel and single-protein emulsion gel as a vitamin E vector for stomach and intestinal delivery. The *in vitro* digestion experiments revealed that the mixed protein emulsion gel improves the bioavailability of vitamin E. Yan *et al.* showed that zein–sodium alginate double-cross-linked emulsion gels exhibited higher light stability and bioaccessibility of curcumin and resveratrol than the single-cross-linked emulsion gels.[126]

6.3.3 Bioactive Proteins in Food Preservation

6.3.3.1 Role of Proteins in Food Preservation

Chemical, physical, and biological deterioration of foods is common for all food products during storage. It is necessary to prevent food deterioration to maintain the quality of foods while having a longer shelf life. Effective preservation can be obtained by controlling different biological, physical, and chemical factors that affect food deterioration. Generally, microbial, enzyme, and other deterioration-related factors are controlled with thermal or non-thermal preservation, adding certain preservatives, and protective packaging against deteriorative factors, such as moisture, dryness, air, and light.[127] Protein-based bioactive compounds have emerged in the food industry to preserve numerous food products due to their antimicrobial and antioxidant properties. The consumer demand for chemical preservative-added food products is reducing drastically due to their adverse health effects on humans with prolonged consumption. Therefore, the use of natural biological sources of preservatives and active packaging concepts in enhancing the shelf life of foods has attracted current attention. Consequently, it was reported that the use of protein hydrolysates of plant, fish, and milk protein and biologically active peptides with antimicrobial and antioxidant properties is possible as a promising alternative to chemical preservatives and biologically active ingredients in active packaging.[128]

6.3.3.1.1 Antimicrobial Properties. The antimicrobial ability of peptides manifests as antibacterial, antiviral, or antifungal activities, mostly attributed to their α-helix and β-fold secondary structures.[25] Mainly, the electrostatic interactions between the peptide and positively charged or anionic lipids on the membrane of the target microorganism and the hydrophobicity and flexibility of the peptide affect the antibacterial activity.[129] Peptides exert antimicrobial action either by acting on microbial cell membranes, causing damage, or by moving into microbial cells without damaging the cell membrane of the target microbe. When the antimicrobial peptide acts on the microbial cell membrane, initially unstructured peptides contained in an aqueous solution become a highly stabilized secondary structure and get distributed on the surface of the cell membrane forming pores while disturbing the integrity of the cell membrane.[25,129] The mode of different actions in forming cell membrane pores has been explained using the barrel stave, the carpet, the aggregate, and the toroidal pore models.[129] Despite the membrane damage mechanism, few other peptides affect important cellular functions in microbial cells, including inhibitions of nucleic acid synthesis, protein synthesis, cell wall synthesis, and enzyme activity, leading to the death of microbial cells.[130]

Some antimicrobial peptides show antifungal activities through different mechanisms of action. Due to the amphipathic nature, peptides bind to the membrane surface, causing fungal cell lysis. Synthetic peptides, such as SMAP-29 and Ib-AMPs, have this mode of action as antifungal agents. These peptides may or may not permeate the cell membrane. Some peptides, including omiganan, can pass into the cellular membrane to interfere with cellular functions, such as cell wall synthesis and essential macromolecule synthesis. Omiganan is effective against *Candida albicans*. A halocidin analog peptide (di-K19Hc) isolated from sea peach indicates promising antifungal effects against strains of *Aspergillus* and *Candida*.[129] Some peptides are discovered to have potential antiviral activities. These peptides block the entry of viruses through interaction with either the host cell or the virus. Lactoferrin acts as an antiviral agent *via* interacting with heparin sulfate molecules on the surface of the host cell.[129]

Although antimicrobial peptides have gained more attention due to their antibacterial and antifungal properties, they also exhibit antiviral activity, benefitting in their application as chemical preservatives. Therefore, the interest in protein-based antimicrobial compounds as a preservative in food and some cosmetic products has been developed recently. However, more attention should be paid to the safety assessment of these new preservative agents. The U.S. Food and Drug Administration has given approval only for two antimicrobial peptides (nisin and ε-poly-L-lysine) to use as food preservatives.[25]

6.3.3.1.2 Antioxidant Properties. Scientific evidence reveals the antioxidant activity of amino acids and peptides *via* inhibition of lipid peroxidation, free radical-scavenging ability, and metal ion chelation. Therefore, these natural compounds can be an alternative to synthetic antioxidants. The structure of amino acids or peptides determines their antioxidant potency. Hydroxyl, carboxyl, and aromatic reactive groups in free amino acids are responsible for their antioxidant activity. However, the mechanism of antioxidant potency is pH-dependent. At pH values below the acid dissociation constant (pKa), non-ionized forms of reactive groups of amino acids quench free radicals *via* donating hydrogen atoms to them, whereas ionized reactive groups, which carry excess electrons at pH > pKa, neutralize free radicals.[131] Some whole proteins and protein hydrolysates derived from different plant and animal origins also have antioxidant properties. On a nanomole basis, patatin, a purified storage protein of potato, showed similar antiradical activity compared to butylated hydroxytoluene (BHT) and higher than that of reduced glutathione.[132] Moreover, amino acid side-chain modifications revealed that cysteine and tryptophan residues in patatin are responsible for its antioxidant potency against radicals. BPs with antioxidant activities are typically composed of 2–12 amino acids.[133] Structural features of peptides, such as the type and position of amino acid residues, peptide chain length, secondary structure, and presence of hydrophobic and aromatic amino acids, are closely related to specific antioxidant properties.[131] The molecular weight of the BP is one of the major factors that determine its free radical-scavenging, metal chelation, and lipid peroxidation inhibition activities. Studies have reported that shorter peptides have higher antioxidant activities than long-chain peptides.[134,135] Besides the peptide size, peptides with amide, acidic, aromatic, hydrophobic, and positively charged amino acids usually have higher antioxidant properties.[131]

6.3.3.2 Applications of Bioactive Proteins in Food Preservation

Protein-based bioactive compounds, especially antimicrobial and antioxidant peptides, can be considered effective natural preservatives widely applicable in food and cosmetics. However, when it comes to practical applicability, more attention should be given to the stability of these compounds in various food matrixes. The stability of BPs can be affected by the structural properties of peptides and several external and internal environmental factors, such as processing and storage temperature, salt concentration, pH, food matrix, additives in food products, and other processing and storage conditions.[136–138] Modern technological initiatives, including the molecular design and optimization of peptides, encapsulation, and delivery systems based on nanotechnology, have been developed to enhance the stability and reduce the toxicity of these bioactive compounds in food applications. However, the application of BPs in food preservation is still in the theoretical stage, and more researches need to be carried out to explore the structure–activity relationship of peptides, further improve stability, and address potential side effects.[25]

6.3.3.2.1 Applications as Preservatives. Researchers are now focusing on using bioactive proteins and peptides as natural preservatives in starchy foods, fish and meat products, beverages, and fruit- and vegetable-based products. Besides food preservatives, some peptides can be used as preservatives in cosmetic products.

At present, the only bacteria-derived antimicrobial peptide that is widely used in food applications as a natural preservative is nisin. As a preservative, nisin can extend the shelf life of dairy products, such as cheese and yogurt, canned vegetables and juice, fish and meat products, and alcoholic beverages, such as beer and wine.[139] In these applications, nisin is known to inhibit the growth of many foodborne pathogens, including thermophilic, acid-tolerant, and aerobic microorganisms. However, being less stable in narrow pH ranges and at higher temperatures limits the applications of nisin in food preservation.[140] Therefore, a trend has emerged to discover new preservatives for food preservation. Results of *in vitro* studies conducted by Hyldgaard *et al.*[141] showed that the blend of ε-poly-L-lysine (a cationic peptide produced by *Streptomyces albulus*) and isoeugenol had a distinct antibacterial effect against *Pseudomonas putida* and inhibited the growth of six different Gram-negative bacterial strains from three different genera. The findings of this study also suggested the use of the said antimicrobial blend in the preservation of turkey meat products. It was found that antibacterial peptides mytichitin-A and mytichitin-CB derived from *Pichia pastoris* are potent preservatives to extend the shelf life and maintain the quality attributes of fresh pork and related products during cold storage.[142,143] Mytichitin-A effectively restrained the growth of primary spoilage microorganisms in pork, including *Pseudomonas fluorescens* and *Pseudomonas tolaasii*, during storage at 4 °C. Furthermore, it was evident that mytichitin-A treatment at a concentration of 25 μg ml^{-1} was excellent in inhibiting pork spoilage, lowering drip loss, reducing lipid oxidation, and significantly hindering the increase in pH while maintaining better sensory qualities of pork meat.[142] Mytichitin-A peptide has higher thermal stability at a temperature range of 4–90 °C and pH stability at a range between 2 and 8. Having low toxicity to animal cells with hemolytic activity of ≤1.0% at a concentration of 60 μg ml^{-1} is a positive aspect of using this peptide as a food preservative. Similarly, mytichitin-CB displayed antibacterial activity against a number of foodborne and food spoilage bacteria, including *Staphylococcus aureus*, *Escherichia coli* O157, and *Pseudomonas fluorescens*, at inhibition concentrations of 0.23–1.24 μg ml^{-1} in pork preservation.[143] Moreover, the combination of 4 mg l^{-1} of mytichitin-CB with 10 g l^{-1} of chitosan and 300 mg l^{-1} of nisin has confirmed its potency as a preservative blend in pork preservation. Therefore, the authors suggest using mytichitin-A and mytichitin-CB as novel preservatives in meat products. In addition, glycinin basic peptide extracted from soybean can effectively inhibit the growth of several fungi, such as *Aspergillus niger* and *Penicillium* sp., in starchy foods, such as fresh wet noodles.[144] Antimicrobial peptides, such as Ac-AMP2 and pediocin DT016 derived from *Amaranthus caudatus* and *Pediococcus pentosaceus* DT016, respectively, can also be used to preserve several fruits and vegetables, including pears, lettuce, parsley, rock salad, and spinach.[145,146] Some peptides extracted from edible insects have a broad antimicrobial activity that can be used as natural food preservatives. These insect antimicrobial peptides are heat-stable, low-molecular-weight compounds commonly found in insect immunity systems. There are several antibacterial peptides discovered from flies, such as cecropins, ceratotoxin, andropin, sapecins, phormia, defensing, drosocin, metchnikowin, diptericin, attacins, and sarcotoxins and antifungal peptides, such as AFP and drosomycin.[147] Enterocin AS-48 is another type of antimicrobial peptide used in cider, fruit, and vegetable juices as a preservative, while enterocin CCM4231 peptide is used to preserve soy milk.[148] These antimicrobial peptides can be used in preservation applications alone or in combination with other BPs, antibiotics, essential oils, and polymeric nanoparticles to enhance their effectiveness.

Moreover, certain BPs can be considered a potent preservative in cosmetic products besides being a preservative in food products. Yan *et al.*[130] have discussed the ability of peptides, such as KR-12-pa, nisin, and active peptides, produced through hydrolysis of bovine lactoferrin to use in the field of cosmetic preservatives, as they are harmless to the human body.

6.3.3.2.2 Packaging Applications. Recently, consumer preference for safe food consumption has drastically increased, and this trend has led the focus on food preservation toward innovative solutions in packaging technologies, such as active packaging and biodegradable food packaging. Active packaging is one of the emerging technologies that refers to incorporating various active ingredients into the package to maintain or extend the product quality and shelf life while improving food safety and sensory properties.[149] In this context, BPs have gained attention for being used as an active ingredient in active packaging. Lactoferrin is a bioactive protein having promising antimicrobial properties found in cheese whey, and it can be incorporated into edible film composite to enhance the shelf life of various perishable food products. Lactoferrin is a potent antimicrobial protein with a wide range of antimicrobial activity against several Gram-positive and Gram-negative bacteria, yeast, fungi, some parasites, and viruses, whereas the BP lactoferricin, a hydrolyzed product of lactoferrin, is also recognized for its antimicrobial activity against several food spoilage microbes, including *E. coli*, *S. aureus*, and *Candida* spp.[17] This study further reported that lactoferrin could be incorporated into chitosan or bacterial cellulose-based film composites in active packaging. The best antimicrobial packaging performance against *L. monocytogenes* was observed when the chitosan film was incorporated with a combination of lactoferrin, nisin, and lysozyme. Bovine lactoferrin contained bacterial cellulose film-coated fresh sausage pig's meat and also had a higher bactericidal efficiency.

A detailed study by Di Pierro *et al.*[150] indicated several applications of whey protein-based edible coatings and films as a bio-based packaging solution to preserve perishable food products. Biodegradable biopolymer films use proteins, lipids, and polysaccharides as base sources. In addition, these biofilms can be incorporated with bioactive compounds, such as antimicrobial peptides or other active ingredients, to enhance the shelf life and improve the safety and organoleptic properties while delaying or preventing the growth of spoilage microorganisms. Moreover, potential applications of dairy whey protein isolate-based edible coatings or films are effective and green packaging options to preserve different foods, including fresh and processed fruits and vegetables, baked goods, cheese, peanuts, and meat products.[150,151] In addition to using whey protein isolates, the protein base for these biofilms can be sourced from corn, soybean, rice, wheat, peanut, cottonseed, casein, collagen, gelatin, and egg albumin.[17,152] Biodegradable soy protein isolate, a valuable by-product from the soybean oil industry, has gained attention for food preservation and packaging technology.[153] Soy protein isolate typically contains more than 90% of proteins, facilitating its film-forming ability. Although soy protein isolate-based biofilms have significant gas barrier properties, bulk or surface modifications are still required to improve their mechanical and water-resistant properties. Silk fibroin, another remarkable protein derived from living organisms including spiders, has been developed into an effective preservative edible coating for fruits, such as strawberry, banana, and apricot.[154–156] Silk fibroin coatings preserve the freshness of fruits by protecting them from fungal attacks, decelerating the respiration of fruits, extending fruit firmness, and preventing fruit from dehydration. Furthermore, water annealing treatment results in structural modification in silk fibroin films and has been identified as an effective technique for edible silk fibroin coatings to preserve fruits.[156]

6.3.4 Bio-based Plastics from Proteins and Lipids

Plastics are highly valued materials with versatile use in almost every industry, including packaging, consumer products, manufacturing, construction, health care, textile, and transportation, because of their low cost, extraordinary versatility, high strength, and durability. However, the growing global production of plastics simultaneously raises plastic waste, a serious global issue.[157] Plastic pollution is related to the accumulation of stable and durable plastics in the environment, and it results in long-term harmful effects on nature, human health, and other living organisms, causing physical damage, chemical harm, and biological threats.[158] Therefore, a growing concern has been paid to eco-friendly bio-based plastics that are biodegradable and based on renewable sources. There are three types of bioplastics: (1) bio-based and biodegradable, (2) only biodegradable, and (3) solely bio-based plastics.[159] These bioplastics can be used as a cost-effective alternative to conventional petroleum-based plastics in food packaging, agricultural applications, and various medical applications, which can minimize the risk of contamination and other potential infections caused by conventional plastics. Biomaterials from polysaccharides, proteins, and lipids, such as corn, potato, vegetable oils, cereal crops, food waste, wood, and seaweed, are used to synthesize bioplastics.[160]

6.3.4.1 Protein-based Bioplastics

A recent study on albumin, soy, and whey protein-based bioplastics produced with the use of different plasticizers, such as water, glycerol, and natural rubber latex, has reported good thermal and viscoelastic properties and promising antimicrobial properties against Gram-positive *Bacillus subtilis* and Gram-negative *Escherichia coli*.[161] Moreover, both albumin and whey protein-based bioplastics with the glycerol plasticizer had the strongest antibacterial activity. The authors suggested the possible applications of these protein-based bioplastics in medicinal and food packaging sectors, emphasizing the need for more research toward their intended use. It is found that the mechanical properties, water-uptake capacity, and rheological characteristics of bioplastic materials; the effect of other additives used in bioplastics; and bioplastic production conditions mostly depend on the type of protein.[162] Felix *et al.*[162] developed biodegradable plastic materials with significant mechanical properties for various applications using albumen protein isolate rich in globular proteins and crayfish flour rich in myofibrillar proteins. Several other studies also focused on developing protein-based bioplastics from different protein sources, such as potato, rice, wheat gluten, pea protein, albumen, and albumen/gluten blends, which exhibit improved mechanical properties to be used in various applications, including biodegradable films and food packaging.[163–165] Thammahiwes *et al.*[166] developed a novel wheat gluten/fish scale composite bioplastic through compression molding at 130 °C for 10 min with a pressure applied to 250 kg cm^{-2} molds. The incorporation of fish scales, which are rich in collagen, into wheat gluten-based bioplastic improved the tensile strength. The composite bioplastic showed a higher tensile strength ranging from 6.5 to 7.5 MPa, while the neat wheat gluten-based bioplastic showed a tensile strength of 3.40 MPa. The good interaction and strong adhesion between the fish scale powder and wheat gluten matrix in the composite bioplastic prevent water penetration into the gluten matrix, thereby controlling the degradation rate of wheat-based biofilms used in the agricultural sector.

Recently, more attention has been paid to keratin-based biofunctional applications due to the increased generation of keratin-rich waste biomass from the poultry, meat, and wool industries. Keratin is an insoluble, durable, and fibrous protein with a high amount of cysteine and is mainly found in mammals, birds, and reptiles.[167] Keratin, which can be potentially used in medicinal, pharmaceutical, cosmetic, agricultural, and biotechnological applications, is generally extracted from waste biomass using acidic hydrolysis, alkaline hydrolysis, enzymatic hydrolysis, ionic liquid hydrolysis, and alkaline–enzymatic hydrolysis techniques.[168] The keratin is usually composed of 90–100 amino acids with 10.2–10.4 kDa molecular weight, and its chemical structure can be α-helix, β-helix, or β-pleated sheet. Acid hydrolysis of wool waste precipitated two different keratin polypeptides at pH 3.22 and 5.55 based on isoelectric precipitation.[169] Several studies reported that keratin extracted from chicken feather wastes had been successfully developed into keratin-based biodegradable bioplastics having significant mechanical and thermal properties for application in the food packaging, biopolymer, biomedical, and pharmaceutical industries.[170–172] Dou *et al.*[172] prepared hydrolyzed feather keratin-based biofilm with uniform, translucent, and tough features that can potentially be used in packaging and agricultural fields. The process included an economical method of alkali extraction and acid precipitation with urea and sodium sulfide to collect hydrolyzed feather keratin from feather biomass, followed by biofilm formation using the compression molding method. Furthermore, adding glycerol as a plasticizer in the biofilm formation provides flexibility to the keratin-based biofilm, as pure keratin may cause fragility in the biofilm.[170]

6.3.4.2 Lipid-based Bioplastics

Plant oil-based bioplastics have promising mechanical and physicochemical properties for their intended applications. A recent study reported that coffee oil could be sourced to produce bioplastics.[173] The fatty acid profile of coffee oil extracted from spent coffee grounds is rich in cafestol and kahweol. In addition, coffee oil is a rich source of triglycerides, which can be used to produce lipid epoxides, precursors for bioplastics. Triglycerides in various vegetable oils, such as soybean, jatropha, linseed, sunflower, palm, castor, Nahar seed, and canola oil, can be considered potential precursors to produce biopolymers.[174] These biopolymers are usually prepared using different polymerization reactions. Epoxidized soybean and linseed oils have been used to synthesize green nanocomposites by acid-catalyzed curing of the epoxy-containing triglyceride in the presence of organically modified clay.[175] As a promising alternative to petrochemical polymers, novel epoxidized soybean oil/clay nanocomposites have also been developed with a curing agent, triethylenetetramine, by Liu *et al.*[176] Characterization of these nanocomposites exhibits greater thermal stability (<180 °C), mechanical properties, and barrier properties, further confirming their potential for use in diversified applications as biodegradable bioplastics, coating materials, and insulation materials. The triglycerides in soybean oil cannot be homo- or copolymerized by radical-initiated polymerization. Therefore, soybean oil has been functionalized as a monomer for radical polymerization using maleinization, epoxidation, acrylation, and hydroxymethylation.[177]

Pan *et al.*[178] synthesized a castor oil-based bioplastic material capable of functioning as a fluorescent probe and carbon nanotube dispersion stabilizer by polyesterification. In another study, methyl hydroxyl branched fatty acid (MHBFA) derivatives from waste cooking oil have been engineered to produce bioplastic with improved mechanical properties, such as higher tensile strength and toughness.[179] Plant oil-derived bioplastics have significant characteristics to use as alternatives to conventional plastics. Moreover, metal–ligand coordination has also been proven to improve the thermo-mechanical properties of plant oil-based bioplastics by creating physical cross-links, thereby accelerating the chain entanglements for plant oil-derived polymers.[180]

6.3.5 Both Protein and Lipid as 3D Printing Inks for the Biomedical and Food Industries

6.3.5.1 Functional Protein for 3D Printing

Proteins are sustainable replacements for dietary staples used for 3D printing. The viscosity of most protein-based 3D printing inks decreases as the shear rate increases. This non-Newtonian behavior, called shear thinning, is a desired attribute as it lowers the pressure required to extrude the ink and supplements the flow of paste-like material during extrusion. Shear thinning behavior is essential to ensure smooth extrusion and retain the desired shape of printed structures. Proteins can form emulsion gels and protein-concentrated emulsions as 3D printing inks.

6.3.5.1.1 Animal Proteins. Animal proteins [*e.g.*, whey protein isolate (WPI), cod proteins, and gelatin] have become the major components in the 3D printing of emulsion gels. Liu *et al.*[181] analyzed the rheological and mechanical behavior of milk protein composite gels for extrusion-based 3D food printing. The results showed that increasing the total protein content was beneficial for 3D printing. They also prepared the 3D printing inks by fabricating gel-like emulsions with whey protein isolate.[182] To pave the way for developing high-protein food inks, Sager *et al.*[183] explored the extrusion-based 3D printability of heat-induced high-protein WPI gels. They found that the substitution of a fraction of WPI by whey protein resulted in good printability and stable shapes after deposition. The high internal phase emulsions (HIPEs) stabilized with 50 mg ml^{-1} of cod proteins exhibited the highest printing resolution, gel strength, hardness, adhesiveness, and chewiness during 3D printing. These food-grade HIPE inks have the potential to diversify the applications of 3D printing in foods, cosmetics, drug delivery systems, and packaging materials.[184] Gelatin was also applied to stabilize food-grade HIPEs as 3D printing inks; subsequent 3D printing on colored cards and food confirmed that the obtained HIPEs had good extrudability and printing performance.[185] Moreover, due to adding nutritional and functional value to printed products, ovalbumin and high-density and low-density lipoprotein-based food inks seem promising to produce 3D printed structures of complex shapes.

6.3.5.1.2 Plant Protein. Compared with the studies on animal-derived proteins as extrusion 3D printing inks, plant proteins have attracted a greater interest due to the safety of food ingredients and the high price of animal-based proteins.

Soy protein isolate (SPI) exhibited great potential in 3D printing due to its ability to form a gel with good water-holding capacity. Many researchers investigated the use of SPI gel for 3D printing.[186,187] This is probably because SPI gel exhibits higher viscosity and rheological properties unsuitable for 3D printing. Appropriate modification is therefore needed before SPI gel can be successfully printed. Another study investigated the 3D-printed geometrical texture of SPI mixtures with sodium alginate and gelatin.[188] Compared with SPI, SPI mixture with gelatin improved the hardness and chewiness of 3D-printed geometries. The overall results suggested that the food matrix of SPI, sodium alginate, and gelatin will be a promising material in 3D food printing. Phuhongsung *et al.*[189] used the combination of a formulation comprising SPI, κ-carrageenan, and vanilla flavor as a printing material. The results could conclude that adding carrageenan and post-printing microwave heating at an optimum level of microwave power enhanced the printability and flavor of an SPI gel.

Pea protein is rich in essential amino acids except methionine. The anti-nutritional factor in pea protein is much less than in soybean protein; thus, it gets digested more easily. Feng *et al.*[190] suggested the potential of food-grade high internal phase Pickering emulsions prepared with pea protein complexes as new materials suitable for 3D printing, broadening the application of plant protein-based emulsions to food 3D printing. Álvarez-Castillo *et al.* partially replaced porcine plasma protein with pea protein concentrate (PPC) and SPI in the prepared dough.[191] They found that doughs containing PPC and SPI experienced a noticeable increase in viscoelastic moduli and viscosity compared to those containing only porcine plasma, which was beneficial for 3D printing. Overall, adding proteins to the formulations of food inks increases the content of highly bioavailable proteins and improves their printability and stability.

6.3.5.2 Functional Lipid for 3D Printing

Lipids are widely found in food and are responsible for the texture, taste, and flavor of food, which plays a vital role in determining the palatability of diet and can be used to improve the hardness, juiciness, and lubricity of products.[192] Many studies have shown that the application of lipid additives in 3D printing inks significantly improved the poor printing characteristics of the system, making unsuitable natural food gels successful for 3D printing.

6.3.5.2.1 Animal Fat. Lecithin is used as an emulsifier in many food emulsions, confectionery, and chocolate products. When used in 3D inks, it can reduce the wear and tear of the nozzle as a lubricant. Chocolate, when added with fat and lecithin, has more suitable rheological properties for printing due to the wrapping effect of fat and lecithin on solid chocolate particles, which reduces the interaction between them.[193,194] Sager *et al.*[183] found that adding whipping cream (10% w/w) to WPI gel formulations forming an emulsion gel increased the gel stiffness and improved printability. It was reported that curcumin has better solubility in medium-chain triglycerides than coconut, olive, and corn oils. Curcumin emulsion was prepared using medium-chain triglycerides and added to lotus root powder gel for 3D printing.[195] Butter was used in 3D-printed cookies to enhance heat, light, and oxygen stability during baking and further storage.[196] Le Tohic *et al.*[197] found that the molecular size of the added fat and its distribution in cheese also affect the color of the 3D-printed products, and the color of the sample becomes deeper with the increase in fat molecules.

6.3.5.2.2 Vegetable Oil. The most reported oil phases of emulsion-based 3D printing inks were vegetable and corn. A high internal phase Pickering emulsion (HIPPE) was prepared by corn oil as 3D printing inks, demonstrating that HIPPE could 3D-print selected shapes with high resolution and shape fidelity at pH 7 and 50 mM NaCl.[198] Canola oil was also used to prepare high-quality, low-fat 3D-printed foods.[199]

Similarly, flavor oils, such as olive oil, have a long history of application as condiments to provide food with increased flavor. Olive oil was added to the dough for printing, making it smoother and more liquid due to the plasticizing and lubrication of the fat in the ingredients.[181] In addition, olive oil was added to a 3D-printed functional chicken meat-based snack to tailor complex 3D-printed functional muscle food.[200] Bi *et al.*[201] investigated the formation mechanism and 3D printing properties of Pickering HIPEs fabricated by casein/pectin hybrid particles with flavor oil (the *Litsea cubeba* oil diluted 1 : 1 with soybean oil). For the nutritional deficiency of the population, the ink can be a nutritional enhancement, such as emulsions and emulsion gels, as the matrix of ink can be introduced to different oils, which can improve the printing performance of emulsions and emulsion gel ink and further enhance its nutritional value.

All in all, in recent years, while customizing delicate structures and simplifying food supply chains, the formulations of printable inks have often been adjusted to achieve personalized nutritional design. By adding specific nutrients or functional compounds (*e.g.*, protein and fat) in 3D printing, food formulas can help people, such as athletes, pregnant women, or children, to promote health and prevent diseases.

6.3.6 Applications of Proteins and Lipids in Molecular Imaging

Molecular imaging (MI) can be defined as *in vivo* visualization and quantitative measuring of functions of biological and cellular processes.[202] It plays a key role in the medical industry for early disease diagnosis, characterization, targeted therapy, personalized therapy, and real-time treatment monitoring, including molecular-specific effects. MI involves imaging techniques, such as positron-emission tomography (PET), single-photon emission computed tomography (SPECT), magnetic resonance spectroscopy (MRS), magnetic resonance imaging (MRI), ultrasound, and optical and acoustic imaging techniques. Currently, novel molecular imaging probes/contrast agents have been developed using peptide- or lipid-related biomolecules to improve sensitivity, accuracy, and target specificity in molecular imaging.

Xue *et al.*[203] designed a novel class of protein-based MRI contrast agents with higher sensitivity and specificity for the MI of cancer biomarkers. As the binding protein, side chains from the scaffold protein have been used, and it also serves as a chelator, which can strongly bind to contrast agents. Currently, the application of MRI is limited primarily due to the lack of sensitivity of contrast agents. The sensitivity of contrast agents is improved by increasing the relaxivity of contrast agents, which is attributed to the relaxation of water in different tissues and organs. The sensitivity could be enhanced by linking the lanthanide gadolinium chelates (Gd^{3+}), which is a frequently used metal ion for MRI contrast agents, to polymers, proteins, particles, and liposomes; encapsulation of Gd^{3+} chelates into micelles and caged structures; and targeting to receptors.[204,205] Several other studies reported that Gd^{3+} chelates have been linked with albumin and lysine-containing random-coil polymers to increase the relaxivity of MRI contrast agents.[206,207] Charron *et al.*[208] broadly discussed the discovery, development, and radiolabeling of several peptides as molecular imaging agents based on the target specificity of protein molecules toward particular endogenous receptors. Natural peptides, such as bombesin, gastrin, cholecystokinin, ghrelin, glucagon-like peptide-1, a melanocyte-stimulating hormone, neuropeptide Y, neurotensin, somatostatin, substance P, tachyplesin, polyphemusin, and vasoactive intestinal peptide, serve as precursors to develop molecular imaging agents as targeting vectors. With different structural modifications to enhance *in vivo* stability and affinity to the receptor target, these peptides can be used as molecular imaging probes in different MI techniques. Similarly, another study reported potential applications of peptide- and peptide hormone-based imaging probes for targeted molecular imaging in PET, SPECT, MRI, optical imaging, and multimodality imaging.[209]

Besides protein-based MI probes, another MRI contrast agent that can precisely measure the tissue pH has been designed by loading paramagnetic gadolinium complex into liposomes.[205] Liposome vesicles comprise an aqueous inner cavity surrounded by a phospholipid bilayer. The pH measurements by this lipid-based contrast agent are related to the relaxometric properties of the paramagnetic complex at different magnetic fields. Additionally, magnetoliposomes, liposomes combined with nanomaterials such as superparamagnetic iron oxide nanoparticles, are emerging as multimodal contrast agents and cancer theragnostics with various imaging methods.[210,211] Magnetoliposomes can be developed as multimodal contrast agents because of the flexibility to incorporate cationic lipids, fluorescent lipid conjugates, targeting ligands, and polyethylene glycol either in the core of liposome or in the lipid bilayer to design all-in-one nanosystem.[212,213] It is found that there is a strong correlation between the fluidity of the lipid bilayer and the relaxivity value of magnetoliposomes in which magnetic nanoparticles are embedded in the bilayer.[213] Furthermore, the important physicochemical characteristics of phospholipids, such as biocompatibility and strong binding on the particle's surface, facilitate the development of liposome-based MRI contrast agents.[212]

6.3.7 Protein-based Bioactive Hydrogels for Biological Applications

The recent focus on cell biology and tissue engineering has emerged toward manipulating bioactive materials to understand cellular functionalities by mimicking tissue stiffness, immobilizing biomolecules, and biochemical patterning.[214] Hydrogels have been recognized as successful *in vitro* platforms in cell culturing and tissue engineering applications. DeForest and Anseth[215] have published a comprehensive review of recent advancements in bioactive hydrogel systems in assessing 2D and 3D cell functions. It was further reported that user-defined tailorability of chemical and mechanical properties of hydrogels enables precise biofunctionality in understanding cellular behavior. Recently, patterning BPs or proteins on these hydrogels has been catapulted as promising biomaterials that allow the identification of cell behavior, including biochemical cues. Typical cell culturing methods depend on soluble factors within the medium to elicit the biofunctionality of cells. They cannot mimic cellular signaling mechanisms, such as matrix-bound growth factors, cell–cell signaling, and spatially specific biochemical cues, as precisely as *in vivo* behaviors.[216] Dorsey *et al.*[214] reported on photochemistry reaction kinetics to pattern vascular endothelial growth factor and ephrinB2 on hydrogels to demonstrate the effects on endothelial cell behavior while retaining the bioactivity of proteins or peptides. Lysozyme bioactivity assay illustrated that the photochemistry reaction parameters, including the type of photoinitiator, concentration, and ultraviolet exposure, affect the bioactivity of protein and cell viability. Therefore, a compressive understanding of photochemistry reaction kinetics in patterning proteins on various hydrogels is required to maintain protein bioactivity and thereby accurately control biochemical and biophysical signals for biological applications. The immobilization of BPs and proteins, such as growth factors or signaling molecules, on biomimetic hydrogels can be applicable in biotechnological applications to investigate cell behavior, including cell migration, signaling mechanisms, and wound healing, as well as to develop clinically relevant tissue in the medical industry.[217,218] In various tissue engineering applications, complex structures such as microvasculature can be designed using 3D micropatterning techniques, as the bioactivity of immobilized biomolecules can be controlled.[217] Another novel approach has been presented to immobilize site-specific functional proteins, which are modified through a sortase-mediated transpeptidation modification technique, within gels to probe and direct cell fate in 4D.[219] Photo-reversible patterning of bioactive site-specifically modified fluorescent proteins, enzymes, and growth factors within gel biomaterials was reported to have excellent spatiotemporal resolution while maintaining the bioactivity of native proteins. Additionally, patterning approaches to immobilize predefined spatiotemporally controlled bioactive proteins on surfaces, including site-specific protein immobilization, are also useful in developing biosensors and protein microarrays for biomedical and biotechnological research fields.[220,221]

6.4 Conclusions and Future Prospects

Natural bioactive proteins and lipids can be sourced from various plant, animal, marine, and microbial origins. The significance of these bioactive compounds is mainly due to their ability to provide health-promoting and therapeutic effects in humans beyond their basic nutritional importance. The structure of proteins or lipids plays a key role in determining their biological functionalities through different mechanisms of action. Recently, attention has emerged in many studies toward bioactive compounds, including BPs, protein hydrolysates, and lipid-based compounds, in developing functional foods, nutraceuticals, pharmaceuticals, cosmeceuticals, biomedicine, and applications in other sectors. Advanced technologies, such as nanotechnology and encapsulation, can be combined with the production of functional ingredients to improve their stability and bioavailability within the food matrix and during gastrointestinal digestion until reaching the target tissue/cell. The interactions between proteins and polyphenols have turned the focus of scientists toward producing protein–polyphenol aggregates as novel stable, functional ingredients. However, more *in vivo* and *in vitro* studies must be performed to understand the safety considerations, long-term effects on humans, possible side effects, and toxic reactions involved with consuming bioactive protein- and lipid-based foods and drugs. Furthermore, the antimicrobial and antioxidant activity of BPs indicates a great potential to use them as alternative chemical preservatives and antimicrobial agents in active packaging. Besides the bioactivity, different proteins and plant oils are used in bioplastic production, and these bio-based plastics can be applied in food packaging, agricultural, and various medical applications, minimizing the risk of contamination and other potential infections caused by conventional plastics. Functional proteins and lipids could also be used as inks for 3D printing applications to enhance printed products' nutritional and functional values in the food and biomedical industry. Additionally, peptide- or lipid-based novel molecular imaging probes show improved sensitivity, accuracy, and target specificity in molecular imaging. BP-based hydrogels are also important in the applications of biomedical and biotechnological research fields.

Moreover, strong scientific evidence is required to determine the optimal levels of the bioactive compounds to be applied as functional ingredients in various applications, considering sensory properties and health and safety concerns.[222] In addition, the research focus could be extended to human-based clinical trials to explore disease reduction claims, thereby increasing the development and utilization of value-added applications with promising health claim benefits.[79] A systematic approach should also be explored to scale up the commercial applications of these bioactive protein- and lipid-based bioactive compounds.

References

1. X. Fan, L. Bai, L. Zhu, L. Yang and X. Zhang, *J. Agric. Food Chem.*, 2014, **62**, 9211–9222.
2. C. C. Udenigwe and R. E. Aluko, *J. Food Sci.*, 2012, **77**, R11–R24.
3. M. Segura-Campos, L. Chel-Guerrero, D. Betancur-Ancona and V. M. Hernandez-Escalante, *Food Rev. Int.*, 2011, **27**, 213–226.
4. Y. Zhang, T. Zhang, Y. Liang, L. Jiang and X. Sui, *J. Agric. Food Chem.*, 2021, **69**, 8929–8943.
5. V. Chiurchiu and M. Maccarrone, *Curr. Opin. Pharmacol.*, 2016, **29**, 54–62.
6. Y. Tintut and L. L. Demer, *Trends Endocrinol. Metab.*, 2014, **25**, 53–59.
7. R. E. Aluko, *Functional Foods and Nutraceuticals. Food Science Text Series*, Springer, New York, 2012, vol. **42**, pp. 23–36.
8. V. Chauhan and S. S. Kanwar, in *Biotechnological production of bioactive compounds*, Elsevier, 2020, pp. 107–137.
9. C. M. Urista, R. A. Fernández, F. R. Rodriguez, A. A. Cuenca and A. T. Jurado, *Food Sci. Technol. Int.*, 2011, **17**, 293–317.
10. Z. F. Bhat, S. Kumar and H. F. Bhat, *Crit. Rev. Food Sci. Nutr.*, 2017, **57**, 566–578.

11. J. T. Ryan, R. P. Ross, D. Bolton, G. F. Fitzgerald and C. Stanton, *Nutrients*, 2011, **3**, 765–791.
12. C. S. F. Bah, A. E.-D. A. Bekhit, A. Carne and M. A. McConnell, *Compr. Rev. Food Sci. Food Saf.*, 2013, **12**, 314–331.
13. B. M. Kimatu, L. Zhao, Y. Biao, G. Ma, W. Yang, F. Pei and Q. Hu, *Food Chem.*, 2017, **230**, 58–67.
14. X. Geng, G. Tian, W. Zhang, Y. Zhao, L. Zhao, H. Wang and T. B. Ng, *Sci. Rep.*, 2016, **6**, 24130.
15. Y.-Z. Xiang, Y.-M. Zhang, Y.-Y. Liu, M. Zhang, L.-B. Lin and Q.-L. Zhang, *Food Control*, 2021, **127**, 108110.
16. P. Ye, J. Wang, M. Liu, P. Li and Q. Gu, *LWT--Food Sci. Technol.*, 2021, **143**, 111125.
17. I. Dinika, D. K. Verma, R. Balia, G. L. Utama and A. R. Patel, *Trends Food Sci. Technol.*, 2020, **103**, 57–67.
18. G. M. Suarez-Jimenez, A. Burgos-Hernandez and J. M. Ezquerra-Brauer, *Mar. Drugs*, 2012, **10**, 963–986.
19. Jumeri and S. M. Kim, *Food Sci. Biotechnol.*, 2011, **20**, 1075–1085.
20. S. Y. Naqash and R. A. Nazeer, *J. Aquat. Food Prod. Technol.*, 2010, **19**, 180–192.
21. X. Fan, L. Bai, L. Zhu, L. Yang and X. Zhang, *J. Agric. Food. Chem.*, 2014, **62**, 9211–9222.
22. T. Lafarga, A. Sánchez-Zurano, S. Villaró, A. Morillas-España and G. Acién, *Trends Food Sci. Technol.*, 2021, **116**, 176–185.
23. S. P. Patil, A. Goswami, K. Kalia and A. S. Kate, *Int. J. Pept. Res. Ther.*, 2020, **26**, 955–968.
24. R. E. Aluko, A. T. Girgih, R. He, S. Malomo, H. Li, M. Offengenden and J. Wu, *Food Res. Int.*, 2015, **77**, 10–16.
25. S. Zhang, L. Luo, X. Sun and A. Ma, *J. Agric. Food Chem.*, 2021, **69**, 12369–12384.
26. F. Roy, J. I. Boye and B. K. Simpson, *Food Res. Int.*, 2010, **43**, 432–442.
27. X. Y. Ye and T. B. Ng, *J. Life Sci.*, 2002, **70**, 1129–1138.
28. H. X. Wang and T. B. Ng, *Peptides*, 2007, **28**, 547–552.
29. P. B. Pelegrini, E. F. Noronha, M. A. R. Muniz, I. M. Vasconcelos, M. D. Chiarello, J. T. A. Oliveira and O. L. Franco, *Biochim. Biophys. Acta, Proteins Proteomics*, 2006, **1764**, 1141–1146.
30. J. Banerjee, R. Singh, R. Vijayaraghavan, D. MacFarlane, A. F. Patti and A. Arora, *Food Chem.*, 2017, **225**, 10–22.
31. A. Sánchez and A. Vázquez, *Food Qual. Saf.*, 2017, **1**, 29–46.
32. A. Alemán, E. Pérez-Santín, S. Bordenave-Juchereau, I. Arnaudin, M. Gómez-Guillén and P. Montero, *Food Res. Int.*, 2011, **44**, 1044–1051.
33. S.-K. Kim and I. Wijesekara, *J. Funct. Foods*, 2010, **2**, 1–9.
34. K. F. Chai, A. Y. H. Voo and W. N. Chen, *Compr. Rev. Food Sci. Food Saf.*, 2020, **19**, 3825–3885.
35. S. Sanjukta and A. K. Rai, *Trends Food Sci. Technol.*, 2016, **50**, 1–10.

36. Y. Hou, Z. Wu, Z. Dai, G. Wang and G. Wu, *J. Anim. Sci. Biotechnol.*, 2017, **8**, 1–13.
37. B. L. Nilsson, M. B. Soellner and R. T. Raines, *Annu. Rev. Biophys. Biomol. Struct.*, 2005, **34**, 91–118.
38. A. Ialenti, V. Santagada, G. Caliendo, B. Severino, F. Fiorino, P. Maffia, A. Ianaro, F. Morelli, B. Di Micco and M. Cartenì, *Eur. J. Biochem.*, 2001, **268**, 3399–3406.
39. X. Xu, H. Yan, J. Chen and X. Zhang, *Biotechnol. Adv.*, 2011, **29**, 667–674.
40. A. Büttner-Mainik, J. Parsons, H. Jérôme, A. Hartmann, S. Lamer, A. Schaaf, A. Schlosser, P. F. Zipfel, R. Reski and E. L. Decker, *Plant Biotechnol. J*, 2011, **9**, 373–83.
41. Y. Li, D. He, B. Li, M. N. Lund, Y. Xing, Y. Wang, F. Li, X. Cao, Y. Liu, X. Chen and J. Yu, *Trends Food Sci. Technol.*, 2021, **110**, 470–482.
42. R. K. Saini and Y.-S. Keum, *Life Sci.*, 2018, **203**, 255–267.
43. E. Sokoła-Wysoczanska, T. Wysoczanski, J. Wagner, K. Czyz, R. Bodkowski, S. Lochynski and B. Patkowska-Sokoła, *Nutrients*, 2018, **10**, 1561.
44. R. E. Aluko, in *Functional Foods and Nutraceuticals*, Springer, New York, NY, USA, 2012, vol. **42**, pp. 23–36.
45. V. Verardo, P. Garcia-Salas, E. Baldi, A. Segura-Carretero, A. Fernandez-Gutierrez and M. F. Caboni, *Food Res. Int.*, 2014, **65**, 445–452.
46. C. Soares, S. Sousa, S. Machado, E. Vieira, A. P. Carvalho, M. J. Ramalhosa, S. Morais, M. Correia, T. Oliva-Teles, V. F. Domingues and C. Delerue-Matos, *Foods*, 2021, **10**, 1366.
47. M. A. Sayeda, G. H. Ali and F. K. El-Baz, *Int. J. Pharm. Sci. Rev. Res.*, 2015, **34**, 210-215.
48. B. R. Kumar, G. Deviram, T. Mathimani, P. A. Duc and A. Pugazhendhi, *Biocatal. Agric. Biotechnol.*, 2019, **17**, 583–588.
49. J. B. German and C. J. Dillard, *Crit. Rev. Food Sci. Nutr.*, 2006, **46**, 57–92.
50. R. Rombaut and K. Dewettinck, *Int. Dairy J.*, 2006, **16**, 1362–1373.
51. C. Lopeza, M. Blota, V. Briard-Biona, C. Ciriéb and B. Grauletc, *Food Res. Int.*, 2017, **100**, 864–872.
52. S. S. AbuMweis, C. P. Marinangeli, J. Frohlich and P. J. Jones, *Can. J. Cardiol.*, 2014, **30**, 1225–1232.
53. N. Meshginfar, H. Tavakoli, K. Dornan and F. Hosseinian, *Crit. Rev. Food Sci. Nutr.*, 2021, **61**, 1394–1403.
54. J.-H. Lin, L.-Y. Liu, M.-H. Yang and M.-H. Lee, *J. Agric. Food Chem.*, 2004, **52**, 4984–4986.
55. R. K. Saini, P. Prasad, X. Shang and Y.-S. Keum, *Int. J. Mol. Sci.*, 2021, **22**, 13643.
56. M. Guderjan, S. Töpfl, A. Angersbach and D. Knorr, *J. Food Eng.*, 2005, **67**, 281–287.

57. F. Bezerra, M. De Oliveira, P. Bezerra, V. Cunha, M. Silva, W. Da Costa, R. Pinto, R. Cordeiro, J. Da Cruz and A. C. Neto, in *Green Sustainable Process for Chemical and Environmental Engineering and Science*, Elsevier, 2020, pp. 149–167.
58. S. P. J. N. Senanayake and F. Shahidi, in *Bailey's industrial oil and fat products*, 2005, pp. 1–29.
59. T. Yuan, W. Wei, X. Wang and Q. Jin, *LWT*, 2020, **128**, 109255.
60. S. Ji, F. Xu, N. Zhang, Y. Wu, X. Ju and L. Wang, *Food Biosci.*, 2021, **44**, 101396.
61. B. H. Kim and C. C. Akoh, *J. Food Sci.*, 2015, **80**, C1713–C1724.
62. L. Han, Z. Xu, J. Huang, Z. Meng, Y. Liu and X. Wang, *J. Agric. Food Chem.*, 2011, **59**, 12635–12642.
63. M. Hossen and E. Hernandez, *Eur. J. Lipid Sci. Technol.*, 2005, **107**, 730–736.
64. H. T. Osborn and C. C. Akoh, *Compr. Rev. Food Sci. Food Saf.*, 2002, **1**, 110–120.
65. Y. Li and J. Yu, *J. Med. Food*, 2015, **18**, 147–156.
66. B. A. Murray and R. J. FitzGerald, *Curr. Pharm. Des.*, 2007, **13**, 773–791.
67. L. A. Dave, C. A. Montoya, S. M. Rutherfurd and P. J. Moughan, *PLoS One*, 2014, **9**, e98922.
68. D. Giordano, M. Costantini, D. Coppola, C. Lauritano, L. N. Pons, N. Ruocco, G. di Prisco, A. Ianora and C. Verde, *Adv. Microb. Physiol.*, 2018, **73**, 171–220.
69. R. J. FitzGerald and H. Meisel, *Br. J. Nutr.*, 2000, **84**, 33–37.
70. J. Tan, C. McKenzie, M. Potamitis, A. N. Thorburn, C. R. Mackay and L. Macia, in *Advances in Immunology*, Academic Press, Boston, MA, 2014, vol. **121**, pp. 91–119.
71. U. N. Das, *Arch. Med. Res.*, 2020, **51**, 282–286.
72. E. E. Kooijman, V. Chupin, B. de Kruijff and K. N. J. Burger, *Traffic*, 2003, **4**, 162–174.
73. E. E. Kooijman, K. M. Carter, E. G. v. Laar, V. Chupin, K. N. J. Burger and B. d. Kruijff, *Biochem.*, 2005, **44**, 17007–17015.
74. S. S. AbuMweis, C. P. F. Marinangeli, J. Frohlich and P. J. H. Jones, *Can. J. Cardiol.*, 2014, **30**, 1225–1232.
75. M. Ledoux, P. Juanéda and J. L. Sébédio, *Eur. J. Lipid Sci. Technol.*, 2007, **109**, 891–900.
76. K. J. Shingfield, Y. Chilliard, V. Toivonen, P. Kairenius and D. I. Givens, in *Bioactive components of milk*, Springer, 2008, vol. **606**, pp. 3–65.
77. V. Dhaka, N. Gulia, K. S. Ahlawat and B. S. Khatkar, *J. Food Sci. Technol.*, 2011, **48**, 534–541.
78. L. López-Barrios, J. A. Gutiérrez-Uribe and S. O. Serna-Saldívar, *J. Food Sci.*, 2014, **79**, R273–R283.
79. D. Bhandari, S. Rafiq, Y. Gat, P. Gat, R. Waghmare and V. Kumar, *Int. J. Pept. Res. Ther.*, 2020, **26**, 139–150.

80. Z. Karami and B. Akbari-Adergani, *J. Food Sci. Technol.*, 2019, **56**, 535-547.
81. S. Y. Lee, H. J. Kang, J. H. Kang and S. J. Hur, *J. Funct. Foods*, 2021, **81**, 104477.
82. B. Yang, T.-t. Wang, Y.-s. Yang, H.-l. Zhu and J.-h. Li, *J. Drug Deliv. Sci. Technol.*, 2021, **66**, 102880.
83. M. Betz, B. Steiner, M. Schantz, J. Oidtmann, K. Mäder, E. Richling and U. Kulozik, *Food Res. Int.*, 2012, **47**, 51–57.
84. A. Helal, D. Tagliazucchi, E. Verzelloni and A. Conte, *J. Funct. Foods*, 2014, **7**, 506–516.
85. P. Chanphai and H. A. Tajmir-Riahi, *Food Hydrocolloids*, 2019, **89**, 461-467.
86. J. Xiong, Y. H. Chan, T. Rathinasabapathy, M. H. Grace, S. Komarnytsky and M. A. Lila, *Food chem.*, 2020, **331**, 127279.
87. B. Jin, X. Zhou, Y. Liu, X. Li, Y. Mai, Y. Liao and J. Liao, *Int. J. Biol. Macromol.*, 2018, **116**, 1–7.
88. N. Kaur, V. Chugh and A. K. Gupta, *J. Food Sci. Technol.*, 2014, **51**, 2289–2303.
89. K. E. Lane and E. J. Derbyshire, *Crit. Rev. Food Sci. Nutr.*, 2018, **58**, 62-69.
90. J. C. Ruiz Ruiz, E. D. L. L. Ortiz Vazquez and M. R. Segura Campos, *Crit. Rev. Food Sci. Nutr.*, 2017, **57**, 1423–1434.
91. M. Geranpour, E. Assadpour and S. M. Jafari, *Trends Food Sci. Technol.*, 2020, **102**, 71–90.
92. P. J. Jones and S. S. AbuMweis, *Curr. Opin. Clin. Nutr. Metab. Care*, 2009, **12**, 147–151.
93. A. Poli, F. Marangoni, A. Corsini, E. Manzato, W. Marrocco, D. Martini, G. Medea and F. Visioli, *Nutrients*, 2021, **13**, 2810.
94. C. W. Pouton and C. J. H. Porter, *Adv. Drug Deliv. Rev.*, 2008, **60**, 625-637.
95. M. Corzo-Martínez, L. Vázquez, P. Arranz-Martínez, N. Menéndez, G. Reglero and C. F. Torres, *Food Bioprod. Process.*, 2016, **100**, 311–322.
96. A. Pezeshki, H. Hamishehkar, B. Ghanbarzadeh, I. Fathollahy, F. K. Nahr, M. K. Heshmati and M. Mohammadi, *Food Biosci.*, 2019, **27**, 11-17.
97. A. Beloqui, M. Á. Solinís, A. Rodríguez-Gascón, A. J. Almeida and V. Préat, *Nanomedicine*, 2016, **12**, 143–161.
98. M. Gautam and D. Santhiya, *Colloids Surf., A*, 2019, **577**, 637–644.
99. I. Katouzian and S. M. Jafari, *Trends Food Sci. Technol.*, 2016, **53**, 34–48.
100. C. Chang, J. Li, Y. Su, L. Gu, Y. Yang and J. Zhai, *Food Hydrocolloids*, 2022, **123**, 106963.
101. F. Iemma, U. G. Spizzirri, F. Puoci, R. Muzzalupo, S. Trombino, R. Cassano, S. Leta and N. Picci, *Int. J. Pharm.*, 2006, **312**, 151–157.
102. Y. Wang, Y. Chang, Y. Xue, Z. Li, Y. Wang and C. Xue, *Food Hydrocolloids*, 2016, **52**, 510–519.

103. A. Abaee, M. Mohammadian and S. M. Jafari, *Trends Food Sci. Technol.*, 2017, **70**, 69–81.
104. Z.-L. Wan, J. Guo and X.-Q. Yang, *Food Funct.*, 2015, **6**, 2876–2889.
105. A. Maltais, G. E. Remondetto and M. Subirade, *Food Hydrocolloids*, 2010, **24**, 518–524.
106. Q. Zhang, L. Gu, Y. Su, C. Chang, Y. Yang and J. Li, *Int. J. Biol. Macromol.*, 2021, **172**, 281–288.
107. F. Alavi, Z. Emam-Djomeh, M. S. Yarmand, M. Salami, S. Momen and A. A. Moosavi-Movahedi, *Food Hydrocolloids*, 2018, **85**, 267–280.
108. X. Ding and P. Yao, *Langmuir*, 2013, **29**, 8636–8644.
109. W. Yang, X. Liang, L. Xu, C. Deng, W. Jin, X. Wang, Y. Kong, M. Duan, Y. Nei and J. Zeng, *Food Res. Int.*, 2020, **131**, 109048.
110. A. Shi, X. Feng, Q. Wang and B. Adhikari, *Food Hydrocolloids*, 2020, **109**, 106117.
111. M. Destribats, M. Rouvet, C. Gehin-Delval, C. Schmitt and B. P. Binks, *Soft Matter*, 2014, **10**, 6941–6954.
112. F. Liu and C.-H. Tang, *Food Hydrocolloids*, 2016, **60**, 606–619.
113. I. Tavernier, W. Wijaya, P. Van der Meeren, K. Dewettinck and A. R. Patel, *Trends Food Sci. Technol.*, 2016, **50**, 159–174.
114. Y. Yang, Z. Fang, X. Chen, W. Zhang, Y. Xie, Y. Chen, Z. Liu and W. Yuan, *Front. Pharmacol.*, 2017, **8**, 287.
115. A. Magri, M. Petriccione, M. A. Cerqueira and T. J. Gutiérrez, *Adv. Colloid Interface Sci.*, 2020, **285**, 102279.
116. J. Xiong, Y. H. Chan, T. Rathinasabapathy, M. H. Grace, S. Komarnytsky and M. A. Lila, *Food Chem.*, 2020, **331**, 127279.
117. F. Diarrassouba, G. Remondetto, G. Garrait, P. Alvarez, E. Beyssac and M. Subirade, *Food Chem.*, 2015, **173**, 203–209.
118. C. Chang, T. G. Meikle, Y. Su, X. Wang, C. Dekiwadia, C. J. Drummond, C. E. Conn and Y. Yang, *Food Chem.*, 2019, **280**, 65–72.
119. H. Hoffmann and M. Reger, *Adv. Colloid Interface Sci.*, 2014, **205**, 94–104.
120. A. Sarkar and E. Dickinson, *Curr. Opin. Colloid Interface Sci.*, 2020, **49**, 69–81.
121. W. Lu, A. L. Kelly and S. Miao, *Trends Food Sci. Technol.*, 2016, **47**, 1–9.
122. X. Zhang, S. Zhang, M. Zhong and B. L. Y. Qi, *Food Chem.*, 2022, **380**, 132212.
123. W. W. Mwangi, H. P. Lim, L. E. Low, B. T. Tey and E. S. Chan, *Trends Food Sci. Technol.*, 2020, **100**, 320–332.
124. D. Xu, F. Yuan, Y. Gao, A. Panya, D. J. McClements and E. A. Decker, *Food Chem.*, 2014, **156**, 374–379.
125. Y. Ji, C. Han, E. Liu, X. Li, X. Meng and B. Liu, *Food Chem.*, 2022, **378**, 132090.
126. J. Yan, X. Liang, C. Ma, D. J. McClements, X. Liu and F. Liu, *Food Hydrocolloids*, 2021, **113**, 106473.

127. N. N. Potter and J. H. Hotchkiss, in *Food Science*, Springer, Chapman & HalI, New York, 5th edn, 1995.
128. J. Tkaczewska, *Trends Food Sci. Technol.*, 2020, **106**, 298–311.
129. P. J. P Espitia, N. d. F. a. F. Soares, J. S. e. d. R. Coimbra, N. e. J. e. d. Andrade, R. S. Cruz and E. A. A. Medeiros, *Compr. Rev. Food Sci. Food Saf.*, 2012, **11**, 187–204.
130. Y. Yan, Y. Li, Z. Zhang, X. Wang, Y. Niu, S. Zhang, W. Xu and C. Ren, *Colloids Surf., B*, 2021, **202**, 111682.
131. R. E. Aluko, in *Handbook of Antioxidants for Food Preservation*, Woodhead Publishing, 2015, pp. 105–140.
132. Y.-W. Liu, C.-H. Han, M.-H. Lee, F.-L. Hsu and W.-C. Hou, *J. Agric. Food Chem.*, 2003, **51**, 4389–4393.
133. D. Takahashi, S. K. Shukla, O. Prakash and G. Zhang, *Biochimie*, 2010, **92**, 1236–1241.
134. L. Xing, R. Liu, X. Gao, J. Zheng, C. Wang, G. Zhou and W. Zhang, *Food Chem.*, 2018, **266**, 420–426.
135. J. Ren, M. Zhao, J. Shi, J. Wang, Y. Jiang, C. Cui, Y. Kakuda and S. J. Xue, *Food Chem.*, 2008, **108**, 727–736.
136. S. Mandel, J. Michaeli, N. Nur, I. Erbetti, J. Zazoun, L. Ferrari, A. Felici, M. Cohen-Kutner and N. Bachnoff, *Sci Rep.*, 2021, **11**, 6603.
137. X. Wang, H. Yu, R. Xing, X. Chen, S. Liu and P. Li, *BioMed Res. Int.*, 2017, **2017**, 9746720.
138. R. Lucas, M. J. Grande, H. Abriouel, M. Maqueda, N. Ben Omar, E. Valdivia, M. Martinez-Canamero and A. Gálvez, Application of the broad-spectrum bacteriocin enterocin AS-48 to inhibit Bacillus coagulans in canned fruit and vegetable foods, *Food Chem. Toxicol.*, 2006, **44**, 1774–1781.
139. M. Rai, R. Pandit, S. Gaikwad and G. Kovics, *J. Food Sci. Technol.*, 2016, **53**, 3381–3394.
140. L. H. Deegan, P. D. Cotter, C. Hill and P. Ross, *Int. Dairy J.*, 2006, **16**, 1058–1071.
141. M. Hyldgaard, R. L. Meyer, M. Peng, A. A. Hibberd, J. Fischer, A. Sigmundsson and T. Mygind, *Int. J. Food Microbiol.*, 2015, **215**, 131–142.
142. D. M. Meng, X. M. Yang, X. Q. Sun, L. Cheng and Z. C. Fan, *J. Food Process. Preserv.*, 2021, **45**, e15404.
143. D.-M. Meng, S.-N. Sun, L.-Y. Shi, L. Cheng and Z.-C. Fan, *Food Control*, 2021, **125**, 108041.
144. J. Hou, Y.-Q. Li, Z.-S. Wang, G.-J. Sun and H.-Z. Mo, *LWT--Food Sci. Technol.*, 2017, **83**, 267–274.
145. Y. Huang, L. Gao, M. Lin and T. Yu, *Postharvest Biol. Technol.*, 2021, **171**, 111298.
146. B. Ramos, T. R. S. Brandão, P. Teixeira and C. L. M. Silva, *Food Microbiol.*, 2020, **85**, 103282.

147. W. Zhao, L. Lu and Y. Tang, *Int. J. Food Eng.*, 2010, **6**, DOI: 10.2202/1556-3758.1943.
148. L. Settanni and A. Corsetti, *Int. J. Food Microbiol.*, 2008, **121**, 123–138.
149. K. B. Biji, C. N. Ravishankar, C. O. Mohan and T. K. Srinivasa Gopal, *J. Food Sci. Technol.*, 2015, **52**, 6125–6135.
150. P. Di Pierro, L. Mariniello, V. L. Giosafatto, M. Esposito, M. Sabbah and R. Porta, in *Food Packaging and Preservation*, Elsevier, 2018, pp. 439–456.
151. H. Jooyandeh, *Pak. J. Nutr.*, 2011, **10**, 296–301.
152. N. S. Said and N. M. Sarbon, Protein-Based Active Film as Antimicrobial Food Packaging: A Review, in *Active Antimicrobial Food Packaging*, ed. I. Var and S. Uzunlu, IntechOpen, 2019, https://doi.org/10.5772/intechopen.80774.
153. F. Song, D. L. Tang, X. L. Wang and Y. Z. Wang, *Biomacromolecules*, 2011, **12**, 3369–3380.
154. B. Marelli, M. A. Brenckle, D. L. Kaplan and F. G. Omenetto, *Sci. Rep.*, 2016, **6**, 1–11.
155. N. Jaramillo-Quiceno and A. Restrepo-Osorio, *J. Appl. Polym. Sci.*, 2020, **137**, 48505.
156. H. Tahir, N. Pervez, J. Nadeem, A. Khan and Z. Hassan, *Braz. J. Biol.*, 2019, **80**, 115–121.
157. M. Mittal, D. Mittal and N. K. Aggarwal, *Environ. Sci. Pollut. Res.*, 2022, 1–15.
158. P. Li, X. Wang, M. Su, X. Zou, L. Duan and H. Zhang, Characteristics of plastic pollution in the environment: a review, *Bull. Environ. Contam. Toxicol.*, 2021, **107**, 577–584.
159. T. D. Moshood, G. Nawanir, F. Mahmud, F. Mohamad, M. H. Ahmad and A. AbdulGhani, *Curr. Res. Green Sustainable Chem.*, 2022, 100273.
160. C. Lim, S. Yusoff, C. G. Ng, P. E. Lim and Y. C. Ching, *J. Environ. Chem. Eng.*, 2021, **9**, 105895.
161. A. Jones, A. Mandal and S. Sharma, *J. Appl. Polym. Sci.*, 2015, **132**, DOI: 10.1002/app.41931.
162. M. Felix, V. Perez-Puyana, A. Romero and A. Guerrero, *J. Appl. Polym. Sci.*, 2017, **134**, 45430.
163. D. Gómez-Heincke, I. Martínez, M. Stading, C. Gallegos and P. Partal, *Food Hydrocolloids*, 2017, **73**, 21–29.
164. A. Jerez, P. Partal, I. Martínez, C. Gallegos and A. Guerrero, *Rheol. Acta*, 2007, **46**, 711–720.
165. V. Perez, M. Felix, A. Romero and A. Guerrero, *Food Bioprod. Process.*, 2016, **97**, 100–108.
166. S. Thammahiwes, S.-A. Riyajan and K. Kaewtatip, *J. Cereal Sci.*, 2017, **75**, 186–191.
167. S. Sharma and A. Gupta, *Braz. Arch. Biol. Technol.*, 2016, **59**, e16150684.
168. S. Sharma, A. Gupta and A. Kumar, in *Keratin as a Protein Biopolymer*, Springer, 2019, pp. 1–18.

169. J. Zhang, Y. Li, J. Li, Z. Zhao, X. Liu, Z. Li, Y. Han, J. Hu and A. Chen, *Powder Technol.*, 2013, **246**, 356–362.
170. S. Sharma, A. Gupta, A. Kumar, C. G. Kee, H. Kamyab and S. M. Saufi, *Clean Techn. Environ. Policy*, 2018, **20**, 2157–2167.
171. N. Ramakrishnan, S. Sharma, A. Gupta and B. Y. Alashwal, *Int. J. Biol. Macromol.*, 2018, **111**, 352–358.
172. Y. Dou, B. Zhang, M. He, G. Yin and Y. Cui, *Chin. J. Chem. Eng.*, 2016, **24**, 415–420.
173. K. Williamson and E. Hatzakis, *Food Res. Int.*, 2019, **119**, 683–692.
174. M. R. Islam, M. D. H. Beg and S. S. Jamari, *J. Appl. Polym. Sci.*, 2014, **131**, DOI: 10.1002/app.40787.
175. H. Uyama, M. Kuwabara, T. Tsujimoto, M. Nakano, A. Usuki and S. Kobayashi, *Chem. Mater.*, 2003, **15**, 2492–2494.
176. Z. Liu, S. Z. Erhan and J. Xu, *Polymer*, 2005, **46**, 10119–10127.
177. L. Fu, L. Yang, C. Dai, C. Zhao and L. Ma, *J. Appl. Polym. Sci.*, 2010, **117**, 2220–2225.
178. J.-L. Pan, C.-R. Xu, F.-R. Zeng, Y. Liang, T. Zhang, J. Xu, Z.-L. Li and Z.-C. Li, *Appl. Polym. Mater.*, 2021, **3**, 2054–2062.
179. J.-H. Kim, Y.-R. Oh, J. Hwang, J. Kang, H. Kim, Y.-A. Jang, S.-S. Lee, S. Y. Hwang, J. Park and G. T. Eom, *Waste Manage.*, 2021, **124**, 195–202.
180. M. E. Lamm, L. Song, Z. Wang, B. Lamm, L. Fu and C. Tang, *Polym. Chem.*, 2019, **10**, 6570–6579.
181. Y. Liu, X. Liang, A. Saeed, W. Lan and W. Qin, *Innovative Food Sci. Emerging Technol.*, 2019, **54**, 9–18.
182. Y. Liu, W. Zhang, K. Wang, Y. Bao, J. M. Regenstein and P. Zhou, *Food Bioprod. Process.*, 2019, **12**, 1967–1979.
183. V. F. Sager, M. B. Munk, M. S. Hansen, W. L. P. Bredie and L. Ahrné, 2021, **10**, 8.
184. X. Li, X. Xu, L. Song, A. Bi, C. Wu, Y. Ma, M. Du and B. Zhu, *ACS Appl. Mater. Interfaces*, 2020, **12**, 45493–45503.
185. C. Wu, X. Na, W. Ma, C. Ren, Q. Zhong, T. Wang and M. Du, *Chem. Eng. J.*, 2021, **412**, 128724.
186. P. Phuhongsung, M. Zhang and S. Devahastin, *LWT*, 2020, **122**, 109019.
187. M. Shahbazi, H. Jäger, J. Chen and R. Ettelaie, *Food Hydrocolloids*, 2021, **121**, 107054.
188. J. Chen, T. Mu, D. Goffin, C. Blecker, G. Richard, A. Richel and E. Haubruge, *J. Food Eng.*, 2019, **261**, 76–86.
189. P. Phuhongsung, M. Zhang and B. Bhandari, *Food Res. Int.*, 2020, **137**, 109605.
190. T. Feng, C. Fan, X. Wang, X. Wang, S. Xia and Q. Huang, *Food Hydrocolloids*, 2022, **124**, 107265.
191. E. Álvarez-Castillo, S. Oliveira, C. Bengoechea, I. Sousa, A. Raymundo and A. Guerrero, *J. Food Eng.*, 2021, **288**, 110255.
192. Y. Chen, M. Zhang, Y. Sun and P. Phuhongsung, *Food Hydrocolloids*, 2022, **123**, 107160.

193. M. Lanaro, D. P. Forrestal, S. Scheurer, D. J. Slinger, S. Liao, S. K. Powell and M. A. Woodruff, *J. Food Eng.*, 2017, **215**, 13–22.
194. S. Mantihal, S. Prakash, F. C. Godoi and B. Bhandari, *Food Res. Int.*, 2019, **119**, 161–169.
195. C. Chen, M. Zhang, C. Guo and H. Chen, *Innovative Food Sci. Emerging Technol.*, 2021, **68**, 102605.
196. M. V. Vieira, S. M. Oliveira, I. R. Amado, L. H. Fasolin, A. A. Vicente, L. M. Pastrana and P. Fuciños, *Food Hydrocolloids*, 2020, **107**, 105893.
197. C. Le Tohic, J. J. O'Sullivan, K. P. Drapala, V. Chartrin, T. Chan, A. P. Morrison, J. P. Kerry and A. L. Kelly, *J. Food Eng.*, 2018, **220**, 56–64.
198. T. Ma, R. Cui, S. Lu, X. Hu, B. Xu, Y. Song and X. Hu, *Food Hydrocolloids*, 2022, **125**, 107418.
199. M. Shahbazi, H. Jäger, R. Ettelaie and J. Chen, *Food Hydrocolloids*, 2021, **120**, 106967.
200. E. G. Bulut and K. Candoğan, *LWT*, 2022, **154**, 112768.
201. A.-Q. Bi, X.-B. Xu, Y. Guo, M. Du, C.-P. Yu and C. Wu, *Food Chem.*, 2022, **371**, 131349.
202. M. A. Pysz, S. S. Gambhir and J. K. Willmann, *Clin. Radiol.*, 2010, **65**, 500–516.
203. S. Xue, J. Qiao, F. Pu, M. Cameron and J. J. Yang, *Wiley Interdiscip. Rev.: Nanomed. Nanobiotechnol.*, 2013, **5**, 163–179.
204. J. L. Major and T. J. Meade, *Acc. Chem. Res.*, 2009, **42**, 893–903.
205. E. Gianolio, S. Porto, R. Napolitano, S. Baroni, G. Giovenzana and S. Aime, *Inorg. Chem.*, 2012, **51**, 7210–7217.
206. L. S. Karfeld-Sulzer, E. A. Waters, N. E. Davis, T. J. Meade and A. E. Barron, *Biomacromolecules*, 2010, **11**, 1429–1436.
207. P. Caravan, G. Parigi, J. M. Chasse, N. J. Cloutier, J. J. Ellison, R. B. Lauffer, C. Luchinat, S. A. McDermid, M. Spiller and T. J. McMurry, *Inorg. Chem.*, 2007, **46**, 6632–6639.
208. C. L. Charron, J. L. Hickey, T. K. Nsiama, D. R. Cruickshank, W. L. Turnbull and L. G. Luyt, *Nat. Prod. Rep.*, 2016, **33**, 761–800.
209. S. Lee, J. Xie and X. Chen, *Biochemistry*, 2010, **49**, 1364–1376.
210. S. J. Soenen, M. Hodenius and M. De Cuyper, *Nanomedicine*, 2009, **4**, 177–191.
211. S. J. H. Soenen, J. Cocquyt, L. Defour, P. Saveyn, P. V. d. Meeren and M. D. Cuyper, *Mater. Manuf. Processes*, 2008, **23**, 611–614.
212. H. Fattahi, S. Laurent, F. Liu, N. Arsalani, L. V. Elst and R. N. Muller, *Nanomedicine*, 2011, **6**, 529–544.
213. N. Kostevšek, C. C. L. Cheung, I. Serša, M. E. Kreft, I. Monaco, M. Comes Franchini, J. Vidmar and W. T. Al-Jamal, *Nanomaterials*, 2020, **10**, 889.
214. T. B. Dorsey, A. Grath, A. Wang, C. Xu, Y. Hong and G. Dai, *Bioact. Mater.*, 2018, **3**, 64–73.
215. C. A. DeForest and K. S. Anseth, *Annu. Rev. Chem. Biomol. Eng.*, 2012, **3**, 421–444.

216. T. B. Dorsey, A. Grath, C. Xu, Y. Hong and G. Dai, *J. Visualized Exp.*, 2017, **127**, e55873.
217. S.-H. Lee, J. J. Moon and J. L. West, *Biomaterials*, 2008, **29**, 2962–2968.
218. M. S. Hahn, J. S. Miller and J. L. West, *Adv. Mater.*, 2006, **18**, 2679–2684.
219. J. A. Shadish, G. M. Benuska and C. A. DeForest, *Nat. Mater.*, 2019, **18**, 1005–1014.
220. M. Humenik, A. Winkler and T. Scheibel, *Biopolymers*, 2021, **112**, e23412.
221. D. Weinrich, P. Jonkheijm, C. M. Niemeyer and H. Waldmann, *Angew. Chem., Int. Ed.*, 2009, **48**, 7744–7751.
222. J. A. M. Prates and C. M. R. P. Mateus, *Rev. Med. Vet.*, 2002, **153**, 155–160.

CHAPTER 7

Nanoliposomes for Promising Encapsulation and Delivery of Potential Nutraceutical Ingredients

S. ANULEKSHMI[a], K. R. RESHNA[a], PREETHA BALAKRISHNAN[a] AND SREERAG GOPI*[a,b]

[a]Molecules BioLabs Private Limited, Building No: III/634, First Floor, Commercial Building, Kinfra Small Industrial Park, Koratty, Kerala 680309, India; [b]Academy of Material Research, Mahatma Gandhi University Innovation Foundation, Building No. 572, Priyadarshini Hills, Athirampuzha, Kottayam, Kerala 686560, India
*E-mail: sreeraggopi@gmail.com

7.1 Introduction

Lipids are particularly attractive in the formation of closed spherical lipid bilayers, known as liposomes. The self-assembly of building blocks into nanoparticles or microstructures is an area of intense interest.[1] Liposomes are very diverse structures for research, treatment, and analytical applications. They consist of a lipid bilayer with hydrophobic chains of lipids forming the bilayer, while the polar head groups of lipids face both the extracellular solution and the inner chamber.[2] Their structures are similar to those of cells and, hence, can be used as a more easily characterized

Smart Materials No. 41
Functional Materials from Lipids and Proteins
Edited by Nandika Bandara and Aman Ullah

Published by the Royal Society of Chemistry, www.rsc.org

vessel to explain the permeability of biological molecules, such as membrane lipids, DNA, and proteins, to ions and drugs. Liposomes are classified according to their lamellarity and size, such as multilamellar vesicles (MLVs), unilamellar vesicles (ULVs), and smaller unilamellar vesicles (SUVs, 1 μm).[3] It can be formulated using cholesterol and natural phospholipids. Liposomes are also preferred as an advanced delivery system for nutraceutical products.[4] Depending on the membrane structure of the liposome, the vesicles can be classified into large multilamellar (0.1 to 6 mm in size) or small and large unilamellar vesicles.[5]

The stability of liposomes has been reported as a function of the change in their average diameter and distribution, and the confirmed potential value indicates the strength of repulsive interactions between liposomes, preventing particle localization or aggregation. Therefore, the relationship of each variable with other factors and the theory of curvature are discussed. The production of nanoliposomes is an effective strategy for the encapsulation and controlled release of bioactive molecules. Liposomes have been shown to efficiently encapsulate and stabilize bioactive molecules against a range of environmental changes and protect them from degradation.[6]

Nanoliposome-carried encapsulation systems can be achieved by two methods, such as passive loading encapsulation, which involves the entrapment of compounds during the process of vesicle formation, and active or remote loading encapsulation. Hydrophilic compounds are encapsulated within the aqueous phase, while hydrophobic compounds are located in the lipid bilayer of the liposomes. On the other hand, amphiphilic molecules are located in their lipid-soluble region between the liposomal lipid bilayers.[7] Active or remote loading encapsulation involves the incorporation of bioactive compounds into intact vesicles. This method has been found to be very effective in formulating drugs or compounds that are high in lipid ratios using a pumping or forcing mechanism. In addition, the product regulates the internal locking of bioactive compounds into the liposome, which may enhance regulatory release activity[8] (Figure 7.1).

The word "nutraceutical" is a combination of nutrition and pharmaceutical.[9] Nutraceuticals are dietary components that play an important role in modifying and maintaining normal physiological functions in healthy human beings. They detoxify our body and restore healthy digestion and eating habits. Nutraceuticals can be classified into many types, mainly traditional and non-traditional. Substances with established nutritional values, herbs or botanical products, reagents derived from other sources, functional foods, probiotics and prebiotics, polyunsaturated fatty acids, antioxidant vitamins, polyphenols, and spices are among the other classifications.[9]

Traditional nutraceuticals are natural foods that contain many natural ingredients that provide benefits in addition to basic nutrition, such as the lycopene in tomatoes, the omega-3 fatty acids in salmon, the resveratrol in

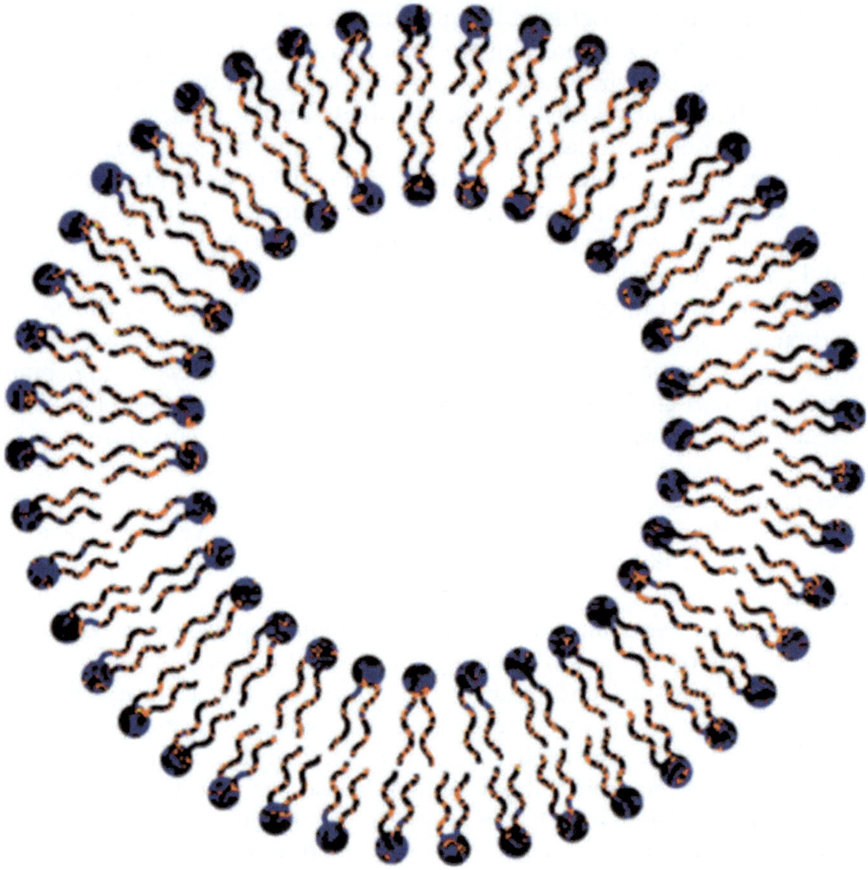

Figure 7.1 Structure of liposomes.

grapes, or the saponins in soybeans.[10] Food that enhances the nutritional value of non-traditional nutraceuticals by combining agricultural reproduction or nutrients and ingredients. Examples include β-carotene-enriched rice, soybeans, orange juice fortified with calcium, and cereals with added vitamins or minerals.[11]

Nutraceuticals are naturally occurring oral dietary components found in foods that are believed to have medical or health benefits.[12] Nutraceuticals as adjuvants can enhance the therapeutic effect of certain drugs, increasing the number of pathways, such as enhanced regeneration of blocked monoamines, thereby providing extraordinary neurobiological effects.[13] A nutraceutical has been shown to have physiological properties, protecting against chronic diseases. Bioactive ingredients and phytochemicals in nutraceuticals maintain or promote health, and they occur at the intersection of the pharmaceutical and food industries. Such substances can range from isolated nutrients to dietary supplements. They play a vital role in maintaining a healthy immune system, *i.e.*, low or excessive consumption can adversely affect health.[14]

Nutraceuticals are currently being studied as a medicine that works as both a preventive and a curative medicine. They provide nutritional and therapeutic benefits. In addition, they are natural and biologically compatible, so there are no side effects. This facilitates their ready acceptance as dietary supplements without special dosage and worries about long-term use. Nutraceuticals can be produced from natural resources, such as spices, fruits, vegetables, and plants. However, nutraceuticals are subject to environmental pressures that require encapsulation for long-term storage and adequate bioavailability. Although nutraceuticals are usually affected by environmental factors, their inclusion is a general recommendation, especially for therapeutic purposes.[15]

The term "nutraceutical" was coined from words "nutrition" and "pharmaceuticals" in 1989 by Stephen DeFelice.[9] In his opinion, a nutraceutical is often defined as a dietary component that offers encapsulation systems, which are excellent choices for preserving bioactive compounds and as additives in food applications. It facilitates the precise and controlled release of hydrophobic or lipophilic and hydrophilic or lipophobic molecules into the food preparation system.[16] Encapsulation techniques come in various forms, giving food processors the benefit of protecting sensitive food components from damage and exposure to environmental factors, preventing loss of nutritional and sensory characteristics and facilitating time release systems and fluid synthesis. This encapsulation procedure offers several advantages, such as high encapsulation efficiency, ability to load lipophilic and hydrophilic drugs together, biological compatibility, receptor-mediated site-specific targeting, and controlled release triggering.[17] Liposomal encapsulation has been shown to be the most appropriate form of nutraceuticals, especially when given in cancer therapy. Researchers have found that liposomal encapsulation is a prominent and reliable technique for enhancing the stability of nutraceuticals, targeting delivery, and enhancing bioavailability.[15]

7.1.1 Importance of Nutraceuticals in Life

The medical or health benefits of nutraceuticals include the prevention or treatment of diseases.[14] Compared to other therapeutic agents, the use of nutraceuticals as an attempt to achieve desirable treatment results with fewer side effects has been a great financial success. Nutraceuticals detoxify our body and restore healthy digestion and eating habits.[18] Nutraceuticals are marketed in concentrated forms, such as tablets, pills, powders, and tinctures, as a single substance or a combined preparation. Nutraceuticals are nutritious foods that contain a specific ingredient, such as omega-3 oils from spirulina, garlic, soy, or salmon. They are also known as medical diets, nutritional supplements, or dietary supplements.[9]

Nutraceuticals have enormous chances for growth and expansion in terms of health benefits.[19] Nutraceutical formulations are taken not as drugs but as food. Therefore, unlike pharmaceutical preparations, nutraceutical preparations need to meet more requirements. As nutrient supplements and functional foods are contained in nutraceuticals, the ingredients used in the formulation must be food grade. This limits the choice of researchers and reduces the scope of innovations in the nutraceutical domain. Therefore, choosing the right material to prepare the formulation is a major challenge.[20]

Nutraceuticals are products that are used as medicine rather than for nutrition. A nutraceutical product is defined as a substance that has physiological properties or protects against chronic diseases. Currently, there is considerable interest in nutraceuticals because of their nutritional safety and therapeutic effects.[21] Nutraceuticals are medical foods that play a role in maintaining and enhancing health, modifying the immune system, and thereby preventing and treating specific diseases. Thus, the nutraceutical sector can be envisioned as one of the missing blocks in an individual's health benefits.[22]

Nutraceuticals might be defined as substances that have physiological benefits and also provide protection against chronic diseases. It may be used to improve health, delay the aging process, prevent chronic diseases, increase life expectancy, or support the body's structure or function. Nowadays, there is considerable interest in nutraceuticals due to their nutritional, safety, and therapeutic effects. Nutraceuticals can be used to improve health, delay the aging process, prevent chronic diseases, increase life expectancy, or support the structure and function of the body.[23] Nutraceuticals are one type of nutritional supplement that can be used for health purposes other than nutrition.[24]

Most nutraceuticals have multiple therapeutic properties, and most of them have antioxidant activity capable of counteracting this diseased pathogenic condition. They are therefore considered healthy sources of health promotion, especially for the prevention of life-threatening diseases.[21]

Nutraceuticals are not drugs because they have not gone through an approval process. Nutraceuticals have no official meaning and are not a specific category of food. Most of the time, they are natural, whole foods that consumers have been eating for thousands of years. The importance of nutraceuticals to the human body is that they provide all the essential nutrients that a healthy diet should have. Too often, the daily rush and our diet led to unhealthy lifestyles. Proper nutrition administration provides a better quality of life, a healthier life, a better mood, confidence, better functioning, and a better social environment. Therefore, in the future, we will see the urgency of nutraceutical soups, nutraceutical processed meat, bread, and sausage. In addition, many of these foods may be genetically engineered. Compared with other therapeutic agents, the use of nutraceuticals as an attempt to achieve desirable treatment results with fewer side effects has been a great financial success. Nutraceuticals should contribute to the prevention of such diseases. The world's research strategy toward nutraceuticals should be to live healthier lives and improve quality of life in the future.[25]

Most nutraceuticals have many therapeutic benefits for various diseases. Nutraceuticals claim to have a physiological benefit or protect against various diseases, such as heart disease, anti-diabetics, anti-cancer agents, chronic inflammation, immune boosters, and disorders. Nutraceuticals play a key role in creating greater confidence in product quality and effectiveness, a better market for nutrient products, public awareness, and increased healthcare industry awareness; setting an autonomous agenda; and improving the health value of our food. They also help us live longer and avoid particular medical conditions. They also have a psychological benefit from doing something for oneself and are perceived to be more "natural" than traditional medicine and less likely to produce unpleasant side effects. They may provide food for populations with special needs.[19]

Nutraceuticals are intended to play an important role in future therapeutic development. Community health authorities consider nutrition prevention and treatment to be an excellent tool for maintaining good health and combating acute and chronic diseases caused by malnutrition, thereby promoting the most beneficial health, endurance, and quality of life. A class of nutraceuticals is emerging in clinical practice, but important pharmaceutical and clinical issues need to be addressed through prior research.[20]

According to the theory of nutraceutical and nutritional therapy, the goal is to use the effectiveness of such nutraceuticals to eliminate toxins from the body, avoid vitamin and mineral deficiencies, and restore healthy digestion and eating habits. Although many health problems require specific medications, nutritional therapy can prevent many other conditions, such as certain diseases and chronic diseases. These disorders may differ from chronic fatigue, energy loss, insomnia, and osteoarthritis, to backaches, skin complaints, and asthma. Comprehensive nutritional therapy always tries to use as many health-affecting factors as possible to improve the overall health condition of an individual. It depends not only on balanced dietary nutrients but also on enough exposure to sunlight, fresh air, pure water, proper exercise and rest, emotions, and self-control to build up healthy human immunity. The immune system is the basic and essential part of the body's defense system. Even in clinical settings, nutritional therapy is getting more and more attention for helping the recovery of various kinds of patients.[23]

7.2 Classification of Nutraceuticals

The term "nutraceuticals" itself is a broad category that includes many categories and subcategories. The contribution of nutraceuticals to public health is one of the most careful approaches in the healthcare system. Although the system claims to have advanced technologies for treating diseases, the nutritional approach is considered due to its power to treat diseases through diet. For the best explanation of the comparison, pharmaceuticals include drugs for the treatment of diseases, but nutraceuticals are prescribed to prevent diseases. There is a functional diet that incorporates vitamins, fats, proteins, and carbohydrates for healthy survival. However, those foods are called nutraceuticals when they help prevent and treat diseases in addition to meeting deficiency conditions. In recent years, dietary and lifestyle disorders have become major health concerns. With the availability of the state-of-the-art methodology for determining the nutritional quality of food, increased attention and awareness of healthcare issues need to be studied. It can be reviewed in various studies that nutraceuticals provide non-specific biological therapy for the treatment and prevention of various diseases.[26] To understand their applications, nutraceuticals need to be classified based on their intended use (Figure 7.2).

7.3 Traditional Nutraceuticals

This category includes foods that are not subject to voluntary modification. The ingredients are natural and have some potential to be actively involved in health benefits. Lycopene, a component of tomatoes, is an example of this category.[27]

7.3.1 Phytochemicals

Phytochemicals are the chemical components of plants that have different biological functions. These are reported to contain active ingredients that have an effect on the metabolism and biochemistry of organisms, thus providing health benefits.[28]

7.3.2 Herbals

The class includes herbs of medicinal value to be included in the treatment and prevention of diseases. Botanical products may contain fresh herbs or parts of dried leaves, fruits, stalks, seeds, roots, or any concentrated extract.[18]

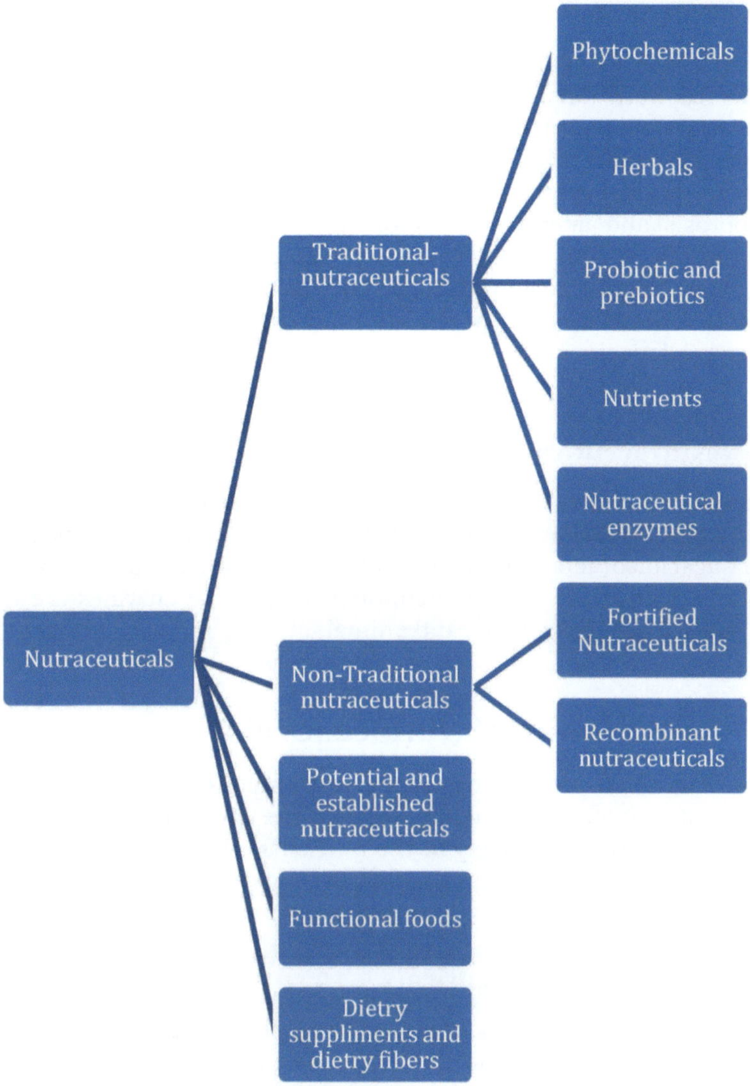

Figure 7.2 Classification of nutraceuticals.

7.3.3 Probiotics and Prebiotics

The probiotic category includes food components that are real-time micronutrients beneficial to health. Their function is to stay close to the gastrointestinal tract in specific places, and their survival leads to the elimination of germs. They are the fertilizing agents of probiotic bacteria in the large intestine. They are not affected by gastric pH and digestive acids.[29]

7.3.4 Nutrients

Nutrients such as vitamins, minerals, amino acids, and fatty acids have established nutritional activity. Most vegetables, grains, dairy products, fruits, meat, and poultry products contain vitamins that can help cure heart disease, stroke, cataracts, osteoporosis, diabetes, and cancer. Minerals found in plants, animals, and dairy products are useful for treating osteoporosis and anemia and improving strong bones, teeth, muscles, nerve impulses, and heart rhythm. Flaxseed and salmon contain the fatty acids omega-3 polyunsaturated fatty acids (PUFAs), which are strong regulators of inflammatory processes, maintain brain function, and reduce cholesterol deposits.[30]

7.3.5 Nutraceutical Enzymes

Enzymes are an integral part of life, without which our body would stop working. People suffering from health problems, such as hypoglycemia, blood sugar disorders, digestive problems, and obesity, can eliminate the symptoms with dietary enzyme supplements. These enzymes are derived from microorganisms, plants, and animals.[31]

7.4 Non-traditional Nutraceuticals

In non-traditional nutraceuticals, the nutritional content is increased by adding nutrients, and this component of the diet includes food components that improve the quality of the nutrients. An example of this species is rice rich in beta-carotene.[27]

7.4.1 Fortified Nutraceuticals

Strengthening of food components is the process of adding micronutrients (essential nutrients and vitamins) to the diet to increase the effectiveness and nutritional value. An example of this is milk containing cholecalciferol, which is used for vitamin D deficiency.[26]

7.4.2 Recombinant Nutraceuticals

Biotechnology and genetic engineering can be utilized to extract bioactive components using enzymatic or fermentation technology, which enhances foods, such as yoghurt and cheese. Gold kiwifruit is genetically modified for high levels of ascorbic acid, carotenoids, lutein, and zeaxanthin.[26]

7.5 Potential and Established Nutraceuticals

Potential nutraceuticals guarantee medicinal properties. These nutraceuticals became established drugs only after adequate data demonstration and clinical testing for their effectiveness and safety. All nutraceuticals are potential agents, although not all of them are established.[30]

7.6 Functional Foods

Functional foods are sources of absolutely essential nutrients that provide more than the amount needed for maintenance, growth, and development. In addition to basic nutrition, the term refers to a particular type of food or food component that carries evidence of providing health-promoting ingredients. The functional food class includes several subtypes, including cereals, legumes, and fermented foods. The potential for a functional diet, including whole grains, such as rice, corn, wheat, millet, sorghum, and buckwheat, has been identified in a number of ways to eliminate coronary heart disease, lower blood pressure, and reduce tumor incidence. Similar to grains, legumes are another subtype of functional food that is an important component of both traditional and modern diets. They are highly nutritious and rich in biologically valuable proteins, bioactive peptides, and amino acids, and include kidney beans, split beans, legumes, and soybeans. They have been shown to have deep antioxidant and protective effects against cardiovascular diseases and diabetes.[32]

7.7 Dietary Supplements and Dietary Fibers

A dietary supplement is a product that contains a supplement that is added as a remedy for deficiencies or diseases. The inclination toward dietary supplements has been multiplied many times over to improve health, fitness, and performance and bodybuilding, and as a tonic to delay aging. A dietary supplement is something that enhances food and its nutritional value. Vitamins and minerals are available in the diet as multiple ingredients or as single ingredients in the market. Dietary supplements, other than vitamins and minerals, include concentrates or ingredients of herbs, botanicals, amino acids, pure extract, glandular extract, and organ cells. It should not be used as a traditional or conventional food ingredient or as a regular item in food or diet. Dietary fiber and high-fiber products are of great interest, as they have significant health benefits.[26,27]

7.8 Lipid Carrier System: Nanoliposomes

Lipid- or phospholipid-based carrier systems are one of the most promising encapsulation technologies used in the rapidly growing field of nanotechnology. They include lipidic nanoemulsions, liposomes, nanoliposomes, cochleates, archaeosomes, solid lipid nanoparticles (SLNs), and nanostructured lipid carriers (NLCs).[33] Lipid-based nanoencapsulation systems have different advantages, including the ability to entrap substances with different solubility, the ability to produce natural ingredients on an industrial scale, and targetability. Lipid-based carriers can protect synthesized substances, especially in high-water-activity applications. They provide stability to water-soluble substances, especially in high-water-activity applications. Lipid-based carriers can encapsulate not only water-soluble substances but also lipid-soluble molecules, separately or simultaneously (Figure 7.3), providing a synergistic effect. Additionally, lipid-based nanocarriers can be targeted *in vivo* to the active site within the body *via* active (*e.g.* by the incorporation of antibodies) or passive (*e.g.*, targeting based on the particle size) mechanisms.[34]

Lipid-based carrier systems, such as liposomes, nanoliposomes, and proteoliposomes (*i.e.*, liposomes containing one or more proteins), have been shown to be highly adaptable to deliver biological, biochemical, and pharmacological targets. Targetability is a very useful property of liposomes and nanoliposomes. Bioactive molecules aim for their effect *in vitro* or *in vivo* to achieve a sufficient concentration of bioactive compounds at the target site for their optimal efficacy. The timely and targeted release enhances the effectiveness of bioactive materials, broadens their application range, and ensures optimal dosing, thereby improving

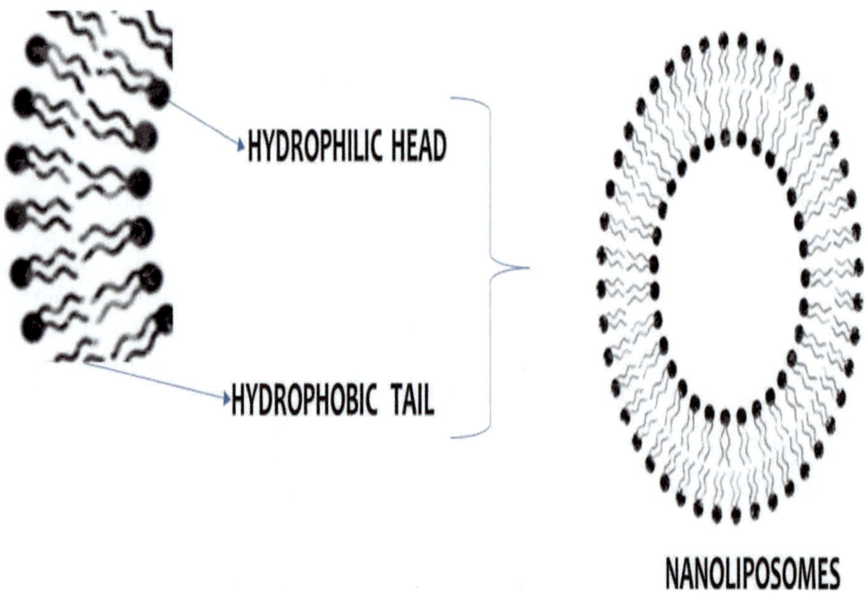

Figure 7.3 Simultaneous encapsulation of lipid-soluble molecules (hydrophobic), inside the lipid phase, and water-soluble compounds (hydrophilic), inside the aqueous phase, of a nanoliposome.

the cost-effectiveness of the product. Lipid vesicles mainly target active (*e.g.*, synthesis of antibodies) or inactive (*e.g.*, particle size-based targeting) systems.[35]

The rapidly evolving lipid-based nanoencapsulation bodies are ideal carriers for a variety of purposes. Other nanocarriers containing carbohydrate or protein building blocks are not suitable for full-scale processing due to the involvement of complex chemical and thermal processes that cannot be fully observed. However, lipid-based nanocarriers can be applied on an industrial scale and provide high entrapment efficiencies and low toxicity. Most bioactive ingredients are internally hydrophobic, including polyphenols, fatty acids, lipophilic vitamins, nutraceuticals, aromas, and preservatives. For the accommodation of these lipophilic agents, oil is usually prepared in water supply. In addition, the availability of digestible lipids facilitates the intestinal absorption of bioactives because they are able to absorb and carry hydrophobic molecules. Therefore, the abovementioned advantages hold lipid-based nanocarriers as robust and flexible delivery vectors.[36]

Nanostructured lipid carriers are used for further applications in various pharmaceutical and food particles. Lipid-based nanocarriers in the food sector include SLNs and NLCs.[7] SLNs are derived from oil–water emulsions with some modifications, using solid–liquid instead of lipid–lipid (oil) to maintain solidity at body temperature. The emergence and advancement of SLNs represent a major breakthrough in the field of nanoscience. SLNs, compared to conventional nanoparticles (emulsions and polymeric nanoparticles), provide high stability, biocompatibility, targeted delivery, sustainable release of bioactive molecules, bioactive components, and affordable and simple scale-up.[37]

Therefore, to eliminate the shortcomings of SLNs, in the 1990s, NLCs were introduced as next-generation SLNs. Unlike the structure of SLNs, NLCs have enhanced bioactive retention capabilities with controlled levels of solid and liquid lipids as well as enhanced levels of released attributes. The melting point of the final structure decreases compared to pure solid lipids, and the final particles do not retain their solid structure at ambient and body temperatures.[37] NLCs have a lot of amorphous crystals in their structures, which gives a lot of space to the bioactive molecules and reduces the ejection rate of the bioactive compounds. The amount of liquid lipids in the matrix plays a dynamic role in inhibiting the crystallization process by combining spatially different compounds. This procedure gives tangible differences in the particle matrix. The three classes of NLCs are the incomplete structured type, the amorphous type, and the multiple types. All models allow for compact particle scattering (>30–95%) and increased payload concentrations.[36]

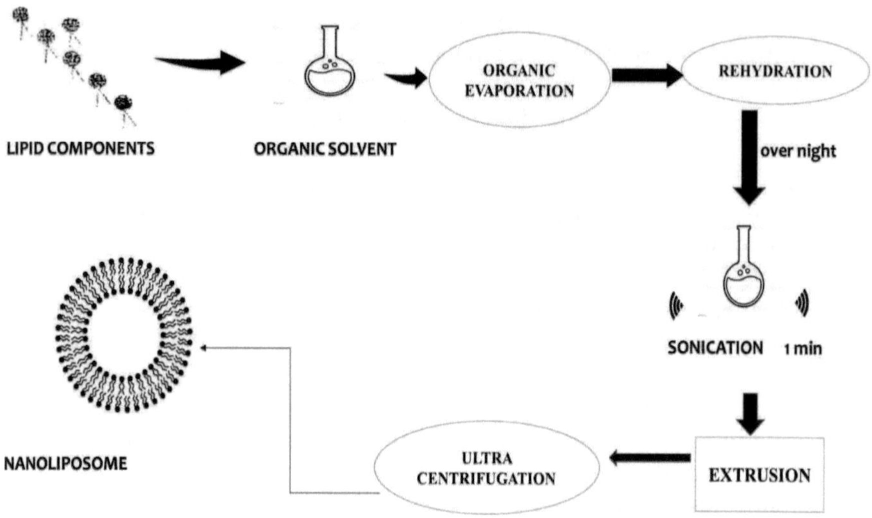

Figure 7.4 Schematic representation of nanoliposome preparation.

7.9 Method of Preparation of Nanoliposomes

During liposome preparation (Figure 7.4), phospholipid vesicles were generated at concentrations ranging from 100 to 250 mg mL^{-1}. Sodium chloride includes 0.2 mg of sodium acid and 10 mg of ascorbic acid per milliliter of Q water. Ascorbic acid has been used as a model molecule for encapsulation efficiency. Solutions are integrated with a high-speed blender into a one-pass homogenizer after 1 min at 8000 rpm at a pressure of 69 MPa.

Different levels of phospholipids and sterol supplements were tested, and the results showed a total concentration of 150 mg mL^{-1} lipids (approximately up to 200 mM for lecithin liposomes only). Two different commercial sterol preparations were tested (Archer Daniels Midland Company, Decatur, IL, USA): phosphate, distilled monoglycerides, silicon dioxide, and antioxidants, and an oil-soluble preparation containing Cardio-Aid-S and plant sterol esters and antioxidants. The amount of the total plant sterol esters is >90 g 100 g^{-1} (depending on the manufacturer's specifications), which is a mixture of B-cytosterol, stigmasterol, and composterol. It has previously been shown that these plant sterols are well incorporated into soybean phosphatidylcholine, forming large unilamellar vessels.[38]

Three different methods of mixing were tested in this study. Plant sterols were added to the phospholipid mixture in two different phospholipid/plant sterol preparation ratios: 14/1 and 13/2 (g/g) (approximately 1 : 0.1 and 1 : 0.3 lecithin ratios), up to a total of 150 mg mL^{-1} of suspension. Before homogenization, the mixtures were prepared by simply mixing at room temperature, by mixing with heat, or by heating in the presence of ethanol before dissolving the plant sterols. In short, for the "basic" mixing method, the plant sterols were mixed with phospholipids and imidazole buffer containing ascorbic acid at pH 7.0 and then stirred for 45 min before consolidation. An alternative method of mixing plant sterols and phospholipids was used, in which the mixture was dispersed in imidazole buffer containing ascorbic acid at pH 7, then heated to 45 °C for 45 min, and cooled for 45 min before consolidation. In the case of the "ethanol" method, the phospholipids and plant sterols are mixed with about 5 mL of ethanol at room temperature and then heated to 60 °C for 45 min. Ethanol was then evaporated at 80 °C, and the paste was hydrated in a buffer containing 10 mg mL^{-1} ascorbic acid at pH 7, as described in the basic method. Cholesterol-containing liposomes were also prepared using the ethanol and heat methods.[38]

7.9.1 Microfluidization

Microfluidization is based on microfusion/homogenization/microemulsification and is designed for the organization of liposomes. Encapsulation efficiencies can reach up to 75%. The aqueous dispersion of liposomes often tends to merge or coalesce, leading to hydrolysis or oxidation. First, an appropriate suspension medium and the components of the nanoliposomes (depending on the planned application) were selected. In the suspension medium, after the correct mixing of the sample by continuous stirring and setting the correct procedure parameters, such as the flow rate, air intake, operating pressure, the size of the interaction chamber as we achieved in the previous step of suspension in a reservoir, and the number of passes after the procedure parameters were set, the air valve was opened, leading to the liquid dispersal, which flowed through a filter into the interaction chamber. Subsequently, the obtained suspension can be recycled through the equipment. Before the next passage, the suspension had to be cooled as the temperature in the interaction chamber increased, in an atmosphere containing nitrogen or argon for 1 h, which eventually led to the annealing and stability of the sample.[39]

7.9.2 Extrusion

In this technique, a diluted suspension of liposomes (containing liquid crystalline lipids) can pass continuously through filters, which can reduce the size of the pores. In addition to the pressure that can be applied manually, this method limits the back pressure that the syringe and filter holder can withstand.[40] First, a sample containing micrometrical liposomes (MLVs) was prepared, one or two stacked polycarbonate filters modified for the next stages of the liposome were transferred to a hot spot on the stand or heating block, and then a thermometer was inserted. The phase of liposome suspension lipids should be above the transition temperature. The hot plate should be changed according to this principle and the buffer was filled completely in the syringe and then passed through the extractor.

The goal is to reduce the dead volume. In the next step, the liposome suspension removes the buffer that was inserted into one of the extractor's syringes (donor syringe). Then, the syringe was carefully placed on one end of the device, applying a gentle twist. In the next step, the second syringe (receiver syringe) was placed on the end of the extruder, making sure that the receiver syringe plunger was set to zero before starting the next step. The stainless-steel hexagonal nut was added, so that two opposite ends were allowed to fall vertically. To ensure good thermal contact between the syringes and the heating block, the swing-arm clips should be exposed to a cosmetic laser at approximately 5 J for 10 min, allowing the suspension of nanoliposomes to reach the temperature of the heating block. The next step involves gently pushing the plunger of the alternate syringe to transfer the suspension to the original syringe. To obtain nanoliposomes with the desired properties, the extrusion process must be repeated. The number of passages through the filters is always odd. The minimum number of passes after the extraction is seven. The extractor must be removed from the heating block, and then the filled syringe must be removed from the extruder. For example, the obtained nanoliposomes should be injected into a clean bottle, and in the next step, the extruder parts can be washed first with ethanol and then with purified water. The final stage involves keeping the nanoliposomes in an inert atmosphere, such as nitrogen or argon for 1 h at temperatures above the transition stage, which ultimately leads to sample analysis and stability.[34]

7.9.3 Sonication Technique

Ultrasonic techniques can be used to form liposomes. The basic materials required for liposome formation are amphiphilic molecules derived from or based on biological membrane lipids. For the configuration of SUVs, lipid dispersion is gently sonicated – for example, handheld ultrasonic device UP50H (50 W and 30 kHz) and VialTweeter or UTR200 ultrasonic reactor – in an ice bath. The duration of such ultrasonic handling will be roughly doubled, ranging from 5 to 15 min. Sonication of multilamellar vesicle liposomes is a different technique for making small unilamellar vesicles. Sonication of MLVs occurs at room temperature. In addition, ultrafiltration is used to reduce or eliminate pyrogens. The solution was transferred to a flask (pear-shaped or round bottom) and placed in a rotary evaporator. The aim is to remove the solvents at room temperature throughout the phase under negative pressure. This procedure is carried out until a thin layer of dry lipid components is obtained from the flask. The residues of the biological solvents were removed or flushed with a vacuum pump (*e.g.*, nitrogen or argon). Glass beads (*e.g.*, 500 m in diameter) were added to a flask containing dry lipids, and then, a suitable aqueous phase, such as distilled water or buffer, was added. Dry lipids were stirred for 1–5 min, or the flask was shaken by hand to disperse them into the hydration liquid. At this stage, microscopic (MLV)-type liposomes were obtained and transferred to the MLV-containing flask probe (tip) sonicator. For targeted investigation sonication, the tip of the sonicator was placed in the MLV flask, turning the sample on for 20 s, at 20 s intervals, for a total of 10–15 min. Once the sonication was properly executed, we got nanoliposomes as the final product.[1]

7.9.4 Heating Methods

Most nanoliposome-manufacturing techniques use either toxic solutions (*e.g.*, chloroform, methanol, diethyl ether, or acetone) or high-shear-force procedures. The residues of these toxins are thought to persist in the final liposome or nanoliposome preparation, causing toxicity and influencing the stability of the lipid vesicles.[35,41] Although there are methods to reduce the concentration of the residual solvents in liposomes (*e.g.*, gel filtration, dialysis, and vacuum), these are difficult and time-consuming procedures in practice. In addition, the amount of these solvents in the final formulations should be evaluated to ensure the clinical suitability of the products. Therefore, it is best to avoid the utilization of these solvents in the manufacture of nanoliposomes, as this will reduce the time and cost of preparation, especially on industrial scales. There have been reports on the use of high pressure or high shear forces during nanoliposome formation (*e.g.*, as occurring during microfluidization) and the adverse effects these procedures have on the structure of the coated material. In the absence of toxic solvents, these barriers can be overcome by alternative preparation methods, such as heating, which can be used to prepare liposomes and nanoliposomes (in addition to some other carrier systems) using a single apparatus.[39]

7.10 Liposome-encapsulated Nutraceuticals

Liposomes have been used and applied in many fields in the pharmaceutical industry. Liposomes are spherical in shape (15–1000 nm), made up of a bilayer of phospholipids, and within the bilayer, the hydrophobic tails of the phospholipid groups face each other, while the hydrophilic heads face the inner core and the outer border of the liposome. This structure allows the hydrophilic material to be synthesized within the nucleus of the liposome, while the hydrophobic material is split into two layers, one of which is partitioned within the bilayer. The ability to carry hydrophilic and lipophilic components, in addition to their biocompatibility and biodegradability, is an important factor in the use of liposomes for pharmaceutical applications.[42] Liposomes increase the efficacy and therapeutic index of the carrying drugs, increase the stability and reduce the toxicity of encapsulated agents, and improve the pharmacokinetic effects (reduced elimination and increased circulation lifetimes). Their use is essential to increase the bioavailability of natural compounds.[43]

7.10.1 Nanoliposome-encapsulated Vitamins

Liposomes are microscopic, hollow, spherical vesicles made up of a lipid bilayer. When loaded with pharmaceuticals and dietary supplements, liposomes are a very effective method of drug delivery. When ingested, the pharmacokinetic benefits of liposomal intestinal absorption outweigh those of the normal absorption of the coated drug. That is, the delivery of a drug or supplement with a slow or controlled absorption pattern, such as vitamin C, may be accelerated when encapsulated into a liposome. With this concept, the purpose of the current investigation was to compare the circulatory concentrations of vitamin C contained within liposomes after the oral intake of vitamin C with that after the intravenous administration of an equivalent dose. The intake of vitamin C encapsulated within liposomes results in greater circulatory density than that of unencapsulated vitamin C.[44] Given the increased bioavailability, it is hypothesized that liposomal vitamin C may significantly improve oxidative stress caused by ischemia–reperfusion compared with unencapsulated vitamin C.[45]

Vitamin D3 nanoliposomes were successfully fabricated using thin-film hydration–sonication. Adding a certain amount of cholesterol improved the electrostatic stability of the empty liposomes but did not significantly change the zeta potential when the vitamin was incorporated into the liposomes. The effect of the nanoliposome is shown under dark conditions at 4 °C. The effects of temperature and light show that stable gel-phase lipids provide the greatest protection against vitamin D3 degradation. These results suggest that nanoliposomes containing bioactive materials may have the potential to improve the shelf life of nutraceuticals and drug delivery systems.[46]

Vitamin E as a nutraceutical is an essential vitamin for the human body, which provides health and stability to the cells and cell membranes. The encapsulation of vitamin E is necessary for increasing its solubility in food and cell membranes. Several delivery systems have been developed for vitamin E encapsulation, depending on the price, ease of use, and biocompatibility of liposomes. Nanoliposomes have been used for the encapsulation of different food bioactives and nutraceutical compounds. Vitamin E encapsulation in liposomes was carried out by the destruction method. The behavior of vitamin E is similar to other hydrophobic stabilizers, so it is used in liposome formulation. In addition, some stabilizers, such as gamma oryzanol (GO), improve encapsulation efficiency and stability.[47]

The vitamin A-encapsulated nanoliposome formulation enhances the aqueous dispensability and encapsulation efficiency of vitamin A-loaded nanoliposomes. It is prepared by the thin-layer hydration method.[48]

7.10.2 Nanoliposome-encapsulated Minerals

Deficiencies of certain essential minerals can adversely affect human health. However, fortifying foods with certain minerals (*e.g.*, calcium, magnesium, and iron) is usually not possible because of drawbacks, such as complexity when combined with other foods and increased risk of oxidation if added directly to the diet.[49] Today, iron deficiency is recognized as one of the major nutritional deficiencies worldwide. This condition is mainly due to iron deficiency in the diet, lack of its bioavailability, or both.[50] Iron deficiency in the blood should not be ignored, as it can cause anemia when the hemoglobin level in the blood is below normal. Due to the contradiction between iron and calcium in milk, iron deficiency in milk can lead to iron deficiency in young children.[51] Therefore, it is necessary to protect the milk from iron–calcium resistance, fat oxidation consequences, and metallic flavor and enrich it with iron. To achieve this, supplementation with iron milk coated in safe and non-toxic carriers is recommended. To this end, the highly soluble form of iron (*i.e.*, ferrous sulfate) was chosen because of its cost-effectiveness and high bioavailability. Generally, iron salts are divided into three groups based on their solubility: (1) water-soluble, (2) acid-soluble, and (3) poorly soluble salts. Water-soluble iron salts (*e.g.*, ferrous sulfate and ferric chloride) have the highest bioavailability and the lowest cost. Studies have shown that water-soluble ferrous salts have a higher bioavailability than ferric salts. However, water-soluble iron salts can cause undesirable changes in fortified food products, especially lipid oxidation, as well as changes in odor, color, and texture. Iron sulfate changes color and taste rapidly when added directly to food products. However, yogurt containing ferrous sulfate did not change color after 2 weeks. In iron-rich nanoliposome formulations, the use of antioxidants, such as vitamin E or ascorbic acid, in the structure of nanoliposomes is a common method of protecting ferrous ions from oxidation.[52]

7.10.3 Nanoliposome-encapsulated Plant Extract

Phenolic compounds are sensitive to environmental factors. Encapsulation protects food-sensitive components from temperature, humidity, and pH processes. Encapsulation, on the other hand, eliminates the extreme and unnecessary benefits of compounds produced in foods.[53] Smaller-sized nanocrystals are more stable due to their higher resistance to gravity (Brownian motion).[54] The liposome structure consists of two hydrophilic and hydrophobic segments; the hydrophilic compounds are enclosed within the aqueous phase of the liposome, while the hydrophobic compounds are encapsulated between the two phospholipid layers, thus acting as a reservoir of phospholipid bilayer extract.[55]

7.11 Encapsulation of Enzymes Within Liposomes

Enzymes are biological macromolecules that can accelerate various biochemical reactions. These biocatalysts play an important role in the establishment and development of industrial bioprocesses. Because enzymes are mostly proteins, they are sensitive to various physiological changes that can be triggered by certain factors, such as heat, pH, ionic strength, and other denaturation agents.[56] In 1970, Sessa and Weissmann conducted the first investigation into the encapsulation of enzymes by liposomes. The purposes of using immobilized enzymes include their application at a wide range of temperatures and pH values to intensify their catalytic activity and achieve higher productivity.[57] Enzymes can be incorporated into the core or membranes of liposomes based on their hydrophilic or hydrophobic properties so that a layer of the liposome can protect them from environmental stresses. Several studies have demonstrated the ability of liposomal carriers to load various enzymes. Most of the enzymes encapsulated in food-relevant liposomal carriers belong to the oxidoreductase and hydrolase classes.[58] Due to their food compatibility, high efficiency, and applications in the food industry, nanoliposomal enzymes can easily be used in food formulations due to their high efficiency and applications in the dairy sector, especially in cheese production.[46]

7.12 Pre-clinical and Clinical Study of Liposome-encapsulated Nutraceuticals

The oral administration of nutraceutical encapsulation has evolved into nanocarrier-based platforms as a promising drug delivery system to overcome pharmacokinetic limitations.[59] In recent years, a wide variety of nanocarriers, including polymeric nanoparticles, solid lipid nanoparticles, polymeric conjugates, liposomes, micelles, and carbon nanotubes, have been explored to modulate pharmacokinetics. As chitosan-based nanocarriers have the ability to cross tight junctions *via* the paracellular route and modulate P-gp in the epithelial cells, several mechanisms have been proposed to increase the oral bioavailability of these nanocarriers. Polymeric nanoparticles made of poly-lactic-*co*-glycolic acid (PLGA) are absorbed by special Peyer's patches (M-cells). Lipid-based formulations can modulate P-gp and CYP450, increase intracellular concentration, and stimulate lipoprotein/chylomicron production. In addition, the encapsulation of nutraceuticals into these systems protects against adverse gastrointestinal (GI) interactions. For example, a recent study showed that the encapsulation of epigallocatechin gallate (EGCG) in nanoliposomes developed using phospholipids, cholesterol, and Tween 80 reduced its absorption in stimulated intestinal fluids by approximately 10-fold.[23] Nutraceuticals, such as vitamin E and fatty acids, are also used as adjuvants in the manufacture of nanocarriers and play a crucial role in increasing the bioavailability of anti-cancer drugs. For example, D-α-tocopheryl polyethylene glycol 1000 succinate (Vitamin E TPGS), a non-ionic surfactant, inhibits P-glycoprotein *via* ATPase inhibition and thus modulates the pharmacokinetics of the administered P-gp substrate.[60]

These nanocarriers exert an "enhanced permeation and retention (EPR)" effect through passive and active targeting. Due to the leaky vasculature of tumor endothelial cells, these nanocarriers can effectively penetrate into tumors through "passive targeting". Furthermore, these nanocarriers have the ability to control the agent targeting a specific site by attaching appropriate ligands to the surface of nanocarriers through active targeting. A wide range of ligands, such as hyaluronic acid, folic acid, and arginylglycylaspartic acid (RGD) peptides, have been used to deliver drugs to the tumor site due to their excessive expression in cancer cells. These nanocarriers can be classified into three categories based on the excipients used to make them: polymer-based nanocarriers, lipid-based nanocarriers, and inorganic nanocarriers.[61]

As an alternative approach to the current clinical management of neurodegenerative diseases, derivatives of various natural products are useful in patient care. These include food additives, herbal compositions, and nutrients found in abundance in nature. Most of these extracts contain antioxidants, such as α-tocopherol, ascorbic acid, reactive oxygen agents, transition metals, chelating agents, enzymes, and other polyphenols, which act as receptor activators, influencing the potential genetic and epigenetic mechanisms in neurodegenerative disorders. In general, nutraceuticals are both nutritious and medicinal.[62] Nutraceuticals are defined as a substance that protects against physical or chronic diseases. Many nutrients and their components exhibit healing effects in their natural environment. For example, flaxseed and salmon contain omega-3 polyunsaturated fatty acids, which are effective regulators of the inflammatory response, brain function, and cholesterol deposition. The tannin (*Lavandula angustifolia*) contained in lavender has a therapeutic effect on depression, blood pressure, and stress. Tannin also has stimulant properties that are good for the nervous system. Tannin is also good for hysteria and paralysis. In human experiments, aromatherapy with lavender and rosemary has been shown to be effective in non-pharmacological therapy of dementia and excellent cognitive function due to their free radical scavenging activity in Alzheimer's patients.[63]

The encapsulation of nutraceuticals into nanocarrier-based platforms (nano-chemo-prevention) has emerged as a promising drug delivery system for overcoming the pharmacokinetic limitations of oral administration of nutraceuticals.[64] In recent years, a wide variety of nutraceuticals on nanocarriers, including polymeric nanoparticles, solid lipid nanoparticles, polymeric conjugates, liposomes, micelles, and carbon nanotubes, have been explored to modulate pharmacokinetics. As chitosan-based nanocarriers have the ability to cross the tight junctions *via* the paracellular route and modulate the existing P-gp in epithelial cells, several mechanisms have been suggested to increase the oral bioavailability of the nanocarriers. Polymeric nanoparticles made of PLGA are absorbed by special Peyer's patches (M-cells). Lipid-based formulations have the ability to modulate P-gp and CYP450, increase intracellular concentrations, and stimulate lipoprotein or chylomicron production. In addition, the encapsulation of nutraceuticals into these systems protects against adverse GI interventions. For example, a recent study demonstrates that the encapsulation of EGCG in nanoliposomes developed using phospholipids, cholesterol, and Tween 80 reduced its degradation in simulated intestinal fluids by approximately 10-fold.[65] Nutraceuticals, such as vitamin E and fatty acids, are also used in the manufacture of nanocarriers and play a crucial role in enhancing the bioavailability of anticancer drugs. For example, the use of D-α-tocopheryl polyethylene glycol 1000 succinate (Vitamin E TPGS), a non-ionic surfactant, inhibits P-glycoprotein *via* ATPase inhibition and thus modulates the pharmacokinetics of the administered P-gp substrate.[61]

Scyllo-inositol and endogenous inositol inhibit the aggregation of the β-amyloid stereoisomer. They bind to the amyloid-beta peptide (Aβ) and prevent Aβ aggregation and fibrillation. The neuroprotective effects of turmeric and curcumin have been well documented. Resveratrol plays an important role in res-amyloid-induced memory disorder by neuronal inflammation and apoptosis by cyclic adenosine monophosphate (CAMP) reaction elemental-binding (CREB) protein signaling. Although the chemistry of some natural nutraceuticals is well documented, the study of nutraceutical compositions presents some challenges to the medicinal chemistry and, therefore, their therapeutic efficacy in relation to clinical trials. Nutraceuticals have various beneficial effects and are therefore used to treat cancer, inflammation, hypertension, heart disease, atherosclerosis, obesity, diabetes, and neurodegenerative disorders. Many nutrients, such as gallic acid, caffeine, quercetin, and curcumin, exhibit anti-aging and antioxidant activity. In addition, nutraceuticals exhibit anti-hypercholesterolemic and anti-inflammatory properties, which increase cholinergic production and inhibit acetylcholinesterase.[66]

Fish oils, containing polyunsaturated fatty acid methyl esters, reduce the risk of heart diseases and improve brain function. Nutraceuticals inhibit neurodegeneration by altering cellular and molecular pathways and by incorrect protein return. Despite their benefits, nutraceuticals have not been successful in human therapy due to poor absorption, low bioavailability, and limited access to the blood–brain barrier (BBB). Poor absorption rate is related to the laboratory characteristics and insolubility of active ingredients. The interaction of nutraceuticals with other chemicals or drugs can interfere with the treatment results or produce undesirable physical effects. Low solubility and limited accessibility of bioactive compounds in the gastrointestinal tract likely affect their stability and bioavailability. For example, in the case of curcumin, oral bioavailability is low due to decreased serum levels, partial tissue distribution, first-pass metabolism, and short lifespan, influencing potential genetic and epigenetic mechanisms in neurodegenerative disorders. At the same time, coenzyme Q10 exhibits low aqueous solubility and a slow dissolution rate in gastrointestinal fluids. In the case of large movements, tissue permeability, P-glycoprotein excretion, and active transport are challenged. Therefore, it is important to develop methods to increase the bioavailability and distribution of nutraceuticals at the target site with appropriate approaches.[29]

Although many approaches, including chemical modifications and liposome-medicated delivery, have been tried with reasonable success, nanoparticles in conjunction with nutraceuticals promise to increase drug delivery throughout the blood–brain barrier. However, while many nano-formulations are expected to improve bioavailability and increase accessibility, nutraceuticals are a late research topic in cancer treatment. As nutraceuticals are generally subject to environmental factors, their inclusion is a general recommendation, especially for therapeutic purposes. Liposomal encapsulation has proven to be the most suitable form of nutraceuticals, especially when given in cancer therapy. The researchers agreed that liposomal encapsulation is a leading and reliable technology for improved stability of nutraceuticals, targeted delivery, and improved bioavailability. The key nutrients in liposomes, including curcumin, resveratrol, fisetin, crocin, and sterols, have anticancer potential. The scientific consensus was clear about the ability of the nutrients contained as liposomes to regulate and inhibit the growth of cancer cells. Nutraceutical liposomes can act as curative and preventive drugs. This is crucial in gaining more scientific consensus, especially due to the diversity of cancer types, the new types of cancers identified, their complexities, the variable needs, and the sustainability of the targeted delivery. Nutraceuticals containing liposomes can significantly help control cancer and its mitigation, and there is evidence that it is an alternative to conventional chemotherapy.[15]

Anti-cancer nutraceuticals loaded into nanoparticles show remarkable results, showing improved solubility, absorption, bioavailability, and anticancer potential in comparison to nutraceuticals used alone. However, it should be noted that until recent years, relatively few studies effectively tested these nanostructures on real food systems.[67] Probiotic cells in their encapsulated form are mainly used for the commercialization of nutraceutical products. However, the need to develop new foods as viable carriers for these microorganisms has been widely discussed.[68] The incorporation of probiotics into food products has some challenges, such as the survival of living cells during processing and storage and undesirable interactions with the food matrix. Thus, probiotic encapsulation aims to increase their stability and function and provides a controlled release to effectively adhere and colonize the intestine. In this sense, polymers used for microencapsulation should be able to protect cells from stomach acidity and release them from neutralization in the alkaline conditions of the small intestine.[69] Therefore, to ensure that microcapsules maintain the viability of probiotics, it is essential to take into account certain physical and chemical properties, such as coating material, the type of culture, the number of initial cells, the size of particle size, and water solubility. Although different techniques and wall materials have been reported for efficient probiotic encapsulation, maintaining the viability of probiotics is still a major challenge, and further research on this topic is necessary.[70]

7.13 Conclusion and Future Perspectives

Increasing studies are being conducted on the nanoliposomal encapsulation of technically and biologically active compounds with the aim of incorporating them into nutrition and food products. This will increase the bioavailability of such compounds to nanoliposomes, increase their solubility and dispersal in food systems, and stabilize them against stress in food processing and digestion. The protection and controlled sustainable release of sensitive bioactives enable manufacturers to prepare high-quality, renewable, and safe product formulations, functional foods, and nutraceuticals. Future food or nutraceutical formulations of nanoliposome scattering should focus on one or more step-by-step activities in liposome formation and the possibility of large-scale and continuous production processes without the use of solvents or detergents. The clean labels that consumers demand the most can be further strengthened by emphasizing bioactive ingredients (*e.g.*, sphingolipids and phospholipids). Microfluidization and especially high-pressure integration will help to provide the infrastructure, large capacity, and continuous production potential that are widely available in the food processing industry. In addition, innovative and sustainable sources of liposomes are currently being suggested. In the future, more variants of liposome-forming lipids may be expected to remain commercially available at a reasonable cost. Response surface methodology (RSM)-based process optimization methods or similar technologies may be increasingly used to enhance the effectiveness and reproducibility of liposome production systems. Spray-dried liposome systems or proliposomes may be more prevalent in active food production. If this approach can reduce the number of steps of bioactive incorporation, reduce manufacturer weight, and increase the profitability of innovative functional foods, this may be a step in this direction. Especially from the pharmaceutical field, there is a lot of information about the interactions of liposomes with active cells or organisms. We hope that in the near future, there will be an increase in investigations into cellular adaptations and interactions of bioactive-bearing nanoliposomes in cell culture environments or *in vivo* investigations whether such liposomes were designed for analysis and research development or integrated into models or real-world food systems. Pharmaceutical advances, surface modification to enhance targeted delivery, formation of decorated liposomes, prevention of mucus entrapment, and better stability during storage and digestion need to be properly addressed in the food and nutraceutical areas. Similarly, the importance of phytosomes can be safely increased, while the bioavailability of active compounds from phytosomes needs to be closely investigated.

Nutraceuticals are intended to play important roles in future therapeutic developments. Community health authorities consider the use and treatment of nutraceuticals as powerful tools for maintaining good health and acting against acute and chronic diseases caused by malnutrition, thereby promoting the most beneficial health, endurance, and quality of life. Liposome technology has many applications in the pharmaceutical and cosmetic industries. Although there is a lot of information on liposome production and behavior, most of it is based on highly purified phospholipids, food-unsafe production methods, and highly resource-intensive technologies. However, to extend the utilization of liposomes to food applications, future research will have to focus on the production of lipid vesicles through safe and up-scalable methods and with less expensive ingredients. Nanoencapsulation technologies for food have been studied over the past decade for increased bioavailability and the effectiveness of the bioactive compound in treating diseases or maintaining the well-being of the human population, as well as the stability and storage of food products due to their antioxidant and anti-microbiological properties. Bioactive behavior masks the unpleasant taste or odor of certain harsh compounds. Some of those products have already entered the market and are poised to make a big impact in the coming years. Clinical trials, basic health claims, targeted population based on geographical conditions, and the cost and affordability of nutraceutical products all influence economic and market growth.

Declaration of Interest

The authors state no conflict of interest and have received no payment in preparation of this manuscript.

Acknowledgements

We would like to express our special thanks of gratitude to ADSO Naturals Pvt. Ltd. who gave us the golden opportunity to do this review, which also helped us in doing a lot of research and we came to know about many new things. We are really thankful to all of them.

References

1. Z. Fakhravar, P. Ebrahimnejad, H. Daraee and A. Akbarzadeh, Nanoliposomes: Synthesis Methods and Applications in Cosmetics, *J. Cosmet. Laser Ther.*, 2016, **18**(3), 174–181.
2. T. Subramani and H. Ganapathyswamy, An Overview of Liposomal Nano-Encapsulation Techniques and Its Applications in Food and Nutraceutical, *J. Food Sci. Technol.*, 2020, **57**(10), 3545–3555.
3. R. Genç, M. Ortiz and C. K. O'Sullivan, Curvature-Tuned Preparation of Nanoliposomes, *Langmuir*, 2009, **25**(21), 12604–12613.
4. C. I. Price, J. W. Horton and C. R. Baxter, Topical Liposomal Delivery of Antibiotics in Soft Tissue Infection, *J. Surg. Res.*, 1990, **49**(2), 174–178.
5. D. Zucker, A. V. Andriyanov, A. Steiner, U. Raviv and Y. Barenholz, Characterization of PEGylated Nanoliposomes Co-Remotely Loaded with Topotecan and Vincristine: Relating Structure and Pharmacokinetics to Therapeutic Efficacy, *J. Controlled Release*, 2012, **160**(2), 281–289.
6. H. P. Klug and L. E. Alexander, in *X-Ray Diffraction Procedures for Polycrystalline and Amorphous Materials*, John Wiley & Sons, New York, 1974, p. 960.
7. J. Y. Fang, C. L. Fang, C. H. Liu and Y. H. Su, .Lipid Nanoparticles as Vehicles for Topical Psoralen Delivery: Solid Lipid Nanoparticles (SLN) versus Nanostructured Lipid Carriers (NLC), *Eur. J. Pharm. Biopharm.*, 2008, **70**(2), 633–640.
8. K. A. Edwards and A. J. Baeumner, Analysis of Liposomes, *Talanta*, 2006, **68**(5), 1432–1441.
9. K. A. Gaidhani, M. Harwalkar and P. S. Nirgude, World Journal of Pharmaceutical Research Seed Extracts, *World J. Pharm. Res.*, 2014, **3**(3), 5041–5048.
10. D. Paolino, A. Mancuso, M. C. Cristiano, F. Froiio, N. Lammari, C. Celia and M. Fresta, Nanonutraceuticals: The New Frontier of Supplementary Food, *Nanomaterials*, 2021, **11**(3), 1–20.
11. V. Shekhar, A. K. Jha and J. S. Dangi, in *Nutraceuticals: A re-emerging health aid*, Proceedings of the International Conference on Food, Biological and Medical Sciences (FBMS-2014), Bangkok, Thailand, 2014, pp. 28–29.
12. M. Levine, C. Conry-Cantilena, Y. Wang, R. W. Welch, P. W. Washko, K. R. Dhariwal, J. B. Park, A. Lazarev, J. F. Graumlich, J. King and L. R. Cantilena, Vitamin C Pharmacokinetics in Healthy Volunteers: Evidence for a Recommended Dietary Allowance, *Proc. Natl. Acad. Sci. U. S. A.*, 1996, **93**(8), 3704–3709.
13. K. P. van der Burg, L. Cribb, J. Firth, D. Karmacoska and J. Sarris, Nutrient and Genetic Biomarkers of Nutraceutical Treatment Response in Mood and Psychotic Disorders: A Systematic Review, *Nutr. Neurosci.*, 2021, **24**(4), 279–295.
14. E. K. Kalra, in *Nutraceutical-Definition and Introduction*, 2003, vol. **5**.

15. D. Yadav, K. Sandeep, D. Pandey and R. K. Dutta, Liposomes for Drug Delivery, *J. Biotechnol. Biomater.*, 2017, **07**(4), 1–8.
16. T. Toniazzo, I. F. Berbel, S. Cho, C. S. Fávaro-Trindade, I. C. F. Moraes and S. C. Pinho, β-Carotene-Loaded Liposome Dispersions Stabilized with Xanthan and Guar Gums: Physico-Chemical Stability and Feasibility of Application in Yogurt, *LWT--Food Sci. Technol.*, 2014, **59**(2P2), 1265–1273.
17. W. Aadinath, A. Bhushani and C. Anandharamakrishnan, Synergistic Radical Scavenging Potency of Curcumin-in-β-Cyclodextrin-in-Nanomagnetoliposomes, *Mater. Sci. Eng., C*, 2016, **64**, 293–302.
18. B. Chauhan, G. Kumar, N. Kalam and S. H. Ansari, Current Concepts and Prospects of Herbal Nutraceutical: A Review, *J. Adv. Pharm. Technol. Res.*, 2013, **4**(1), 4–8.
19. M. Faruque Ahmad, F. Ali Ahmad, Z. Azaz Ahmad Azad, M. Sarfaraz Alam and S. Amir Ashraf, Nutraceutical Is the Need of Hour, *World J. Pharm. Pharm. Sci.*, 2013, **2**(5), 2516–2525.
20. R. Daenicke, Advances in Equine Nutrition II, *Anim. Feed Sci. Technol.*, 2002, **96**(3–4), 245.
21. H. Nasri, A. Baradaran, H. Shirzad and M. R. Kopaei, New Concepts in Nutraceuticals as Alternative for Pharmaceuticals, *Int. J. Prev. Med.*, 2014, **5**(12), 1487–1499.
22. L. Das, E. Bhaumik, U. Raychaudhuri and R. Chakraborty, Role of Nutraceuticals in Human Health, *J. Food Sci. Technol.*, 2012, **49**(2), 173–183.
23. J. Zhao, Nutraceuticals, Nutritional Therapy, Phytonutrients, and Phytotherapy for Improvement of Human Health: A Perspective on Plant Biotechnology Application, *Recent Pat. Biotechnol.*, 2008, **1**(1), 75–97.
24. S. H. Zeisel, Regulation of "Nutraceuticals", *Science*, 1999, **285**(5435), 1853–1855.
25. M. C. Braithwaite, C. Tyagi, L. K. Tomar, P. Kumar, Y. E. Choonara and V. Pillay, Nutraceutical-Based Therapeutics and Formulation Strategies Augmenting Their Efficiency to Complement Modern Medicine: An Overview, *J. Funct. Foods*, 2014, **6**(1), 82–99.
26. S. Ruchi, K. Amanjot, T. Sourav, B. Keerti and B. Sujit, Role of Nutraceuticals in Health Care: A Review, *Int. J. Green Pharm.*, 2017, **11**(3), S385–S394.
27. S. S. Lokhande, Role of Nutraceuticals in Various Diseases: A Comprehensive Review, *Asian J. Pharm. Res.*, 2018, **8**(4), 236.
28. C. J. Dillard and J. B. German, Phytochemicals: Nutraceuticals and Human Health, *J. Sci. Food Agric.*, 2000, **80**(12), 1744–1756.
29. C. Ramaa, A. Shirode, A. Mundada and V. Kadam, Nutraceuticals - An Emerging Era in the Treatment and Prevention of Cardiovascular Diseases, *Curr. Pharm. Biotechnol.*, 2006, **7**(1), 15–23.

30. P. Panwar, B. Pandey, P. C. Lakhera and K. P. Singh, Preparation, Characterization, and in Vitro Release Study of Albendazole-Encapsulated Nanosize Liposomes, *Int. J. Nanomedicine*, 2010, **5**(1), 101–108.
31. J. Singh and S. C. Sinha, Classification, Regulatory Acts and Applications of Nutraceuticals for Health, *Int. J. Pharm. Biol. Sci.*, 2012, **2**(1), 177–187.
32. M. Spano, Functional Foods, Beverages, and Ingredients in Athletics, *Strength Cond. J.*, 2010, **32**(1), 79–86.
33. H.-F. Wen, C. Yang, D.-G. Yu, X.-Y. Li and D.-F. Zhang, Electrospun Zein Nanoribbons for Treatment of Lead-Contained Wastewater, *Chem. Eng. J.*, 2016, **290**, 263–272.
34. M. R. Mozafari, C. Johnson, S. Hatziantoniou and C. Demetzos, Nanoliposomes and their applications in food nanotechnology, *J. Liposome Res.*, December 2008, 309–327.
35. S. Khorasani, M. Danaei and M. R. Mozafari, Nanoliposome Technology for the Food and Nutraceutical Industries, *Trends Food Sci. Technol.*, 2018, **79**(July), 106–115.
36. S. Akhavan, E. Assadpour, I. Katouzian and S. M. Jafari, Lipid Nano Scale Cargos for the Protection and Delivery of Food Bioactive Ingredients and Nutraceuticals, *Trends Food Sci. Technol.*, 2018, **74**, 132–146.
37. S. M. Pyo, R. H. Müller and C. M. Keck, in *Encapsulation by Nanostructured Lipid Carriers*, Elsevier Inc, 2017, https://doi.org/10.1016/B978-0-12-809436-5.00004-5.
38. M. Alexander, A. Acero Lopez, Y. Fang and M. Corredig, Incorporation of Phytosterols in Soy Phospholipids Nanoliposomes: Encapsulation Efficiency and Stability, *LWT--Food Sci. Technol.*, 2012, **47**(2), 427–436.
39. M. R. Mozafari, Nanoliposomes: Preparation and Analysis, *Methods Mol. Biol.*, 2010, **605**, 29–50.
40. M. J. Hope, M. B. Bally, G. Webb and P. R. Cullis, Production of Large Unilamellar Vesicles by a Rapid Extrusion Procedure. Characterization of Size Distribution, Trapped Volume and Ability to Maintain a Membrane Potential, *Biochim. Biophys. Acta, Biomembr.*, 1985, **812**(1), 55–65.
41. H. Kikuchi, H. Yamauchi and S. Hirota, A Polyol Dilution Method for Mass Production of Liposomes, *J. Liposome Res.*, 1994, **4**(1), 71–91.
42. A. Akbarzadeh, R. Rezaei-Sadabady, S. Davaran, S. W. Joo, N. Zarghami, Y. Hanifehpour, M. Samiei, M. Kouhi and K. Nejati-Koshki, Liposome: Classification, Preparation, and Applications, *Nanoscale Res. Lett.*, 2013, **8**(1), 1–9.
43. C. Bonechi, A. Donati, G. Tamasi, A. Pardini, H. Rostom, G. Leone, S. Lamponi, M. Consumi, A. Magnani and C. Rossi, Chemical Characterization of Liposomes Containing Nutraceutical Compounds: Tyrosol, Hydroxytyrosol and Oleuropein, *Biophys. Chem.*, 2019, **246**, 25–34.
44. S. Gopi and P. Balakrishnan, Evaluation and Clinical Comparison Studies on Liposomal and Non-Liposomal Ascorbic Acid (Vitamin C)

and Their Enhanced Bioavailability, *J. Liposome Res.*, 2020, DOI: 10.1080/08982104.2020.1820521.

45. J. L. Davis, H. L. Paris, J. W. Beals, S. E. Binns, G. R. Giordano, R. L. Scalzo, M. M. Schweder, E. Blair and C. Bell, Liposomal-Encapsulated Ascorbic Acid: Influence on Vitamin C Bioavailability and Capacity to Protect against Ischemia–Reperfusion Injury, *Nutr. Metab. Insights*, 2016, **9**, NMI.S39764.
46. M. Mohammadi, B. Ghanbarzadeh and H. Hamishehkar, Formulation of Nanoliposomal Vitamin D3 for Potential Application in Beverage Fortification, *Adv. Pharm. Bull.*, 2014, **4**(Suppl 2), 569–575.
47. S. Amiri, B. Ghanbarzadeh, H. Hamishehkar, M. Hosein, A. Babazadeh and P. Adun, Vitamin E Loaded Nanoliposomes: Effects of Gammaoryzanol, Polyethylene Glycol and Lauric Acid on Physicochemical Properties, *Colloids Interface Sci. Commun.*, 2018, **26**(July), 1–6.
48. A. Pezeshky, B. Ghanbarzadeh, H. Hamishehkar, M. Moghadam and A. Babazadeh, Vitamin A Palmitate-Bearing Nanoliposomes: Preparation and Characterization, *Food Biosci.*, 2016, **13**, 49–55.
49. S. M. T. Gharibzahedi and S. M. Jafari, The Importance of Minerals in Human Nutrition: Bioavailability, Food Fortification, Processing Effects and Nanoencapsulation, *Trends Food Sci. Technol.*, 2017, **62**, 119–132.
50. S. Horton and J. Ross, The Economics of Iron Deficiency, *Food Policy*, 2003, **28**(1), 51–75.
51. E. E. Ziegler, Consumption of Cow's Milk as a Cause of Iron Deficiency in Infants and Toddlers, *Nutr. Rev.*, 2011, **69**(Suppl 1), 37–42.
52. S. Xia and S. Xu, Ferrous Sulfate Liposomes: Preparation, Stability and Application in Fluid Milk, *Food Res. Int.*, 2005, **38**(3), 289–296.
53. M. Gibis, E. Vogt and J. Weiss, Encapsulation of Polyphenolic Grape Seed Extract in Polymer-Coated Liposomes, *Food Funct.*, 2012, **3**(3), 246–254.
54. M. Fathi, M. R. Mozafari and M. Mohebbi, Nanoencapsulation of Food Ingredients Using Lipid Based Delivery Systems, *Trends Food Sci. Technol.*, 2012, **23**(1), 13–27.
55. C. Sebaaly, H. Greige-Gerges, G. Agusti, H. Fessi and C. Charcosset, Large-Scale Preparation of Clove Essential Oil and Eugenol-Loaded Liposomes Using a Membrane Contactor and a Pilot Plant, *J. Liposome Res.*, 2016, **26**(2), 126–138.
56. L. A. Prado Barragán, J. J. Buenrostro-Figueroa, C. N. A. González and I. Marañon, in *Production, Stabilization, And Uses Of Enzymes From Fruit and Vegetable Byproducts*, Elsevier Inc., 2016, https://doi.org/10.1016/B978-0-12-803622-8.00010-0.
57. V. L. Dos Santos and M. V. Dias-Souza, Strategies Based on Microbial Enzymes and Surface-Active Compounds Entrapped in Liposomes for Bacterial Biofilm Control, in *Nanobiomaterials in Antimicrobial Therapy*, William Andrew Publishing, 2016, ch. 11, pp. 385–418.

58. S. Raveendran, B. Parameswaran, S. B. Ummalyma, A. Abraham, A. K. Mathew, A. Madhavan, S. Rebello and A. Pandey, Applications of Microbial Enzymes in Food Industry, *Food Technol. Biotechnol.*, 2018, **56**(1), 16–30.
59. G. Multhoff, M. Molls and J. Radons, Chronic Inflammation in Cancer Development, *Front. Immunol.*, 2011, **2**(JAN), 1–17.
60. M. S. Muthu, S. A. Kulkarni, J. Xiong and S. S. Feng, Vitamin e TPGS Coated Liposomes Enhanced Cellular Uptake and Cytotoxicity of Docetaxel in Brain Cancer Cells, *Int. J. Pharm.*, 2011, **421**(2), 332–340.
61. D. Arora and S. Jaglan, Nanocarriers Based Delivery of Nutraceuticals for Cancer Prevention and Treatment: A Review of Recent Research Developments, *Trends Food Sci. Technol.*, 2016, **54**, 114–126.
62. R. Meenambal and M. M. Srinivas Bharath, Nanocarriers for Effective Nutraceutical Delivery to the Brain, *Neurochem. Int.*, 2020, **140**, 104851.
63. S. Ghaffari and N. Roshanravan, The Role of Nutraceuticals in Prevention and Treatment of Hypertension: An Updated Review of the Literature, *Food Res. Int.*, 2020, **128**(May 2019), DOI: 10.1016/j.foodres.2019.108749.
64. N. P. Aditya, Y. G. Espinosa and I. T. Norton, Encapsulation Systems for the Delivery of Hydrophilic Nutraceuticals: Food Application, *Biotechnol. Adv.*, 2017, **35**(4), 450–457.
65. T. Wang, Y. Deng, Y. Geng, Z. Gao, J. Zou and Z. Wang, Preparation of Submicron Unilamellar Liposomes by Freeze-Drying Double Emulsions, *Biochim. Biophys. Acta, Biomembr.*, 2006, **1758**(2), 222–231.
66. K. L. Cheung and J. F. R. Robertson, Preoperative Hormone Therapy Trials for Breast Cancer, *Breast*, 2001, **10**(1), 1–5.
67. N. A. Lopes and A. Brandelli, Nanostructures for Delivery of Natural Antimicrobials in Food, *Crit. Rev. Food Sci. Nutr.*, 2018, **58**(13), 2202–2212.
68. L. K. Sarao and M. Arora, Probiotics, Prebiotics, and Microencapsulation: A Review, *Crit. Rev. Food Sci. Nutr.*, 2017, **57**(2), 344–371.
69. H.-S. Kwak, Overview of Nano- and Microencapsulation for Foods, in *Nano- and Microencapsulation for Foods*, ed. H.-S. Kwak, 2014, ch. 1, https://doi.org/10.1002/9781118292327.
70. M. Yao, J. Xie, H. Du, D. J. McClements, H. Xiao and L. Li, Progress in microencapsulation of probiotics: A review, *Compr. Rev. Food Sci. Food Saf.*, 2020, **19**(2), 857–874.

CHAPTER 8

Lipid- and Protein-based Nanodelivery Systems

ANUJIT GHOSAL[a,b], THILINI DISSANAYAKE[a,b] AND NANDIKA BANDARA*[a,b]

[a]Department of Food and Human Nutritional Sciences, Faculty of Agricultural and Food Sciences, University of Manitoba, Winnipeg R3T 2N2, Canada; [b]Richardson Centre for Food Technology and Research (RCFTR), 196, Innovation Drive, Winnipeg, Manitoba R3T 6C5, Canada
*E-mail: Nandika.Bandara@umanitoba.ca

8.1 Introduction

Technological advancements have influenced the biomedical industry by upgrading theragnostic tools, designing new drug-like molecules and their fabrication, and elucidating the delivery mechanisms for bioactive or functional macromolecules.[1,2] Although designing new bioactive molecules is a complex process, scientists, pharmaceutical industries, and research centers may have developed many potentially bioactive molecules that are sparingly soluble in water due to their higher hydrophobicities and larger molecular weights. Several Food and Drug Administration (FDA)-approved therapeutic drug molecules with high efficacies against many diseases are known for their insolubility in water. Therefore, they are either used in high dosages that cause severe side effects or modified to increase the solubility.[3]

The pharmacokinetics and pharmacodynamics of drug molecules have been influenced by their bioavailability, retention time, stability, and solubility (aqueous) within the host.[4] It has been observed that the partial

Smart Materials No. 41
Functional Materials from Lipids and Proteins
Edited by Nandika Bandara and Aman Ullah

Published by the Royal Society of Chemistry, www.rsc.org

solubility of many bioactive molecules, such as curcumin, resveratrol, paclitaxel, vitamins (A, D, and E), fibrous proteins (keratins and collagens), lipids, essential oils, and other macromolecules, diminishes their effectiveness and broad-spectrum applications.[3,5] The reduced solubility in the aqueous system causes shorter bioavailability, restricts the bioactivity, and results in unwanted precipitation, higher dosage concentrations, side effects, and increased toxicity. Therefore, the production of water-soluble bioactive molecules, formulation of novel techniques to improve the solubility of partially soluble drug molecules, and finding effective drug delivery mechanisms are always in great demand.[6,7] Adsorption of the drug molecules on biodegradable or metal oxide nanoparticles, surface functionalization (chemical modification), conjugation (pro-drug formulations), ionization, and encapsulation are a few such routes that enhance the solubility and, hence, the potency of the drug molecules.[1,8] The enhancement in solubility also improves the storage life, bioactivity, and uniform distribution of bioactive molecules through various modes, such as solution (oral/injection), food products, pills, fumes (nasal), and permeable creams/films/patches. Among these, the delivery of the bioactive molecules after encapsulating *via* suitable vehicles/platforms/polymeric agents is the most common, facile, and preferred method, particularly for the delivery of ultrasensitive, unstable, and reactive drug molecules, which are protected by layering or stealth coatings of biodegradable and compatible green polymers, proteins, or other biomolecules.[9–11] Encapsulation of bioactive molecules has the most negligible impact on the original chemical structure and does not alter the bioactivity of the molecule. However,

Table 8.1 A few examples of bioactive molecules delivered using the encapsulation technique.

Bioactive molecules	Encapsulating materials	Ref.
Co-delivery of curcumin and resveratrol	Core–shell nanoparticles (electrospraying); zein as a core and PEG or ethyl cellulose as a shell; 3D-printed gelatin and gelatin-based hydrogel	15
Curcumin	Lipid-based nanoparticles (cubosomes and liposomes); monoolein, monopalmitolein, phytantriol, and distearoylphosphatidylcholine	16
Melittin	Lipid-based nanoparticles (niosomes); cholesterol, Span 60, and Tween 60	17
Prodigiosin	Chitosan microspheres; Tween 20 and glutaraldehyde	18
Insulin	Silk fibroin-based hydrogel; ethylene glycol and triethylene glycol	19
Histidine; WENGE (Trp-Glu-Asn-Gly-Glu) and KHIQK (Lys-His-Ile-Gln-Lys) peptides	Cold-set whey protein microgels; calcium chloride	20

selecting the encapsulating material, such as polyethylene glycol (PEG), polyethyleneimine, polyvinyl alcohol (PVA), carbohydrates, cellulose, lipids, and proteins, greatly influences the overall drug efficacy. Mostly, environmentally benign, biocompatible, biodegradable, and stimuli-responsive materials are preferred over the other categories of materials. Nowadays, due to numerous value-added functionalities, proteins and lipids are the most on-demand biomaterials for delivering bioactive molecules.[12–14] The need for such materials and their modification and prerequisite engineering of biomolecules are discussed in the subsequent sections. Table 8.1 lists a few examples of bioactive molecules encapsulated using lipids or proteins to deliver and eradicate diseases.

8.2 Need for Innovative Biomaterials

The constant rise in population, pollution, and consumption of materials has generated a demand for the use of plant- and animal-based biodegradable alternatives in most sectors.[21–23] Applying biodegradable materials improves the understanding of their chemical, physical, and mechanical properties. Scalable materials with value-added functionalities are necessary for the vast application of biomass derivatives in composite materials, the textile industry, tissue engineering, scaffolds, stimuli-responsive substances, biosensing, photochromatic panels, imaging, and bioelectronics.[24–26] Figure 8.1 presents selected applications of bioderivatives around the world. Being biodegradable and derived from natural resources is not the only criterion for selecting the material for biomedical or food applications. The selection depends on the bio-interactions and response, known composition, and purity of biomaterials, *i.e.*, to ascertain low or zero side effects of the material.[27,28] During metabolism, the complex biological processes, cellular events, and immune system responses may consider any foreign biodegradable materials (synthetic or natural) as harmful or toxic. Therefore, it is essential to understand and predict the probable interactions as well as the response between the dynamic bioprocess (protein interactions, macrophage fusion, cell proliferation/growth, and mineralization) and biomaterials used for biomedical (scaffolds, delivery agents, bio-additives, biosensors, and other bio-implants) as well as food applications (packaging, food product development, and delivery of nutraceuticals).[29,30]

Biomaterials could be engineered by altering the surface functional groups and available chemical, ionic, or physical linkages, separating allergens/toxins, suitably processing to retain all the beneficial properties, such as protein purity, functionality, or activity through enzymatic or catalytic modification, nonthermal processing, and cold gelation.[31,32] The literature reports several biomaterials modified using traditional and new innovative techniques. Innovative materials with predictable surface

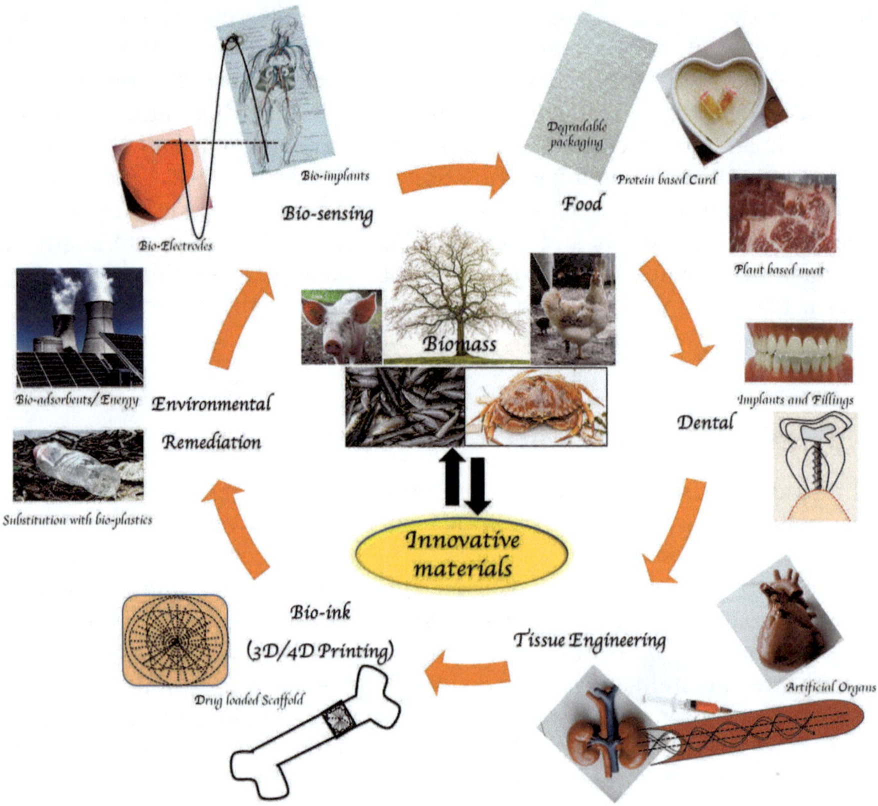

Figure 8.1 Schematic presentation of the avenues where innovative materials are needed.

response can be developed by following the below-mentioned routes and using knowledge of detailed functional characterizations.[24,33–35]

1. Engineering the surface functional properties (denaturation, enzymatic alterations, *etc.*)
2. Conjugation with active molecules, such as proteins, lipids, carbohydrates, and peptides
3. Fabrication of pro-drug-like molecules
4. Formulation of metal–organic framework or metal–chelating complexes
5. Generation of exfoliated structure through chemical, physical, or mechanical treatments
6. Using state-of-the-art innovative techniques, such as electrospinning, electrodeposition, 3D printing, and microfluidizers.

The delivery of bioactive molecules in humans through invasive or non-invasive routes, such as oral, intravenous/muscular injection, dispersion, inhalation, ocular, transdermal, pulmonary, vaginal, and others, has been challenging, and innovations are in progress.[36,37] Particularly, the dispersion of hydrophobic drug-like macromolecules, vitamins, minerals, and phenolic species in food or as a supplement has drawn much attention. Fabrication of functional materials for delivering specific bioactive molecules at targeted regions with certain concentrations and maintaining a sustainable time-bound release under physiological conditions are in high demand.[38,39] The biomaterials obtained from plants and animals and derivatization of biowaste products, such as extracted protein, a variety of lipids, fat molecules, and other bioactive species that can mimic natural or biological functions, are preferred in the encapsulation.

As the integral components of biological systems, protein- and lipid-based drug delivery motifs or agents have higher biocompatibility, allowing them to bypass the immune response and supporting the targeted delivery of loaded molecules.[40–42] These biomaterials, individually or in combination, could inherit various value-added properties promising for their biomedical and food applications. The utilization of these biomaterials improves the beneficial properties of other engineered biopolymers. The modification gives better control over the rate of degradability, encapsulation efficiency (hydrophilic, hydrophobic, and amphiphilic drug molecules), release kinetics, and biocompatibility. The stimuli-responsive biopolymers could exhibit dynamic and responsive properties toward dynamic environmental conditions, *i.e.*, self-repair, responsive toward change in pH, temperature, osmotic pressure, ionic concentration, and bioreagents, and have lower or zero immune responsiveness, which are a few other beneficial properties expected in the innovative biomaterials.

8.3 Protein-based Formulations as Nanodelivery Agents

Apart from being a major macronutrient in humans or animals, proteins from plant and animal sources have been exploited to develop delivery platforms for drugs, DNA, small interfering RNA (siRNA), vitamins, and other bioactive molecules. They are known for enhancing the permeability and retention effect of the intended biomolecules.[43,44] The proteins used in such platforms could become one of the sources to produce essential amino acids within the body after partial digestion or hydrolysis of the drug delivery platforms.[45] Furthermore, they could be used in various compositions to optimize drug-loading efficiency and release mechanisms as they are inherently digestible and biodegradable. They could bypass the immune systems, occasionally evading the enzyme and other biological barriers in human systems. The proteins could protect the active molecules from degradation and unwanted interactions with other biomolecules in biological systems. Combining drugs with certain proteins further improves the targetability and bioactivity against diseases, such as cancer.[46,47] Proteins have multifunctional surfaces and could combine/mix with polymers, lipids, carbohydrates, phenolics, and other macromolecules to formulate stable drug delivery vehicles. Drug delivery platforms from protein or their derivatives have been developed primarily due to their ability to self-assemble under the surrounding conditions. Most proteins respond to pH, temperature, metallic ions, charged species, and ionic concentration changes. Proteins, such as gelatin, albumin, keratin, collagen, elastin, silk, ferritin/apoferritin, capsid, soy, and whey proteins, have been reported to form assemblies, such as spheres, cages, rods, hydrogels, films, and others. Due to these reasons, in many instances, proteins or derived amino acids have been reported to be delivered in a cocktail mixture of vaccines or therapy to boost immunity (generation of antibodies in response to COVID-19 and hepatitis B vaccine) or to support the biological processes (enzymatic actions).[48–50] Protein-based systems have been actively studied in several clinical trials and research laboratories to formulate highly compatible and efficient drug delivery platforms.[42,51] Table 8.2 provides a few of these protein-based drug delivery systems and the acquired geometry.

The demand for specific proteins can easily be fulfilled by the traditional/innovative extraction procedures from biomass (plant or animal biomasses) or programmed to be harvested from bacterial or yeast cultures followed by purification and storage under optimum conditions.[27,63] The extraction of proteins from bioresources could easily be carried out *via* traditional/modified/advanced or by combining several extraction techniques.[64,65] A suitable extraction process would result in products with high protein yield and purity, with a lower impact on the techno-functional properties. Alkaline, acid, salt, and solution (organic/inorganic/deep eutectic solvents) extraction; isoelectric and micellar precipitation; dry fractionation; ultrasonication; and supercritical fluid extraction are a few examples of the preferred techniques for biomass-to-protein conversion. The properties of proteins are greatly affected by the processing technique/s selected and other processing or modification treatments (sonication; thermal, plasma, and UV treatments; and enzymatic action).[66,67]

Furthermore, during the processing of proteins, new or modified steps could eradicate the existing allergens, polyphenols, and natural toxins to create hypoallergenic proteins for biomedical and food applications.[68] All these procedures could influence (positively or negatively) the final properties of the extracted proteins, such as thermal stability, emulsification, solubility, available amino acids, degree of denaturation, and biodegradability. Hence, better control over all these parameters would channel the abilities of the formulated protein-based drug carriers. A schematic presentation has been shown in Figure 8.2, elaborating a protein cycle

Table 8.2 Protein-based drug carriers.

Protein used	Drug/properties	Platform	Ref.
Caseinate/kappa-carrageenan and zein	Quercetin/higher *in vitro* antioxidant activities and intracellular reactive oxygen species	Nanoparticles (d <150 nm)	52
Pea protein-carboxymethylated corn fiber gum	Curcumin/high loading efficiency, good water dispersibility, and antioxidant and radical-scavenging activities	Nanoparticles/composite materials/homogenizer and lyophilization	53
Soy protein isolate	Ciprofloxacin/noncytotoxic, polycrystalline, increased thermal stability, and pH responsiveness	Hydrogel/chemical grafting with acrylic copolymer	54
α-Lactalbumin/whey protein	Ampicillin/biocompatibility with TR146 human buccal epithelium and NIH 3T3 murine fibroblast cells	Nanofibers/electrospinning	55
Pea, lentil, and faba bean protein concentrates	Omega-3 polyunsaturated fatty acids (fish oil)	Oil-in-water emulsions (d_{32} <500 nm)/homogenization	56
Ferritin, small heat shock proteins, and vault proteins	Various hydrophobic and hydrophilic bioactive molecules	Nanocages	57
Zein and sodium caseinate	Simvastatin/improved hypolipidemic activities	Nanoparticles/liquid–liquid phase separation method	58
Whey proteins	Heat-sensitive nutraceutical polyphenols	Nanoparticles/cold gelation	59
Engineered spider silk protein	Methyl violet or ethacridine lactate (model cationic drugs)/biodegradable and sustainable drug release profile	Drug-eluting coatings or implantable films	60
Soy protein isolate, zein, and SPI/zein blends	Riboflavin/near-zero-order release kinetics	Microsphere/cold gelation	61
Gliadin (wheat protein) and lectin-conjugated gliadin	Amoxicillin, clarithromycin, and omeprazole/antibacterial (*Helicobacter pylori*)	Nanoparticles/desolvation procedure	62

Figure 8.2 Schematic presentation of the basic steps involved in converting biomass into proteins, formulation of delivery agents, and the final degradation.

(source to source) in the development of protein-based drug carriers. The basic steps involved are acquiring, extracting, formulating, and degrading the final products into the building components while releasing the loaded bioactive molecules.

8.4 Lipid-based Formulations as Nanodelivery Agents

Lipids from plant and animal sources have been used progressively in recent years for various biomedical applications, including the fabrication of drug delivery platforms. Lipids can be extracted from oil seeds (soybean, canola, sunflower, flaxseed, and hemp), oils (castor oil, coconut oil, and olive oil), and various animal sources (fish, meat, eggs, and milk). Various lipid-based platforms used for the sustainable delivery of bioactive molecules are shown in Figure 8.3. Technological advancements and nanonization strategies have improved the advantageous properties of lipid-based drug delivery systems by many folds.[69,70] ONPATTRO™ is the first FDA-approved lipid-based delivery (siRNA) system in 2018 for polyneuropathy due to genetic disorders. Similar delivery vehicles have been utilized to treat COVID-19 through the delivery of mRNA and other cocktails of therapeutics.[70,71] The hydrophobic characteristic of lipids supports the dispersion or dissolution of most water-fearing bioactive molecules (resveratrol, quercetin, curcumin, and essential oils), drugs (ciprofloxacin, triclosan, camptothecin, taxol, and diazepam), DNA, RNA, and polyolefins. The dispersion enables the easy transportation of the active molecules within the aqueous system. However, it has always been challenging to form a thermodynamically stable lipid-based dispersion system due to its incompatibility with an aqueous system, and systematic structuring or functionalization is necessary.[72]

Various lipid-based colloidal structures have been reported, such as liposomes, solid lipid nanoparticles, nanostructured lipid carriers, organogels, nanoemulsions, Pickering emulsions, micellar formulations, and hydrogels.[73–75] The formulation of an efficient lipid-based drug delivery system depends on the characteristic properties of lipids, such as molecular weight, melting point, pH response, chemical structure, available charges, targeting ligands, and associated co-polymers/stabilizers. Lipid-based nanocarriers bypass one of the most complex biological

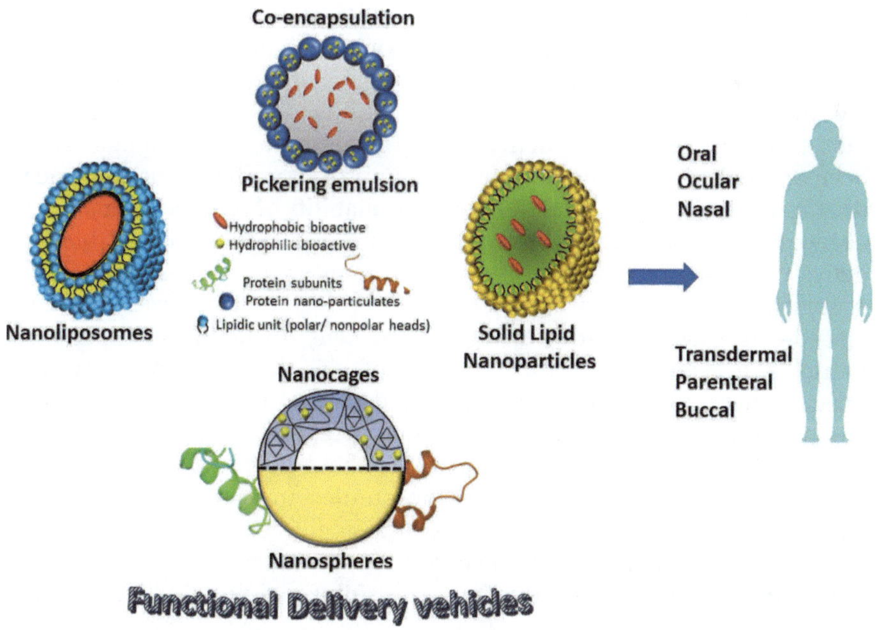

Figure 8.3 Lipid-based functional drug delivery platforms.

barriers in neuro-medicine applications, *i.e.*, blood–brain and gut barriers.[74] Generally, chemical stabilizers, such as surfactants (sodium dodecyl sulfate, sodium stearate, and Triton X), surface functionalization (generation of –OH, –COOH, and $-NH_2$ groups), high-energy techniques (ultra-homogenization, sonication, microfluidization, and supercritical fluid assistance), or template materials (hydrophilic polymers, proteins, and water-soluble biomolecules), are used.[76–78] One of the lipids obtained from plants and animals with huge applications in drug delivery is phospholipids. The phospholipids consist of two fatty acid chains and a hydrophilic head with a phosphoric acid unit with choline, ethanolamine, and other organic molecules. They are one of the key components in all cell membranes, are amphiphilic and biocompatible, and have been explored extensively in drug delivery applications.[79,80] The biomimetic response of bio-formulations from lipids makes them influential and more suitable for clinical trials.

8.5 Protein–Lipid Conjugates as Nanodelivery Systems

Protein–lipid conjugation is introduced as a better solution for most of the challenges associated with single-carrier-based delivery systems, as the presence of both proteins and lipids results in mutual benefits.[81] Lipids show excellent characteristics as nanodelivery systems due to their low toxicity, biodegradability, biocompatibility, easiness of industrial production, excellent functionality in emulsification, and the possibility of delivering both hydrophilic and lipophilic bioactive compounds.[82,83] Proteins also exhibit such characteristics including, but not limited to, low toxicity, biocompatibility, biodegradability, the possibility of surface modifications, and target delivery.[84,85] The protein–lipid conjugation has been applied to nondelivery systems to improve the therapeutic efficacy of drugs and genes and to increase the bioavailability of nutraceuticals.[81,86,87]

Lipid-based nanodelivery systems face many issues during bioactive delivery, such as the instability of liposomes in the presence of serum proteins, incompatibility between the solubility of bioactives and carriers, low drug accumulation in the tumor cells, instability of lipid nanoparticles and emulsions, the difficulty of skin penetration of proteins, toxicities due to the accumulation of drugs in healthy cells, nuclease degradation of genes, incomplete gene suppression, and undesirable side effects.[81] Therefore, numerous approaches have been developed to conjugate proteins and lipids in scientific studies to give all the desirable characteristics of proteins and lipids in a single nanodelivery system and address the challenges of single-carrier-based nanodelivery systems. The published literature has shown the successful use of protein–lipid conjugation in bioactive delivery applications. Figure 8.4 shows several techniques

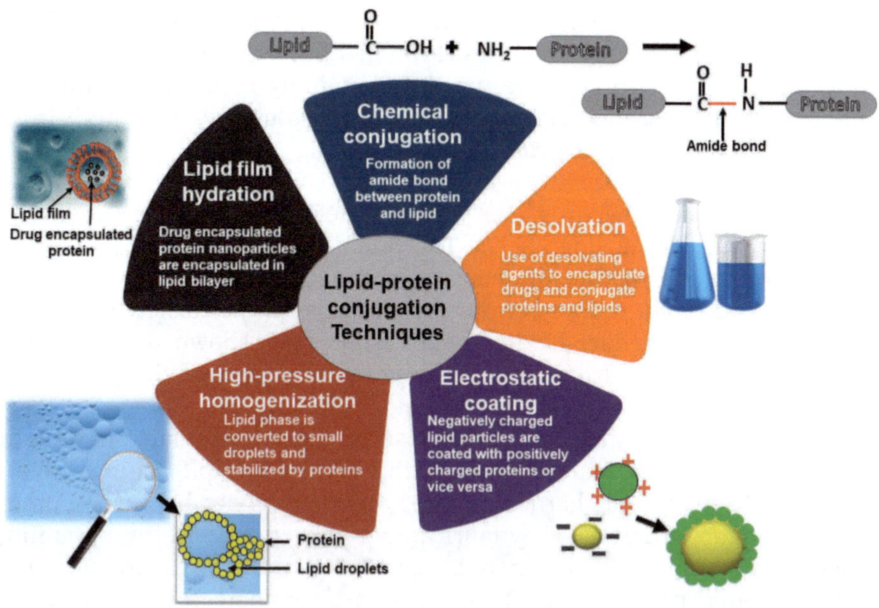

Figure 8.4 Illustration of different techniques used for protein–lipid conjugation. Reproduced from ref. 81 with permission from Elsevier, Copyright 2022.

used for protein–lipid conjugation, mainly chemical conjugation, desolvation, lipid film hydration, homogenization, and electrostatic coating.[88]

Chemical conjugation involves the formation of an amide bond between lipid and protein, which is contributed by carboxyl groups and amine groups, respectively. Chemical conjugation differs from other methods due to the formation of a covalent bond between protein and lipid,[89,90] whereas the other conjugation techniques include electrostatic interactions,[91] hydrophilic/hydrophobic interactions,[92] hydrophobic interactions,[93] and physical mixing.[94]

Protein is a better solution to overcome the two main issues of drug delivery systems in cancer therapy: toxicity to normal tissues and low accumulation in tumor sites. Albumin is an excellent example, which can be conjugated to lipids to form delivery systems with an enhanced permeation and retention effect due to its nano-scale size. Furthermore, albumin selectively binds to proteins, such as gp60 and secreted protein acidic and rich in cysteine (SPARC) that are overexpressed in tumor cells.[95] Other than the target delivery, improving the stability of the delivery system by protein or lipid coatings is another benefit of protein–lipid conjugation. The physicochemical stability of antioxidants can be protected during processing and storage conditions when incorporated with lipids. Chuacharoen and Sabliov (2016) have shown that lecithin-incorporated zein nanoparticles had improved chemical stability against UV radiation compared to control.[96] Reduced oral bioavailability is one of the main issues related to most drugs. Protein–lipid conjugation has proved its excellent ability to enhance oral bioavailability of bioactives through various approaches, including increasing drug solubility, protection from acidic pH and digestive enzymes, and enhancing intestinal permeability.[88] Cytotoxic drugs create toxicity against not only cancer cells but also normal cells. Even though the drug has proved its efficiency against carcinoma cells, it causes several side effects. The protein–lipid conjugated nanoparticles improve the therapeutic efficacy of the drugs due to the increased solubility and target delivery of the drugs. Thereby, the toxicity to normal body cells is also reduced.[87] Research results have shown the enhanced controlled release of drugs by lipid–protein conjugated delivery systems, thereby high drug circulation time and bioavailability.[97]

The bioaccessibility of the bioactive compounds encapsulated in protein nanoparticles can be increased by mixing them with lipid nanoparticles. Such a delivery system harnesses the synergistic effect between protein and lipid nanoparticles.[98] They fabricated a delivery system for curcumin, where curcumin-loaded zein nanoparticles are mixed with curcumin-free digestible lipid nanoparticles. Lipid nanoparticles increased the bioaccessibility of curcumin by forming mixed micelles in the intestine by solubilizing and transporting hydrophobic curcumin.[98] Furthermore, a study conducted by Liu *et al.* (2018) has shown that the aqueous dispersion and chemical stability of resveratrol and curcumin could be improved by core–shell nanoparticles that have been designed using zein–epigallocatechin gallate conjugates and rhamnolipid coating.[99] Freire *et al.* (2018) prepared an emulsion gel using different proteins to deliver n-3 fatty acids and tannins. Sodium caseinate, whey protein isolate, and soy protein were used in the emulsion, and they showed a solid-like structure.[100] Moreover, the prepared gel emulsion provided adequate stability to lipids during storage and showed excellent binding properties. Chen *et al.* (2016) encapsulated indocyanine green (ICG), an amphiphilic compound used in near-infrared (NIR) fluorescence imaging for tumor diagnosis.[91] However, its efficient use is limited by aggregation in the aqueous medium; degradation due to the high temperature, lights, and oxidants; and rapid clearance by the liver. Even though encapsulation in liposomes helps to overcome the issues mentioned earlier, liposomes also suffer from their inherent instability. Coating with human serum albumin improved the stability of liposomes, and its nontoxicity to normal cells and internalization in the tumor cells were proved by *in vitro* and *in vivo* studies.[91]

Likewise, several protein–lipid conjugated delivery systems have been reported in the literature, including, but not limited to, chemically conjugated gelatin–cholesterol lipid microparticle scaffolds to maintain the controlled release of encapsulated curcumin and to prevent post-engraftment complications,[89] rice bran–soybean oil nanoemulsions to enhance the bioavailability and cell permeability of encapsulated quercetin

Table 8.3 Recent applications of protein–lipid conjugation in nanoencapsulation of bioactive compounds.

Protein	Lipid	Conjugation method	Bioactive compound	Objective of conjugation	Ref.
Whey protein	Corn oil	Homogenization	β-Carotene	Increasing the β-carotene stability	101
Zein	Lecithin	Antisolvent co-precipitation	Curcumin	Increasing encapsulation efficiency, loading capacity, and stability of curcumin	102
Zein	Lecithin	High-pressure homogenization	Lutein	Increasing lutein stability by preventing photo- and thermal degradation	96
Albumin	A mixture of cholesterol, phosphatidylcholine, and DSPE-PEG2000 carboxylic acid	EDC/NHS chemical conjugation	Paclitaxel and vorinostat	Increasing target delivery, co-encapsulation, and colloidal stability of paclitaxel	103
Gelatin	Oleic acid	EDC/NHS chemical conjugation	Irinotecan hydrochloride	Developing an amphiphilic and stable delivery system	104
Albumin	Lecithin	Desolvation	Paclitaxel and borneol	Increasing target delivery and bioavailability of paclitaxel	105
Wheat germ agglutinin lectin	Glyceryl monostearate	EDC/NHS chemical conjugation	Paclitaxel	Increasing bioavailability and target delivery of paclitaxel	90
Gelatin	Oleic acid	EDC/NHS chemical conjugation	Sesamol	Increasing skin permeation of sesamol	106
Gelatin	Oleic acid	EDC/NHS chemical conjugation	Doxorubicin	Increasing target delivery of doxorubicin	107
Albumin	Phosphatidylethanolamine	Electrostatic interaction	Indocyanine green (ICG)	Increasing target delivery and stability of ICG	91
Transferrin	Phosphatidylcholine and phosphoethanolamine	Chemical conjugation	α-Mangostin	Increasing penetration through the blood–brain barrier	91

(continued)

Table 8.3 (*continued*)

Protein	Lipid	Conjugation method	Bioactive compound	Objective of conjugation	Ref.
Gelatin	Oleic acid	Physical mixing	Zaltoprofen	Increasing skin permeation of zaltoprofen	94
Gelatin	Soybean phosphatidylcholine	Lipid film hydration	Paclitaxel	Increasing physical stability of paclitaxel against lyophilization stress	108
Albumin	Lecithin	High-pressure homogenization	Teniposide	Reducing systemic toxicity and enhancing antitumor activity	87

and overall stability of the delivery system,[92] transferrin–phosphatidylcholine and phosphoethanolamine-conjugated liposomes to enhance the penetration of α-mangostin through the blood–brain barrier,[96] and bovine serum albumin–lecithin-conjugated nanoparticles to increase the circulation time.[93] Table 8.3 illustrates some recent applications of protein–lipid conjugation in drug and nutraceutical delivery.

Even though protein–lipid conjugated delivery systems have been introduced to solve the issues related to single-polymer delivery systems, some limitations are also related to protein–lipid conjugated systems. In some cases, toxicities of the conjugated delivery systems have been reported. For example, toxicity results from some proteins, such as bovine serum albumin and zein, were reported due to triggering the immune system and resisting digestion.[85,88] In addition, toxicities can be recorded due to crosslinkers such as glutaraldehyde.[104] Penetration through biological barriers is another concern where nanocarriers penetrate the mucus layer and epithelium cells.[109] Furthermore, conjugation should be able to provide new properties that are not present in either component alone. In that case, there are complexities in developing new delivery systems as it is difficult to predict the behavior of the system based only on the *in vitro* results. Therefore, the delivery systems should be synthesized with a better understanding of the behavior of the delivery systems inside the body.[88]

8.6 Future Perspectives

The continuous increase in the population, consumption of products, and pollution worldwide demands policy changes, such as green alternatives, sustainable development, and circular economy. With suitable modification and chemical engineering of biomass derivatives, such as lipids and proteins, they have the potential to satisfy most of the physical, chemical, and mechanical requirements of smart biomaterials in the biomedical and food industries. In recent years, a huge demand for these biomaterials has resulted, and in the coming years, more technological advancements in their extraction, purification, and storage procedures are expected. Furthermore, the bio-resemblance, compatibility, and biodegradability of such greener excipients, particularly the protein–lipid conjugates, innovate delivery mechanisms and efficacy of bioactive molecules.

Protein–lipid conjugates have been used prominently in drug and gene delivery applications. However, there is a high potential for developing protein–lipid conjugation systems to enhance the target delivery and controlled release. Utilizing proteins, such as transferrin, whey proteins, cruciferin, and others, to coat liposomes or nanoparticles to bypass biological barriers is currently tested in clinical trials as a second-line treatment for several diseases and targeting ligands in cancer cells.[110–113] Protein-based nanoparticles are known for treating cancers with enhanced target delivery and safety profiles of the drugs.[114–116] Most of the studies have been performed in laboratory scales, and there is a clear gap between laboratory and commercial platform applications. Although they are not widely used in food products with their proven advantages, they are becoming popular for encapsulating bioactive compounds in food-based products and biomimicking protein–lipid-based materials.[117,118]

Overall, the relatively poor physico-mechanical properties compared to synthetic or natural polymers, variation in properties of the proteins depending on the extraction and purification methodology, and source of protein (soy, oats, canola, quinoa, sunflower, pea, peanuts, fava beans, rice, legumes, chicken feathers, and fish scales) and economic aspects have caused certain reservation in the broader application of protein-based platforms. Other shortcomings associated with the proteins are their short-time stability in solution and at elevated temperatures. Furthermore, when present in high concentrations, few proteins could trigger immune responses and enzymatic actions, which may cause untimely unloading of loaded drug molecules. A better understanding of protein chemistry, advancement in purification and extraction techniques, and conceptualization of the associated problems with protein-based drug delivery platforms would possibly result in advanced delivery systems with cell-specific delivery abilities.[119]

Cytotoxicity caused by the delivery system itself has been a big concern. Animal proteins, such as bovine serum albumin, can trigger immune responses due to their exogenous nature when used in conjugated systems. The resistance of proteins, such as zein in digestion or other allergens in bio-components (proteins/lipids), could also trigger immunogenic responses.[85,88] Therefore, research studies on these shortfalls and minimizing the toxicities of the loaded drug molecules *via* sheath coating or encapsulation are needed. Even though research studies have proved the synergistic effects of co-encapsulation of bioactive compounds, there is a shortage of marketed protein–lipid-based delivery systems that can co-encapsulate contrary soluble bioactive compounds in one cocktail. As proteins and lipids can be chemically tethered, they would act as good media for contrary solubilities. There is a potential to fabricate protein–lipid conjugated delivery systems for co-encapsulation of contrary soluble bioactive compounds. In the same direction, rather than providing them as capsules, there is a high potential to increase the consumption of nutraceuticals through daily consumption of functional foods and beverages or bio-supplements. Future research should be designed to develop new food products containing proteins and lipids to satisfy the required nutritional and pharmaceutical aspects.

Abbreviations

DNA	Deoxyribonucleic acid
FDA	Food and Drug Administration
ICG	Indocyanine green
NIR	Near-infrared
PVA	Polyvinyl alcohol
PEG	Polyethylene glycol
RNA	Ribonucleic acid
siRNA	Small interfering RNA
SPARC	Secreted protein acidic and rich in cysteine

References

1. S. Jain, N. Patel, M. K. Shah, P. Khatri and N. Vora, Recent Advances in Lipid-Based Vesicles and Particulate Carriers for Topical and Trans dermal Application, *J. Pharm. Sci.*, 2017, **106**, 423.
2. C. Corti, P. P. M. B. Giachetti, A. M. M. Eggermont, S. Delaloge and G. Curigliano, Therapeutic vaccines for breast cancer: Has the time finally come?, *Eur. J. Cancer*, 2022, **160**, 150.
3. Y. Agrawal and V. Patel, Nanosuspension: An approach to enhance solubility of drugs, *J. Adv. Pharm. Technol. Res.*, 2011, **2**, 81.
4. S. Zhang, *et al.*, Lecithins-Zein nanoparticles for antifungal treatment: Enhancement and prolongation of drug retention in skin with reduced toxicity, *Int. J. Pharm.*, 2020, **590**, 119894.
5. S. Kalepu and V. Nekkanti, Insoluble drug delivery strategies: Review of recent advances and business prospects, *Acta Pharm. Sin. B*, 2015, **5**, DOI: 10.1016/j.apsb.2015.07.003.
6. L. Motelica, D. Ficai, O. C. Oprea, A. Ficai and E. Andronescu, Smart food packaging designed by nanotechnological and drug delivery approaches, *Coatings*, 2020, **10**, DOI: 10.3390/COATINGS10090806.
7. E. R. Ghomi, *et al.*, Innovations in Drug Delivery for Chronic Wound Healing, *Curr. Pharm. Des.*, 2021, **28**, 340.
8. D. Pooja, *et al.*, Improving efficacy, oral bioavailability, and delivery of paclitaxel using protein-grafted solid lipid nanoparticles, *Mol. Pharm.*, 2016, **13**, 3903.
9. H. A. Dad, T. W. Gu, A. Q. Zhu, L. Q. Huang and L. H. Peng, Plant Exosome-like Nanovesicles: Emerging Therapeutics and Drug Delivery Nanoplatforms, *Mol. Ther.*, 2021, **29**, DOI: 10.1016/j.ymthe.2020.11.030.
10. H. M. G. Barriga, M. N. Holme and M. M. Stevens, Cubosomes: The Next Generation of Smart Lipid Nanoparticles?, *Angew. Chem., Int. Ed.*, 2019, **58**, DOI: 10.1002/anie.201804067.
11. J. Lu, *et al.*, One-Step Protein Conjugation to Upconversion Nanoparticles, *Anal. Chem.*, 2015, **87**, 10406.

12. Y. Xu, *et al.*, Novel Solid Lipid Nanoparticle with Endosomal Escape Function for Oral Delivery of Insulin, *ACS Appl. Mater. Interfaces*, 2018, **10**, 9315.
13. A. E. Pottash, C. Kuffner, M. Noonan-Shueh and S. M. Jay, Protein-based vehicles for biomimetic RNAi delivery, *J. Biol. Eng.*, 2019, **13**, 1.
14. A. Ghosal and N. Bandara, Lipid-based nanostructures in food applications, in *Food, Medical, and Environmental Applications of Nanomaterials*, 2022, DOI: 10.1016/B978-0-12-822858-6.00005-4.
15. M. M. Leena, T. Anukiruthika, J. A. Moses and C. Anandharamakrishnan, Co-delivery of curcumin and resveratrol through electrosprayed core-shell nanoparticles in 3D printed hydrogel, *Food Hydrocolloids*, 2022, **124**, 107200.
16. C. Chang, T. G. Meikle, C. J. Drummond, Y. Yang and C. E. Conn, Comparison of cubosomes and liposomes for the encapsulation and delivery of curcumin, *Soft Matter*, 2021, **17**, 3306.
17. F. Dabbagh Moghaddam, *et al.*, Delivery of melittin-loaded niosomes for breast cancer treatment: an in vitro and in vivo evaluation of anti-cancer effect, *Cancer Nanotechnol.*, 2021, **12**, 14.
18. S. O. Dozie-Nwachukwu, *et al.*, Extraction and encapsulation of prodigiosin in chitosan microspheres for targeted drug delivery, *Mater. Sci. Eng., C*, 2017, **71**, 268.
19. B. Maity, S. Samanta, S. Sarkar, S. Alam and T. Govindaraju, Injectable Silk Fibroin-Based Hydrogel for Sustained Insulin Delivery in Diabetic Rats, *ACS Appl. Bio Mater.*, 2020, **3**, 3544.
20. T. Egan, D. O'Riordan, M. O'Sullivan and J. C. Jacquier, Cold-set whey protein microgels as pH modulated immobilisation matrices for charged bioactives, *Food Chem.*, 2014, **156**, 197.
21. G. Kaur, S. Sharma, S. A. Mir and B. N. Dar, Nanobiocomposite Films: a "Greener Alternate" for Food Packaging, *Food Bioprocess Technol.*, 2021, **14**, DOI: 10.1007/s11947-021-02634-x.
22. A. M. Elbagory, R. M. Marima and Z. Dlamini, Role and merits of green based nanocarriers in cancer treatment, *Cancers*, 2021, **13**, DOI: 10.3390/cancers13225686.
23. A. Ghosal, O. U. Rahman and S. Ahmad, High-Performance Soya Polyurethane Networked Silica Hybrid Nanocomposite Coatings, *Ind. Eng. Chem. Res.*, 2015, **54**, 12770.
24. P. Rozhin, C. Charitidis and S. Marchesan, Self-assembling peptides and carbon nanomaterials join forces for innovative biomedical applications, *Molecules*, 2021, **26**, DOI: 10.3390/molecules26134084.
25. M. Agrawal, *et al.*, Biomaterials in treatment of Alzheimer's disease, *Neurochem. Int.*, 2021, **145**, 105008.
26. F. Zafar, A. Ghosal, E. Sharmin, R. Chaturvedi and N. Nishat, A review on cleaner production of polymeric and nanocomposite coatings based on waterborne polyurethane dispersions from seed oils, *Prog. Org. Coat.*, 2019, **131**, 259.

27. P. Ghaffari-Bohlouli, *et al.*, Protein by-products: Composition, extraction, and biomedical applications, *Crit. Rev. Food Sci. Nutr.*, 2022, **63**, 9436.
28. J. A. C. Pérez, *et al.*, Bioinspired biomaterials and enzyme-based biosensors for point-of-care applications with reference to cancer and bio-imaging, *Biocatal. Agric. Biotechnol.*, 2019, **17**, DOI: 10.1016/j.bcab.2018.11.015.
29. J. R. Smith, V. Kholodovych, D. Knight, W. J. Welsh and J. Kohn, QSAR models for the analysis of bioresponse data from combinatorial libraries of biomaterials, *QSAR Comb. Sci.*, 2005, **24**, 99.
30. J. M. Anderson, Biocompatibility and Bioresponse to Biomaterials, in *Principles of Regenerative Medicine*, 2019, https://doi.org/10.1016/b978-0-12-809880-6.00039-4.
31. M. Venkateswara Rao, C. K. Sunil, A. Rawson, D. v. Chidanand and N. Venkatachlapathy, Modifying the plant proteins techno-functionalities by novel physical processing technologies: a review, *Crit. Rev. Food Sci. Nutr.*, 2021, DOI: 10.1080/10408398.2021.1997907.
32. J. Jovanović, *et al.*, Enzymatic treatment of soy protein concentrate: Influence on the potential techno-functional and antioxidant properties,*J. Hyg. Eng. Des.*, 2020, **30**, 58.
33. H. Zhou, *et al.*, Injectable biomaterials for translational medicine, *Mater. Today*, 2019, **28**, DOI: 10.1016/j.mattod.2019.04.020.
34. D. Grinberg, *et al.*, 4D Printing based piezoelectric composite for medical applications,*J. Polym. Sci., Part B: Polym. Phys.*, 2019, **57**, 109.
35. J. M. Anderson, A. Rodriguez and D. T. Chang, Foreign body reaction to biomaterials, *Semin. Immunol.*, 2008, **20**, DOI: 10.1016/j.smim.2007.11.004.
36. T. Tong, L. Wang, X. You and J. Wu, Nano and microscale delivery platforms for enhanced oral peptide/protein bioavailability, *Biomater. Sci.*, 2020, **8**, DOI: 10.1039/d0bm01151g.
37. C. R. Garcia, *et al.*, Nanoemulsion delivery systems for enhanced efficacy of antimicrobials and essential oils, *Biomater. Sci.*, 2022, **10**, DOI: 10.1039/d1bm01537k.
38. C. A. Peng, L. Kozubowski and W. R. Marcotte, Advances in Plant-Derived Scaffold Proteins, *Front. Plant Sci.*, 2020, **11**, DOI: 10.3389/fpls.2020.00122.
39. S. Sankaran, J. Becker, C. Wittmann and A. del Campo, Optoregulated Drug Release from an Engineered Living Material: Self-Replenishing Drug Depots for Long-Term, Light-Regulated Delivery, *Small*, 2019, **15**, 1804717.
40. S. Gomes, I. B. Leonor, J. F. Mano, R. L. Reis and D. L. Kaplan, Natural and genetically engineered proteins for tissue engineering, *Prog. Polym. Sci.*, 2012, **37**, DOI:10.1016/j.progpolymsci.2011.07.003.

41. W. Kim, Y. Wang and C. Selomulya, Dairy and plant proteins as natural food emulsifiers, *Trends Food Sci. Technol.*, 2020, **105**, DOI: 10.1016/j.tifs.2020.09.012.
42. M. Kumar, *et al.*, Plant-based proteins and their multifaceted industrial applications, *LWT*, 2022, **154**, 112620.
43. M. Mahdavi, A. Fattahi and S. Nouranian, Doxorubicin Stability and Retention on PEGylated Graphene Oxide Nanocarriers Adjacent to Human Serum Albumin, *ACS Appl. Bio Mater.*, 2020, **3**, 7646.
44. F. Khan, *et al.*, Synthesis, classification and properties of hydrogels: Their applications in drug delivery and agriculture, *J. Mater. Chem. B*, 2022, **10**, DOI: 10.1039/d1tb01345a.
45. S. H. M. Gorissen, *et al.*, Protein content and amino acid composition of commercially available plant-based protein isolates, *Amino Acids*, 2018, **50**, 1685.
46. G. Labib, Overview on zein protein: a promising pharmaceutical excipient in drug delivery systems and tissue engineering, *Expert Opin. Drug Delivery*, 2018, **15**, DOI: 10.1080/17425247.2017.1349752.
47. H. Liu, *et al.*, Predicting effective drug combinations using gradient tree boosting based on features extracted from drug-protein heterogeneous network, *BMC Bioinf.*, 2019, **20**, 1.
48. B. A. Witika, *et al.*, Nano-Biomimetic Drug Delivery Vehicles: Potential Approaches for COVID-19 Treatment, *Molecules*, 2020, **25**, DOI: 10.3390/MOLECULES25245952.
49. J. Schwestka and E. Stoger, Microparticles and nanoparticles from plants—the benefits of bioencapsulation, *Vaccines*, 2021, **9**, DOI: 10.3390/vaccines9040369.
50. M. G. Corigliano, *et al.*, Plant heat shock protein 90 as carrier-adjuvant for immunization against a reporter antigen, *Vaccine*, 2013, **31**, 5872.
51. S. Hong, *et al.*, Protein-based nanoparticles as drug delivery systems, *Pharmaceutics*, 2020, **12**, DOI: 10.3390/pharmaceutics12070604.
52. Y. Zou, *et al.*, Encapsulation of quercetin in biopolymer-coated zein nanoparticles: Formation, stability, antioxidant capacity, and bioaccessibility, *Food Hydrocolloids*, 2021, **120**, 106980.
53. Y. Wei, *et al.*, Core-shell pea protein-carboxymethylated corn fiber gum composite nanoparticles as delivery vehicles for curcumin, *Carbohydr. Polym.*, 2020, **240**, 116273.
54. S. Mehra, S. Nisar, S. Chauhan, V. Singh and S. Rattan, Soy Protein-Based Hydrogel under Microwave-Induced Grafting of Acrylic Acid and 4-(4-Hydroxyphenyl)butanoic Acid: A Potential Vehicle for Controlled Drug Delivery in Oral Cavity Bacterial Infections, *ACS Omega*, 2020, **5**, 21610.
55. M. B. Stie, *et al.*, Waterborne Electrospinning of α-Lactalbumin Generates Tunable and Biocompatible Nanofibers for Drug Delivery, *ACS Appl. Nano Mater.*, 2020, **3**, 1910.

56. C. E. Gumus, E. A. Decker and D. J. McClements, Gastrointestinal fate of emulsion-based ω-3 oil delivery systems stabilized by plant proteins: Lentil, pea, and faba bean proteins, *J. Food Eng.*, 2017, **207**, 90.
57. E. J. Lee, N. K. Lee and I. S. Kim, Bioengineered protein-based nanocage for drug delivery, *Adv. Drug Deliv. Rev.*, 2016, **106**, DOI: 10.1016/j.addr.2016.03.002.
58. O. A. A. Ahmed, K. M. Hosny, M. M. Al-Sawahli and U. A. Fahmy, Optimization of caseinate-coated simvastatin-zein nanoparticles: Improved bioavailability and modified release characteristics, *Drug Des., Dev. Ther.*, 2015, **9**, 655.
59. S. Sadeghi, A. Madadlou and M. Yarmand, Microemulsification–cold gelation of whey proteins for nanoencapsulation of date palm pit extract, *Food Hydrocolloids*, 2014, **35**, 590.
60. J. G. Hardy, A. Leal-Egaña and T. R. Scheibel, Engineered spider silk protein-based composites for drug delivery, *Macromol. Biosci.*, 2013, **13**, 1431.
61. L. Chen and M. Subirade, Elaboration and characterization of soy/zein protein microspheres for controlled nutraceutical delivery, *Biomacromolecules*, 2009, **10**, 3327.
62. S. Ramteke, N. Ganesh, S. Bhattacharya and N. Jain, Triple therapy-based targeted nanoparticles for the treatment of Helicobacter pylori, *J. Drug Target.*, 2008, **16**, 694.
63. S. Yu, *et al.*, Optimizing light sources for selective growth of purple bacteria and efficient formation of value-added products, *J. Cleaner Prod.*, 2021, **280**, 124493.
64. M. Kumar, *et al.*, Advances in the plant protein extraction: Mechanism and recommendations, *Food Hydrocolloids*, 2021, **115**, DOI: 10.1016/j.foodhyd.2021.106595.
65. O. D. Fagbemi, B. Sithole and T. Tesfaye, Optimization of keratin protein extraction from waste chicken feathers using hybrid pre-treatment techniques, *Sustainable Chem. Pharm.*, 2020, **17**, 100267.
66. J. Yang and L. M. C. Sagis, Interfacial behavior of plant proteins — novel sources and extraction methods, *Curr. Opin. Colloid Interface Sci.*, 2021, **56**, DOI: 10.1016/j.cocis.2021.101499.
67. E. A. Oke and S. P. Ijardar, Advances in the application of deep eutectic solvents based aqueous biphasic systems: An up-to-date review, *Biochem. Eng. J.*, 2021, **176**, DOI: 10.1016/j.bej.2021.108211.
68. F. G. Chizoba Ekezie, J. H. Cheng and D. W. Sun, Effects of nonthermal food processing technologies on food allergens: A review of recent research advances, *Trends Food Sci. Technol.*, 2018, **74**, DOI: 10.1016/j.tifs.2018.01.007.
69. L. M. Ickenstein and P. Garidel, Expert Opinion on Drug Delivery Lipid-based nanoparticle formulations for small molecules and RNA drugs Lipid-based nanoparticle formulations for small molecules and RNA drugs, *Expert Opin. Drug Delivery*, 2019, **16**, 1205.

70. Y. Sato, T. Nakamura, Y. Yamada and H. Harashima, The nanomedicine rush: New strategies for unmet medical needs based on innovative nano DDS, *J. Controlled Release*, 2021, **330**, 305.
71. S. Tiwari, *et al.*, Antibacterial and antiviral high-performance nanosystems to mitigate new SARS-CoV-2 variants of concern, *Curr. Opin. Biomed. Eng.*, 2022, **21**, 100363.
72. A. Ashkar, A. Sosnik and M. Davidovich-Pinhas, Structured edible lipid-based particle systems for oral drug-delivery, *Biotechnol. Adv.*, 2022, **54**, DOI: 10.1016/j.biotechadv.2021.107789.
73. A. Beloqui, M. Á. Solinís, A. Rodríguez-Gascón, A. J. Almeida and V. Préat, Nanostructured lipid carriers: Promising drug delivery systems for future clinics, *Nanomedicine*, 2016, **12**, 143.
74. M. Amiri, *et al.*, Engineered Solid Lipid Nanoparticles and Nanostructured Lipid Carriers as New Generations of Blood-Brain Barrier Transmitters, *ACS Chem. Neurosci.*, 2021, **12**, DOI: 10.1021/acschemneuro.1c00540.
75. O. Taratula, A. Kuzmov, M. Shah, O. B. Garbuzenko and T. Minko, Nanostructured lipid carriers as multifunctional nanomedicine platform for pulmonary co-delivery of anticancer drugs and siRNA, *J. Controlled Release*, 2013, **171**, 349.
76. Y. Huang, *et al.*, In vitro and in vivo correlation for lipid-based formulations: Current status and future perspectives, *Acta Pharm. Sin. B*, 2021, **11**, DOI: 10.1016/j.apsb.2021.03.025.
77. E. Elizondo, J. Veciana and N. Ventosa, Nanostructuring molecular materials as particles and vesicles for drug delivery, using compressed and supercritical fluids, *Nanomedicine*, 2012, **7**, DOI: 10.2217/nnm.12.110.
78. N. Kimura, *et al.*, Development of a Microfluidic-Based Post-Treatment Process for Size-Controlled Lipid Nanoparticles and Application to siRNA Delivery, *ACS Appl. Mater. Interfaces*, 2020, **12**, 34011.
79. R. P. Singh, H. v. Gangadharappa and K. Mruthunjaya, Phospholipids: Unique carriers for drug delivery systems, *J. Drug Delivery Sci. Technol.*, 2017, **39**, DOI: 10.1016/j.jddst.2017.03.027.
80. M. Rahimnejad, *et al.*, Emerging Phospholipid Nanobiomaterials for Biomedical Applications to Lab-on-a-Chip, Drug Delivery, and Cellular Engineering, *ACS Appl. Bio Mater.*, 2021, **4**, DOI: 10.1021/acsabm.1c00932.
81. T. Dissanayake, X. Sun, L. Abbey and N. Bandara, Recent advances in lipid-protein conjugate-based delivery systems in nutraceutical, drug, and gene delivery, *Food Hydrocolloids Health*, 2022, **2**, 100054.
82. L. De Souza Simões, *et al.*, Micro- and nano bio-based delivery systems for food applications: In vitro behavior, *Adv. Colloid Interface Sci.*, 2017, **243**, 23.

83. M. Fathi, M. R. Mozafari and M. Mohebbi, Nanoencapsulation of food ingredients using lipid based delivery systems, *Trends Food Sci. Technol.*, 2012, **23**, 13.
84. D. Verma, N. Gulati, S. Kaul, S. Mukherjee and U. Nagaich, Protein based nanostructures for drug delivery, *J. Pharm.*, 2018, **2018**, 9285854.
85. A. O. Elzoghby, W. M. Samy and N. A. Elgindy, Protein-based nanocarriers as promising drug and gene delivery systems, *J. Controlled Release*, 2012, **161**, 38.
86. A. A. Khan, K. S. Allemailem, S. A. Almatroodi, A. Almatroudi and A. H. Rahmani, Recent strategies towards the surface modification of liposomes: an innovative approach for different clinical applications, *3 Biotech*, 2020, **10**, 1.
87. X. He, *et al.*, Encapsulation of teniposide into albumin nanoparticles with greatly lowered toxicity and enhanced antitumor activity, *Int. J. Pharm.*, 2015, **487**, 250.
88. M. Gaber, *et al.*, Protein-lipid nanohybrids as emerging platforms for drug and gene delivery: challenges and outcomes, *J. Controlled Release*, 2017, **254**, 75.
89. P. C. Li, *et al.*, Gelatin scaffold with multifunctional curcumin-loaded lipid-PLGA hybrid microparticles for regenerating corneal endothelium, *Mater. Sci. Eng., C*, 2021, **120**, 111753.
90. D. Pooja, *et al.*, Improving efficacy, oral bioavailability, and delivery of paclitaxel using protein-grafted solid lipid nanoparticles, *Mol. Pharm.*, 2016, **13**, 3903.
91. S. Chen, *et al.*, Human serum albumin (HSA) coated liposomal indocyanine green for in vivo tumor imaging, *RSC Adv.*, 2016, **6**, 15220.
92. W. Chen, *et al.*, Rice bran protein-based nanoemulsion carrier for improving stability and bioavailability of quercetin, *Food Hydrocolloids*, 2020, **108**, 106042.
93. Q. Peng, *et al.*, Nanocomplex based on biocompatible phospholipids and albumin for long-circulation applications, *ACS Appl. Mater. Interfaces*, 2014, **6**, 13730.
94. S. Pawar and V. Pande, Oleic acid coated gelatin nanoparticles impregnated gel for sustained delivery of zaltoprofen: formulation and textural characterization, *Adv. Pharm. Bull.*, 2015, **5**, 537.
95. C. Tao, Y. J. Chuah, C. Xu and D.-A. Wang, Albumin conjugates and assemblies as versatile bio-functional additives and carriers for biomedical applications, *J. Mater. Chem. B*, 2019, **7**, 357.
96. T. Chuacharoen and C. M. Sabliov, Stability and controlled release of lutein loaded in zein nanoparticles with and without lecithin and pluronic F127 surfactants, *Colloids Surf., A*, 2016, **503**, 11.
97. P. H.-L. Tran, T. T.-D. Tran, T. V. Vo, C. L.-N. Vo and B.-J. Lee, Novel multifunctional biocompatible gelatin-oleic acid conjugate: self-assembled nanoparticles for drug delivery, *J. Biomed. Nanotechnol.*, 2013, **9**, 1416.

98. L. Zou, *et al.*, Enhancing the bioaccessibility of hydrophobic bioactive agents using mixed colloidal dispersions: Curcumin-loaded zein nanoparticles plus digestible lipid nanoparticles, *Food Res. Int.*, 2016, **81**, 74.
99. F. Liu, *et al.*, Fabrication and characterization of protein-phenolic conjugate nanoparticles for co-delivery of curcumin and resveratrol, *Food Hydrocolloids*, 2018, **79**, 450.
100. M. Freire, *et al.*, Emulsion gels containing n-3 fatty acids and condensed tannins designed as functional fat replacers, *Food Res. Int*, 2018, **113**, 465.
101. Y. Fan, J. Yi, Y. Zhang, Z. Wen and L. Zhao, Physicochemical stability and in vitro bioaccessibility of β-carotene nanoemulsions stabilized with whey protein-dextran conjugates, *Food Hydrocolloids*, 2017, **63**, 256.
102. L. Dai, *et al.*, Structural characterization, formation mechanism and stability of curcumin in zein-lecithin composite nanoparticles fabricated by antisolvent co-precipitation, *Food Chem.*, 2017, **237**, 1163.
103. H. B. Ruttala, *et al.*, Molecularly targeted co-delivery of a histone deacetylase inhibitor and paclitaxel by lipid-protein hybrid nanoparticles for synergistic combinational chemotherapy, *Oncotarget*, 2017, **8**, 14925.
104. C. Park, C. L.-N. Vo, T. Kang, E. Oh and B.-J. Lee, New method and characterization of self-assembled gelatin-oleic nanoparticles using a desolvation method via carbodiimide/N-hydroxysuccinimide (EDC/NHS) reaction, *Eur. J. Pharm. Biopharm.*, 2015, **89**, 365.
105. B. Tang, *et al.*, Lipid-albumin nanoassemblies co-loaded with borneol and paclitaxel for intracellular drug delivery to C6 glioma cells with P-gp inhibition and its tumor targeting, *Asian J. Pharm. Sci.*, 2015, **10**, 363.
106. S. R. ElMasry, R. M. Hathout, M. Abdel-Halim and S. Mansour, In Vitro transdermal delivery of sesamol using oleic acid chemically-modified gelatin nanoparticles as a potential breast cancer medication, *J. Drug Delivery Sci. Technol.*, 2018, **48**, 30.
107. N. M. Meghani, *et al.*, Design and evaluation of clickable gelatin-oleic nanoparticles using fattigation-platform for cancer therapy, *Int. J. Pharm.*, 2018, **545**, 101.
108. P. Guan, *et al.*, Solidification of liposomes by freeze-drying: The importance of incorporating gelatin as interior support on enhanced physical stability, *Int. J. Pharm.*, 2015, **478**, 655.
109. D. J. McClements, Advances in edible nanoemulsions: Digestion, bioavailability, and potential toxicity, *Prog. Lipid Res.*, 2021, **81**, 101081.
110. J. D. Heidel and M. E. Davis, Clinical developments in nanotechnology for cancer therapy, *Pharm. Res.*, 2011, **28**, 187.
111. L. Pan, *et al.*, Whey Protein Isolate Coated Liposomes as Novel Carrier Systems for Astaxanthin, *Eur. J. Lipid Sci. Technol.*, 2020, **122**, 1900325.

112. S. Woods, *et al.*, Glucosamine-NISV delivers antibody across the blood-brain barrier: Optimization for treatment of encephalitic viruses, *J. Controlled Release*, 2020, **324**, 644.
113. A. Akbari, A. Lavasanifar and J. Wu, Interaction of cruciferin-based nanoparticles with Caco-2 cells and Caco-2/HT29-MTX co-cultures, *Acta Biomater.*, 2017, **64**, 249.
114. R. van der Meel, L. J. C. Vehmeijer, R. J. Kok, G. Storm and E. V. B. van Gaal, Ligand-targeted particulate nanomedicines undergoing clinical evaluation: Current status, *Adv. Drug Deliv. Rev.*, 2013, **65**, 1284.
115. A. Jain, S. K. Singh, S. K. Arya, S. C. Kundu and S. Kapoor, Protein Nanoparticles: Promising Platforms for Drug Delivery Applications, *ACS Biomater. Sci. Eng.*, 2018, **4**, DOI: 10.1021/acsbiomaterials.8b01098.
116. A. S. Pitek, H. Hu, S. Shukla and N. F. Steinmetz, Cancer Theranostic Applications of Albumin-Coated Tobacco Mosaic Virus Nanoparticles, *ACS Appl. Mater. Interfaces*, 2018, **10**, 39468.
117. A. Ashfaq, K. Jahan, R. U. Islam and K. Younis, Protein-based functional colloids and their potential applications in food: A review, *LWT*, 2022, **154**, 112667.
118. K. Han, T. Y. Park, K. Yong and H. J. Cha, Combinational Biomimicking of Lotus Leaf, Mussel, and Sandcastle Worm for Robust Superhydrophobic Surfaces with Biomedical Multifunctionality: Antithrombotic, Antibiofouling, and Tissue Closure Capabilities, *ACS Appl. Mater. Interfaces*, 2019, **11**, 9777.
119. M. D. Levasseur, *et al.*, Cell-Specific Delivery Using an Engineered Protein Nanocage, *ACS Chem. Biol.*, 2021, **16**, 838.

CHAPTER 9

Hydrophobic and Superhydrophobic Protein-based Materials for Functional Applications

BOON PENG CHANG[a], JIAN ZHOU[b] AND TIZAZU H. MEKONNEN*[a]

[a]Department of Chemical Engineering, University of Waterloo, 200 University Avenue, Engineering 6, N2L 3G1 Waterloo, ON, Canada; [b]School of Material Science and Engineering, Key Laboratory for Polymeric Composite & Functional Materials of Ministry of Education, Guangzhou Key Laboratory of Flexible Electronic Materials and Wearable Devices, Sun Yat-sen University, Guangzhou, Guangdong 510275, China
*E-mail: tizazu.mekonnen@uwaterloo.ca

9.1 Introduction

Water contact angle (WCA) provides us with an indication of the wettability or the interface interaction behavior of a liquid spread over a solid surface. When water droplets fall on a solid flat surface, the WCA typically ranges between 0 and 90° (hydrophilic behavior). Hydrophobic (water-hating) surface produces a WCA in the range of 90–150°. A higher WCA indicates a lower wettability of the surface. Superhydrophobic materials show an almost perfect sphere with a WCA equal to or greater than 150°, which

Smart Materials No. 41
Functional Materials from Lipids and Proteins
Edited by Nandika Bandara and Aman Ullah

Published by the Royal Society of Chemistry, www.rsc.org

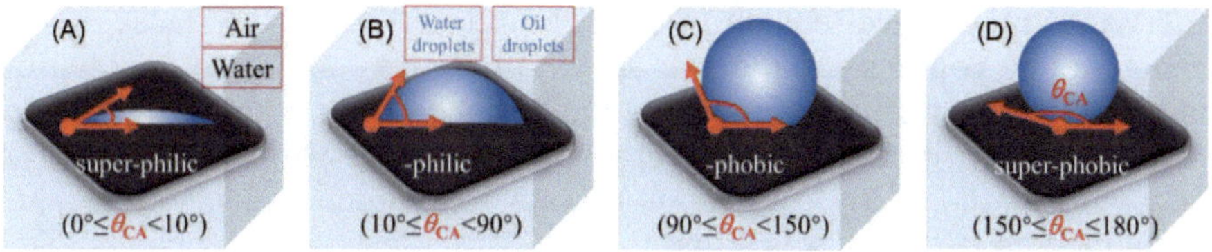

Figure 9.1 Graphical images of the degree of contact angles θ_{CA} and water or oil droplets on four types of surfaces. Adapted from ref. 1 with permission from Elsevier, Copyright 2021.

are non-wetting (see Figure 9.1). A sliding angle (SA) is another common parameter reported along with a WCA. SA represents the angle at which a water droplet starts to roll off from the contacting surfaces when tilted. Water droplets roll off from superhydrophobic surfaces effortlessly when tilted at a small SA due to poor wettability and a high WCA. These contact angle parameters provide important information for developing ingredient formulations and modifications to the coating, ink or paint that spreads over the substrate surface.

Superhydrophobic materials are a relatively new and emerging advanced class of materials that are in high demand due to their potential applications in a variety of advanced products, such as self-cleaning materials; oil/water separation; drug delivery; tissue engineering; high-sensitive sensors; liquid-repellent and stain-resistant textiles; anti-icing, anti-fouling and anti-corrosion and optical devices.[2,3] These applications play a vital role in solving our daily life problems and could bring the human lifestyle to the next level. For example, applying a hydrophobic coating to ship hulls can reduce water resistance, thereby increasing cruising speed and decreasing fuel consumption. Similarly, hydrophobic coatings on airplane or car windshields can repel rain and dirt, keeping them clean.

The wetting behavior of solid materials is usually measured by evaluating the contact angle (θ) between a water droplet and the contact surface, which is defined as the angle produced by solid–liquid interface (γ_{SL}) and the liquid–vapor interface (γ_{LV}) (see Figure 9.2A). The relationship between

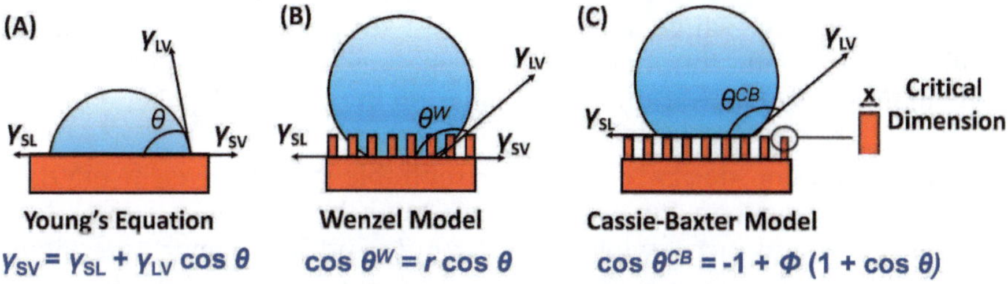

Figure 9.2 Schematic illustration of wetting behavior models: (A) Young's model, (B) Wenzel's model and (C) Cassie–Baxter's model.

the WCA and surface tension of solid, liquid and vapor interfaces can be expressed by Young's equation, as shown in Figure 9.2A. Young's equation is based on the wetting of perfectly flat surfaces at an equilibrium energy state. However, it cannot be directly applied to real-world material surfaces, which are mostly rough. Another two extended models were later developed to address this limitation of Young's equation, which takes material surface energies and surface roughness into consideration. These two models, *i.e.*, Wenzel and Cassie–Baxter's models, describe the WCA with various degrees of heterogeneous surfaces for real-world objects. The Wenzel model assumes that a liquid droplet completely fills up the rough grooves of the surface at the area of contact and becomes unable to slide on the surface[4] (see Figure 9.2B). The Wenzel model is presented in eqn (9.1):

$$\cos\theta^{w} = r\cos\theta \tag{9.1}$$

where θ^{w} is the apparent contact angle (Wenzel), θ is the actual contact angle and r is the surface roughness factor. Therefore, r is unity ($r = 1$) when the surface is completely smooth or flat, while $r > 1$ represents a rough surface.

On the other hand, Cassie–Baxter's model assumes that water droplets contact the top asperities without penetrating the valleys of surfaces.[4] A "composite" surface is then formed between the trapped air packets and the tip of the surface (see Figure 9.2C). The Cassie–Baxter model is presented in eqn (9.2):

$$\cos\theta^{CB} = -1 + \Phi(1 + \cos\theta) \tag{9.2}$$

where θ^{CB} is the apparent contact angle (Cassie–Baxter), θ is the actual contact angle and Φ is the fraction of the solid surface in contact with the liquid.

The Cassie–Baxter model is effective in predicting the WCA of highly rough surfaces. For instance, the water droplet does not penetrate the hierarchical surface of the lotus leaf. Air is entrapped in the convex cell grooves of the lotus leaf, resulting in the Cassie–Baxter state. The water droplet rests mostly on the entrapped air and has minimum contact with the leaf structure, which makes it superhydrophobic.[5,6] The wettability models were briefly introduced in this chapter, and for a more detailed explanation of the derivation and explanation of the models, please refer to ref. 3, 4 and 6–9.

Oftentimes, the design of superhydrophobic materials is inspired by nature. The most popular examples are lotus leaves, water strider legs, duck feathers and insect wings (*i.e.*, butterfly and leafhopper) (see Figure 9.3). The unique surface topography and microstructure of the lotus leaf enable superior water repellency and self-cleaning properties (a.k.a. the

Figure 9.3 Superhydrophobic surfaces found in nature: (A) lotus leaf, (B) water strider legs, (C) duck feather and (D) leafhopper wing.

lotus effect). Hierarchical structures and fibrous, micro-fibrous, micro-pattern and nano-pattern surface structures are typical hydrophobic surfaces found in nature. To date, many research works have focused on mimicking and recreating these micro- and nanometric scales of hierarchical structure surfaces (bioinspired) to develop functional materials with superhydrophobic behavior.[10–15]

Low surface energy and rough surface texture with hierarchical micro- and nanoscales are the two critical factors to produce hydrophobic and superhydrophobic surfaces. Typical design methods to produce superhydrophobic surface materials include spray-drying, coating, casting, topping, and chemical deposition of hydrophobic functional groups onto a substrate. These include fatty acids, fluorinated alkanes, organic silanes, long alkyl chains, aromatic azides and alkanethiols.[16–18] Besides the various types of hydrophobic surface modifications, the lotus effects inspired by the lotus leaf are the widely adopted methods to produce superhydrophobic surfaces. Zhang *et al.*[11] reported a general route to produce core/shell/satellite composite particles with a micro/nano hierarchical structure that mimics the lotus leaf surfaces. Polydopamine was used as an adhesive medium between the silica (SiO_2) microparticle substrate and metal nanoparticles, which was inspired by the strong adhesion ability of mussel protein adhesives (see Figure 9.4A). The polydopamine can completely encapsulate the hydrophilic SiO_2 substrate (see Figure 9.4B and C). The developed SiO_2/

polydopamine microparticles react with silver ions in the silver nitrate solution to form nanoparticles on the surfaces (Figure 9.4D). Surfaces constructed with this method presented a WCA of greater than 150° (Figure 9.4E). This versatile fabrication method of core/shell/satellite composite

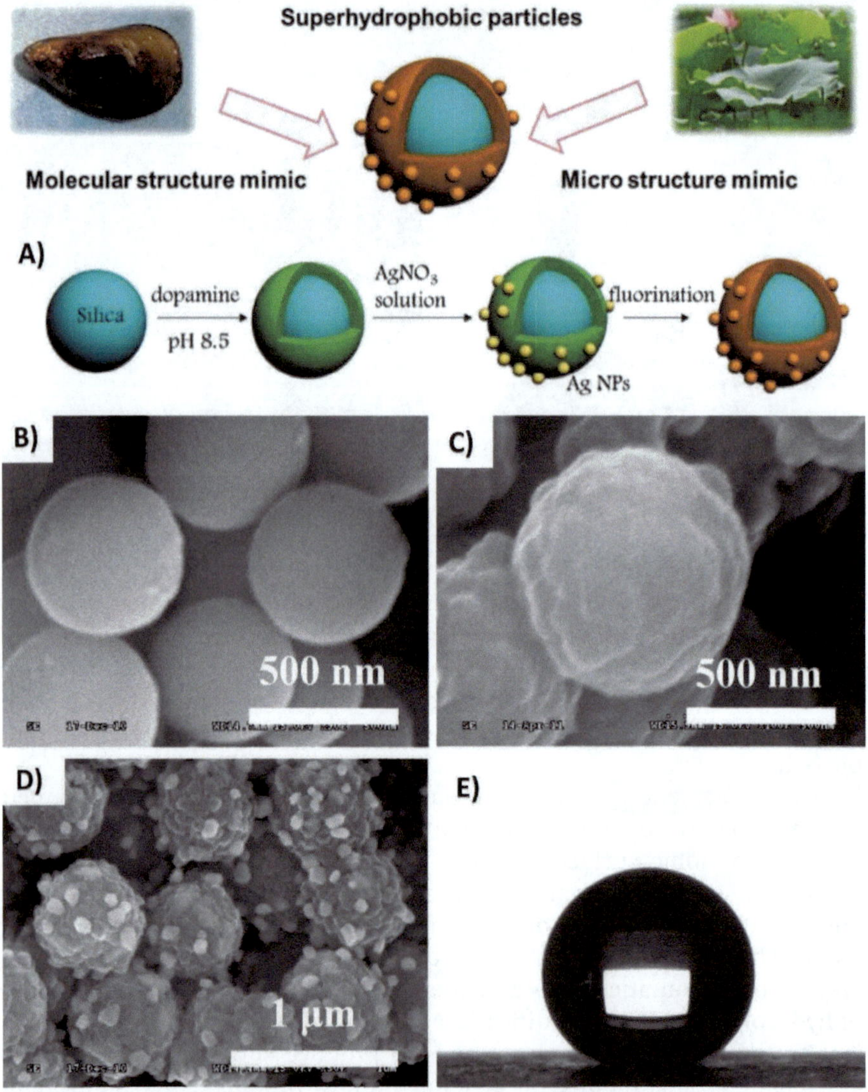

Figure 9.4 (A) Synthesis of superhydrophobic particles inspired by the lotus leaf and the mussel adhesive protein. (B) SiO_2, (C) SiO_2 coated with polydopamine, (D) SiO_2/polydopamine/nano-silver core/shell/satellite composite particles, and (E) an image of a water droplet sitting on the surface composed of superhydrophobic particles. Reproduced from ref. 11 with permission from American Chemical Society, Copyright 2012.

particles was demonstrated to be applied to any other microparticles as core materials, such as polycaprolactone particles and iron particles.[11]

Various methods can be used to produce superhydrophobic surfaces, including chemical etching, sol–gel process, electrodeposition, chemical functionalization, lithography, chemical vapor deposition templating, layer-by-layer deposition, phase separation, colloidal aggregation and assembling, electrospinning and electrospraying.[3] Most of the current superhydrophobic materials are usually produced from various synthetic polymers with fluorinated compounds. Exposure to fluorinated substances causes potential adverse health effects on the human body and is harmful to the environment. This problem has triggered the development of better, safer and environment-friendly materials from bio-based resources.

Protein-based materials derived from renewable resources, such as plants (*e.g.*, soybeans and corn) or animals (*e.g.*, milk and collagen), remain popular choices for various emerging bioproduct applications, especially biomedical and eco-friendly packaging. This is due to their biocompatibility with the human body, non-toxicity, sustainability and biodegradability. Protein is a natural biopolymer composed of diverse types of amino acids linked together *via* peptide bonds. Proteins can be categorized into 4 types based on their structure, *i.e.*, primary, secondary, tertiary and quaternary protein. Primary protein presents the linear linkage of amino acids, while secondary protein exhibits some degrees of folding (*i.e.*, α-helix and β-pleated sheet). Tertiary and quaternary proteins have complex three-dimensional shapes and structures with various bonds, including hydrogen bonds, ionic bonds, Van der Waals forces, hydrophobic bonds and disulfide bonds. Proteins derived from plants and animals are renewable, low cost and eco-friendly.[19] Figure 9.5 shows the current emerging applications of plant- and animal-based proteins in food and other industries.

While protein-based materials are attractive for green biodevices and functional biomaterial development, native proteins typically lack mechanical strength and are highly sensitive to moisture, which limits their practical applications.[21] Lipid addition, plasticization, pH alteration and cross-linking procedures induced by heat, enzymes, chemicals and irradiation are usually carried out on proteins to produce the desired performance attributes.[22] Studies have reported modifying various protein resources for developing active packaging materials[23–25] and self-healing and UV-resistant protective films.[26] The broad interest in sustainable protein-based biomaterials as an alternative to synthetic petro-based material suggests the value of further exploration of these biomaterials for various applications. More recently, several studies have focused on the development of hydrophobic and superhydrophobic protein-based materials. This chapter reviewed the advances in the fabrication of hydrophobic and superhydrophobic materials and applications from various protein-based resources. The hydrophobic and superhydrophobic properties and the related performance of proteinaceous

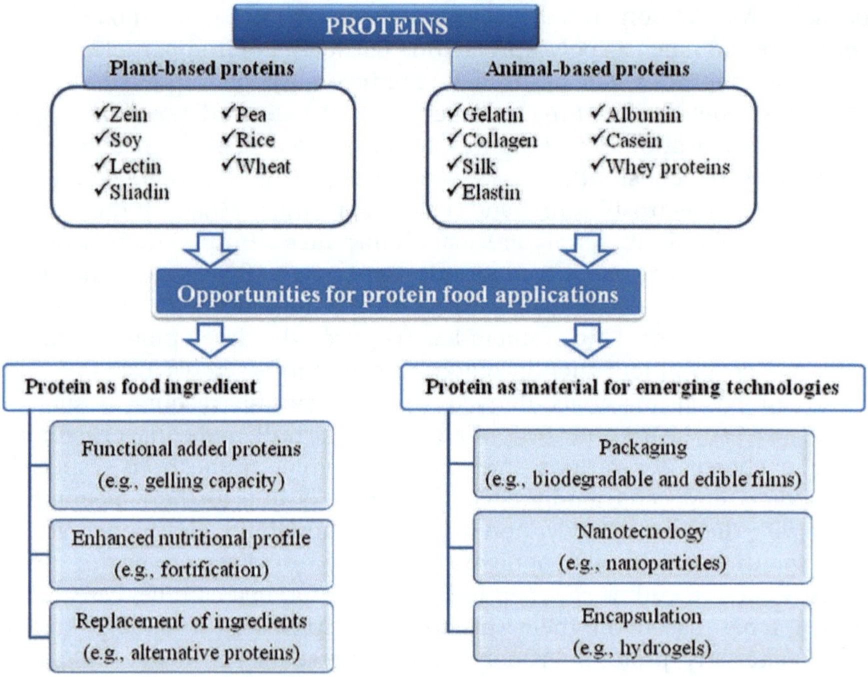

Figure 9.5 Potential applications of plant- and animal-based proteins in food and other emerging industries. Adapted from ref. 20, https://doi.org/10.3389/fsufs.2018.00077, under the terms of the CC BY 4.0 license, https://creativecommons.org/licenses/by/4.0/.

materials from plants and animals were systematically summarized and covered. Finally, the challenges and prospects of these materials in superhydrophobic applications were discussed.

9.2 Proteinaceous Superhydrophobic Materials

9.2.1 Plant-based Proteins

9.2.1.1 Soy Proteins

Soy protein isolate (SPI), obtained through various purification processes from soymeal, a co-product of the soybean oil industry, is an excellent choice for the development of bio-based advanced materials. Its great film-forming ability and ease of processability made it a popular selection for developing various advanced green products, such as hydrogels,[27,28] coatings,[29] antimicrobial films,[30] self-healing and antibacterial films,[31] wound dressings[32] and packaging materials.[23,33,34] However, SPI is relatively brittle, highly sensitive to moisture and has hydrophilicity (WCA around 40–60°), which limit its practical applications.[35] Despite these shortcomings, various studies have reported the modifications of SPI films for developing superhydrophobic materials with promising results.

SPI contains both hydrophobic (benzene rings, –SH) and hydrophilic (–OH, –COOH and $–NH_2$) groups.[36] It was reported that the hydrophobicity of oxidized corn starch films was improved after blending with 10–30 wt.% of SPI.[37] The WCA of oxidized corn starch film increased from ~37° to ~105° at 30 : 70 ratios of SPI : starch when cast at a pH 10 solution. This may be due to the increase in hydrophobic amino acids on the film surface structure as the SPI content increases. Hydrophobicity of SPI can be improved by reducing the hydrophilic groups, increasing the hydrophobic groups and increasing the inter- and intramolecular hydrogen bonding in the protein.[38] Moreover, hydrophilicity and water sensitivity of SPI can be adjusted and reduced by blending with more hydrophobic materials. It was found that the WCA and water resistance of the SPI films increase with an increasing content of poly(butylene adipate)-based waterborne polyurethanes (PUs).[39] A modified SPI film with hydrophobic copolymers, poly(styrene-co-glycidyl methacrylate), showed a WCA of ~115° compared to the unmodified SPI film (~62°).[30] The arylation of ramie fiber-reinforced SPI protein with 2,2-diphenyl-2-hydroxyethanoic acid through a simple dip-coating process improved the water resistance of the composites and increased the WCA from 33° for non-arylated to 70° after arylated.[40] Furthermore, the SPI reinforced with vertically aligned ramie fibers significantly enhances the mechanical properties of the SPI-based composites.

The manufacturing process also plays an essential role in determining the surface properties of the protein film. Garrido *et al.*[41] found that the SPI films produced by casting (wet process) were more hydrophobic than those produced by compression molding (dry process). The WCA of the casted SPI films was observed to be 99° compared to that of the compressed SPI films of only 35°. This is due to the longer times required during the film-forming process by casting, which leads to the formation of hydrophobic films. However, the films processed by compression were more thermally stable and exhibited more extraordinary tensile properties. This can be due to a higher degree of denaturation when the films were compressed at high temperatures and, thus, a better interaction between SPI and the plasticizing agents, such as glycerol.[41]

Various studies have reported the development of superhydrophobic and self-cleaning SPI films through the synthesis of hierarchical micro-/nano-structures. Xie *et al.*[42] demonstrated a facile method to develop a durable, superhydrophobic, self-cleaning photoluminescent film from SPI, polyethylene imide (adhesive medium) and nano-SiO_2 particles through the immersion coating process (Figure 9.6A). The modified soy protein films exhibited excellent self-cleaning performance without deteriorating their tensile properties and flexibility (Figure 9.6B and C). Furthermore, the film exhibited bright fluorescence even after immersion in different organic solvents for up to 3 days (Figure 9.6D). Figure 9.6E and F shows the superior WCA of more than 150° of the developed superhydrophobic film. It was noted that the WCA remained unchanged after the 72 h environmental

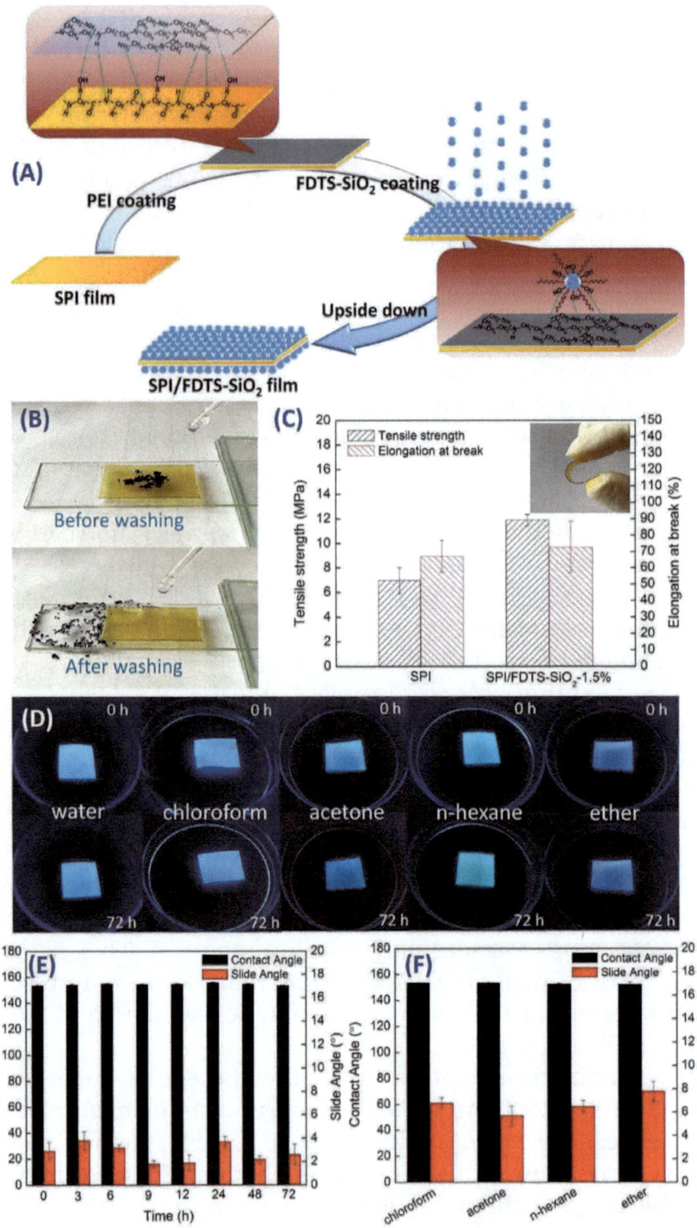

Figure 9.6 (A) The fabrication of soy protein isolate/perfluorodecyltrichlorosilane/SiO_2 nanoparticle (SPI/PFDTS-SiO_2) films. (B) Self-cleaning performance, (C) tensile properties, (D) photographs of the films under UV irradiation at 365 nm before and after immersion in different solvents for 3 days, (E) WCAs and SAs as a function of immersion time and (F) contact angles of the films after immersion in different organic solvents for 3 days. Reproduced from ref. 42 with permission from Elsevier, Copyright 2017.

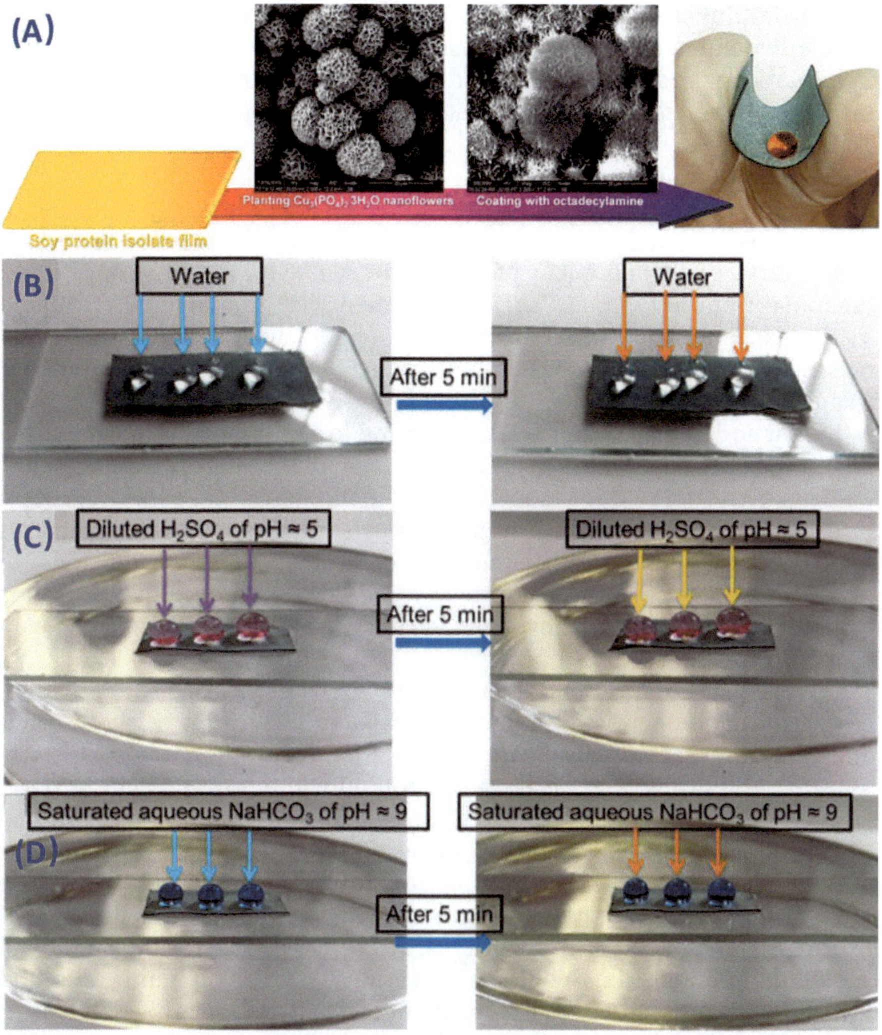

Figure 9.7 (A) The $Cu_3(PO_4)_2{\cdot}3H_2O$ nano-flower on SPI hydrophobic films. (B) The hydrophobicity of the developed SPI film toward deionized water, (C) acidic solution and (D) alkaline solution. Reproduced from ref. 43 with permission from American Chemical Society, Copyright 2017.

exposure and after immersion in different organic solvents for up to 3 days. Growing nano-flowers of $Cu_3(PO_4)_2{\cdot}3H_2O$ on SPI to obtain superhydrophobic and self-cleaning materials was also reported from the same research group[43] (see Figure 9.7A). The developed hydrophobic SPI films repelled deionized water as well as acidic or alkaline droplets (Figure 9.7B–D). The film displayed self-cleaning properties when tilted by as low as 5°.

A hydrophilic/superhydrophobic material was also produced through simple dynamic control of the crystal growth of the zeolitic imidazolate framework (a porous metal–organic framework) on the SPI film, which is inspired by the *Stenocara* beetle's ability to collect water from moist air.[44] This biomimetic film with controllable pattern surfaces displays suitable water collection properties in moist air, holding a potential application in atmospheric water collection in the future, such as capturing water fog in desert areas.

The surface hydrophobicity of SPI can further be improved by optimizing microbial transglutaminase surface treatment.[45–47] This may be attributed to the exposure of hydrophobic groups of SPI, which are initially suppressed in the interior of the molecules, before the transglutaminase treatment.[46]

Kumar and Zhang[38] reported a simple and low-cost method of generating hydrophobic groups on SPI surfaces, which mimic plant leaves. For this, an acrylation of SPI films was carried out through a dip-coating process in a benzilic acid solution. The process generated microphase structures through the non-covalent interactions of aromatic groups, resulting in phase-separated hydrophobic surfaces. The developed modified SPI films exhibited enhancement in hydrophobicity, water resistance and mechanical properties.

In practical applications, superamphiphobic materials are even more attractive than superhydrophobic materials, as they can repel both water and oils. These materials possess unique surface properties that cause water and oil droplets to bead up and roll off without wetting the surface. Liu *et al.*[35] produced durable, self-cleaning, superamphiphobic multifunctional SPI films through a facile surface-coating strategy. The SPI films were first dip-coated with polydimethylsiloxane (PDMS) as an adhesive medium and subsequently spray-coated with pre-synthesized fluorinated hybrid halloysite nanotubes/silica particles (HNTs/SiO_2). The developed superamphiphobic films could repel various liquids, including oils and low-surface-tension liquids (*e.g.*, *n*-heptane). In addition, these SPI films can withstand various harsh conditions without losing their integrity or performance, which include 20 times tape-peeling tests, strong acid/basic and organic solvent immersion and blade scratch.

9.2.1.2 Corn Zein

Corn has served as a significant staple in both food and feed industries, as well as in various industrial applications, for a considerable period. Biorefinery products from corn (*e.g.*, starch and protein) make essential contributions to the development of a wide range of bio-based products (bioproducts), biofuels and the overall bioeconomy, fostering global sustainable development. Corn starch or meal is one of the main precursors for the fermentation of sugars to produce lactic acid, which is the monomer for producing polylactic acid (PLA). The production of biofuels as a complement and an alternative to fossil fuels is growing due to environmental concerns associated with fossil fuels. The intensive expansion of the bioethanol industry that uses corn resources as a precursor has also increased the popularity of corn and its derivatives as an important industrial crop. During bioethanol production, the fermentation and distillation processes produce large amounts of organic waste by-products in the form of dried distillers' grains with solubles (DDGS). Depending on the source of grain and the bioethanol production method, DDGS mainly consists of crude protein (~25–35%), lipids and fibers. Currently, DDGS is a low-value co-product typically used as animal feed.[48] Further research is looking at the possibilities of utilizing DDGS for other value-added applications.

Zein is a protein obtained from corn, which can be extracted from corn endosperm. Recently, zein by-products from corn starch production and DDGS from bioethanol production have gained much attention for their value-added applications. Zein is a tasteless and odorless protein comprising mixtures of different peptides with various molecular sizes, charges and solubility.[49,50] Zein is generally recognized as a safe (GRAS) bio-based material due to its low toxicity, biocompatibility and biodegradability.[51] Unlike soy proteins, zein is insoluble in water and is considered an amphiphilic protein. A large amount of hydrophobic amino acid domains, rather than hydrophilic domains, in the zein molecular structures make it more hydrophobic than other types of proteins. Its hydrophobicity makes it an appealing material to tune the hydrophilicity of other bio-based materials and obtain predominantly hydrophobic functional materials.

Zein can be electrospun from various solvents, such as formic acid[52] and ethanol.[53,54] Wang and Chen[55] reported improved wet stability of electrospun hordein fibers (prolamin protein fibers) after the incorporation of zein. The developed protein fabric exhibited a strong network interaction and is highly stable in both alcohol and water solutions. Dong *et al.*[54] found that electrospinning and spray-drying of zein could produce superhydrophobic/hydrophobic materials. The authors attribute this phenomenon to the formation of droplet-mediated evaporation-induced self-assembly during the electrospinning or spraying process. The WCA of this material ranges from 135° to 156° by varying the zein concentration (10, 15 and 30 wt.%).

It was demonstrated that the hydrophilic moieties of zein proteins could be modified *via* crosslinking with aldehydes (*e.g.*, glutaraldehyde) and electrospinning. The obtained electrospun-crosslinked zein-based nanofibers displayed a maximum WCA of ~109.85°, which increased with increasing glutaraldehyde concentrations.[21] These crosslinked zein-based nanofibers with superior moisture resistance, mechanical properties and excellent filtration efficiency demonstrated great potential to be used in air filtration applications. Wang *et al.*[56] developed a series of films from zein (2–30%) and konjac glucomannan. The hydrophilicity of the konjac glucomannan film has reduced with the incorporation of zein. The WCA and moisture resistance of konjac glucomannan films were significantly enhanced with the increasing content of zein (from 2% to 10%). Other properties, such as mechanical strength, thermal stability, oxygen and water permeability, were also improved. The hydrophilicity, water vapor permeability (WVP) and water absorption rate of whey protein isolate films were also observed to reduce after the incorporation of sodium caseinate-coated zein nanoparticles, which shows great potential for food packaging materials.[51]

Dong *et al.*[57] produced a hydrophobic zein film through a facile and inexpensive method of monolayer-assisted evaporation-induced self-assembly of zein. The WCA of the zein hydrophobic film reached as high as 126°. Inspired by the lotus leaf surface, Luís *et al.*[58] developed hydrophobic zein-based films modified with licorice essential oil on PDMS substrate to produce the negative template of lotus-leaf-like rugosities to generate the hydrophobic surfaces. The zein-based films (positive replica) with similar hierarchical surface morphology to the lotus leaf were obtained (Figure 9.8). The produced films exhibited great water resistance with a WCA of ~113°. Furthermore, the films possess antibacterial properties against known foodborne pathogens while also maintaining their inherent biodegradability (weight loss ~50–60% after 10 days of the soil burial degradation test).

9.2.1.3 *Seed Proteins*

Protein-based materials from plant resources are attractive to develop edible and biodegradable food packaging films. For instance, protein films from bitter vetch seed protein (*Vicia ervilia*) were fabricated and characterized for biodegradable packaging.[59] The surface hydrophobicity,

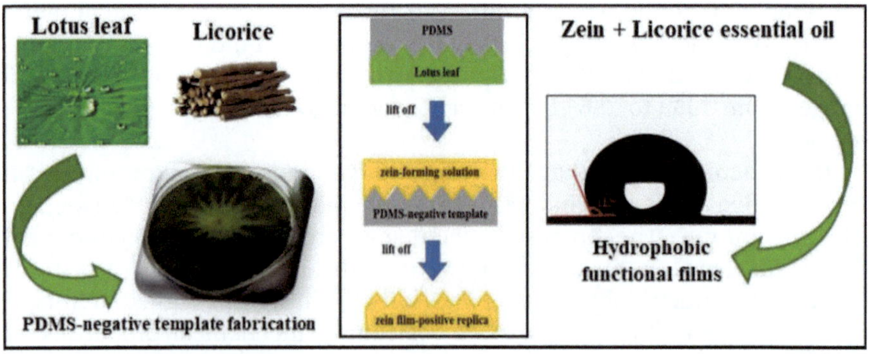

Figure 9.8 Production of hydrophobic zein-based films bioinspired by the lotus leaf surface. Adapted from ref. 58, https://doi.org/10.3390/microorganisms7080267, under the terms of the CC BY 4.0 license, https://creativecommons.org/licenses/by/4.0/.

solubility and WVP of protein films were investigated after modification with calcium chloride ($CaCl_2$). The WCA of the films increased significantly with the addition of $CaCl_2$ from 0.1 to 1.0 wt.%. At 0.7 wt.% $CaCl_2$, the protein films exhibited the highest WCA with ~95° compared to the WCA of 49.8° of the control film. The crosslinking of carboxylic groups of proteins with calcium ions (Ca^{2+}) of proteins reduced the hydrophilic moieties and total soluble matter in the protein, resulting in a marked improvement in the WVP of the films.[59] Overall, the crosslinking of the protein films with calcium salts led to the formation of a rigid three-dimensional structure, resulting in higher mechanical properties and the WVP reduction in the protein films.[60]

It was reported that the hydrophilicity and water absorption of protein films of pumpkin seeds could be reduced after blending with pea starch.[61] The WCA of the protein films of pumpkin seeds increased from ~60° to ~80°. This could be due to the reduction of carboxyl, amino and hydroxyl groups in the protein structure after blending with starch.

9.2.1.4 Gluten

In food packaging-coating applications, it was found that the WCA of the paperboard was decreased after being coated with a wheat gluten-based coating.[62] Moreover, the deposition of wheat gluten coating reduced the water vapor transmission rate of the paperboard. The effect of ultrasound treatment on the surface hydrophobicity of gluten-based film was investigated and reported by Marcuzzo *et al.*[63] It was observed that the WCA of the ultrasound-treated gluten film was less than that of the untreated gluten film. This could be due to the breakdown of the colloidal network of the developed protein-based film, resulting in the exposure of more hydrophilic groups after the ultrasound treatment. The hydrophobicity index of wheat gluten protein increased after planetary ball mill grinding for 20 min.[64] The mechanical force of planetary ball milling breaks the disulfide bonds in gluten protein structures and changes its exposure surface functionality.

9.2.2 Animal-based Proteins

9.2.2.1 Gelatin

Gelatin is a water-soluble animal protein derived from collagen. It is usually obtained *via* boiling animal tissue, such as skin, tendon and bone (by-products from fish and animal processing industries). It is a highly attractive bio-based and biodegradable material for packaging and developing a wide range of other materials because of its excellent film-forming ability. However, it does not have satisfactory water resistance to withstand high moisture when used as packaging materials.[65] Various research works have been reported on modifying gelatin to reduce its hydrophilicity for more practical applications. It was found that the WCA of gelatin/starch blends was elevated by approximately ~40° after the incorporation of graphene oxide (GO) and subsequent crosslinking in a solution-casting process.[66] The thermal stability, mechanical strength and WVP of the GO/gelatin/starch nanocomposites were also improved compared to the gelatin/starch blends. Gelatin (obtained from pigskin) film modified with starch ester (ethylenediamine tetraacetic dianhydride) exhibited hydrophobicity (WCA increased from 77.8° to 126.7°) and flexibility.[67] Deng *et al.*[68] found that the hybridization of gelatin and corn zein nanofibers (a mix ratio of 1 : 1) through the electrospinning method showed an excellent hydrophobicity with a WCA of 118° (larger than the WCA of the independent pure gelatin and zein nanofibers) and good solvent resistance against ethanol.

In a separate study, the WCA of gelatin films was found to reduce after incorporating zein (a more hydrophobic protein) even up to 40 wt.%.[65] The authors attributed this reverse phenomenon to the surface roughness generated by the zein, which affected the film's surface microstructure and wettability. Zein films cast from acetic acid showed much smoother surfaces than those cast from aqueous ethanol solutions.[69] A similar observation was reported by Xia *et al.*[70] on multilayer zein/gelatin films. Therefore, the hydrophilicity of a material does not necessarily change only with the blending with a more hydrophobic material or *via* physical and chemical grafting reactions. Hydrophobicity is a complex material characteristic that depends on surface microstructures, porosity and pore size of the end product in addition to chemical functionalities.

It was reported that the surface hydrophobicity of bovine gelatin-based film has increased after the incorporation of nano-chitin.[71] The hydrophobicity increase was commensurate with the nano-chitin concentration in the studied range (3 to 10 wt.%). The excellent compatibility of chitin nanoparticles with gelatin enhances mechanical properties, thermal stability, solubility reduction and WVP of the gelatin films. Other studies have also demonstrated that the hydrophobicity of the gelatin film can be improved with the incorporation of zinc oxide (ZnO) nanorods,[72] titanium dioxide (TiO_2),[73] clay and silver nanoparticles.[74]

Chen *et al.*[75] developed a mechanically robust, non-toxic durable superhydrophobic protein film from gelatin. For this, the gelatin films were first modified with formaldehyde and glycerol. SiO_2 nanoparticles were then grafted onto gelatin films in the ethanol solvent. The second batch of the films was grafted with modified SiO_2 nanoparticles with a fluorosilane coupling agent (see Figure 9.9A). The results show that the hydrophobicity of films depended on the particle size of the deposited SiO_2 nanoparticles, with a particle size of 303 nm produced the largest WCA compared to particle sizes of 127, 225 and 447 nm (Figure 9.9B and C). In addition, the WCA of these films remains >150° after two days of infiltration in water, which displays excellent long-term water resistance.

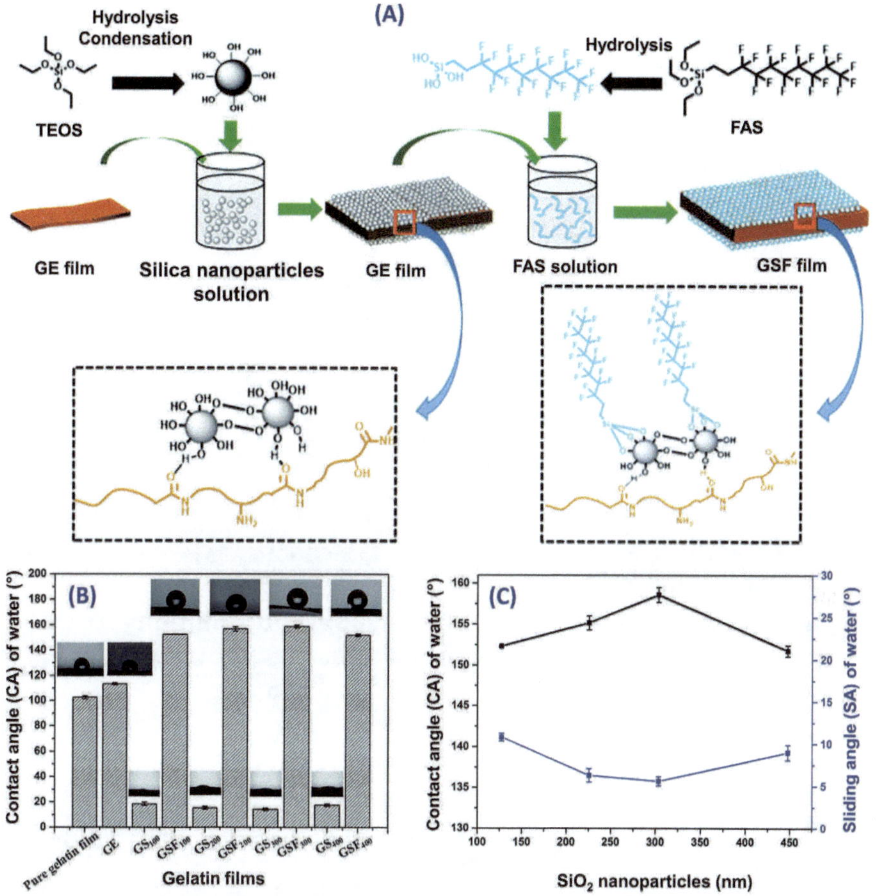

Figure 9.9 (A) Schematic illustration of the fabrication of superhydrophobic gelatin films. (B) and (C) The water contact angles of the pure gelatin film (GE), GE films with silica nanoparticles (sample name: GS) and GE films with silica nanoparticles modified with a fluorosilane coupling agent (FAS) (sample name: GSF) of varying silica nanoparticle sizes. Reproduced from ref. 75 with permission from the Royal Society of Chemistry.

9.2.2.2 *Keratin*

Keratins are cysteine-rich (sulfur-containing amino acid) fibrous animal proteins. The structures, as well as the chemical and physical properties of keratins, vary depending on the source, such as mammals (*e.g.*, hair, horns and wool), reptiles (scale) and birds (feather). α-Keratin and β-keratin are the two primary forms of proteins found in keratin fibers. In recent years, the extraction and utilization of keratin from poultry feathers for material application have drawn much interest in the research community.[76] In general, keratins are insoluble in water and tough due to disulfide bonds

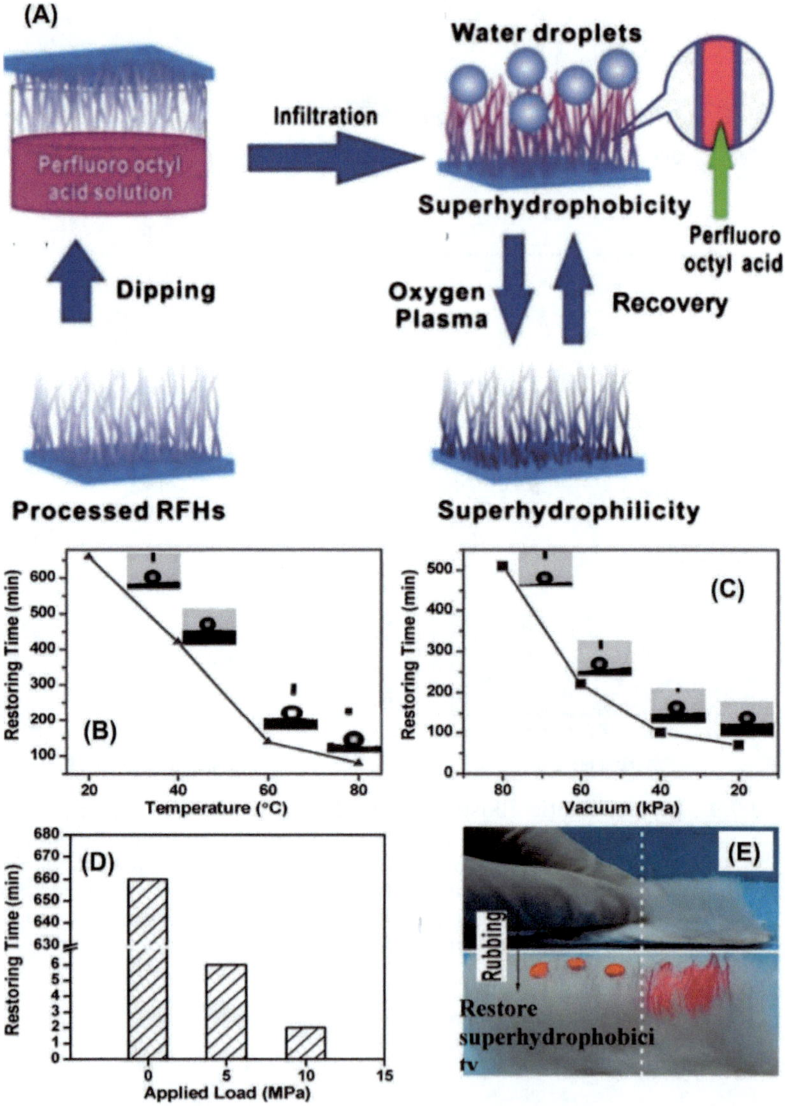

Figure 9.10 (A) Schematic illustration of the fabrication of superhydrophobic rabbit fibrous hairs (RFHs) modified with perfluorooctyl acid (PFA). (B) Recovery time with a function of (B) temperature and (C) vacuum and (D) applied load on the water contact angles of a plasma-treated surface. (E) Top: one side of the superhydrophilic RFHs–PFA rubbed by fingers; bottom: water (red ink) stains formed on the hydrophilic side, and the left side of the RFHs–PFA restored its superhydrophobicity. Adapted from ref. 79 with permission from the Royal Society Chemistry.

and cysteine side chains in the fibers. Esparza *et al.*[77] reported that the scaffolds produced from human hair and chicken feather keratin proteins showed similar WCA (~88°), while the WCA of scaffolds produced from wool keratin was slightly lower (~81°). Active keratin protein nanofiber membranes fabricated from wool fibers *via* an electrospinning process were reported by Figoli *et al.*[78] It was found that the hydrophobicity (>135° of WCA) and roughness of the keratin membranes could be increased by optimization of the electrospinning time (~2 h). However, the water permeability and pore size have decreased as the electrospinning time increased.

Yin *et al.*[79] reported a proof-of-concept of a superhydrophobic material using rabbit fibrous hairs (RFHs) modified with perfluorooctyl acid (PFA). Figure 9.10A shows the dipping process of the RFH in PFA and the regeneration and recovery of superhydrophobic behaviors after plasma treatment. It was noted that the recovery time of the superhydrophobic and plasma-treated RFH–PFA surface could be accelerated with increasing temperatures (Figure 9.10B), evacuation (Figure 9.10C), applying load (Figure 9.10D) and finger rubbing (Figure 9.10E). Overall, studies of the surface wettability modification of keratin proteins are limited in the literature.

9.2.2.3 Silk

Compared to plant-based proteins, silk-based proteins demonstrated superior mechanical properties and durability. Silkworm threads are more robust than collagen[80] and insoluble in water and organic solvents. Fiber-spun silk proteins produced from silkworm (fibroin) and spider (spidroin) are the two main silk proteins that obtained commercial appeal. The tensile strength of silk fibers derived from silkworm ranged from ~300 to 740 MPa, which is comparable or even higher than synthetic fibers, such as aramid fiber.[81] The protein composition and internal structures of silkworms and spiders are different. Fibroin is coated with sericin, while spidroin is coated with glycoprotein. Figure 9.11 shows the scanning electron microscopy (SEM) micrographs and structures of silk fibers produced by *Bombyx mori* and spiders. Fibroin proteins and sericin contain more than 15 different amino acids in various amounts. The amino acids present in the protein structures contribute to different mechanical strengths of the fibers.[80] Spider's dragline shows higher ultimate tensile strength, stiffness, elasticity and toughness than silkworm's thread.[80] However, uncontaminated spider silks are not easy to farm and harvest, such as in the case of silkworms. Scientists have tried to identify silk protein sequences and produce similar protein-based materials for functional material applications.[83,84]

Pre-treatment and a range of extraction processes (a.k.a. degumming process) are typically used to obtain the fibroin protein from the harvested silk cocoons. The sticky sericin protein is removed during the degumming process, and the purified fibroin protein can be collected (see Figure 9.12). Amino acids from silks, such as valine, alanine, phenylalanine, tryptophan,

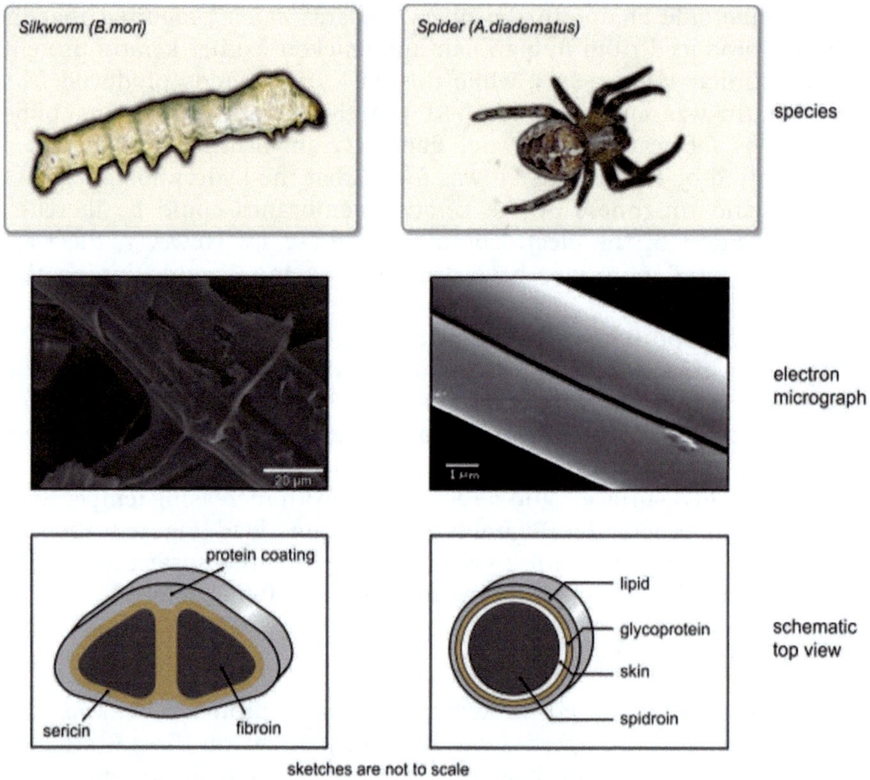

Figure 9.11 Silk fibers produced by *Bombyx mori* silkworms and spiders. Reproduced from ref. 82 with permission from Elsevier, Copyright 2008.

methionine, isoleucine, proline and leucine, are hydrophobic.[81,86] The presence of prominent β-sheets makes silk fibroin hydrophobic (WCA ~90°).[87] The elasticity and degradability of the silk fibroin films can be improved by reducing the β-sheet contents, which is more suitable for film and other biomaterial applications compared to the inherently stiff silk with high β-sheet contents.[88] The content of β-sheet structures and the surface hydrophobicities of silk fiber films vary significantly depending on the starting solvent and the casting substrates used, such as glass, polystyrene (PS) and polytetrafluoroethylene (PTFE).[89] The use of hydrophobic substrates (*e.g.*, PTFE and PS) for silk film casting displays a reduction in the WCA compared to an uncoated template. On the contrary, silk films cast on polar substrates (*e.g.*, glass) increased the WCA. The surface interactions and bondings with the substrate greatly influence the secondary structure of silk proteins, *i.e.*, the change in microphase separation of the silk molecules, the content of β-sheet structures and hence the surface hydrophobicities.[89]

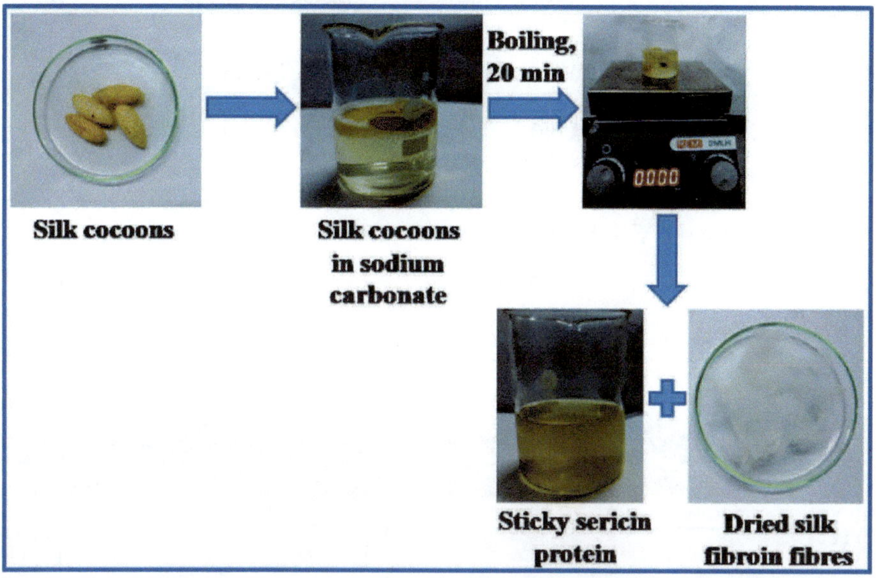

Figure 9.12 Schematic representation showing the degumming process of silk cocoons to remove the sericin protein. Reproduced from ref. 85 with permission from the Royal Society of Chemistry.

Du *et al.*[90] investigated the fabrication of composite films from *Bombyx mori* silkworm cocoon fibroin and hydroxyapatite (HA) with glycerin as a plasticizer. The structure and physico-chemical properties of the composite films were then elucidated. It was found that the WCA of the silk films increases with the increasing content of HA from 2 to 21 wt.%. However, the WCA of silk films decreased significantly by further increasing the HA content to 31 wt.%. This is due to the change in surface roughness of the composite films as a result of the exposure of HA crystals at the film surface.

Water-repellent clothing and textile materials are gaining popularity recently for their ease of cleaning and protection purposes. Water-repellent protein fibers from silk and wool were successfully developed through the chemical modification of their surfaces with octadecenylsuccinic anhydride (ODSA) and dodecenylsuccinic anhydride (DDSA).[91,92] Moreover, both silk and wool protein fibers and membranes modified through these acylation methods show improved thermal stability. Silkworm films coated with fluorinated polypeptide coatings have also presented excellent water repellency.[93] The WCA of the coated silk films was substantially higher than that of the uncoated silk films. Biphenyl groups and perfluoroalkyl chains from the coating imparted the water-repellency effect.

Aslanidou *et al.*[94] fabricated a superhydrophobic and superoleophobic silk fabric (water/oil repellency silk) through a facile spray-coating technique with water-borne siloxane enriched with SiO_2 nanoparticles.

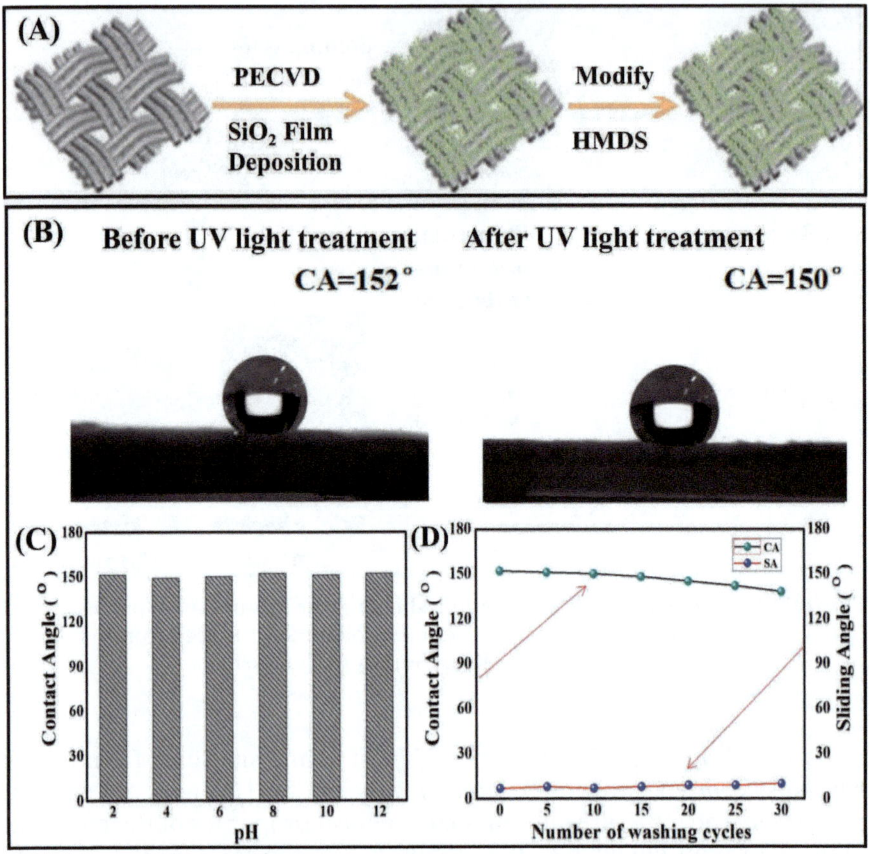

Figure 9.13 (A) Fabrication of superhydrophobic silk fabrics through plasma-enhanced chemical vapor deposition (PECVD) of SiO_2 nanoparticles on the surface of silk followed by hexamethyldisilazane (HMDS) modification. (B) The water contact angle of SiO_2/silk fabrics before and after UV light treatment. (C) Chemical stability and (D) laundering durability of SiO_2/silk fabrics. Adapted from ref. 98, https://doi.org/10.3390/coatings10040349, under the terms of the CC BY 4.0 license, https://creativecommons.org/licenses/by/4.0/.

Water, olive oil and dodecane droplets with contact angle (CA) of more than 150° were observed at an optimum SiO_2 nanoparticle concentration on the treated silk. The produced coating formed microscale clusters with nanostructures, inducing superhydrophobic and superoleophobic surface roughness.

In another study, Patowary *et al.*[85] modified silk fibroin fibers with octadecylamine through a dip-coating method to obtain excellent superhydrophobic and oleophilic silk fibers. A static WCA of ~150° was observed on the modified silk fibers. In addition, the fibers exhibited great oil sorption capacity for motor oil (84.14 g g^{-1}) and crude oil (46.83 g g^{-1}). Maleki *et al.*[95] synthesized robust, porous, multifunctional hybrid aerogels from silk fibroin proteins and polymethylsilsesquioxane (PMSQ). A coupling agent was used to improve the compatibility and grafting between the fibroin and PMSQ. The produced superhydrophobic/oleophilic protein-based aerogels demonstrated a WCA of >150°. The aerogels exhibited a high competency for the selective absorption of various organic pollutants from water and acted as a high-performance filter for continuous water/oil separation.[95]

Chen *et al.*[96] fabricated highly hydrophobic silk fabrics coated with titanium dioxide (TiO_2) nanoparticles through a simple one-step atomic layer deposition process. The increase in the surface roughness of the fabrics after coating with TiO_2 nanoparticles produces a WCA of 135°. Furthermore, the hydrophobic surface of the fabrics was retained even after 30 cycles of washing at optimum TiO_2 coating thickness. Superhydrophilic and underwater superoleophobic silk-based nanofiber membranes, inspired by gills of manta ray fish for oil removal from water, were developed and reported by Li *et al.*[97] The developed silk nanofiber membranes displayed superoleophobicity with oil CA of ~154° and superhydrophilicity with a WCA of ~0°. The silk nanofiber membranes revealed an average oil–water separation efficiency of 99.90% for various oil–water and solvent–water mixtures.

Liang *et al.*[98] produced superhydrophobic silk fabrics with hierarchical roughness structures through plasma-enhanced chemical vapor deposition of SiO_2 nanoparticles on the surface of silk followed by hexamethyldisilazane modification (see Figure 9.13A). The superhydrophobic silk fabric exhibits a superior WCA of ~152°. Furthermore, the superhydrophobic properties obtained on the silk fabrics were able to be maintained after UV treatment (high UV resistance), laundering (numerous washing cycles) and chemical reactions (exposure to different pH solutions up to 6 h) when manufactured with this method (see Figure 9.13B–D).

9.2.3 Effect of Processing Methods

The surface wettability and water solubility of protein-based films are greatly affected by the processing methods and the final film thickness. It was found that the spin-coated films from the amaranth protein isolate showed greater hydrophobicity and less water solubility than those from the cast film method.[99] Incorporating cellulose nanocrystals (CNC) into protein films could enhance the mechanical properties and hydrophobicity and reduce water solubility.[99] Gao *et al.*[16] demonstrated protein self-assembly and deposition of phase-transited lysozyme proteins as a superhydrophobic coating material. The grafting of perfluorooctanyl chloride on lysozyme protein film exhibited superhydrophobic behavior, and the lysozyme coating can also be tailored to obtain a hydrophilic surface through the grafting of a hydrophilic polymer, such as poly(oligoethylene glycol) methacrylate.

9.3 Conclusion

Renewable protein-based materials have recently garnered considerable interest as alternatives to synthetic materials due to their biocompatibility and biodegradability. These materials hold potential for use in biomedical applications as well as eco-friendly food packaging. Protein-based materials display a profound range of properties, depending on the different processing techniques used. A wide range of protein-based materials derived from plants and animals showed promise for their tailorability and functionalization to produce various bio-based hydrophobic and superhydrophobic materials for several functional applications.

Considering the hydrophilic nature of most protein materials, developing hydrophobic and superhydrophobic protein-based materials remains challenging. The hydrophobicity of protein-based materials could be improved by blending and grafting with other more hydrophobic components, such as both synthetic and bio-based polymers, surfactants, natural waxes, oil and fats, nanomaterials and silane. However, achieving long-term hydrophobic and superhydrophobic behaviors (WCA $>150°$) *via* blending methods is difficult. Protein-based materials with superhydrophobic properties can be achieved with an appropriate design of surface roughness patterns and the synthesis and growing of micro/nano hierarchical core/shell structures. These superhydrophobic protein-based materials demonstrated promising applications for self-cleaning coating and self-cleaning fabrics. However, fabrication of these micro/nano hierarchical core/shell structures usually involves various processing steps and intricate chemical reactions. The streamline synthesis of superhydrophobic protein-based materials *via* green chemistry and engineering is needed to ensure that the laboratory product is commercially feasible at an industrial scale.

9.4 Future Prospective

Hydrophobic materials produced from proteins are appealing materials due to their sustainability, biocompatibility and environment-friendly attributes. However, the production of hydrophobic and superhydrophobic protein-based materials is currently limited to laboratory-scale studies. The recovery of protein-based materials from waste streams, especially produced from food industry by-products and co-products, is of utmost importance to upcycle for various value-added and functional bioproducts. It is worthwhile to focus on the molecular crosslinking and effective interface interactions of protein structures to develop mechanically strong, elastic and robust protein-based materials with good barrier performance, low water absorption and hydrophobic properties, which can expand their applications in various industries. Furthermore, durability is always an issue for most green bio-based materials. We need to pay more attention to improving the quality of the protein films, especially their mechanical properties, to ensure that the developed protein-based products remain functional during their service lifetimes.

References

1. Z. He, X. Lan, Q. Hu, H. Li, L. Li and J. Mao, *Prog. Org. Coat.*, 2021, **157**, 106285.
2. T. Darmanin and F. Guittard, *Mater. Today*, 2015, **18**, 273–285.
3. S. Shadmani, M. Khodaei, X. Chen and H. Li, in *Superhydrophobic Surfaces-Fabrications to Practical Applications*, IntechOpen, 2020.
4. M. Ghasemlou, F. Daver, E. P. Ivanova and B. Adhikari, *J. Mater. Chem. A*, 2019, **7**, 16643–16670.
5. C. Neinhuis and W. Barthlott, *Ann. Bot.*, 1997, **79**, 667–677.
6. I. Ahmad and C.-W. Kan, *Materials*, 2016, **9**, 892.
7. E. J. Falde, S. T. Yohe, Y. L. Colson and M. W. Grinstaff, *Biomaterials*, 2016, **104**, 87–103.
8. M. Nosonovsky and B. Bhushan, *Curr. Opin. Colloid Interface Sci.*, 2009, **14**, 270–280.
9. S. Wang,, K. Liu, X. Yao and L. Jiang, *Chem. Rev.*, 2015, **115**, 8230–8293.
10. T. Saison, C. Peroz, V. Chauveau, S. Berthier, E. Sondergard and H. Arribart, *Bioinspir. Biomim.*, 2008, **3**, 046004.
11. L. Zhang, J. Wu, Y. Wang, Y. Long, N. Zhao and J. Xu, *J. Am. Chem. Soc.*, 2012, **134**, 9879–9881.
12. K. Han, T. Y. Park, K. Yong and H. J. Cha, *ACS Appl. Mater. Interfaces*, 2019, **11**, 9777–9785.
13. Y. Wang, L. Wang, S. Wang, R. J. K. Wood and Q. Xue, *Surf. Coat. Technol.*, 2012, **206**, 2258–2264.
14. S. Nishimoto and B. Bhushan, *RSC Adv.*, 2013, **3**, 671–690.
15. T. Darmanin and F. Guittard, *J. Mater. Chem. A*, 2014, **2**, 16319–16359.
16. A. Gao, Q. Wu, D. Wang, Y. Ha, Z. Chen and P. Yang, *Adv. Mater.*, 2016, **28**, 579–587.
17. X. Zhang, F. Shi, J. Niu, Y. Jiang and Z. Wang, *J. Mater. Chem.*, 2008, **18**, 621–633.
18. E. Hosono, S. Fujihara, I. Honma and H. Zhou, *J. Am. Chem. Soc.*, 2005, **127**, 13458–13459.
19. T. H. Mekonnen, *Valorization of waste protein biomass for bio-based plastics, composites, and adhesives development*, PhD Thesis, University of Alberta, 2014, https://doi.org/10.7939/R3WP9TC41.
20. J. T. Martins, A. I. Bourbon, A. C. Pinheiro, L. H. Fasolin and A. A. Vicente, Protein-based structures for food applications: From macro to nanoscale, *Front. Sustain. Food Syst.*, 2018, **2**, 77.
21. X. Yu, C. Li, H. Tian, L. Yuan, A. Xiang, J. Li, C. Wang and A. V. Rajulu, *Chem. Eng. J.*, 2020, **396**, 125373.
22. M. Wihodo and C. I. Moraru, *J. Food Eng.*, 2013, **114**, 292–302.

23. T. Mekonnen, M. Misra and A. Mohanty, *ACS Sustainable Chem. Eng.*, 2016, **4**, 782–793.
24. C. L. Reichert, E. Bugnicourt, M.-B. Coltelli, P. Cinelli, A. Lazzeri, I. Canesi, F. Braca, B. M. Martínez, R. Alonso, L. Agostinis, S. Verstichel, L. Six, S. D. Mets, E. C. Gómez, C. Ißbrücker, R. Geerinck, D. F. Nettleton, I. Campos, E. Sauter, P. Pieczyk and M. Schmid, *Polymers*, 2020, **12**, 1558.
25. C. Johansson, J. Bras, I. Mondragon, P. Nechita, D. Plackett, P. Šimon, D. G. Svetec, S. Virtanen, M. G. Baschetti and C. Breen, *BioResources*, 2012, **7**, 2506–2552.
26. Z. Chang, S. Zhang, F. Li, Z. Wang, J. Li, C. Xia, Y. Yu, L. Cai and Z. Huang, *Chem. Eng. J.*, 2021, **404**, 126505.
27. Y. Zhao, A. Xiao, P. Wu, F. Chen, Q. Zhang, X. Liang, X. Han, X. Shi, Y. Li and Y. Chen, *Tissue Eng., Part A*, 2021, **27**, 788–795.
28. P. Dorishetty, R. Balu, A. Gelmi, J. P. Mata, N. K. Dutta and N. R. Choudhury, *Biomacromolecules*, 2021, **22**, 3668–3678.
29. D. Meng, J. Guo, A. Wang, X. Gu, Z. Wang, S. Jiang and S. Zhang, *Prog. Org. Coat.*, 2020, **148**, 105835.
30. S. Li, E. Donner, H. Xiao, M. Thompson, Y. Zhang, C. Rempel and Q. Liu, *Mater. Sci. Eng. C*, 2016, **69**, 947–955.
31. F. Li, Q. Ye, Q. Gao, H. Chen, S. Q. Shi, W. Zhou, X. Li, C. Xia and J. Li, *ACS Appl. Mater. Interfaces*, 2019, **11**, 16107–16116.
32. Z. Peles and M. Zilberman, *Acta Biomater.*, 2012, **8**, 209–217.
33. T. H. Mekonnen, M. Misra and A. K. Mohanty, Processing, performance, and applications of plant and animal protein-based blends and their biocomposites, in *Biocomposites: Design and Mechanical Performance*, 2015, ch. 8, pp. 201–235, https://doi.org/10.1016/B978-1-78242-373-7.00017-2.
34. T. H. Mekonnen, P. G. Mussone, P. Choi and D. C. Bressler, *Macromol. Mater. Eng.*, 2015, **300**, 198–209.
35. X. Liu, K. Wang, W. Zhang, J. Zhang and J. Li, *Composites, Part B*, 2019, **174**, 107002.
36. R. Kumar, D. Liu and L. Zhang, *J. Biobased Mater. Bioenergy*, 2008, **2**, 1–24.
37. C. Liu, B. Yu, H. Tao, P. Liu, H. Zhao, C. Tan and B. Cui, *LWT*, 2021, **147**, 111529.
38. R. Kumar and L. Zhang, *Ind. Crops Prod.*, 2009, **29**, 485–494.
39. M. Zhang, F. Song, X.-L. Wang and Y.-Z. Wang, *Colloids Surf., B*, 2012, **100**, 16–21.
40. R. Kumar and L. Zhang, *Compos. Sci. Technol.*, 2009, **69**, 555–560.
41. T. Garrido, I. Leceta, S. Cabezudo, P. Guerrero and K. de la Caba, *Eur. Polym. J.*, 2016, **85**, 499–507.
42. W.-Y. Xie, F. Wang, C. Xu, F. Song, X.-L. Wang and Y.-Z. Wang, *Chem. Eng. J.*, 2017, **326**, 436–442.

43. W.-Y. Xie, F. Song, X.-L. Wang and Y.-Z. Wang, *ACS Sustainable Chem. Eng.*, 2017, **5**, 869–875.
44. H. Liu, W.-Y. Xie, F. Song, X.-L. Wang and Y.-Z. Wang, *Chem. Eng. J.*, 2019, **369**, 1040–1048.
45. Y. Jiang, C.-H. Tang, Q.-B. Wen, L. Li and X.-Q. Yang, *Innovative Food Sci. Emerging Technol.*, 2007, **8**, 218–225.
46. C.-H. Tang, Y. Jiang, Q.-B. Wen and X.-Q. Yang, *J. Biotechnol.*, 2005, **120**, 296–307.
47. C.-H. Tang and Y. Jiang, *Food Res. Int.*, 2007, **40**, 504–509.
48. J. Zhao, D. Wang and Y. Li, Proteins in dried distillers' grains with solubles: A review of animal feed value and potential non-food uses, *J. Am. Oil Chem. Soc.*, 2021, **98**(10), 957–968.
49. R. Shukla and M. Cheryan, *Ind. Crops Prod.*, 2001, **13**, 171–192.
50. Y. Chen, R. Ye and J. Liu, *Ind. Crops Prod.*, 2014, **53**, 140–147.
51. P. Oymaci and S. A. Altinkaya, *Food Hydrocolloids*, 2016, **54**, 1–9.
52. C. Yao, X. Li, T. Song, Y. Li and Y. Pu, *Polym. Int.*, 2009, **58**, 396–402.
53. C. Yao, X. Li and T. Song, *J. Appl. Polym. Sci.*, 2007, **103**, 380–385.
54. F. Dong, M. Zhang, W.-W. Tang and Y. Wang, *J. Phys. Chem. B*, 2015, **119**, 5321–5327.
55. Y. Wang and L. Chen, *J. Mater. Chem.*, 2012, **22**, 21592–21601.
56. K. Wang, K. Wu, M. Xiao, Y. Kuang, H. Corke, X. Ni and F. Jiang, *Int. J. Biol. Macromol.*, 2017, **105**, 1096–1104.
57. F. Dong, G. W. Padua and Y. Wang, *Soft Matter*, 2013, **9**, 5933–5941.
58. Â. Luís, F. Domingues and A. Ramos, *Microorganisms*, 2019, **7**, 267.
59. A. Arabestani, M. Kadivar, M. Shahedi, S. A. H. Goli and R. Porta, *Int. J. Biol. Macromol.*, 2013, **57**, 118–123.
60. S. K. Park, C. O. Rhee, D. H. Bae and N. S. Hettiarachchy, *J. Agric. Food Chem.*, 2001, **49**, 2308–2312.
61. X. Xu, H. Liu, S. Duan, X. Liu, K. Zhang and J. Tu, *Mater. Res. Express*, 2020, **6**, 125355.
62. C. Rovera, H. Türe, M. S. Hedenqvist and S. Farris, *Food Packag. Shelf Life*, 2020, **26**, 100561.
63. E. Marcuzzo, D. Peressini, F. Debeaufort and A. Sensidoni, *Innovative Food Sci. Emerging Technol.*, 2010, **11**, 451–457.
64. Z. Liu, Z. Zheng, G. Zhu, S. Luo, D. Zhang, F. Liu and Y. Shen, *Food Chem.*, 2021, **363**, 130251.
65. S. Ahammed, F. Liu, M. N. Khin, W. H. Yokoyama and F. Zhong, *Food Hydrocolloids*, 2020, **105**, 105804.
66. S. Afshar and H. Baniasadi, *Int. J. Biol. Macromol.*, 2018, **109**, 1019–1028.
67. C. Shi, F. Tao and Y. Cui, *Int. J. Biol. Macromol.*, 2018, **109**, 863–871.
68. L. Deng, X. Zhang, Y. Li, F. Que, X. Kang, Y. Liu, F. Feng and H. Zhang, *Food Hydrocolloids*, 2018, **75**, 72–80.
69. K. Shi, J. L. Kokini and Q. Huang, *J. Agric. Food Chem.*, 2009, **57**, 2186 2192.

70. C. Xia, W. Wang, L. Wang, H. Liu and J. Xiao, *Food Packag. Shelf Life*, 2019, **19**, 76–85.
71. S. Sahraee, J. M. Milani, B. Ghanbarzadeh and H. Hamishehkar, *Int. J. Biol. Macromol.*, 2017, **97**, 373–381.
72. A. Mohammadi Nafchi, M. Moradpour, M. Saeidi and A. K. Alias, *LWT--Food Sci. Technol*, 2014, **58**, 142–149.
73. Q. He, Y. Zhang, X. Cai and S. Wang, *Int. J. Biol. Macromol.*, 2016, **84**, 153–160.
74. P. Kanmani and J.-W. Rhim, *Food Hydrocolloids*, 2014, **35**, 644–652.
75. Y. Chen, W. Lu, Y. Guo, Y. Zhu, H. Lu and Y. Wu, *RSC Adv.*, 2018, **8**, 23712–23719.
76. S. Alahyaribeik and A. Ullah, *Int. J. Biol. Macromol.*, 2020, **148**, 449–456.
77. Y. Esparza, N. Bandara, A. Ullah and J. Wu, *Mater. Sci. Eng., C*, 2018, **90**, 446–453.
78. A. Figoli, C. Ursino, D. O. Sanchez Ramirez, R. A. Carletto, C. Tonetti, A. Varesano, M. P. De Santo, A. Cassano and C. Vineis, *Polym. Eng. Sci.*, 2019, **59**, 1472–1478.
79. X. Yin, D. Wang, B. Yu and F. Zhou, *RSC Adv.*, 2014, **4**, 3611–3614.
80. C. Belbéoch, J. Lejeune, P. Vroman and F. Salaün, *Environ. Chem. Lett.*, 2021, **19**, 1737–1763.
81. P. M. Gore, M. Naebe, X. Wang and B. Kandasubramanian, *Chem. Eng. J.*, 2019, **374**, 437–470.
82. J. G. Hardy, L. M. Römer and T. R. Scheibel, *Polymer*, 2008, **49**, 4309–4327.
83. G. Bhattacharyya, P. Oliveira, S. T. Krishnaji, D. Chen, M. Hinman, B. Bell, T. I. Harris, A. Ghazitabatabaei, R. V. Lewis and J. A. Jones, *Protein Expression Purif.*, 2021, **183**, 105839.
84. C.-F. Hu, Z.-G. Qian, Q. Peng, Y. Zhang and X.-X. Xia, *ACS Biomater. Sci. Eng.*, 2021, **7**, 3608–3617.
85. M. Patowary, K. Pathak and R. Ananthakrishnan, *RSC Adv.*, 2016, **6**, 73660–73667.
86. S. Viju, R. S. Rengasamy, G. Thilagavathi, C. J. Singh and H. A. K. Mohamed, *J. Nat. Fibers*, 2021, 1–11.
87. F. Xu, M. Bao, L. Rui, J. Liu, J. Li, Y. Dou, K. Yang, B. Yuan and Y. Ma, *RSC Adv.*, 2015, **5**, 65684–65689.
88. H.-J. Jin, J. Park, V. Karageorgiou, U.-J. Kim, R. Valluzzi, P. Cebe and D. L. Kaplan, *Adv. Funct. Mater.*, 2005, **15**, 1241–1247.
89. S. Wohlrab, K. Spieß and T. Scheibel, *J. Mater. Chem.*, 2012, **22**, 22050–22054.
90. C. Du, J. Jin, Y. Li, X. Kong, K. Wei and J. Yao, *Mater. Sci. Eng., C*, 2009, **29**, 62–68.
91. T. Arai, G. Freddi, R. Innocenti, D. L. Kaplan and M. Tsukada, *J. Appl. Polym. Sci.*, 2001, **82**, 2832–2841.
92. T. Arai, G. Freddi, R. Innocenti and M. Tsukada, *J. Appl. Polym. Sci.*, 2003, **89**, 324–332.

93. K. Tsuchiya, N. Ifuku, Y. Koyama and K. Numata, *Polym. Degrad. Stab.*, 2019, **160**, 96–101.
94. D. Aslanidou, I. Karapanagiotis and C. Panayiotou, *Prog. Org. Coat.*, 2016, **97**, 44–52.
95. H. Maleki, L. Whitmore and N. Hüsing, *J. Mater. Chem. A*, 2018, **6**, 12598–12612.
96. F. Chen, X. Liu, H. Yang, B. Dong, Y. Zhou, D. Chen, H. Hu, X. Xiao, D. Fan, C. Zhang, F. Cheng, Y. Cao, T. Yuan, Z. Liang, J. Li, S. Wang and W. Xu, *Appl. Surf. Sci.*, 2016, **360**, 207–212.
97. Z. Li, C. M. Tan, W. Tio, J. Ang and D. D. Sun, *Environ. Sci.: Nano*, 2018, **5**, 1466–1472.
98. Z. Liang, Z. Zhou, B. Dong and S. Wang, *Coatings*, 2020, **10**, 349.
99. A. López-Rubio, A. Blanco-Padilla, K. Oksman and S. Mendoza, *Nanomaterials*, 2020, **10**, 2564.

CHAPTER 10

Functional Protein-based Biomaterials

M. AZAM ALI*(0000-0002-0136-6800)[a], MAREE GOULD[a] AND SHUVA BHOWMIK[a,b,c]

[a]Centre for Bioengineering and Nanomedicine, Faculty of Dentistry, Division of Health Sciences, University of Otago, PO Box 56, Dunedin 9054, New Zealand; [b]Department of Food Science, University of Otago, Division of Sciences, PO Box 56, Dunedin 9054, New Zealand; [c]Department of Fisheries and Marine Science, Noakhali Science and Technology University, Noakhali 3814, Bangladesh
*E-mail: azam.ali@otago.ac.nz

10.1 Introduction

Proteins are essential biomacromolecules that exhibit distinct structures and complex amino acid sequences and are biologically active and chemically reactive, along with attractive functionalities.[1–3] Exhibiting multifunctional properties, including biological roles, biocompatibility, and biodegradability, functional proteins have been used to fabricate potential biomaterials for advanced applications in biomedical, agriculture, food, and food packaging sectors.[4–7] Biomaterials are the important elements of drug delivery systems, tissue engineering, smart material development, and encapsulation agents, which are derived naturally or synthesized artificially (*e.g.*, recombinant or genetically engineered techniques).[8,9] Several naturally derived functional proteins from animals (*e.g.*, collagen, gelatin, albumin, elastin, keratin, silk, and resilin), dairy (*e.g.*, casein and whey

Smart Materials No. 41
Functional Materials from Lipids and Proteins
Edited by Nandika Bandara and Aman Ullah

Published by the Royal Society of Chemistry, www.rsc.org

protein), plant (*e.g.*, soy protein, corn zein, and wheat gluten), and marine (*e.g.*, fish collagen, sponge collagen, and marine lectins) sources have been used for the development of biomaterials.[3,10–12] Protein-based biomaterials have several advantages, such as low immunogenicity, structural flexibility, biocompatibility, low antigenicity, relative abundance, hydrophilicity, cost-effectiveness, biomechanical responses, easy processing, and higher degradability, over synthetic materials (*e.g.*, polylactic acid [PLA], polycaprolactone, poly-*p*-dioxanone, and polyglycolic acid).[13–16] In contrast, inadequate site specificity during chemical conjugation stages, polydispersity in molecular weights (MWs), environmental unfriendliness, and accumulation in the human body resulting in deleterious effects, including allergenicity, are the major drawbacks of using synthetic materials.[1,17] Similarly, some limitations, such as moderate mechanical properties, shorter degradation rate, and lower thermal stability, have been detected for the use of protein biomaterials in biomedical settings (*e.g.*, cell differentiation, drug delivery, and wound dressing).[18,19] However, these drawbacks can easily be overcome by developing protein biomaterial composites and using advanced manufacturing technologies that enhance functional bioactivity and mechanical and thermal stability.[20,21] In particular, cell-to-cell and cell-to-matrix interactions and three-dimensional structure of protein biomaterials have emerged as potential candidates for use in the biomedical industry. Additionally, protein biomaterials are attractive due to their hierarchical assembly that can easily be tuned to form a customized structure.[22] This chapter provides an overview of the structure, function, properties, and manufacturing process of biomaterials derived from functional proteins. Thereafter, the potential application of protein-based biomaterials will be discussed, and finally, future challenges in the development of smart biomaterials and advanced applications will be explored.

10.2 Multi-hierarchical Structure and Protein Chemistry

The multi-hierarchical structure of protein is demonstrated in Figure 10.1.

10.2.1 Primary Structure

The simplest form of a protein that is developed from the amino acid sequence in a polypeptide chain is called the primary structure of protein. There are twenty amino acids, mostly used for the formation of proteins. On the basis of side chains, amino acids are classified into four groups, such as nonpolar amino acids (alanine, glycine, isoleucine, leucine, methionine, phenylalanine, proline, tryptophan, and valine), uncharged amino acids (asparagine, cysteine, glutamine, serine, threonine, and

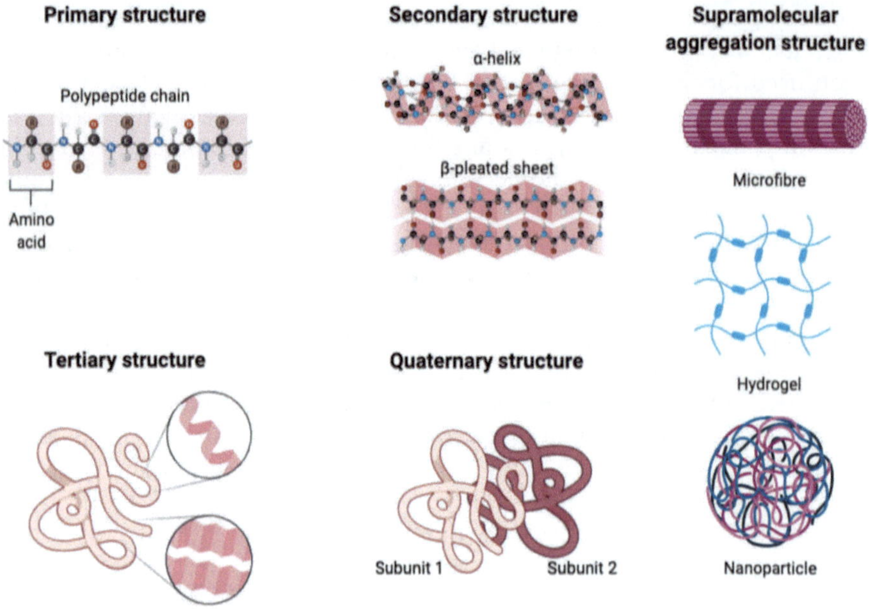

Figure 10.1 Multi-hierarchical structure of protein. Created with BioRender.com.

tyrosine), basic amino acids (arginine, histidine, and lysine), and acidic amino acids (aspartic acid and glutamic acid).[23,24] In general, the carboxyl group of one amino acid is connected with the amino group of another amino acid to form a polypeptide chain. The length of amino acid chains determines the type of protein and its biological functions.[25,26]

10.2.2 Secondary Structure

The formation of hydrogen bonds between the carbonyl and amino groups within different parts of the polypeptide backbone leads to a local folded structure, known as the secondary structure of protein. The most fundamental types of secondary structure of proteins are α-helix and β-sheets, while β-helix is the more unusual structure.[1,27] A hydrogen bond formed between the oxygen of the carbonyl group and the hydrogen of the amino group in the amino acid chain leads to the formation of the helix, whereas sheets are formed by a hydrogen bond between the sequence of amino and carbonyl groups that causes them to separate on the backbone of the polypeptide chain. The β-sheets can be parallel (possess the same direction, N- and C-terminus match up) or antiparallel (possess the opposite direction, N-terminus of one strand and C-terminus of another strand).[28] The secondary structure of proteins has a great role in the functional and mechanical characteristics, such as β-sheets demonstrating high structural rigidity, while α-helix providing strong strain-hardening and elasticity capacity.[29] Additionally, the α-helix can assemble into a coiled-coil for the desired application due to the significant hydrophobic interaction between specific amino acids that are positioned on the outer surface of the coiled-coil.[30] Coiled-coil structures can respond to pH and temperature due to the deprotonation or protonation of residues at the surrounding sites of their hydrophobic core.[31–34] The main advantages of the secondary structure of proteins are that their properties can be tuned easily by altering the sequence of amino acids in peptide molecules.[35]

10.2.3 Tertiary Structure

The three-dimensional structure of a polypeptide chain is called the tertiary structure of protein. This structure is usually formed by different types of interactions, including disulphide bridges between the sulfhydryl groups on the side groups of amino acid chains, which play a significant role in generating the three-dimensional structure of proteins. Furthermore, ionic bonds, hydrogen bonds, and hydrophobic interactions between the nonpolar amino acid side chains are essential to develop the tertiary structure of protein.[9,36]

10.2.4 Quaternary Structure

Proteins are made up of more than one polypeptide chain, known as protein subunits. These subunits arrange themselves to generate a complex structure that is referred to as the quaternary structure of protein. This structure is further modified through diverse interactions, including disulphide bridges, ionic bonds, and hydrogen bonding. This type of structure is commonly detected in larger proteins, and sometimes, their catalytic activity depends on prosthetic groups that are necessary for functional activity.[37]

10.2.5 Supramolecular Aggregation Structure

While the secondary, tertiary, and quaternary structures of proteins are useful for several physicochemical, prime structural, and functional properties of proteins, in general, it is possible to convert these structures into higher-order supramolecular aggregation structures that enhance the functional activity of proteins.[1] This structure of protein is essential for several potential applications, including drug delivery. In this regard, the most common groups of supramolecular structures, such as hydrogels, fibres, and particles/nanoparticles, are considered for application in medical sectors.[38–40] Some researchers reported using gels and fibres as most appropriate for drug release and conceivable for distributed delivery *in vivo*. The major attributes, especially mechanical characteristics and thermal stability, of different hierarchical structures of proteins exhibit a significant discrepancy.[41,42] As a result, the hierarchical arrangement of proteins at different levels allows for the interactions of desirable design, which is consequential for their target applications.[28,36]

10.3 Tailoring of Functional Proteins as Active Biomaterials

The functional properties of protein-based active biomaterials can be enhanced through hybridization techniques with bioactive molecules to improve the biological function *in vivo* or *in vitro*.[9,18] Furthermore, the functional protein biomaterials have been customized by physical adsorption and encapsulation, chemical modification, and physical and chemical crosslinking methods, which render them able to develop different active molecules, including antibiotics and growth factors, improve cell viability and elicit strong tissue responses.[13,36,43,44] Some straightforward approaches, such as the addition of amino acids to the structural protein domain, suggest developing functional protein biomaterials through reaction with their side chains.[45,46] For instance, cysteine and lysine can be crosslinked with maleimide and *N*-hydroxysuccinimide, respectively, to develop active biomaterials from functional proteins. Therefore, the sequence of amino acids in β-sheets can be strategically modified to improve the gelation properties, stimuli responsiveness, and nanofiber morphology for selective applications. The functional protein biomaterials are well suited for various applications, including in biomedical sectors, due to their metabolized natural amino acid chains, which are often biocompatible with tissues and cells, and minimal inflammation to the host.[47–50]

Physical adsorption is an effective and simple immobilization technique that is commonly applied to attach bioactive molecules. The efficiency of the adsorption method depends on some physical characteristics, such as surface topography, wettability, and functional group, and chemical properties including electrical charge and pH.[51,52] Most of the biomaterials derived from functional proteins are hydrophobic in nature; hence, the introduction of several effective methods is needed to enhance their wettability and develop them with desirable hydrophilic properties. In this regard, some physical methods, such as UV light, bombardment with ions, and plasma modification, have been used to hinder chemical bonds between carbon and non-carbon atoms that render unsaturated bonds and radicals, which react with oxygen to enhance hydrophilicity and reactivity towards biological molecules.[18,53] The main advantages of natural functional protein biomaterials are the presence of strong reactive chemical groups, such as amide, hydroxyl, and carboxyl, which cause an increase in hydrophilic properties and powerful interactive capacity with bioactive molecules. However, the stability of binding adsorbed bioactive molecules depends on several conditions, such as ionic strength, pH, and the concentration of adsorbed species.[51] Some studies have reported that changes in environmental conditions alter the binding stability of molecules, which causes uncontrolled release of immobilized species.[18] To overcome these limitations, covalent immobilization has been applied for the stable attachment of molecules in the target application. Covalent conjugation is an effective strategy and has demonstrated controlled release of immobilized agents for desirable time periods and target sites compared to the adsorption method.[54] In recent years, the carbodiimide coupling method has gaining more attention in protein chemistry to interact activated carboxylic acid groups with the amines that exist in the peptide chain.[55,56] The use of the coupling method has some limitations, such as difficulty in identifying new functional peptide–protein biomaterials due to their background noise and reactive amine groups.[57] The reactive side groups need to be managed through deprotection for their sustainable functional activity.

Encapsulation of molecules within functional protein biomaterials is an emerging technique to control the release of bioactive agents in the desirable area. In most cases, chemical modifications are needed for the controlled and effective release of encapsulated bioactive molecules. Additionally, several factors, such as topology and structure of proteins during the incorporation of molecules into functional protein biomaterials, can be considered for inhibiting aggregation and denaturation of protein that lead to loss of bioactivity.[58] In this context, novel techniques have been required for the functionalization of structures of proteins and the sustainable release of bioactive compounds.[59] Several researchers have stated that recombinant DNA strategies can be good options to address the limitations noted above, which may help to formulate the next generation of functional protein-based biomaterials.[9]

The physical crosslinking technique has been used to improve the mechanical and physical properties of functional proteins without bringing in cross-linker residues or toxic compounds compared to the chemical crosslinking method. In general, dehydrathermal, ultraviolet radiation, and gamma irradiation processes are used in the physical crosslinking method.[60–62] Functional proteins could be degraded through long exposure to physical treatment. Furthermore, high strength and uniform crosslinking degree could not be achieved in functional protein biomaterials by using the physical crosslinking method.[63] Therefore, it is estimated that the properties of functional proteins can benefit from a synergistic effect between the physical treatment and other crosslinking techniques.[64]

Apart from these, the functional properties of protein-based biomaterials could be efficiently enhanced through chemical modification or chemical crosslinking methods. Crosslinking agents, such as isocyanates (*e.g.*, hexamethylene diisocyanate), aldehydes (*e.g.*, glutaraldehyde), inorganic crosslinkers, photoreactive agents, and plant extracts (*e.g.*, tannic acid and genipin), have been used for improving the properties of functional protein biomaterials.[65–67] There are various types of chemical crosslinkers that exist, but the number of nontoxic crosslinkers is limited. Hence, modification or a combination of chemical crosslinkers with other compounds is needed to fabricate low-toxicity or nontoxic crosslinkers. In recent years, some low-toxic chemical crosslinking agents, such as dialdehyde cellulose, dialdehyde alginate, oxidized starch, and dialdehyde carboxymethyl cellulose, have been applied for the modification of functional protein-based biomaterials.[68,69] Additionally, some synthetic polymers, such as polyurethane, polyvinyl alcohol (PVA), PLA, and polylactic acid-co-glycolic acid (PLGA), and natural carbohydrate polymers, including chitosan, cellulose, and glycosaminoglycans, have been used to improve the properties of functional protein-based biomaterials.[70–73]

10.4 Protein-based Biomaterials and Their Sources, Structure, Properties, and Function

The types, sources, structure, function, and properties of functional protein-based biomaterials are presented in Table 10.1.

10.4.1 Animal Sources

10.4.1.1 Collagen

In animals and humans, collagen is the most prominent and abundant structural protein, constituting up to almost one-third of the dry weight of all proteins in the body and is an essential component of the extracellular matrix, cells, and connective tissue.[3,87] The major sources of collagen include pigs, sheep, cows, rats, and nerve tissue or placenta of humans.[88–90] To date, approximately 30 different types of collagens have been selected and characterized. Among them, most of the collagens exhibit a triple-helical structure, which is built up based on three α-chains, consisting of more than 1000 amino acids. Of the 30 types of collagen, only types I, II, III, V, and XI are considered fibre-forming collagen, and these types of collagen are useful for developing functional biomaterials.[76] Collagens have excellent functions, such as cell migration, tissue repair, tissue scaffolding, and cell adhesion.[28]

10.4.1.2 Gelatin

Gelatin is a popular, biocompatible, biodegradable, and nonimmunogenic functional protein-based biomaterial derived through alkaline or acid hydrolysis or by enzymatic or thermal denaturation of type I collagen, mostly derived from porcine or bovine origins.[8,91] The type I collagen consists of approximately 85–92% of protein, minerals, and water. The

Table 10.1 Types, sources, structure, function, and properties of functional protein-based biomaterials.[a]

Types	Sources	Structure	Properties and function	Ref.
Collagen	• Human collagen • Porcine • Bovine • Marine organisms • Frog skin • Duck feet • Chicken • Bird's feet • Sheep skin • Alligator skin/bone	• Triple helix • Three polypeptide α-chains • Gly-X-Y triplet units • Quaternary structure • Supramolecular form	• Low antigenicity • Good compatibility • Strong biodegradability • Hydrophilicity • Cellular adhesion capacity • Tissue regeneration ability • Desirable thermal and mechanical stability	74
Gelatin	• Type I collagen • Porcine skin • Bovine ossein	• Triple-helical structure • (Gly-X-Pro)*n* triplet units	• Biodegradable • Biocompatible • Nonimmunogenic • Highly soluble • High mechanical and thermal stability of crosslinking gelatin	8
Albumin	• Eggs • Chicken • Bovine serum • Human serum	• Globular • Monomeric protein • Crystal	• Soluble • Biodegradable • Nontoxic • pH- and temperature-sensitive • Ligand-binding capacity	75
Elastin	• Bovine • Cattle • Goats • Chicken	• α-Helices (3%) • β-Sheets (41%) • β-Turns (21%)	• Second abundant macromolecules in the extracellular matrix • Thermal responsiveness • Nontoxic • Biodegradable	76

(continued)

Table 10.1 (*continued*)

Types	Sources	Structure	Properties and function	Ref.
Keratin	• Animal skin • Hairs • Nails • Feathers • Scales	• High cysteine content • α-Helical structure • β-Sheets structure	• Good mechanical, thermal, and chemical stability • Highly tuneable • Nontoxic • Biocompatible • Cell adhesion motifs • Biological stimulus responsive	10
Silk	• Silkworms • Spiders	• Light chain (MW 26 kDa) • Heavy chain (MW 390 kDa) • β-Sheet structure • High-tuneable molecular weight	• Response to external stimuli • High mechanical and thermal stability • High immunogenic tolerance • High biodegradation rate • Low inflammatory	77
Resilin	• Exoskeletons of insects • Arthropods	• Secondary structure • Polypeptide chain rich in glycine	• Low stiffness • High resilience • Efficiently store energy • pH and temperature responsive	78
Casein	• Bovine milk	• Secondary structure	• Favourable hydrophilicity • Desirable biocompatibility • Swelling property • Cell attachment property • Highly sensitive to pH and temperature	79
Whey protein	• Milk	• Globular structure	• High thermal stability	80

(*continued*)

Table 10.1 (*continued*)

Types	Sources	Structure	Properties and function	Ref.
			• Cell-binding ability • Biodegradable in nature	
Soy protein	• Soybean	• Globular structure	• Aggregation ability • pH sensitivity • Strongly interact with other macromolecules	81
Corn zein	• Maize or corn	• Triple super helix • Rich in nonpolar residues	• Presence of hydrophobic and hydrophilic regions • Insoluble in aqueous environment • Biodegradable • Nontoxic	82
Wheat gluten	• Wheat grain	• High percentage of glutamine amino acids present in their structure	• Good water-holding capacity • High thermal stability • Biocompatible • Nontoxic	83
Fish collagen	• Cod skin • Cuckoo ray skin • Lantern shark skin • Atlantic grenadier • Rabbitfish skin • Catshark cartilage • Tuna scales	• Low amino acid content in their structure • Triple-helical α-domain structure	• Low denaturation temperature • Low molecular weight • Antimicrobial activity • Biodegradability • Water solubility • Nontoxic • Cell-binding ability	84
Marine sponge collagen	• Sponge • Demosponge	• 3D matrix network	• Low immunogenicity • Good mechanical properties • Antioxidant activity	85

(*continued*)

Table 10.1 (*continued*)

Types	Sources	Structure	Properties and function	Ref.
Fish lectin	• Turbot • Perch • Sea lamprey • Sea bass • Zebrafish	• β-Sheet structure	• Biocompatibility • Cell attachment ability • Carbohydrate-binding sites • Agglutination capacity • Immobilization ability • Cell growth control • Enhanced immunity	86

[a]Gly: glycine; Pro: proline; X: any type of amino acids.

gelatin has a triple-helical structure, repeating (Glycine-X-Proline)*n* triplets, where X demonstrates the type of amino acid. In most cases, arginine, valine, methionine, and lysine are commonly present in the chain of gelatin.[92] In recent years, two types of commercial natural gelatin, such as type A (cationic in nature and derived from the skin of porcine) and type B (anionic in nature and obtained from ossein), have been available in the international markets.[93] Gelatin is nontoxic to cells and represents no or less antigenicity due to the absence of tryptophan and tyrosine in its structure.[94] Additionally, it can be co-fabricated with other less toxic or nontoxic synthetic or natural polymers to enhance its functional properties.

10.4.1.3 Albumin

Albumin is the most familiar and abundant protein in human blood plasma and has several outstanding properties, such as nontoxicity, biodegradability, and ligand-binding capacity, including metal ions, hormones, bilirubin, and amino acids, which serve different physiological and functional roles.[95–97] The bovine serum albumin (molecular weight 69 kDa), egg white ovalbumin (molecular weight 47 kDa), and human serum albumin (molecular weight 66.5 kDa) are the major sources of albumin in nature.[75,98,99] The amino acid chains, structure, and differently charged groups of albumin allow it to bind with other compounds by specific mechanisms, such as electrostatic interaction with negatively and positively charged groups or dual compounds. Additionally, the presence of amine, thiol, and carboxyl groups in the albumin facilitates the attachment of several types of ligand molecules for target applications.[100,101]

10.4.1.4 Elastin

Elastin is the second most abundant structural protein in the connective tissue and extracellular matrix, which contributes to the resilience and elasticity of several living tissues, such as lungs, skin, blood vessels, and ligaments.[102,103] It is extracted from different animal sources by applying the partial hydrolysis technique using various hydrolysis agents, such as oxalic acid, alkali, potassium hydroxide, and guanidine.[104,105] Elastin contains both hydrophilic and hydrophobic domains, and its structure can be changed during the enzymatic crosslinking process *via* lysine residues in the extracellular area. Elastin has the ability to bind to multiple molecules at a time as targeting agents to certain locations or tissues in the body.[106,107] Furthermore, environmental factors, including temperature, can change the phase of elastin, which helps to design specific biocomposite polymers for enhancing functional activity.

10.4.1.5 Keratin

Keratin is a fibrous structural protein prevalent in the epithelial cells of humans and animal nails, feathers, skin, hair, scales, and other epidermal appendages.[3] There are three different molecular structures of keratin such as alpha-, beta-, and gamma-keratin. Alpha-keratin has a low concentration of sulphur and is abundantly detected in soft tissues. However, beta-keratin is abundant in histidine, lysine, serine, glycine, and tryptophan. It creates a beta-sheet structure supported by hydrogen bonds, which gives nails, scales, and skin stiffness. The glycine, cysteine, and tyrosine are abundant in the third configuration of keratin, known as gamma-keratin. Its amorphous structure generates many intra- and intermolecular disulphide bonds. It is an essential matrix protein that keeps keratin fibres together, giving hair with mechanical strength. Additionally, negatively charged keratin nanoparticles or hydrogels can readily stick with positively charged therapeutics for drug delivery applications. Furthermore, the protein's amine, carbonyl, and carboxyl groups offer advantageous drug attachment sites through the creation of chemical conjugation or hydrogen bonds. In addition, keratin has been used as a "smart" material for its stimulus-responsive properties at different pH, and could be a helpful material during the treatment of targeted injured tissues. Therefore, keratin-based hydrogels, nanoparticles, and films have been extensively explored for food packaging and biomedical applications.

10.4.1.6 Silk

Silk is a hydrophobic, stimuli-responsive, block copolymeric, fibrous, and β-sheet structural protein derived from domestic silkworms (*Bombyx mori*) and spiders (*Nephila clavipes* and *Araneus diadematus*).[77,108] Silk fibroin and sericin are the two main proteins of silk that represent unique biological (low inflammatory, controllable biodegradation, and non-toxic) and physicochemical (swelling, hydration resistance, high network density, thermal stability, and high mechanical ability) properties for certain applications, including the biomedical industry.[109] The silk fibroin is used for further processing through material fabrication, chemical modification, and sterilization. This is possible due to its significant mechanical and thermal stability.[110] Both sericin and silk fibroin have different functional groups that can easily interact with other biomacromolecules, demonstrating strong immunogenic tolerance behaviour.

10.4.1.7 Resilin

Resilin is an abundant elastomeric protein in nature extracted from the exoskeleton of arthropods and insects. It is remarkable due to its high resilience and strain, fatigue lifetime, good mechanical properties, high biodegradability, low stiffness, and capacity to store energy.[78,111,112] The first sequence of resilin was detected in fruit fly (*Drosophila melanogaster*), followed by flea (*Ctenocephalides felis*), African mosquito (*Anopheles gambiae*), and buffalo fly (*Haematobia irritans exigua*), which led to the development of the molecular cloning method to generate resilin-type protein with high functional properties.[113–115] The main advantage of resilin-type biomaterials is the specific sequence and function constructed through genetic modifications or engineering at a molecular level. The cell-adhesion ability, sequences of matrix metalloproteinase, and controlled cell response property of resilin make up tuneable attributes to develop multifunctional biomaterials.[111]

10.4.2 Dairy Sources

10.4.2.1 Casein

Casein is a main fraction of milk protein used as a functional biomaterial due to its unique physicochemical and structural features, such as stabilizing ability, emulsification capacity, ion-binding ability, water-binding activity, and ability to easily interact with other macromolecules while also providing phosphate and calcium.[116,117] Furthermore, casein has UV absorbance properties, making it act as a safeguard against harmful radiation, particularly UV radiation. Casein demonstrates high thermal stability due to the lack of disulphide linkages in their chain that render to form secondary and/or tertiary structures.[118,119] There are four different types of caseins detected in bovine milk, such as αs_1-casein, αs_2-casein, β-casein, and κ-casein.[120,121] The αs_1-casein and αs_2-casein are highly phosphorylated, containing 7–8 phosphate groups and almost 12 carboxyl groups. Two variants, such as A1 β-casein (histidine position at 67) and A2 β-casein (proline position at 67), are identified in cow milk. κ-Casein is a relatively minor component of casein protein and is soluble in the presence of calcium ions.

10.4.2.2 Whey Protein

Whey protein is a coproduct of casein generally derived from milk, is globular in structure, and contains high sulfhydryl amino acid groups, allowing it to form intermolecular covalent bonds during the processing at high temperatures.[80,122] Two main proteins, such as α-lactalbumin (30–65%) and β-lactalbumin (10–25%), constitute approximately 65% of the total dry weight of whey protein.[123,124] α-Lactalbumin is considered a high-thermal stable whey protein, with a molecular weight of 14 kDa, containing 123 amino acid residues, especially leucine, lysine, asparagine, aspartic acid, and cysteine, along with four disulphide bonds. The MW of β-lactalbumin is 18.3 kDa, consisting of 162 amino acid residues, particularly leucine, glutamic acid, lysine, alanine, glutamine, and asparagine. The nontoxic nature, high binding capacity, and biodegradability properties of whey protein make it an outstanding biomaterial for sustainable applications.

10.4.3 Plant Sources

10.4.3.1 Soy Protein

Soy protein, one of the key plant-derived functional proteins from soybean, is an abundant, sustainable, low-cost, eco-friendly biomaterial.[125,126] Soy protein is separated and purified by applying the proper separation and verification technique, which consists of 91% protein.[127] Two proteins, namely glycinin (60%) and β-conglycinin (40%), are reported to exist in soy protein isolate.[128] In general, soy proteins exist as globular structures in water and strongly aggregate during the addition of a solvent or cross-linking agent. Additionally, soy protein is enriched with different amino acids, especially leucine, glutamine, and aspartic acid, which are rendered to form secondary bonding structures and interact with various specific macromolecules to enhance their functionality.[129]

10.4.3.2 Corn Zein

Zein is the main storage protein of corn, containing amino acid residues, such as proline (10%), leucine (19%), alanine (12%), glutamic acid (20%), and glutamine (10–15%).[130,131] Both the hydrophobic and hydrophilic regions are present in the zein structure but insoluble in water and generally soluble in urea, alkaline or anionic surfactants, or alcohol-enriched water.[132] There are four types of zein identified, including α-zein (MW 19–22 kDa), β-zein (MW 14 kDa), γ-zein (MW 16–27 kDa), and δ-zein (MW 10 kDa). The zeins easily interact with biological composites based on their properties, such as molar mass, conformation, and surface charge. The heterogeneity in the structure of zein makes up multiple isoelectric points, depending on different proportions of amino acid residues that enhance their functionality along with binding capacity.

10.4.3.3 Wheat Gluten

Gluten is a cohesive and dough-forming protein extracted from wheat grain.[83] Gluten consists of 75% protein on a dry weight basis. The water-holding capacity and thermostability of wheat gluten make it regarded as a functional biomaterial. The structure of gluten contains a high percentage of the amino acid glutamine. Gluten is insoluble in water due to the presence of highly stable covalent and non-covalent bonds but soluble in 70% ethanol solution.[133] The modification of gluten through enzymatic and chemical techniques can improve its solubility and enhance its functional properties.[134] The presence of disulphide bonds in the structure of gluten improves its elasticity and allows it to interact with other macromolecules through hydrogen bonds.

10.4.4 Marine Sources

10.4.4.1 Fish Collagen

Fish collagen is a structural protein extracted from different by-products, such as fish skin, scales, bones, and cartilage, which are generally generated during the processing of fish.[84] Fish collagens have higher absorption ability and are gaining more attention due to their bioavailability compared to porcine and bovine collagens. Fish collagens are denatured at low temperatures at almost 25–30 °C, whereas mammalian collagens are denatured at 35–40 °C.[135] Collagens are primarily obtained from fish, mostly type I collagen, containing β-chain as a large portion and two unique α_1- and α_2-chains in their structure.[136] In addition, fish collagen exhibits a triple-helical structure and strong intra- and inter-molecular hydrogen bonds that make it insoluble in aqueous environment. Therefore, some common techniques, using pepsin-solubilized collagen, acid-solubilized collagen, and supercritical fluid, have been applied for the extraction of collagen from fish to enhance its solubilization rate.[137] The major properties of fish collagen, such as low immunogenicity, biodegradability, and higher binding capacity, have attracted the scientific community for the sustainable use as biomaterials.[138]

10.4.4.2 Marine Sponge Collagen

Marine sponge collagen has low immunogenicity with good mechanical properties, mostly extracted from different marine sponges, such as *Suberites carnosus*, *chondrosia reniformis*, *Geodia cydonium*, and *Axinella cannabina*.[139] Marine sponge collagens have a strong ability to bind with cells, generate pores, and alter osteogenic differentiation, which enhances their functional properties. Additionally, marine sponge collagen is nontoxic and biodegradable and promotes proliferation.[85] The denaturation temperature of marine sponge collagens is lower than that of mammalian collagens, reflecting the environmental temperature at which they live. Some researchers have reported that the thermal stability of marine sponge collagen is closely related to the body and environmental temperatures of sponges.[140] Glycine is the main amino acid in the structure of marine sponge collagen, followed by aspartic acid, glutamic acid, alanine, and proline.[139]

10.4.4.3 *Fish Lectin*

Fish lectin is a complex group of non-catalytic proteins that commonly bind with different carbohydrate moieties, such as glycolipids, oligosaccharides, proteoglycans, glycoconjugates, and monosaccharides.[141] In general, fish lectins are obtained from the mucus of skin, eggs, serum, embryos, plasma, and other tissues of fish, which are used as a guard against infectious diseases. The main types of lectins extracted from fish are X-type lectin, C-type lectin, S-type lectin, F-type lectin, and P-type lectin.[86] The cell agglutination property, antipathogenic activity, and other biological responses make it a potential candidate for functional biomaterial.[142] The most common fish species, such as striped bass (*Morone saxatilis*), sea lamprey (*Petromyzon marinus*), black rockfish (*Sebastes schlegelii*), and sea bass (*Dicentrarchus labrax*), are used for the extraction of lectins.[143]

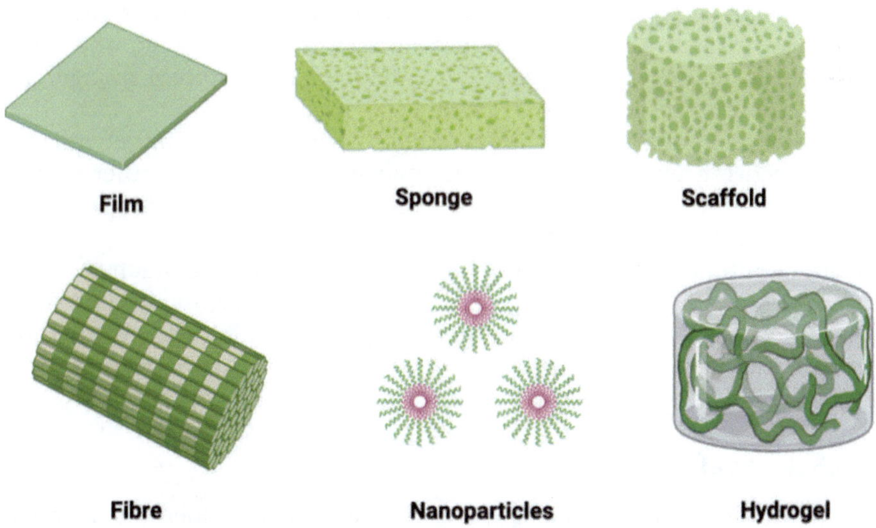

Figure 10.2 Protein-based biomaterials as film, sponge, scaffold, fibre, nanoparticles, and hydrogel. Created with BioRender.com.

10.5 Protein Biomaterials as Films, Sponges, Scaffolds, Fibres, Nanoparticles, and Hydrogels

Protein-based biomaterials used as films, sponges, scaffolds, fibres, nanoparticles, and hydrogels are presented in Figure 10.2.

10.5.1 Film

A film is a well-known barrier membrane, 0.01–0.5 mm in thickness, and developed from functional protein-based biomaterials that are commonly incorporated with other macromolecules through covalent bonding, hydrogen bonding, and simple entrapment.[144] Functional proteins, such as collagen, are commonly used to develop films. Attaching collagen films with other biodegradable membranes has the potential to increase their ability for the controlled release of attached macromolecules.[87] Similarly, keratin-based films are popularly used due to their strong cell attachment sites, large surface area, and biocompatibility nature. In general, solely protein biomaterial-based films have lower mechanical strength; therefore, protein biomaterials that are combined with other proteins or polysaccharides through crosslinking techniques can improve their stiffness, rigidity, and stability. For instance, keratin is blended with silk fibroin for the formation of films that provide strong structural stability, including physical and thermal properties.[145]

10.5.2 Sponge

A sponge is produced from biodegradable protein biomaterials. Some chemical treatments such as iodoacetamide, iodoacetic acid, and 2-bromoethylamine are used to generate amido-sponge, carboxyl-sponge, and amino-sponge, respectively.[146] Chemically modified sponges have been shown to be good mimics of extracellular matrix proteins in general, and highly active groups are present in sponges to allow for further fabrication with other bioactive molecules. In recent years, lyophilization and freezing treatments have been applied for the formation of sponges to avoid harmful chemical crosslinking agents, such as glutaraldehyde.[147] For example, a wool keratin sponge hybrid with other macromolecules shows good mechanical stability, including a homogenous microstructure without crosslinking treatment.

10.5.3 Scaffold

A scaffold is a three-dimensional microstructure developed from protein biomaterials, where pore volume and pore size have been regarded to have desirable effects on the emission of extracellular matrices.[148] The pore size of the protein scaffold plays a key role in cell migration, binding, and growth. The high pore size of the scaffold can allow sustainable gas diffusion, nutrient supply, and removal of metabolic waste but can lead to lower intracellular signalling and cell attachment, whereas small pores demonstrate the opposite.[149] Although most of the protein biomaterial-based scaffolds have outstanding biological properties, they demonstrate low structural stability and mechanical properties.[150] In this regard, some physical and chemical treatments can be applied to obtain crosslinking protein biomaterials, including collagen and gelatin, which are essential to develop functionally active materials with good structural properties.

10.5.4 Fibres

Fibre is a supramolecular structure of a protein that has modulated stiffness, biocompatibility, length, and porosity, and exhibiting these properties, it can be adjusted to different environmental conditions during electrospinning, processing, and synthesis.[38] A fibre has been considered a growth factor and a mimic of the cell extracellular matrix. The triple-helical, α-helical, and β-sheet structures of proteins have been used to fabricate fibres. For example, type I collagen has a triple-helical structure, tropomyosin represents α-helical coiled-coils, and β-structure-rich proteins are used to develop tuneable fibre with good thermal and mechanical properties that lead to enhancing fibre cell attachment capacity.[151] Some researchers have reported that nanofibers have large contact surfaces and greater peptide abilities than fibres.[1,152]

10.5.5 Nanoparticles

Protein biomaterial-based nanoparticles have been gaining more attention in recent years due to their excellent cell attachment ability, which carry target macromolecules *via* hydrogen bonding, disulphide bond generation, electrostatic interactions, and chemical conjugation.[1,153,154] Positively charged molecules electrostatically bind with the negatively charged keratin nanoparticles. This type of interaction demonstrates sustainable and long-term release of the molecules from their carrier.[154] The range of nanoparticle sizes is 1 to 1000 nm. The size and charge of nanoparticles are important parameters for effective colloidal stability, tissue distribution, molecule release, cellular uptake, and macromolecule encapsulation.[155] For example, the size and dispersion of gelatin nanoparticles are significantly affected by pH, with cationic type A nanoparticles formed from gelatin at pH 3 and anionic type B nanoparticles formed at pH 11.[3]

10.5.6 Hydrogel

Hydrogels are highly swellable, holding over 80% water of their weight, biomaterials derived through chemical or physical crosslinking with ionic, dynamic, or covalent bonding, demonstrating unique functional properties.[156,157] Functional proteins have been tailored to generate hydrogels with a high diversity of properties, including bioactive moieties, strong immune response, degradation, and both hydrophobicity and hydrophilicity for biomolecule encapsulation.[158] Furthermore, hydrogels provide a desirable environment for several small biomolecules, offering protection towards external factors that can threaten the activity of the attached molecules. In recent years, some researchers have been working to develop smart hydrogels that can change their physical and chemical attributes by the change in environmental conditions.[159] In this regard, keratin-based hydrogels have been greatly studied for target applications due to their durability, stability, and a variety of attachment sites for molecules.[146]

10.6 Application of Protein-based Biomaterials

10.6.1 Biomedical

10.6.1.1 Drug Delivery

Protein-based biomaterials have been used as a drug delivery vehicles. For example, subcutaneous injections of hydrogels developed from silk fibroin with encapsulated insulin are capable of the sustainable release of insulin to control the blood glucose of diabetic rats.[160] Hydrogels of keratin have been explored for the controlled delivery of small drug molecules and other macromolecules. The degradation of keratin dictates drug release, and the delivery rate is tuned by controlling the bonding stability along with the number of chemical bonds.[145] For example, gelatin nanoparticles and their hybrids are explored to encapsulate both hydrophilic and hydrophobic bioactive drug molecules, such as anti-cancer drugs (cisplatin and paclitaxel), antimalarial drugs (chloroquine), anti-inflammatory drugs (ibuprofen), and anti-HIV drugs (didanosine).[161–164] Furthermore, gelatin composites incorporated with vancomycin demonstrated sustainable release in several days and reported slow release of drugs, successfully eliminating *Staphylococcus aureus* and *Escherichia coli* levels in co-culture.[165] Drug delivery vehicles through functional protein biomaterials are enhancing the targeting, longevity, and therapeutic efficacy of pharmaceuticals.

10.6.1.2 Tissue Regeneration and Repair

Protein biomaterials have been used as neuroinductive agents to facilitate the regeneration of nerve cells. For instance, keratin hydrogels are used in a mouse nerve injury model to enhance the capacity of Schwann cells in *in vitro* conditions through accelerating cell proliferation and controlling specific gene expression, which is required for target neuronal functions.[166,167] A collagen scaffold is applied for the reconstruction of the skin due to its water absorption ability, biodegradability, and nontoxic nature, which is capable of enhancing the wound healing rate without side effects or inflammation.[168] Furthermore, powdered collagen has the potential to aid in chronic wound healing by promoting cellular recruitment and new tissue growth.[169] Additionally, protein biomaterials are used in bone repair, tendon repair, cartilage repair, cardiac tissue engineering, and cell culture systems and as haemostatic agents.[170–174]

10.6.2 Agriculture

In agriculture, protein biomaterials can be used for sustainable crop production by delivery and controlled release of fertilizers or nutrients, ensuring good-quality soil and water management without threatening the natural resources and environment.[5,175] For example, collagen has been used as a good source of amino acids in plants, particularly proline, which plays an important role as a phytohormone precursor, contributes to pollen fertility, and regulates water balance.[176] In general, the structure of collagens and keratins imparts partial insolubility, leading to high energy consumption; hence, an enzymatic process or crosslinking technique is required to modify their structure and increase their bioactive functions.[177] Additionally, protein biomaterial-based hydrogels can play a significant role in plant growth by accelerating soil aeration, controlling harmful microbial activity, and removing toxic metals.

10.6.3 Food Additives/Functional Foods

In recent years, protein biomaterials have gained more attention in the food industry due to their functional properties and health benefits. For example, marine lectins as biomaterials have been incorporated into functional food or nutraceuticals to avoid calcium deficiency.[11] Marine collagen is an outstanding potential component of functional food due to its functional attributes and good sources of amino acids.[178] To provide a nutritionally enriched balanced diet at the consumer level, collagen could be a good candidate as a food additive or functional ingredient. Similarly, hydrolysed and deaminated zeins have potential applications in puddings, mousses, beverages, and margarine and as dairy substitutes.[179] Furthermore, casein and whey protein in functional food items have demonstrated antimicrobial activity, anti-inflammatory activity, protection against metabolic syndromes (glucose balance, serum lipid levels, body mass index, and hypertension), cancer, and antithrombotic activity.[180]

10.6.4 Food Packaging

In food packaging, protein biomaterials are used as less polluting and environmentally friendly packaging materials in the global food industry. In recent years, biodegradable packaging films have been developed from functional proteins due to their nutritional value, film-forming ability, relative abundance, and health benefits.[7] Some researchers have reported that both collagen and soy protein are used in the development of biodegradable packaging films for the preservation of perishable foods, such as fish and meat.[181] The gelatin incorporated with other macromolecules to develop an antioxidant packaging film reduced the lipid oxidation and extended the shelf-life of beef patties.[182] Additionally, the addition of clove and coriander essential oils to the gelatin film presented a lower oxidation rate for the preservation of cheddar cheese.[183] The shelf-life of fish fillets and fresh shrimp is extended due to the addition of orange leaf essential oils to the gelatin film.[184,185] Milk casein is used to create water-insoluble and opaque films, and their mechanical properties, particularly barrier capacity, can be used to control oxygen permeability.[186] Keratin-based packaging films are sensitive to humidity and are highly brittle, whereas albumin-based films demonstrate high elasticity and transparency, but water vapour permeability is high.[187] Additionally, whey protein-based film has outstanding barrier activity against oxygen along with aroma compounds at moderate relative humidity conditions but has high water vapour permeability and low strength. At the same time, wheat gluten presents good gas barrier attributes and is insoluble in water but demonstrates undesirable humidity sensitivity along with a brittle structure.[188] However, composite or hybrid protein biomaterial-based packaging films showed excellent physical, thermal, and mechanical properties that could be potential candidates for sustainable food packaging.[7]

10.7 Conclusion and Future Outlook

Functional protein-based biomaterials have been considered one of the most effective materials in recent years due to their special tuneable structure, biodegradability, environmentally friendly nature, biocompatibility, nontoxicity or minimal toxicity, and multifunctional properties for the potential applications in biomedical, agriculture, and food industries. However, some shortcomings of protein biomaterials, such as safety issues, lower mechanical strength, and inadequate host response in clinical areas, limit their future sustainable applications. In this regard, researchers should continue to explore suitable new ways of tuning proteins to obtain desirable mechanical and chemical properties. The incorporation of functional proteins with other active proteins or low-toxic synthetic compounds or polysaccharide-based macromolecules can lead to the development of hybrid materials with desirable physicochemical attributes. Besides, several innovative research areas, such as the electrospinning method, biomimetic technique, surface modifications, and 3D-printing process, are needed for the fabrication of effective biomaterials from functional proteins. These techniques will lead to modified protein biomaterials with sustainable mechanical and biological properties to address the different growing fields of the food industry, agriculture, and biomedicine. Additionally, exploring the hierarchical structure of functional proteins is necessary to understand the fundamental properties of biomaterials, which could be useful for predicting and developing new designs of biomaterials as drug delivery vehicles. It is estimated that functional protein biomaterials could be highly customizable through the modification of their physical, chemical, and biological properties for future applications.

Abbreviations

DNA	Deoxyribonucleic acid
MW	Molecular weight
PLA	Polylactic acid
PLGA	Polylactic acid-co-glycolic acid
PVA	Polyvinyl alcohol
UV light	Ultraviolet light

References

1. S. Chu, A. L. Wang, A. Bhattacharya and J. K. Montclare, Protein based biomaterials for therapeutic and diagnostic applications, *Prog. Biomed. Eng.*, 2021, **4**(1), 012003.
2. Y. Han, J. Pan, Y. Ma, D. Zhou and W. Xu, Protein-based biomaterials for combating viral infections: Current status and future prospects for development, *Biosafety and Health*, 2022, **4**(2), 87–94.
3. A. Varanko, S. Saha and A. Chilkoti, Recent trends in protein and peptide-based biomaterials for advanced drug delivery, *Adv. Drug Delivery Rev.*, 2020, **156**, 133–187.
4. H. Agnieray, J. Glasson, Q. Chen, M. Kaur and L. Domigan, Recent developments in sustainably sourced protein-based biomaterials, *Biochem. Soc. Trans.*, 2021, **49**(2), 953–964.
5. J. M. Moreno-Hernández, I. Benítez-García, M. A. Mazorra-Manzano, J. C. Ramírez-Suárez and E. Sánchez, Strategies for production, characterization and application of protein-based biostimulants in agriculture: A review, *Chil. J. Agric. Res.*, 2020, **80**(2), 274–289.
6. S. Portanguen, P. Tournayre, J. Sicard, T. Astruc and P.-S. Mirade, Toward the design of functional foods and biobased products by 3D printing: A review, *Trends Food Sci. Technol.*, 2019, **86**, 188–198.

7. S. J. Calva-Estrada, M. Jiménez-Fernández and E. Lugo-Cervantes, Protein-based films: Advances in the development of biomaterials applicable to food packaging, *Food Eng. Rev.*, 2019, **11**(2), 78–92.
8. A. B. Bello, D. Kim, D. Kim, H. Park and S.-H. Lee, Engineering and functionalization of gelatin biomaterials: From cell culture to medical applications, *Tissue Eng., Part B*, 2020, **26**(2), 164–180.
9. S. M. Choi, P. Chaudhry, S. M. Zo and S. S. Han, Advances in protein-based materials: from origin to novel biomaterials, in *Cutting-edge enabling technologies for regenerative medicine*, 2018, pp. 161–210.
10. F. Costa, R. Silva and A. Boccaccini, Fibrous protein-based biomaterials (silk, keratin, elastin, and resilin proteins) for tissue regeneration and repair, in *Peptides and proteins as biomaterials for tissue regeneration and repair*, 2018, pp. 175–204.
11. C. Zhang, X. Li and S.-K. Kim, Application of marine biomaterials for nutraceuticals and functional foods, *Food Sci. Biotechnol.*, 2012, **21**(3), 625–631.
12. M.-C. Wan, W. Qin and C. Lei, *et al.*, Biomaterials from the sea: Future building blocks for biomedical applications, *Bioact. Mater.*, 2021, **6**(12), 4255–4285.
13. B. K. Shanbhag, C. Liu and G. C. Pradeep, *et al.*, Custom design of protein particles as multifunctional biomaterials, *Adv. Funct. Mater.*, 2022, **32**(2), 2108039.
14. J. E. Gagner, W. Kim and E. L. Chaikof, Designing protein-based biomaterials for medical applications, *Acta Biomater.*, 2014, **10**(4), 1542–1557.
15. Y. Li, H. Meng, Y. Liu and B. P. Lee, Fibrin gel as an injectable biodegradable scaffold and cell carrier for tissue engineering, *Sci. World J.*, 2015, **2015**.
16. P. A. Gunatillake, R. Adhikari and N. Gadegaard, Biodegradable synthetic polymers for tissue engineering, *Eur. Cells Mater.*, 2003, **5**(1), 1–16.
17. S. Banskota, P. Yousefpour, N. Kirmani, X. Li and A. Chilkoti, Long circulating genetically encoded intrinsically disordered zwitterionic polypeptides for drug delivery, *Biomaterials*, 2019, **192**, 475–485.
18. S. Gomes, I. B. Leonor, J. F. Mano, R. L. Reis and D. L. Kaplan, Natural and genetically engineered proteins for tissue engineering, *Prog. Polym. Sci.*, 2012, **37**(1), 1–17.
19. S. A. Maskarinec and D. A. Tirrell, Protein engineering approaches to biomaterials design, *Curr. Opin. Biotechnol.*, 2005, **16**(4), 422–426.
20. L. Li, A. Mahara and Z. Tong, *et al.*, Recombinant resilin-based bioelastomers for regenerative medicine applications, *Adv. Healthcare Mater.*, 2016, **5**(2), 266–275.
21. A. D. Theocharis, S. S. Skandalis, C. Gialeli and N. K. Karamanos, Extracellular matrix structure, *Adv. Drug Delivery Rev.*, 2016, **97**, 4–27.
22. K. Punia, J. B. Kronenberg and J. K. Montclare, Protein biomaterials for theranostic applications, *Mol. Syst. Des. Eng.*, 2019, **4**(6), 1074–1094.

23. C. I. Branden and J. Tooze, in *Introduction to protein structure*, Garland Science, 2012.
24. I. Rehman, M. Farooq and S. Botelho, Biochemistry, secondary protein structure, in *StatPearls*, StatPearls Publishing, 2021.
25. M. N. Belgacem and A. Gandini, in *Monomers, polymers and composites from renewable resources*, Elsevier, 2011.
26. N. H. Silva, C. Vilela, I. M. Marrucho, C. S. Freire, C. P. Neto and A. J. Silvestre, Protein-based materials: from sources to innovative sustainable materials for biomedical applications, *J. Mater. Chem. B*, 2014, **2**(24), 3715–3740.
27. M. Giménez-Andrés, A. Čopič and B. Antonny, The many faces of amphipathic helices, *Biomolecules*, 2018, **8**(3), 45.
28. K. Gelse, E. Pöschl and T. Aigner, Collagens—structure, function, and biosynthesis, *Adv. Drug Delivery Rev.*, 2003, **55**(12), 1531–1546.
29. Y. Li and S. M. Yu, Targeting and mimicking collagens via triple helical peptide assembly, *Curr. Opin. Chem. Biol.*, 2013, **17**(6), 968–975.
30. E. P. DeBenedictis and S. Keten, Mechanical unfolding of alpha- and beta-helical protein motifs, *Soft Matter*, 2019, **15**(6), 1243–1252.
31. L. Ding, Y. Jiang, J. Zhang, H.-A. Klok and Z. Zhong, pH-sensitive coiled-coil peptide-cross-linked hyaluronic acid nanogels: synthesis and targeted intracellular protein delivery to CD44 positive cancer cells, *Biomacromolecules*, 2018, **19**(2), 555–562.
32. L. K. Hill, M. Meleties and P. Katyal, *et al.*, Thermoresponsive protein-engineered coiled-coil hydrogel for sustained small molecule release, *Biomacromolecules*, 2019, **20**(9), 3340–3351.
33. G. Martelli, H. R. Zope, M. B. Capell and A. Kros, Coiled-coil peptide motifs as thermoresponsive valves for mesoporous silica nanoparticles, *Chem. Commun.*, 2013, **49**(85), 9932–9934.
34. L. Feng, Z. Dong, D. Tao, Y. Zhang and Z. Liu, The acidic tumor microenvironment: a target for smart cancer nano-theranostics, *Natl. Sci. Rev.*, 2018, **5**(2), 269–286.
35. E. T. Pashuck, H. Cui and S. I. Stupp, Tuning supramolecular rigidity of peptide fibers through molecular structure, *J. Am. Chem. Soc.*, 2010, **132**(17), 6041–6046.
36. X. Liu, C. Zheng, X. Luo, X. Wang and H. Jiang, Recent advances of collagen-based biomaterials: Multi-hierarchical structure, modification and biomedical applications, *Mater. Sci. Eng., C*, 2019, **99**, 1509–1522.
37. H.-D. Belitz, W. Grosch and P. Schieberle, Food additives, *Food. Chem.*, **2009**, 429–466.
38. E. J. Torres-Martinez, J. M. Cornejo Bravo, A. Serrano Medina, G. L. Pérez González and L. J. Villarreal Gómez, A summary of electrospun nanofibers as drug delivery system: drugs loaded and biopolymers used as matrices, *Curr. Drug Delivery*, 2018, **15**(10), 1360–1374.

39. A. Jain, S. K. Singh, S. K. Arya, S. C. Kundu and S. Kapoor, Protein nanoparticles: promising platforms for drug delivery applications, *ACS Biomater. Sci. Eng.*, 2018, **4**(12), 3939–3961.
40. R. Panahi and M. Baghban-Salehi, Protein-based hydrogels, in *Cellulose-based superabsorbent hydrogels*, Springer, 2019, pp. 1561–1600.
41. J. Sanders and B. Goldstein, Collagen fibril diameters increase and fibril densities decrease in skin subjected to repetitive compressive and shear stresses, *J. Biomech.*, 2001, **34**(12), 1581–1587.
42. L. Sivakumar and G. Agarwal, The influence of discoidin domain receptor 2 on the persistence length of collagen type I fibers, *Biomaterials*, 2010, **31**(18), 4802–4808.
43. D. T. Seroski, A. Restuccia, A. D. Sorrentino, K. R. Knox, S. J. Hagen and G. A. Hudalla, Co-assembly tags based on charge complementarity (CATCH) for installing functional protein ligands into supramolecular biomaterials, *Cell. Mol. Bioeng.*, 2016, **9**(3), 335–350.
44. M. C. Gomes and J. F. Mano, Chemical modification strategies to prepare advanced protein-based biomaterials, *Biomater. Biosyst.*, 2021, **1**, 100010.
45. C.-Y. Lin and J. C. Liu, Modular protein domains: an engineering approach toward functional biomaterials, *Curr. Opin. Biotechnol.*, 2016, **40**, 56–63.
46. T. Heck, G. Faccio, M. Richter and L. Thöny-Meyer, Enzyme-catalyzed protein crosslinking, *Appl. Microbiol. Biotechnol.*, 2013, **97**(2), 461–475.
47. J. Chen, R. R. Pompano and F. W. Santiago, *et al.*, The use of self-adjuvanting nanofiber vaccines to elicit high-affinity B cell responses to peptide antigens without inflammation, *Biomaterials*, 2013, **34**(34), 8776–8785.
48. L. A. Haines-Butterick, D. A. Salick, D. J. Pochan and J. P. Schneider, In vitro assessment of the pro-inflammatory potential of β-hairpin peptide hydrogels, *Biomaterials*, 2008, **29**(31), 4164–4169.
49. P. C. Hsieh, M. E. Davis, J. Gannon, C. MacGillivray and R. T. Lee, Controlled delivery of PDGF-BB for myocardial protection using injectable self-assembling peptide nanofibers, *J. Clin. Invest.*, 2006, **116**(1), 237–248.
50. G. H. Hudalla and W. L. Murphy, in *Mimicking the Extracellular Matrix: The Intersection of Matrix Biology and Biomaterials*, Royal Society of Chemistry, 2015.
51. R. Beutner, J. Michael, B. Schwenzer and D. Scharnweber, Biological nano-functionalization of titanium-based biomaterial surfaces: a flexible toolbox, *J. R. Soc., Interface*, 2010, **7**(Suppl 1), S93–S105.
52. L. Bacáková, E. Filová, F. Rypácek, V. Svorcík and V. Starý, Cell adhesion on artificial materials for tissue engineering, *Physiol. Res.*, 2004, **53**(Suppl 1), S35–S45.

53. V. Karageorgiou, M. Tomkins and R. Fajardo, *et al.*, Porous silk fibroin 3-D scaffolds for delivery of bone morphogenetic protein-2 in vitro and in vivo, *J. Biomed. Mater. Res., Part A*, 2006, **78**(2), 324–334.
54. M. Biondi, F. Ungaro, F. Quaglia and P. A. Netti, Controlled drug delivery in tissue engineering, *Adv. Drug Delivery Rev.*, 2008, **60**(2), 229–242.
55. U. Hersel, C. Dahmen and H. Kessler, RGD modified polymers: biomaterials for stimulated cell adhesion and beyond, *Biomaterials*, 2003, **24**(24), 4385–4415.
56. H. Cao and S.-Y. Xu, EDC/NHS-crosslinked type II collagen-chondroitin sulfate scaffold: characterization and in vitro evaluation, *J. Mater. Sci.: Mater. Med*, 2008, **19**(2), 567–575.
57. A. R. Murphy and D. L. Kaplan, Biomedical applications of chemically-modified silk fibroin, *J. Mater. Chem.*, 2009, **19**(36), 6443–6450.
58. F. Wu and T. Jin, Polymer-based sustained-release dosage forms for protein drugs, challenges, and recent advances, *AAPS PharmSciTech*, 2008, **9**(4), 1218–1229.
59. N. Adhirajan, N. Shanmugasundaram, S. Shanmuganathan and M. Babu, Functionally modified gelatin microspheres impregnated collagen scaffold as novel wound dressing to attenuate the proteases and bacterial growth, *Eur. J. Pharm. Sci.*, 2009, **36**(2–3), 235–245.
60. N. N. Fathima, J. R. Rao and B. U. Nair, Effect of UV irradiation on the physico-chemical properties of iron crosslinked collagen, *J. Photochem. Photobiol., B*, 2011, **105**(3), 203–206.
61. S. Cao, H. Li, K. Li, J. Lu and L. Zhang, A dense and strong bonding collagen film for carbon/carbon composites, *Appl. Surf. Sci.*, 2015, **347**, 307–314.
62. T. Takitoh, M. Bessho, M. Hirose, H. Ohgushi, H. Mori and M. Hara, Gamma-cross-linked nonfibrillar collagen gel as a scaffold for osteogenic differentiation of mesenchymal stem cells, *J. Biosci. Bioeng.*, 2015, **119**(2), 217–225.
63. J. Kozłowska and A. Sionkowska, Effects of different crosslinking methods on the properties of collagen-calcium phosphate composite materials, *Int. J. Biol. Macromol.*, 2015, **74**, 397–403.
64. Y. Hu, L. Liu, W. Dan, N. Dan, Z. Gu and X. Yu, Synergistic effect of carbodiimide and dehydrothermal crosslinking on acellular dermal matrix, *Int. J. Biol. Macromol.*, 2013, **55**, 221–230.
65. X. Fan, K. Chen and X. He, *et al.*, Nano-TiO2/collagen-chitosan porous scaffold for wound repairing, *Int. J. Biol. Macromol.*, 2016, **91**, 15–22.
66. X. Zhou, J. Wang and W. Fang, *et al.*, Genipin cross-linked type II collagen/chondroitin sulfate composite hydrogel-like cell delivery system induces differentiation of adipose-derived stem cells and regenerates degenerated nucleus pulposus, *Acta Biomater.*, 2018, **71**, 496–509.
67. M. H. Lee, S. Y. Kim and H. J. Park, Effect of halloysite nanoclay on the physical, mechanical, and antioxidant properties of chitosan films

incorporated with clove essential oil, *Food Hydrocolloids*, 2018, **84**, 58–67.
68. X. Wang, Y. Wang, L. Li, Z. Gu and X. Yu, Feasibility study of the naturally occurring dialdehyde carboxymethyl cellulose for biological tissue fixation, *Carbohydr. Polym.*, 2015, **115**, 54–61.
69. T. Lu, Q. Li, W. Chen and H. Yu, Composite aerogels based on dialdehyde nanocellulose and collagen for potential applications as wound dressing and tissue engineering scaffold, *Compos. Sci. Technol.*, 2014, **94**, 132–138.
70. L. Jia, M. P. Prabhakaran, X. Qin and S. Ramakrishna, Guiding the orientation of smooth muscle cells on random and aligned polyurethane/collagen nanofibers, *J. Biomater. Appl.*, 2014, **29**(3), 364–377.
71. D. Yin, H. Wu and C. Liu, *et al.*, Fabrication of composition-graded collagen/chitosan–polylactide scaffolds with gradient architecture and properties, *React. Funct. Polym.*, 2014, **83**, 98–106.
72. Y. C. Shin, J. H. Lee and L. Jin, *et al.*, Stimulated myoblast differentiation on graphene oxide-impregnated PLGA-collagen hybrid fibre matrices, *J. Nanobiotechnol.*, 2015, **13**(1), 1–11.
73. R. Huang, W. Li and X. Lv, *et al.*, Biomimetic LBL structured nanofibrous matrices assembled by chitosan/collagen for promoting wound healing, *Biomaterials*, 2015, **53**, 58–75.
74. E. Rezvani Ghomi, N. Nourbakhsh, M. Akbari Kenari, M. Zare and S. Ramakrishna, Collagen-based biomaterials for biomedical applications, *J. Biomed. Mater. Res., Part B*, 2021, **109**(12), 1986–1999.
75. J. A. Huntington and P. E. Stein, Structure and properties of ovalbumin, *J. Chromatogr. B: Biomed. Sci. Appl.*, 2001, **756**(1–2), 189–198.
76. D. Miranda-Nieves and E. L. Chaikof, Collagen and elastin biomaterials for the fabrication of engineered living tissues, *ACS Biomater. Sci. Eng.*, 2017, **3**(5), 694–711.
77. K. Jastrzebska, K. Kucharczyk, A. Florczak, E. Dondajewska, A. Mackiewicz and H. Dams-Kozlowska, Silk as an innovative biomaterial for cancer therapy, *Rep. Pract. Oncol. Radiother.*, 2015, **20**(2), 87–98.
78. L. Li, Z. Tong, X. Jia and K. L. Kiick, Resilin-like polypeptide hydrogels engineered for versatile biological function, *Soft Matter*, 2013, **9**(3), 665–673.
79. C. Holt, J. Carver, H. Ecroyd and D. Thorn, Invited review: Caseins and the casein micelle: Their biological functions, structures, and behavior in foods, *J. Dairy Sci.*, 2013, **96**(10), 6127–6146.
80. M. A. Farooq, M. Aquib and S. Ghayas, *et al.*, Whey protein: A functional and promising material for drug delivery systems recent developments and future prospects, *Polym. Adv. Technol.*, 2019, **30**(9), 2183–2191.
81. Z. Teng, C. Liu, X. Yang, L. Li, C. Tang and Y. Jiang, Fractionation of soybean globulins using Ca2+ and Mg2+: a comparative analysis, *J. Am. Oil Chem. Soc.*, 2009, **86**(5), 409–417.

82. X. Liu, Q. Sun, H. Wang, L. Zhang and J.-Y. Wang, Microspheres of corn protein, zein, for an ivermectin drug delivery system, *Biomaterials*, 2005, **26**(1), 109–115.
83. L. Day, Wheat gluten: Production, properties and application, in *Handbook of food proteins*, Elsevier, 2011, pp. 267–288.
84. H. Jafari, A. Lista and M. M. Siekapen, *et al.*, Fish collagen: extraction, characterization, and applications for biomaterials engineering, *Polymers*, 2020, **12**(10), 2230.
85. M. Pozzolini, S. Scarfì and L. Gallus, *et al.*, Production, characterization and biocompatibility evaluation of collagen membranes derived from marine sponge Chondrosia reniformis Nardo, 1847, *Mar. Drugs*, 2018, **16**(4), 111.
86. P. Elumalai, A. S. Rubeena and J. Arockiaraj, *et al.*, The role of lectins in finfish: a review, *Rev. Fish. Sci. Aquac.*, 2019, **27**(2), 152–169.
87. E. Troy, M. A. Tilbury, A. M. Power and J. G. Wall, Nature-based biomaterials and their application in biomedicine, *Polymers*, 2021, **13**(19), 3321.
88. F. Copes, N. Pien, S. Van Vlierberghe, F. Boccafoschi and D. Mantovani, Collagen-Based Tissue Engineering Strategies for Vascular Medicine, *Front. Bioeng. Biotechnol.*, 2019, **7**, 166.
89. C. Rodrigues, P. Serricella and A. Linhares, *et al.*, Characterization of a bovine collagen-hydroxyapatite composite scaffold for bone tissue engineering, *Biomaterials*, 2003, **24**(27), 4987–4997.
90. M. Spira, B. Liu, Z. Xu, R. Harrell and H. Chahadeh, Human amnion collagen for soft tissue augmentation--biochemical characterizations and animal observations, *J. Biomed. Mater. Res.*, 1994, **28**(1), 91–96.
91. A. Veis, The physical chemistry of gelatin, *Int. Rev. Connect. Tissue Res.*, 1965, **3**, 113–200.
92. A. O. Elzoghby, Gelatin-based nanoparticles as drug and gene delivery systems: reviewing three decades of research, *J. Controlled Release*, 2013, **172**(3), 1075–1091.
93. N. Sahoo, R. K. Sahoo, N. Biswas, A. Guha and K. Kuotsu, Recent advancement of gelatin nanoparticles in drug and vaccine delivery, *Int. J. Biol. Macromol.*, 2015, **81**, 317–331.
94. S. Gorgieva and V. Kokol, Biomaterials and Their Biocompatibility: Review and Perspectives, *Biomater. Appl. Nanomed.*, **2011**, 17–52.
95. E. Kianfar, Protein nanoparticles in drug delivery: animal protein, plant proteins and protein cages, albumin nanoparticles, *J. Nanobiotechnol.*, 2021, **19**(1), 1–32.
96. P. Yousefpour, J. R. McDaniel and V. Prasad, *et al.*, Genetically encoding albumin binding into chemotherapeutic-loaded polypeptide nanoparticles enhances their antitumor efficacy, *Nano Lett.*, 2018, **18**(12), 7784–7793.
97. E. N. Hoogenboezem and C. L. Duvall, Harnessing albumin as a carrier for cancer therapies, *Adv. Drug Delivery Rev.*, 2018, **130**, 73–89.

98. J. Li and P. Yao, Self-assembly of ibuprofen and bovine serum albumin-dextran conjugates leading to effective loading of the drug, *Langmuir*, 2009, **25**(11), 6385–6391.
99. D. C. Carter and J. X. Ho, Structure of serum albumin, *Adv. Protein Chem.*, 1994, **45**, 153–203.
100. A. Maham, Z. Tang, H. Wu, J. Wang and Y. Lin, Protein-based nanomedicine platforms for drug delivery, *Small*, 2009, **5**(15), 1706–1721.
101. M. Jahanshahi and Z. Babaei, Protein nanoparticle: a unique system as drug delivery vehicles, *Afr. J. Biotechnol.*, 2008, **7**, 25.
102. S. Mithieux and A. Weiss, in *Fibrous Proteins: Coiled-Coils, Collagen and Elastomers*, Elsevier, 2005, vol. **70**, pp. 437–461.
103. T. Kurinomaru, T. Maruyama, S. Izaki, K. Handa, T. Kimoto and K. Shiraki, Protein-poly(amino acid) complex precipitation for high-concentration protein formulation, *J. Pharm. Sci.*, 2014, **103**(8), 2248–2254.
104. J. Rosenbloom, [9] Elastin: an overview, *Methods Enzymol.*, 1987, **144**, 172–196.
105. G. C. Yeo, B. Aghaei-Ghareh-Bolagh, E. P. Brackenreg, M. A. Hiob, P. Lee and A. S. Weiss, Fabricated Elastin, *Adv. Healthcare Mater.*, 2015, **4**(16), 2530–2556.
106. X. Zhang, D. Wang, Z. Zhou and Y. Ma, Robust low-rank tensor recovery with rectification and alignment, *IEEE Trans. Pattern Anal. Mach. Intell.*, 2019, **43**(1), 238–255.
107. M. Ozboyaci, D. B. Kokh, S. Corni and R. C. Wade, Modeling and simulation of protein-surface interactions: achievements and challenges, *Q. Rev. Biophys.*, 2016, **49**, 1–45.
108. T. Yucel, M. L. Lovett and D. L. Kaplan, Silk-based biomaterials for sustained drug delivery, *J. Controlled Release*, 2014, **190**, 381–397.
109. S. H. Kim, Y. K. Yeon and J. M. Lee, *et al.*, Precisely printable and biocompatible silk fibroin bioink for digital light processing 3D printing, *Nat. Commun.*, 2018, **9**(1), 1–14.
110. S. Chawla, S. Midha, A. Sharma and S. Ghosh, Silk-Based Bioinks for 3D Bioprinting, *Adv. Healthcare Mater.*, 2018, **7**(8), 1701204.
111. R. S.-C. Su, Y. Kim and J. C. Liu, Resilin: protein-based elastomeric biomaterials, *Acta Biomater.*, 2014, **10**(4), 1601–1611.
112. G. Qin, X. Hu, P. Cebe and D. L. Kaplan, Mechanism of resilin elasticity, *Nat. Commun.*, 2012, **3**(1), 1–9.
113. C. M. Elvin, A. G. Carr and M. G. Huson, *et al.*, Synthesis and properties of crosslinked recombinant pro-resilin, *Nature*, 2005, **437**(7061), 999–1002.
114. R. E. Lyons, E. Lesieur and M. Kim, *et al.*, Design and facile production of recombinant resilin-like polypeptides: gene construction and a rapid protein purification method, *Protein Eng., Des. Sel.*, 2007, **20**(1), 25–32.

115. R. E. Lyons, D. C. Wong and M. Kim, *et al.*, Molecular and functional characterisation of resilin across three insect orders, *Insect Biochem. Mol. Biol.*, 2011, **41**(11), 881–890.
116. F. Rehan, N. Ahemad and M. Gupta, Casein nanomicelle as an emerging biomaterial-A comprehensive review, *Colloids Surf., B*, 2019, **179**, 280–292.
117. J. Trommelen, M. E. Weijzen and J. van Kranenburg, *et al.*, Casein protein processing strongly modulates post-prandial plasma amino acid responses in vivo in humans, *Nutrients*, 2020, **12**(8), 2299.
118. Y. D. Livney, Milk proteins as vehicles for bioactives, *Curr. Opin. Colloid Interface Sci.*, 2010, **15**(1–2), 73–83.
119. S. Rahimi Yazdi, in *Changing the structure of casein micelles to improve the delivery of bioactive compounds*, University of Guelph, 2012.
120. H. E. Swaisgood, Chemistry of the caseins, in *Advanced dairy chemistry —1 Proteins*, Springer, 2003, pp. 139–201.
121. C. Ranadheera, W. Liyanaarachchi, J. Chandrapala, M. Dissanayake and T. Vasiljevic, Utilizing unique properties of caseins and the casein micelle for delivery of sensitive food ingredients and bioactives, *Trends Food Sci. Technol.*, 2016, **57**, 178–187.
122. C. V. Morr and E. Ha, Whey protein concentrates and isolates: processing and functional properties, *Crit. Rev. Food Sci. Nutr.*, 1993, **33**(6), 431–476.
123. S. Patel, Emerging trends in nutraceutical applications of whey protein and its derivatives, *J. Food Sci. Technol.*, 2015, **52**(11), 6847–6858.
124. J. J. Hulmi, C. M. Lockwood and J. R. Stout, Effect of protein/essential amino acids and resistance training on skeletal muscle hypertrophy: A case for whey protein, *Nutr. Metab.*, 2010, **7**(1), 1–11.
125. J. Liu, Z. Li and Q. Lin, *et al.*, A robust, resilient, and multi-functional soy protein-based hydrogel, *ACS Sustainable Chem. Eng.*, 2018, **6**(11), 13730–13738.
126. P. Singh, R. Kumar, S. Sabapathy and A. Bawa, Functional and edible uses of soy protein products, *Compr. Rev. Food Sci. Food Saf.*, 2008, **7**(1), 14–28.
127. J. Kinsella, Functional properties of soy proteins, *J. Crit. Rev. Food Sci. Nutr.*, 1976, 7219–7280.
128. L. Chen, G. Hébrard, E. Beyssac, S. Denis and M. Subirade, In vitro study of the release properties of soy-zein protein microspheres with a dynamic artificial digestive system, *J. Agric. Food Chem.*, 2010, **58**(17), 9861–9867.
129. C. K. Hong and R. P. Wool, Development of a bio-based composite material from soybean oil and keratin fibers, *J. Appl. Polym. Sci.*, 2005, **95**(6), 1524–1538.
130. Y. Wang and G. W. Padua, Formation of zein microphases in ethanol–water, *Langmuir*, 2010, **26**(15), 12897–12901.

131. S. G. Giteru, M. A. Ali and I. Oey, Recent progress in understanding fundamental interactions and applications of zein, *Food Hydrocolloids*, 2021, **120**, 106948.
132. R. Paliwal and S. Palakurthi, Zein in controlled drug delivery and tissue engineering, *J. Controlled Release*, 2014, **189**, 108–122.
133. P. R. Shewry, N. G. Halford, P. S. Belton and A. S. Tatham, The structure and properties of gluten: an elastic protein from wheat grain, *Philos. Trans. R. Soc., B*, 2002, **357**(1418), 133–142.
134. X. Kong, H. Zhou and H. Qian, Enzymatic preparation and functional properties of wheat gluten hydrolysates, *Food Chem.*, 2007, **101**(2), 615–620.
135. M. Raman and K. Gopakumar, Fish collagen and its applications in food and pharmaceutical industry: A review, *EC Nutr.*, 2018, **13**(12), 752–767.
136. S. Benjakul, S. Nalinanon and F. Shahidi, Fish collagen, *Food Biochem. Food Process.*, **2012**, 365–387.
137. F. Subhan, Z. Hussain, I. Tauseef, A. Shehzad and F. Wahid, A review on recent advances and applications of fish collagen, *Crit. Rev. Food Sci. Nutr.*, 2021, **61**(6), 1027–1037.
138. J.-K. Cho, Y.-G. Jin, S.-J. Rha, S.-J. Kim and J.-H. Hwang, Biochemical characteristics of four marine fish skins in Korea, *Food Chem.*, 2014, **159**, 200–207.
139. L.-A. Tziveleka, E. Ioannou, D. Tsiourvas, P. Berillis, E. Foufa and V. Roussis, Collagen from the marine sponges Axinella cannabina and Suberites carnosus: Isolation and morphological, biochemical, and biophysical characterization, *Mar. Drugs*, 2017, **15**(6), 152.
140. R. Pallela, H. Ehrlich and I. Bhatnagar, Biomedical applications of marine sponge collagens, in *Marine Sponges: Chemicobiological and Biomedical Applications.*, Springer, 2016, pp. 373–381.
141. T. Bun Ng, R. Chi Fai Cheung, C. Cheuk Wing Ng, E. Fei Fang and J. Ho Wong, A review of fish lectins, *Curr. Protein Pept. Sci*, 2015, **16**(4), 337–351.
142. S. Russell and J. S. Lumsden, Function and heterogeneity of fish lectins, *Vet. Immunol. Immunopathol.*, 2005, **108**(1–2), 111–120.
143. M. F. Brinchmann, D. M. Patel, N. Pinto and M. H. Iversen, Functional aspects of fish mucosal lectins—interaction with non-self, *Molecules*, 2018, **23**(5), 1119.
144. C. H. Lee, A. Singla and Y. Lee, Biomedical applications of collagen, *Int. J. Pharm.*, 2001, **221**(1–2), 1–22.
145. A. Vasconcelos, A. P. Pêgo, L. Henriques, M. Lamghari and A. Cavaco-Paulo, Protein matrices for improved wound healing: elastase inhibition by a synthetic peptide model, *Biomacromolecules*, 2010, **11**(9), 2213–2220.
146. J. G. Rouse and M. E. Van Dyke, A review of keratin-based biomaterials for biomedical applications, *Materials*, 2010, **3**(2), 999–1014.

147. A. Tachibana, Y. Furuta, H. Takeshima, T. Tanabe and K. Yamauchi, Fabrication of wool keratin sponge scaffolds for long-term cell cultivation, *J. Biotechnol.*, 2002, **93**(2), 165–170.
148. S.-M. Lien, L.-Y. Ko and T.-J. Huang, Effect of pore size on ECM secretion and cell growth in gelatin scaffold for articular cartilage tissue engineering, *Acta Biomater.*, 2009, **5**(2), 670–679.
149. S. H. Oh, I. K. Park, J. M. Kim and J. H. Lee, In vitro and in vivo characteristics of PCL scaffolds with pore size gradient fabricated by a centrifugation method, *Biomaterials*, 2007, **28**(9), 1664–1671.
150. C. Dong and Y. Lv, Application of collagen scaffold in tissue engineering: recent advances and new perspectives, *Polymers*, 2016, **8**(2), 42.
151. K. McGuinness and V. Nanda, Collagen mimetic peptide discs promote assembly of a broad range of natural protein fibers through hydrophobic interactions, *Org. Biomol. Chem.*, 2017, **15**(28), 5893–5898.
152. S. Kajdič, O. Planinšek, M. Gašperlin and P. Kocbek, Electrospun nanofibers for customized drug-delivery systems, *J. Drug Delivery Sci. Technol.*, 2019, **51**, 672–681.
153. A. Vasconcelos and A. Cavaco-Paulo, The use of keratin in biomedical applications, *Curr. Drug Targets*, 2013, **14**(5), 612–619.
154. A. Shavandi, T. H. Silva, A. A. Bekhit and A. E.-D. A. Bekhit, Keratin: dissolution, extraction and biomedical application, *Biomater. Sci.*, 2017, **5**(9), 1699–1735.
155. E. Lee, S. Khan and K.-H. Lim, Gelatin nanoparticle preparation by nanoprecipitation, *J. Biomater. Sci. Polym. Ed.*, 2011, **22**(4–6), 753–771.
156. S. J. Bryant and F. J. Vernerey, Programmable hydrogels for cell encapsulation and neo-tissue growth to enable personalized tissue engineering, *Adv. Healthcare Mater.*, 2018, **7**(1), 1700605.
157. A. Rüter, S. Kuczera, L. Gentile and U. Olsson, Arrested dynamics in a model peptide hydrogel system, *Soft Matter*, 2020, **16**(11), 2642–2651.
158. A. N. Moore and J. D. Hartgerink, Self-assembling multidomain peptide nanofibers for delivery of bioactive molecules and tissue regeneration, *Acc. Chem. Res.*, 2017, **50**(4), 714–722.
159. Y. Wang, M. S. Shim, N. S. Levinson, H. W. Sung and Y. Xia, Stimuli-responsive materials for controlled release of theranostic agents, *Adv. Funct. Mater.*, 2014, **24**(27), 4206–4220.
160. B. Maity, S. Samanta, S. Sarkar, S. Alam and T. Govindaraju, Injectable silk fibroin-based hydrogel for sustained insulin delivery in diabetic rats, *ACS Appl. Bio Mater.*, 2020, **3**(6), 3544–3552.
161. A. Bajpai and J. Choubey, Design of gelatin nanoparticles as swelling controlled delivery system for chloroquine phosphate, *J. Mater. Sci.: Mater. Med.*, 2006, **17**(4), 345–358.
162. S. Amjadi, H. Hamishehkar and M. Ghorbani, A novel smart PEGylated gelatin nanoparticle for co-delivery of doxorubicin and betanin: A

strategy for enhancing the therapeutic efficacy of chemotherapy, *Mater. Sci. Eng., C*, 2019, **97**, 833–841.
163. D. Narayanan, M. Geena, H. Lakshmi, M. Koyakutty, S. Nair and D. Menon, Poly-(ethylene glycol) modified gelatin nanoparticles for sustained delivery of the anti-inflammatory drug ibuprofen-sodium: an in vitro and in vivo analysis, *Nanomedicine*, 2013, **9**(6), 818–828.
164. R. Yarchoan, H. Mitsuya and R. V. Thomas, *et al.*, In vivo activity against HIV and favorable toxicity profile of 2′,3′-dideoxyinosine, *Science*, 1989, **245**(4916), 412–415.
165. Z. Fan, P. Cheng, G. Yin, Z. Wang and J. Han, In situ forming oxidized salecan/gelatin injectable hydrogels for vancomycin delivery and 3D cell culture, *J. Biomater. Sci., Polym. Ed.*, 2020, **31**(6), 762–780.
166. P. Sierpinski, J. Garrett and J. Ma, *et al.*, The use of keratin biomaterials derived from human hair for the promotion of rapid regeneration of peripheral nerves, *Biomaterials*, 2008, **29**(1), 118–128.
167. P. J. Apel, J. P. Garrett and P. Sierpinski, *et al.*, Peripheral nerve regeneration using a keratin-based scaffold: long-term functional and histological outcomes in a mouse model, *J. Hand Surg.*, 2008, **33**(9), 1541–1547.
168. A. A. Mahmoud and A. H. Salama, Norfloxacin-loaded collagen/chitosan scaffolds for skin reconstruction: Preparation, evaluation and in-vivo wound healing assessment, *Eur. J. Pharm. Sci.*, 2016, **83**, 155–165.
169. K. Whitaker, P. Barrow and K. Feree, A case study using chronicure to treat A stage IV chronic wound, *Ostomy Wound Manage.*, 1992, **38**, 36–44.
170. M. T. Islam, R. M. Laing, C. A. Wilson, M. McConnell and M. A. Ali, Fabrication and characterization of 3-dimensional electrospun poly(vinyl alcohol)/keratin/chitosan nanofibrous scaffold, *Carbohydr. Polym.*, 2022, **275**, 118682.
171. M. A. Ali and M. H. Bhuiyan, Types of biomaterials useful in brain repair, *Neurochem. Int.*, 2021, **146**, 105034.
172. S. S. Sali, M. L. Gould, M. Qasim and M. A. Ali, Biodegradable methacrylated casein for cardiac tissue engineering applications, *J. Mater. Chem. B*, 2021, **9**(6), 1557–1567.
173. X. Deng, M. Gould and M. A. Ali, Fabrication and characterisation of melt-extruded chitosan/keratin/PCL/PEG drug-eluting sutures designed for wound healing, *Mater. Sci. Eng., C*, 2021, **120**, 111696.
174. R. Boni, A. Ali, S. G. Giteru, A. Shavandi and A. N. Clarkson, Silk fibroin nanoscaffolds for neural tissue engineering, *J. Mater. Sci.: Mater. Med.*, 2020, **31**(9), 1–17.
175. M. Klein and E. Poverenov, Natural biopolymer-based hydrogels for use in food and agriculture, *J. Sci. Food Agric.*, 2020, **100**(6), 2337–2347.
176. R. E. Häusler, F. Ludewig and S. Krueger, Amino acids--a life between metabolism and signaling, *Plant Sci.*, 2014, **229**, 225–237.
177. N. Muralidharan, R. Jeya Shakila, D. Sukumar and G. Jeyasekaran, Skin, bone and muscle collagen extraction from the trash fish, leather jacket

(Odonus niger) and their characterization, *J. Food Sci. Technol.*, 2013, **50**(6), 1106–1113.
178. G. K. Pal and P. Suresh, Sustainable valorisation of seafood by-products: Recovery of collagen and development of collagen-based novel functional food ingredients, *Innovative Food Sci. Emerging Technol.*, 2016, **37**, 201–215.
179. J. Glusac and A. Fishman, Enzymatic and chemical modification of zein for food application, *Trends Food Sci. Technol.*, 2021, **112**, 507–517.
180. K. Petrotos, E. Tsakali, P. Goulas and A. D'Alessandro, Casein and whey proteins in human health, *Milk Dairy Prod. Funct. Foods*, 2014, **94**, 1–10.
181. M. Ahmad, N. P. Nirmal, M. Danish, J. Chuprom and S. Jafarzedeh, Characterisation of composite films fabricated from collagen/chitosan and collagen/soy protein isolate for food packaging applications, *RSC Adv.*, 2016, **6**(85), 82191–82204.
182. M. G. Gallego, M. H. Gordon, F. Segovia and M. P. Almajano Pablos, Gelatine-based antioxidant packaging containing Caesalpinia decapetala and Tara as a coating for ground beef patties, *Antioxidants*, 2016, **5**(2), 10.
183. J.-H. Lee, M. Won and K. B. Song, Physical properties and antimicrobial activities of porcine meat and bone meal protein films containing coriander oil, *LWT--Food Sci. Technol.*, 2015, **63**(1), 700–705.
184. S. Bahram, M. Rezaie and M. Soltani, *et al.*, Effect of whey Protein Concentrate Coating Cinamon Oil on Quality and Shelf Life of Refrigerated Beluga Sturegeon (Huso huso), *J. Food Qual.*, 2016, **39**(6), 743–749.
185. Y. Alparslan, H. H. Yapıcı, C. Metin, T. Baygar, A. Günlü and T. Baygar, Quality assessment of shrimps preserved with orange leaf essential oil incorporated gelatin, *LWT--Food Sci. Technol.*, 2016, **72**, 457–466.
186. M. Vukić, S. Grujićand B. Odzaković, Application of edible films and coatings in food production, in *Advances in Applications of Industrial Biomaterials*, Springer, 2017, pp. 121–138.
187. N.-B. Song, J.-H. Lee, M. Al Mijan and K. B. Song, Development of a chicken feather protein film containing clove oil and its application in smoked salmon packaging, *LWT--Food Sci. Technol.*, 2014, **57**(2), 453–460.
188. J. R. Rocca-Smith, E. Marcuzzo and T. Karbowiak, *et al.*, Effect of lipid incorporation on functional properties of wheat gluten based edible films, *J. Cereal Sci.*, 2016, **69**, 275–282.

CHAPTER 11

Advances in Protein- and Lipid-based Materials for Cosmetic Applications

MUNIBA IQBAL, SADIA FAROOQ AND AJAZ HUSSAIN*

Institute of Chemical Sciences, Bahauddin Zakariya University, Multan, 60800, Pakistan
*E-mail: drajazhussain@bzu.edu.pk

Proteins and lipids have extensively been used in the field of cosmetics on account of their role as moisturizing, shining and smoothing agents. For instance, a broad range of cosmetic products, including hair cleansing and other cosmeceutical formulations, such as hair-straightening agents, conditioners, shampoos, coloring lotions and many others, are composed of lipid- and protein-based materials. The role of proteins in increasing the softness, elasticity and strength of hair is an established fact now. They also play a role in repairing the split ends of hair and minimizing the aftereffects of bleaching. Moreover, proteins have also been used for a particular type of hair treatment. The addition of proteins to eye-cleansing products increases the tolerability of skin. Skin protection from roughness and dehydration is also achieved by adding proteins to cosmeceutical formulations. In this context, many fatty acids and proteins with high molecular weights have been used for protecting eyes and skin from anionic tensides, as evidenced by red blood cell tests. The role of quaternized protein

Smart Materials No. 41
Functional Materials from Lipids and Proteins
Edited by Nandika Bandara and Aman Ullah

Published by the Royal Society of Chemistry, www.rsc.org

derivatives as anti-irritants, especially for anionic-based cosmeceuticals, has also been reported in the literature.

Lipids constitute another major class of compounds used in the formulation of a variety of cosmeceuticals. Lipids and their modified materials are usually used on the skin to achieve protection from the harmful effects of environmental factors. Moreover, they are used as emulsifiers and smoothing agents for maintaining the smoothness and softness of skin.

11.1 Proteins

In the 1950s, proteins were rationally used in cosmetics for the first time.[1] Proteins are natural macromolecules consisting of amino acid chains linked in a variety of ways and make the structural and functional units of living organisms. The film-forming properties of high-molecular-weight-bearing hydrolysates and native proteins have achieved the attention of the cosmeceutical industry for their preferred use in cosmetics, especially for skincare products, *e.g.*, collagen and its partly deaminated derivatives have been used for many skincare formulations. The capability of protein molecules to form continuous films enables them to maintain softness, smoothness and a shiny, glossy appearance. These proteins should be protected from heat and denaturing agents for several reasons: (1) the length of molecules determines the film-forming properties of these proteins, (2) a large number of exposed H-bonding sites for the linking of water molecules are responsible for hydrating properties of these proteins and (3) properties of these molecules have a direct relationship with the conformation of the triple helix structure.[2]

In the formulation of several skincare products and makeup, different types of proteins are used. Soluble proteins are the main constituent and are suitable for all types of cosmetics, such as gels, lotions, powders and emulsions. In some specific products, insoluble proteins can also be used. For instance, the use of fibrous insoluble collagen (protein) in facial masks ensures moisturization, shine, gloss and smoothness.[1] However, anhydrous decorative products are prepared using silk powder obtained from finely crushed silk fibroin.[2] The molecular binding of proteins with hair has been improved by developing certain protein derivatives containing cationated, silylated and acylated proteins.[3]

There are several bioactive proteins, such as elastin, collagen and fibronectin, which help in repairing wrinkles, skin firming and skin thickening and, hence, can be used commercially as an anti-aging product. Moreover, these proteins can also be used in the regulation of melanin synthesis.[4]

The use of proteins in the cosmetic industry started with using milk proteins. In general, cosmetic products mainly contain five types of hydrolyzed proteins. The hydrolyzed category includes ammonium-hydrolyzed protein, gelatin and sink sialoprotein. These can be used in personal products, such as shampoos, shaving creams and body wash.[5]

Sclerophytin, which is an animal tissue protein, was the first protein used in cosmetics. The most commonly used protein-based cosmetic materials are discussed for each type of protein.

Plant proteins have now gained considerable importance as an alternative to animal proteins obtained from animal by-products. The presence of proteins in wheat, oats, soybean and rice makes them suitable sources. The native proteins can also be used in cosmetic products, but the common practice is to use highly soluble hydrolyzed proteins with intermediate molecular weights. The modifications, such as binding proteins to fatty alkyl quaternary groups, have also been carried out. The film-forming ability of the proteins can be improved by preparing a co-polymer of polyvinylpyrrolidone (PVP) with proteins. A remarkable increase in moisture absorption can be achieved by such modifications. Nevertheless, the major challenges associated with the use of proteins in the cosmetic industry include their odor, change in color with the passage of time and the need for additional preservatives.[6]

11.1.1 Collagen

The use of collagen in cosmetics is based on the fact that this protein has a natural occurrence in the connective tissues of the skin. Considering its occurrence, cosmetic formulations containing collagen are prepared to achieve skin protection, improve skin structure and function and enhance skin appearance. Collagen has the distinct feature of making triple superhelical rods (composed of three polypeptide chains) and insoluble fibers that provide tensile strength. The triple helical structure of collagen has a considerable role in moisture-retaining properties[3] (see Figure 11.1).

Collagen can be obtained from animals and vegetables. Among animal sources, bovine, human collagen, porcine, fish skin and scales are most common. The terrestrial sources include duck feet, kangaroo tails, chicken, rat tail tendons, equine tendon bird's feet, alligator bone/skin, frog skin and sheep skin. All these sources are cheap and easily available, but collagen from these sources may be allergenic and can cause different diseases. Collagen obtained from marine sources, such as fish and jellyfish sponges, can be feasible, and there is no chance of transmission of diseases.

Collagen has been categorized into 26 types, which have been re-grouped into the following eight families:

- Fibril-forming collagen
- Basement membrane collagen

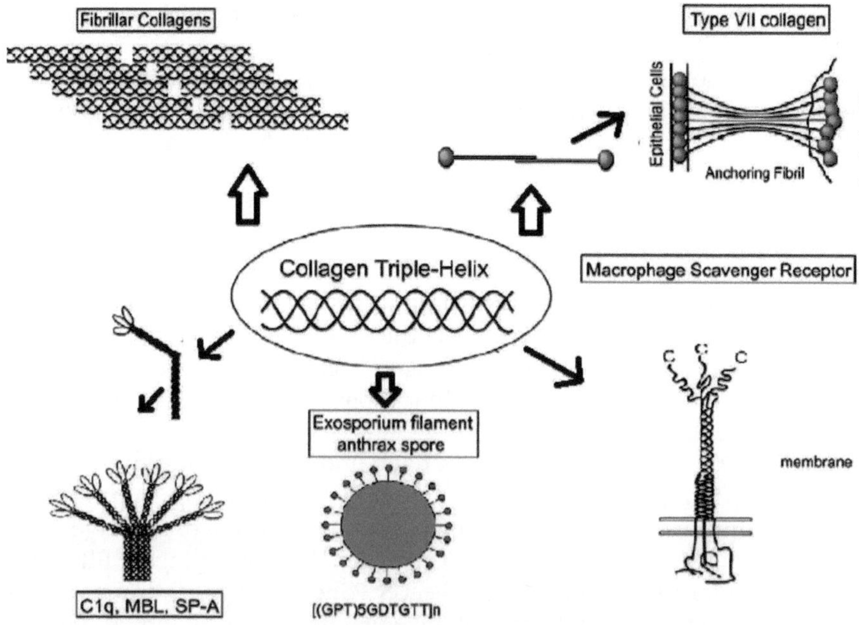

Figure 11.1 Various biological forms of the triple helix of collagen. Reproduced from ref. 7 with permission from Elsevier, Copyright 2005.

- Hexagonal network-forming collagen
- Fibril-associated collagens with interrupted triple helix
- Microfibrillar collagen
- Anchoring fibrils
- Transmembrane collagen
- Multiplexins

Among all, the fibril-forming types are mostly used in the cosmetic industry because of their decomposing ability, recyclability, biocompatibility and easy availability.[8,9]

Type I collagen has potential applications in cosmetics, especially in preparing templates for skin regeneration and dental composites, along with applications in other fields, such as collagen shields (used in the field of ophthalmology) and the preparation of biodegradable matrices. They are mostly used in cosmetics because of their moisturizing activity in making creams and other activities such as anti-aging, anti-wrinkling, protection from UV radiations and wound healing.[10]

Fibrous insoluble collagen is used in making facial masks, which increases skin smoothness and shine. The partially deaminated (desamido collagen) and soluble collagens are able to form a colloidal film with skin, which gives softness, gloss and shine.[1]

For the person suffering from skin burning, collagen can be proved very effective and can be used as an artificial skin substitute. The mixture of collagen with silicon, fibroblast and glycosaminoglycans can be used for plastic surgery of fire victim skin.[11]

Tissue engineers consider skin substitutes as an option for patients suffering from severe burns. This method can be applied to cover the wounds in a situation where the donor skin is not sufficient.[12]

Zyderm, which is a commercial product, is an injectable collagen that is used for the treatment of acne on the skin. Collagen-sodium dodecyl sulfate(C-SDS) has been used in foaming bubble bath gels, shower gels and shampoos due to its high water-binding capability. The SDS is found to be more effective than sodium laurate.[13]

Besides various other uses, such as hand and facial renovation, augmentation of soft tissues and cosmetic surgery for achieving improvement in facial contours, dermal fillers of collagen are used on a large scale to overcome volume deficiencies. Therefore, bovine collagen filler was the first dermal filler approved by the US Food and Drug Administration (FDA) in 1980. The other dermal fillers, such as Bovine Zyderm®, porcine Evolence™ and CosmoDerm®, were introduced at a later stage.[14]

The dermal fillers may also have some disadvantages, such as inflammation, allergic reactions and side effects, which depend on the type of collagen used as a filler. Porcine is proven to be the best for dermal filler purposes.[15,16]

The collagen obtained from the fish skin is considerable for cosmetics compared with bovine collagen and pig collagen to avoid the transmission of animal-derived diseases. Collagen in human skin exists in two forms: type I collagen and type III collagen. The deficiency of collagen in the skin causes wrinkles; therefore, collagen-containing cosmetics could be proved very effective as they form a protective layer on skin and hair due to their good binding ability. These film-forming properties of collagen can prevent the surface from damage and give a smoothing, shining and softening effect.[17]

Collagen is also obtained from marine fish proteins and has been widely used in the formulation of many cosmetic products due to its moisturizing properties. Many skin-firming and skin-hydrating cosmetics have been made using this collagen. Serum is an excellent moisturizer, but it has short-term effects, whereas cream is more active than serum when applied regularly.[18,19]

Proteins and peptides derived from fish have the ability to protect the skin from UV radiation.[20–22] In this context, collagen obtained from jellyfish extract has been used against photoaging. It can be used against UVA and UVB to prevent skin degradation.[20] Collagen hydrolysates from jellyfish also have the ability to moisturize skin and reduce the transepidermal water loss

(TEWL). The collagen obtained from the skin of fish and jellyfish and their collagen hydrolysates exhibit the property of providing effective protection against the harmful and dangerous effects of UV radiation. Besides the degradation of skin lipids, they also stimulate the synthesis of collagen that prevents the skin from photoaging effects.[23]

One of the most flourishing and successful uses of collagen is as a dermal filler as it can reduce the effect of aging. The dermal filler enhances the collagen level in the skin and can also reduce acne scars. The life of collagen filler is 3–7 months. The concentration of collagen in fillers is 3.5–6.5%. These fillers can also be used for lip enhancement.[24–26] The film-forming properties of collagen make it the most important ingredient of many cosmetics that helps in reducing TEWL. This occlusion makes skin smoother, shiny and radiant by preventing skin damage. It also helps in the regeneration of tissues and wound healing.[27–29]

The oral supplementation of collagen hydrolysates also has a positive effect on the skin. After 8 weeks of administering bioactive collagen peptide (BCP), the wrinkles reduced by up to 49.9% and an increase in elastin (18%) and fibrillin (6%) was also observed, which led to an increase in the elasticity and moisture content of the skin, making the appearance better.[30] Some detergents are also used in cosmetics to increase their foaming ability, even though they are associated with adverse effects on skin by making it dry and rough. The addition of soluble collagen to them may reduce the adverse effects of the detergent without lessening its cleansing and foaming ability.[11]

Collagen and collagen hydrolysates are used as conditioning and protective agents. The addition of these ingredients to shampoos improves smoothing, tensioning and manageability with pleasant odor, shine and luster.[31]

The nano-collagens are being used in cosmetics on account of their significant role in increasing skin elasticity, softness and moisture and preventing wrinkles and skin aging. Due to small particle sizes, nano-collagens can penetrate better into the skin and repair the damaged skin very efficiently.[32]

Collagen is the chief and major element of many hydrogels that are basically used as beauty masks. These facial masks help maintain skin elasticity, making it more glossy by reducing skin aging.[33]

A new advancement in the cosmetic industry is the use of collagen blended with other polymeric molecules or biopolymers. The binding of collagen with PVP and polyvinyl alcohol (PVA) increases its film-forming ability. A hydrogen bond is formed between collagen and PVP. Collagen acts as a donor of H, while PVP has a carbonyl group that acts as an H acceptor, facilitating the formation of the H-bond. This hydrogen bonding also enhances the penetration into the skin and hence increases the film-forming properties. The addition of chitosan, keratin, elastin, silk fibroin and hydraulic acid has also been proven very effective for this purpose.[17]

11.1.2 Keratin

Keratin, another intracellular protein, is considered the most specialized epidermal protein. It is regarded as one of the most common proteins in reptiles, birds and mammals. It is the basic part of nails, wool, horns and feathers and also gives strength to the body. Keratin has many applications in the food industry, pharmaceuticals, green chemistry and cosmetics. The major portion of the outer layer of the skin and the specialized structures, such as nails and hairs, are composed of keratin. Therefore, skin and hair formulations usually contain this fibrous film-forming protein. Keratin is composed of polypeptide chains, which frequently contain amino acids, such as arginine, leucine, alanine and cysteine. The intermediate filaments serve as the fundamental units of α-keratin, as shown in Figure 11.2.

Keratin is believed to play a potential role in improving skin and hair by connecting structural proteins, reinstating surface damage, increasing the surface moisture content and acting as a coating. The hydrolyzed form of water-soluble keratin, comprising peptides of 10–20 amino acids, frequently obtained from sheep wool, is considered an active agent of cosmetic products. In this respect, a modified quaternary ammonium group-bearing hydrolyzed keratin has been achieved by hydrolyzing keratin to cocodimonium hydroxypropyl, which acts as an anti-static agent. Most hair formulations contain this polymeric molecule as an important ingredient on account of its damage-preventing and hair-repairing capability. The protein supplements containing keratin are helpful in protecting and repairing hair, as hair is primarily composed of keratin.[3]

Several keratin-based cosmetics have been reported for skin and hair treatment. Keratin with other naturally occurring polymers, such as

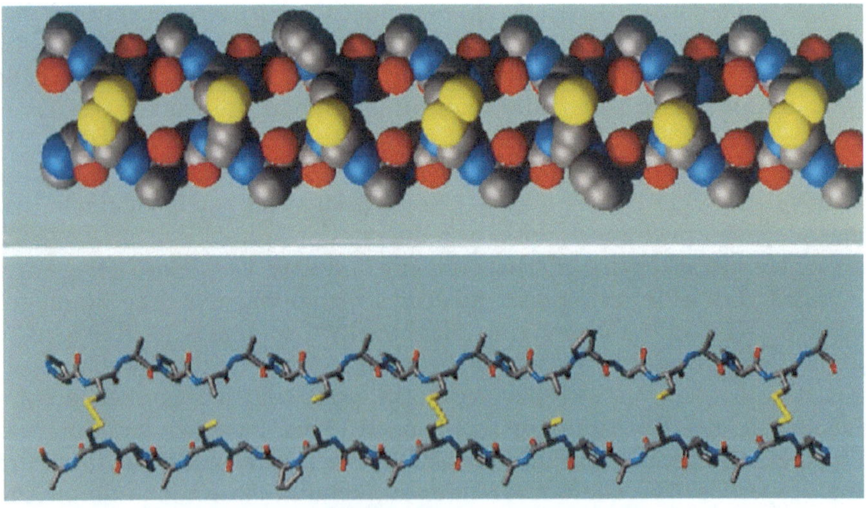

Figure 11.2 Antiparallel chains of dimers of 21mers of keratin. Reproduced from ref. 34 with permission from Elsevier, Copyright 1998.

collagen, chitosan and silk fibroin, has also been used as a component of the blends for cosmetic applications. Keratin provides a glossy, shiny and soft sensation when applied to the skin due to its hydrophilic nature.[35]

Keratin proteins reduce the harmful effects of bleaching creams by developing a protective layer encapsulating the hair.[11]

Keratin from wools, feathers, horns, hoofs and hairs is used in cosmetics due to its complex structure and a high amount of hydrophobic amino acids, causing water insolubility and resistance to the activity of many chemical substances and enzymes.[36]

11.1.3 Milk Proteins

In the past, Eritrean shepherds used camel milk for cleaning hair and skin, Hokkaido Island fishers used soy flour for making facial masks and Cleopatra's legendary used donkey milk for cleansing purposes.[1]

11.1.3.1 Whey Proteins

Whey, a milk protein derived from the liquid portion of milk, contains a blend of proteins and is used for making hair and skincare products, especially lipsticks, dressings, shampoos, hair lotions, emollients, conditioning products, cleansing creams, permanent wave solutions, body lotions, dyes, cold creams, soaps, detergents, makeup products, shaving creams, sun protection agents, depilatories and screening products.[37] Whey protein is composed of mixed globular protein molecules comprising β-lactoglobulin (β-LG; <50% wt/wt), α-lactalbumin (α-LA; <20% wt/wt), immunoglobulins (IgGs; <10% wt/wt) and bovine serum albumin (BSA; <6%, wt/wt), including some other minor peptides or protein components, such as lactoperoxidase, lactoferrin, growth factors and lysozymes. The different fibers of whey protein and its isolates are available for use in cosmetics (see Figure 11.3).

It has recently been reported that the donkey's milk contains a high amount of whey protein and is able to be used for increasing skin hydration. The donkey's milk may play a role in enhancing the skin color homogeneity.[4]

11.1.3.2 Casein

Casein, another important milk protein, makes up 0.1–10% of cosmetics by weight and is used especially in soap stuff, skin creams and softening lotions for achieving improvement in features of cosmetic products, such as skin elasticity maintenance, softening and dryness prevention.[37] Casein is also composed of related phosphoproteins, and various investigations highlighted that leucine is the major amino acid in casein, followed by lysine as the 2nd most frequent amino acid (see Figure 11.4).

Sodium caseinate is a major component at up to 96.9% in the manufacturing of bath oils, tablets and salts. Up to 0.5% of milk protein is used in

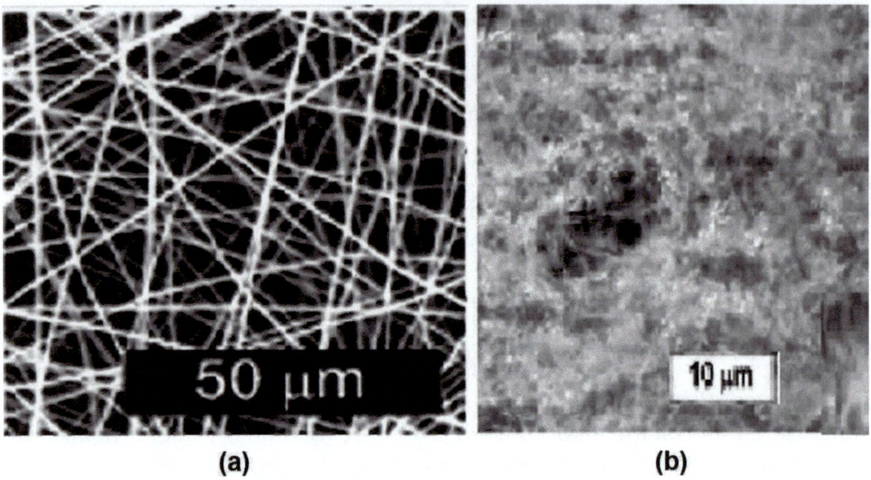

Figure 11.3 (a) β-Lactoglobulin (BLG) protein blended with polyethylene oxide at a ratio of 75 : 25 and (b) whey protein isolate blended with polyethylene oxide at a ratio of 75 : 25. Reproduced from ref. 38 with permission from Elsevier, Copyright 2014.

the preparations of eye makeup. Furthermore, these compounds are used in the formulation of face powders, deodorant sprays, body sprays, fragrances, face and neck sprays, hand sprays and hair sprays. It was reported that 0.013% of casein is used in deodorant sprays and 0.0002% of milk proteins is used in the formulation of face powders.[40]

Caseins are used as active agents in soaps, hand creams and lotions. Casein contains more than 0.1% lactose, giving a milky appearance to the skin, and its hydrolysates enhance skin hydration. They are also added

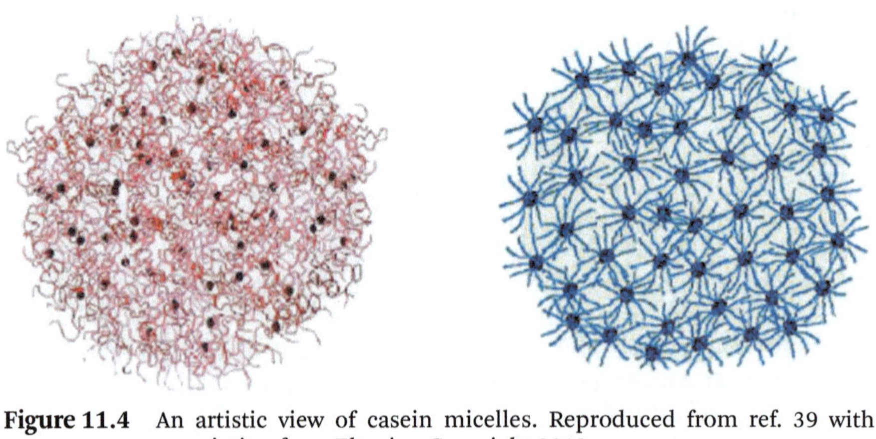

Figure 11.4 An artistic view of casein micelles. Reproduced from ref. 39 with permission from Elsevier, Copyright 2012.

to many other products, such as shampoos, shaving creams, cleansing creams, dressings, and body and hair lotions.[41]

Milk proteins are commonly used in products for showering, bathing and shampoos. In the manufacturing of shampoos, 3–15% milk-derived proteins are added.

11.1.4 Gelatin

Gelatin is used in cosmetics and many healthcare products. It is used as an ingredient in making body lotions, face creams, shampoos, sunscreens, bath salts, bubbles and hair sprays.[42] Gelatin is actually a degraded form of collagen and composed of a variety of amino acids, such as glycine, hydroxyproline, proline, glutamic acid, arginine, alanine, aspartic acid, serine, lysine, leucine, phenylalanine, valine and threonine. The structure of gelatin polypeptide is shown in Scheme 11.1.

11.1.5 *Ganoderma* Proteins

Ganoderma refers to a genus of polypore fungi, and the proteins isolated from its various species are referred to as *Ganoderma* proteins. In nutricosmetics and cosmeceuticals, *Ganoderma* protein can be used as a competitive raw material. It has many biological functions, such as antioxidation, antimelanin (responsible for yellowing and blackening of skin) and anti-inflammatory. *Ganoderma* has a great inhibitory effect

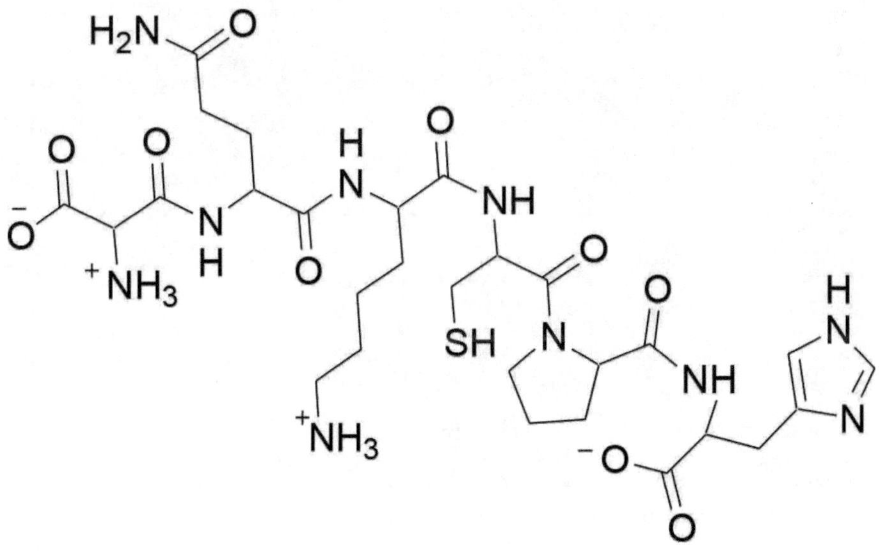

Scheme 11.1 A typical structure of gelatin polypeptide.

on melanoma, and it retards the melanin production, resulting in skin darkening.[43]

The powder obtained from *Ganoderma* spores contains 13 types of amino acids, such as arginine, D-asparagic acid, tryptophan, L-alanine, glycine, L-serine, L-threonine, proline and glutamic acid as major components and small quantities of leucine, methionine, phenylalanine and tyrosine. Laccase enzymes, containing copper ions, have been recently isolated from *Ganoderma lucidum* (see Figure 11.5).

11.1.6 Plant Proteins

In the formulations of skin and hair conditioning creams and lotions, hydrolyzed wheat gluten (HWG) and hydrolyzed wheat proteins (HWPs) are frequently used. The report given by the industry to the FDA revealed that the HWP is the most commonly used protein in skincare products and cosmetics. The reported use of HWP is in approximately 1077 applications,

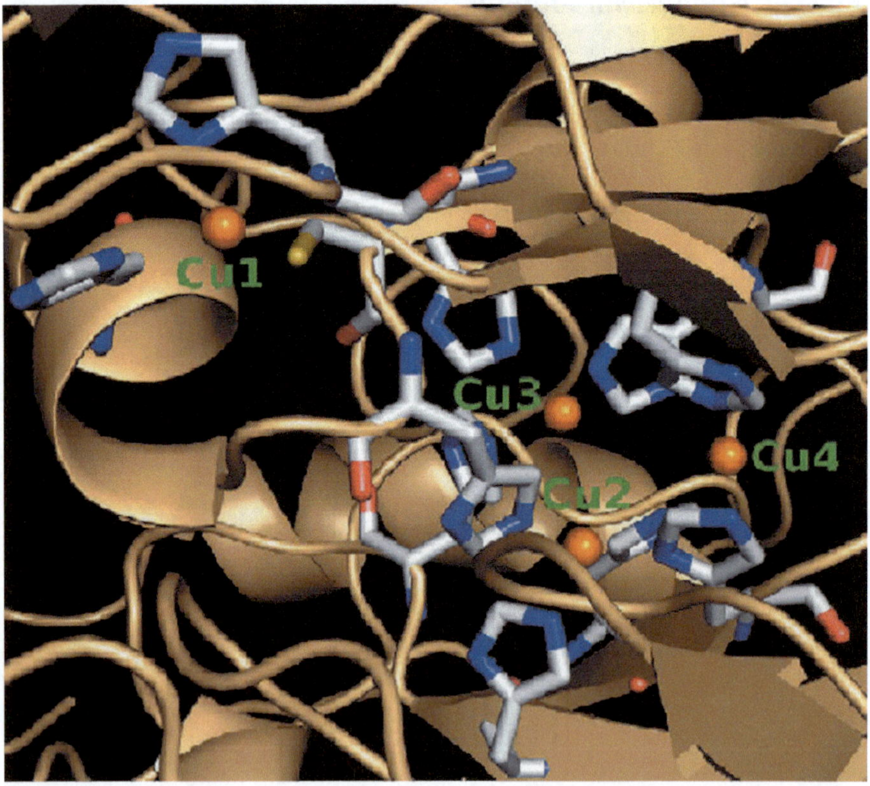

Figure 11.5 Structure of laccase enzyme, a copper-containing enzyme obtained from *Ganoderma lucidum*. Reproduced from ref. 44 with permission from the International Union of Crystallography, Copyright 2011.

with almost half of those reported uses being in non-coloring hair products. There are approximately 78 uses of HWG in skincare products and cosmetics, with almost half of them in hair tints. HWPs are commonly used in hair sprays, hair tonics and aerosols. For the formulations of fresheners and skincare products, both HWG and HWP are used, and these proteins are harmful if they are inhaled.[45]

The hydrolyzed wheat proteins exhibit greater hydrophobicity than collagen hydrolysates and, hence, wheat hydrolysates are more fruitful and effective in cosmetics. In the formulations of shampoos, the hydrolysates of sericin were used and were compared with placebo. The results of sericin were more effective and superior in enhancing skin smoothness and in its ability to be tolerated by the eyes and skin. These proteins also have the ability to form foams, and this property makes these proteins more applicable and beneficial in the formulations of shampoos. There are many examples of these proteins, such as HWP, gelatin, soluble keratin, soluble collagen, sodium caseinate, alkyldiammonium hydroxypropyl-hydrolyzed proteins, protein hydrolysates, serum albumin, polysiloxane co-polymer, desamido collagen, wheat protein and aminomethyl propanolisostearoyl-hydrolyzed protein.[5,46–48]

The protein obtained from wheat gluten is used as a cosmetic raw material due to its low water solubility and the presence of several disulfide bonds. The disulfide bridge-forming ability is the main reason for cosmetic activity. The plant protein derivatives from *A. sativa* (oat) are used in the cosmetic industry. Wild oats (*A. sativa* extract) can be used in hygiene cosmetic products and moisturizing creams for atopic patients' skin due to their antioxidant, anti-inflammatory and anti-pruritic characteristics. However, the use of oat extract in the formulations of creams and other cosmetic products can cause allergic reactions.[49] Rice and bean proteins consist of tyrosine, which plays a crucial role in melanogenesis. Tyrosine is converted into melanin by the tyrosinase enzyme. This is the reason why it is frequently used in cosmetics for tan acceleration.[36]

The rice protein obtained from *Oryza sativa* L. can be used as an anti-aging agent, attributed to its permeability into the skin. The inhibition of the enzyme activity by rice protein may lead to the decomposition of extracellular matrix protein (MCP), which can potentially lead to a reduction in the effects of aging.[50] Wheat-based protein derivatives were examined in aqueous solutions and in oil–water (O/W) emulsions, which provide information on the moisturizing effect on the skin and are also helpful in increasing the viscoelasticity of skin. With the help of emulsion, the film-forming behavior of these proteins can be accelerated.[5,51] Hydrolysates of plant oil proteins, such as sunflower seeds, nuts and almonds, have been frequently used in cosmetics.[36]

Because of many exciting properties, such as emollient and cooling effects, spreadability and moisturizing effect, of bigels (sunflower oil-based bigels/proteins), they are frequently used in making cosmetics.[52]

11.1.7 Silk

Silk extracts and their hydrolyzed forms are applicable up to 0.0036% and 0.024%, respectively, in the formulations of hair sprays. The maximum concentration of silk powder in hair sprays is approximately 0.02%. In the making of perfumes, silk powder and hydrolyzed fibroin are used in small amounts, 0.1% and 0.000047%, respectively. The reported maximum concentration of sericin, silk and silk powder is 0.00047%, 0.1–0.2% and 0.1–1.4%, respectively.[53] The fine crushing and grinding of silk are carried out to achieve fibroin (protein). In the formulations of lipsticks and other beautifying products, this fibroin is extensively used. From these formulations, it has been found that they make skin shiny, smooth and glossy and give lubrication to the formulations. This protein also adds oil–water absorption ability to these formulations. Because of the film-forming behavior of native proteins and having high molecular weight, they are often used in skincare cosmetics.[5,51] The most common sources of silk proteins include silk worms and spider silk. The various types of silk fiber, depicted in the scanning electron microscopy (SEM) and atomic force microscopy (AFM) images, have been reported in the literature (see Figure 11.6).

11.1.8 Eggshell Protein

The membrane of eggshell contains proteins along with other nutrients and is therefore used as a source of proteins. The eggshell proteins can cause a moisturizing effect by transporting and retaining moisture into the stratum corneum (SC) and replenishing the skin, while polypeptides stimulate the growth and the differentiation of new cells, thus furnishing a rejuvenating effect for the skin.[55]

11.1.9 Marine Proteins

A marine protein-based supplement, namely Viviscal, is beneficial in improving the thickness, luster, softness and gloss of hair. It also gives strength to nails, enhances nails' growth and makes skin overall healthy. This supplement can also prevent hair shedding and scalp coverage. The main ingredients of this dietary supplement are AminoMar™ C (protein complex, *Equisetum arvense*), biotin, *Malpighia glabra*, and zinc.[56]

11.1.10 Hydrolyzed Proteins

In order to improve the softness, luster, flexibility, shine and glossy appearance of hair, hydrolyzed proteins are extensively used in conditioning lotions for hair. Most frequently, they act as the main ingredient in

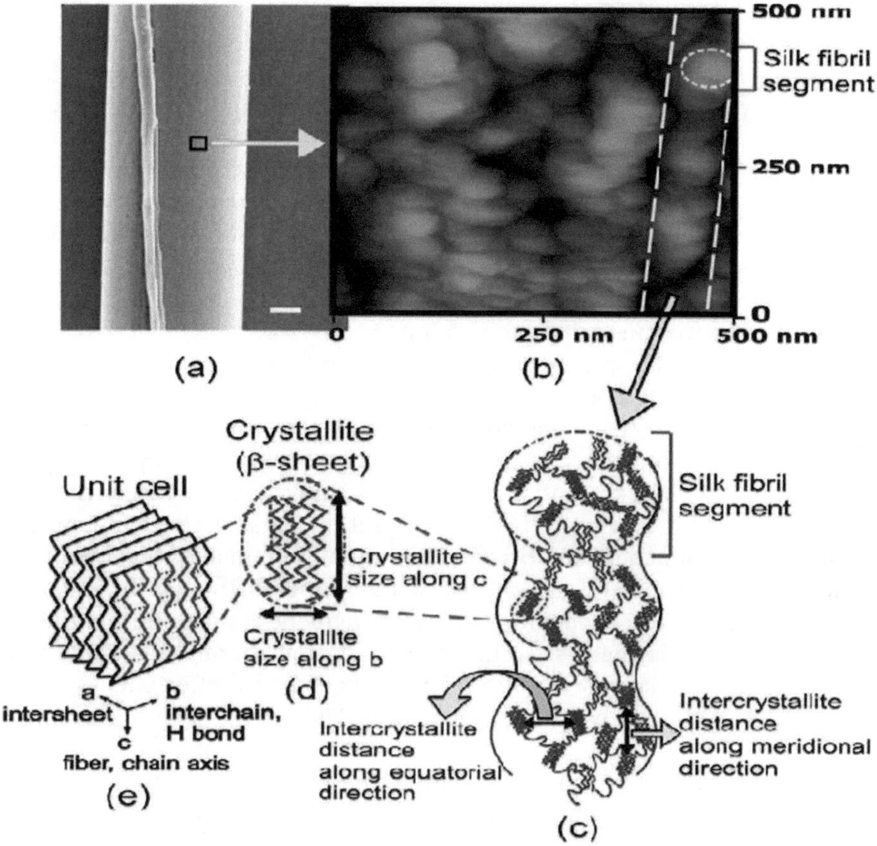

Figure 11.6 Silk fibril structure through (a) SEM and (b) AFM, (c) proposed models for silk β-sheet, (d) crystallite size and (e) anti-parallel β-sheets of crystallite unit cell. Reproduced from ref. 54 with permission from Elsevier, Copyright 2006.

conditioners and shampoos to reduce dryness and improve the appearance of damaged and rough hair. For this formulation, the concentration of hydrolyzed protein is usually 1–2%. These proteins are very essential for the straightening and decolorizing of hair and in permanent curling.[57] In moisturizers and conditioners, the active concentration of the hydrolyzed protein is 0.1–2%.[5,51] Quickly, the proteins were considered to be practical and applicable ingredients for making skin and hair healthier due to their potential to bind water with the SC (skin). Due to buffering and amphoteric characteristics, the hydrolyzed proteins were suggested for lasting waving treatments by preventing damage to hair fibers.[1] For the formulations of many cosmetic products, most of the protein derivatives are derived from simple globular and fibrous proteins, whereas the applications of conjugated proteins are less. Proteins obtained from animals and plants are commonly used for cosmetic purposes, while now the use of proteins derived from lower organisms (fungi and algae) is under consideration.[58]

Hydrolyzed proteins originating from different sources (corn gluten, collagen, wheat proteins, keratin and elastin) have foam-boosting activities. The best activity is shown by the hydrolyzed wheat proteins.[57]

A combination of hydrolyzed proteins with many other moisturizers, such as glycerin, free amino acids and sodium lactate, is also used for cosmetic products. The peptides of polar nature serve as the primary ingredients of the water phase in O/W emulsions. It is believed that polar peptides increase the viscosity of the outer phase due to their hydrophilic nature and contribute to its stabilization.[57]

Hydrolyzed milk protein and soy protein can be used as natural sources for synthesizing conditioners and shampoos.[59] Amino acids belong to the natural moisturizing factors (NMF) and account for 40% of its dry weight. Because of their relatively low molecular weight, they are capable of penetrating the skin and cuticle of the hair more effectively than the higher-molecular-weight protein hydrolysates. Salts of the condensation product of coconut acid and hydrolyzed animal protein and wheat flour and wheat starch are considered safe as cosmetic ingredients by the Cosmetic Ingredient Review (CIR) board. The most frequent clinical presentation of protein contact dermatitis is chronic or recurrent dermatitis. Sometimes an urticarial or vesicular exacerbation has been noted after a few minutes of contact with the causative substance. Hair conditioners containing quaternary hydrolyzed protein or hydrolyzed bovine collagen have induced contact urticaria and respiratory symptoms. Atopic constitution seems to be a predisposing factor in the development of protein contact dermatitis.[6]

11.1.11 Peptide Fragments

In the formulation of anti-aging creams, small fragments of peptides are commonly used because they can easily penetrate through the SC. They also have the ability to influence the synthesis of the extracellular matrix of the skin, and hence, the damaged tissues can easily be healed. During the synthesis of elastin and collagen in the extracellular matrix, many peptide fragments are formed, which can stimulate the production of the extracellular matrix of the skin and improve the skin appearance.[3] The different forms of recently used proteins are listed in Table 11.1.

11.2 Lipid Cosmetics

Lipids are well-defined groups of organic compounds present in all animals, plants and microorganisms. They are natural compounds containing fatty acids as the main component and play pivotal roles in the metabolism and structure of all organisms. Lipids possess hydrophobic

Table 11.1 Proteins used for cosmetics.

S. no	Name of proteins	Source	Cosmetic applications	Cosmetic effects	Ref.
1	Collagen	Animal bones, skin and feet such as duck feet	Facial masks, hair and skincare cosmetic products	Moisturizing and smoothing effect and removing wrinkles	1,10,14,15
2	Keratin	Animal wool, feather, horn, hoof and hair	Hair and skincare/treatment products	Hair repairing and shiny skin and protection from bleaching creams	3,10,33,34
3	Whey	Mammalian milk	Soaps, cleansing creams, lotions, shaving cream and sun screening agents	Skin fairness and smoothness	4,35
4	Casein	Mammalian milk	Face powder, shampoos, creams and lotions	Skin and hair protection	36,37
5	Gelatin	Degradation of collagen	Sun screening agents, hair sprays and lotions	Cosmetic and healthcare effects	38
6	*Ganoderma* protein	Spores of *Ganoderma* species	Various cosmeceuticals such as fairness creams	Antioxidant, anti-inflammatory and skin fairness	39
7	Plant proteins	Plant seeds and other parts	Fresheners, perfumes, creams for atopic patients and aerosol	Antioxidant and anti-inflamma-tory	40,45,47
8	Silk	Silkworm and spider silk	Lipsticks and other makeup products	Shiny and smooth appearance to skin	5,46
9	Eggshell proteins	Eggshell membrane	Skincare products	Moisturizing effect	49
10	Marine	Marine organisms	Protein supplements	Improving the thickness and color of hair and protein structures such as nails	50
11	Hydrolyzed proteins	All the sources of proteins can be used to produce hydro-lyzed proteins	Hair conditioning and skincare products	Moisturising effects	51–53
12	Peptide fragments	Processing of other proteins	Skincare products	Moisturising effect	3

chains, making them hydrophobic in nature, and some lipids may have hydrophilic ends due to the presence of some polar groups, such as glycerol, nitrogen and phosphate group.[60]

Lipid-based cosmetics are applied to boost the appearance and shield the skin and body from external harmful substances by acting as a protective layer and also keeping the skin soft. The major components of these cosmetics are lipids (fatty acid and plant oil) and their derivatives (emollients and emulsifiers).[61]

The oils extracted from plants are more useful than animal fatty acids. The oils from nuts (almonds, walnuts and pistachio), fruits (berries and cherries), vegetables (carrots) and plants (perilla, borage and primrose) are commonly used in personal care products. Vegetable oils possess softening and smoothing effects and, hence, are mostly used to prevent dehydration and reduce skin inflammation.[62] On atopic skin inflammation in the case of dermatoses, marked healing effects of the unsaturated fatty acids have been noticed. They are used in cosmetic milk, creams, emulsions, cosmetic masks, ointments, nail polishes, lipsticks, hair conditioners and bath fluids. Therefore, the shortage of fatty acids may cause extravagant dryness of the skin, inducing various medical and health considerations. Hence, the use of lipids in cosmetics as an important ingredient illustrates their value in hair and skin care.[63]

11.2.1 Essential Oils

Essential oils are volatile and aromatic liquids extracted from plants by the distillation process and named after the plant from which they are extracted. They can be defined as a mixture of fragrant and odorless substances and consist of highly volatile and lipophilic secondary plant metabolites, allyl and isoallyl phenols and principally mono- and sesquiterpenes; in addition to these alkaloids, anthraquinones and coumarin are also present in these volatile oils. Essential oils have wide applications in cosmetics and perfumes.[64] The aroma of these oils and their derivatives is the main reason for their wide applications in cosmetics. Essential oils and fragrances are also used for allergic reactions to the skin. The oils that are used in cosmetics are a complex mixture of hydrocarbons taken from petroleum products with C18–C90 chain lengths.[65,66]

11.2.2 Olive Oil

Olive oil is obtained from ripe olives by pressing. This vegetable oil is used in cosmetics and food as a raw material due to its richness in chemical composition and fascinating utility.

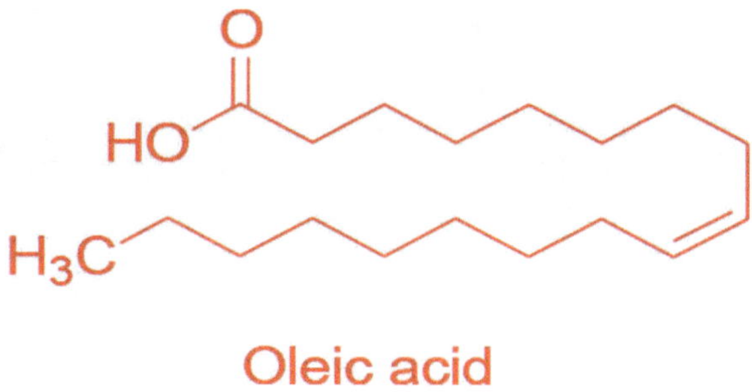

Scheme 11.2 Structure of oleic acid.

Olive oil consists predominantly of oleic acid (65–85%) and is a source of phenolic compounds and tocopherols as bioactive compounds (see scheme 11.2).

In cosmetics, its application may be extended by ozonation, which increases its stability and oil properties. The biological activities of ozonated olive oil increase due to decreases in unsaturation and production of other compounds, such as aldehydes, peroxides and ozonides. The use of ozonated olive oil in cosmetic formulations can introduce novel attractive properties, such as natural aroma, and result in more penetration. Emulsions containing ozonated olive oil detain their properties for a long period in comparison with those with purified olive oil. The antimicrobial activity of olive oil against *Staphylococcus aureus*, *Escherichia coli* and *Candida albicans* after ozonation has been studied, reflecting the enhancement of the antimicrobial potential. Cosmetic emulsion containing ozonated olive oil has been found as more persistent than that of refined olive oil. The growth of the *Aspergillus* sp. mold on the surface of the cosmetic emulsion containing refined olive oil has been studied, showing a visible growth of mold on the cosmetic formulation with a low concentration of ozone (0.04 mole O_3 per 100 g oil) in comparison with ozonated olive oil (0.10 mole O_3 per 100 g oil) with no contamination of mold even after six-month storage (see Figure 11.7).

Ozonated olive oil also makes cosmetic formulations more homogeneous. After application of the formulation, the skin became smooth, providing a silky appearance without a sticky effect.[67]

11.2.3 Kusum Oil

Kusum oil is extracted from a dry forest tree Kusum, which is the host for many insects. It contains 37% of glycerol ester and 91% of fatty acids, having a semisolid state with a bitter almond odor. Normally, mineral oil of petroleum origin is used for the preparation of cosmetic products, causing organic toxicity to the environment because of its non-degradable nature.

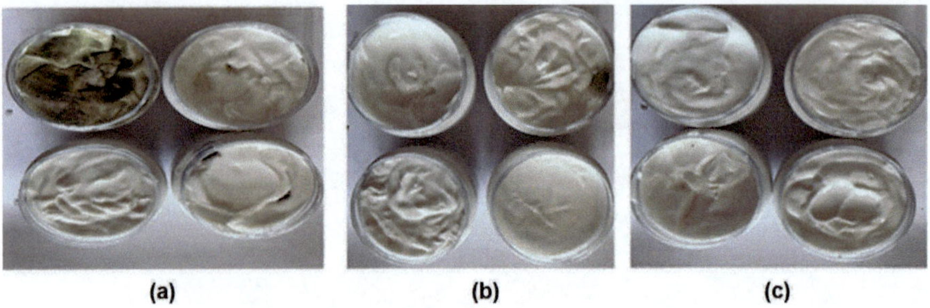

Figure 11.7 Stability of cosmetic emulsions: (a) with refined oil, (b) with 0.04 mole O_3 per 100 g oil and (c) with 0.10 mole O_3 per 100 g oil. Reproduced from ref. 67, https://doi.org/10.3390/molecules26113074, under the terms of the CC BY 4.0 license, https://creativecommons.org/licenses/by/4.0/.

Recently (2018), mineral oil was replaced with Kusum oil in cosmetics, as it is an eco-friendly and renewable raw material. In advanced skin-care formulations, it has been proven an ideal emollient by providing an enhancing skin barrier effect and immediate moisturization. The high melting point and soft consistency offer formulators many ways to optimize stability at high temperatures and skin stability without compromising the attractive appearance and long shelf life. It is not harmful to the environment during disposal. Various cosmetics, such as hair conditioners, moisturizing lotions and shaving creams, can be prepared by using this oil. Physicochemical characteristics of Kusum oil have been explored, and the following results were obtained[68]: acid value, 31.3; iodine value, 60.2; acetyl value, 4.0; saponification value, 234.3; Hehner value, 88.7; unsaponifiable%, 2.1; specific gravity (32 °C), 0.9099 and refractive index (40 °C), 1.46107.

11.2.4 Wax

11.2.4.1 Jojoba Wax

Liquid wax (ester) from sperm whale oil and jojoba seeds is used as additives, lubricants and wax polishes in commercial cosmetics. Jojoba oil is an effective moisturizer for skin and hair, provides oxidative stability and acts as an emollient. It maintains a balanced natural pH and replenishes the lost hair and skin. It also minimizes acne, wrinkles, neurodermatitis and psoriasis and brings back the skin to its normal life and health. The addition of jojoba oil to products enhances the viscosity of lubricants and makes them thermostable even at high temperatures.[69] The modification of jojoba wax by hydroxylation of the double bond and then acylation of the resulting hydroxy moiety to introduce short-chain vicinal acyl substituents have been studied, which improve its cold flow properties and protect the wax liquid from autooxidation and rancidity.[70]

11.2.4.2 Lanolin Wax

The sebaceous secretions (wool fat) of sheep (*Ovis aries*) are a source of lanolin wax. Sebum, which is secreted by the sebaceous glands, has a waxy and oily nature. It provides protection and moisturization to skin and hair, but it may be impoverished by chemicals, sun, pollutants and the aging process, resulting in dryness of skin and hair. It is used in cosmetic products, such as eye products, creams, lipsticks and lotions, as a stabilizer, conditioner and binder for its valuable emulsifying properties. It is an odorless and tasteless wax lipid, useful in bath oil products due to its solubility in mineral oils. Primarily, the lanolin is used in lipsticks and glosses. It has water absorption capability and mineral oil solubility and is used in moisturizers, lotions, creams, cleansing products and emulsions. The use of synthetic rheology modifiers in cosmetics causes allergic reactions, which have been controlled by using lanolin wax in cosmetic emulsions as a modifier. The use of lanolin as a natural product provides the possibility of controlling the texture and consistency of ointments and creams.[71] The lipid part of wool wax (lanolin) consists of free fatty alcohols, free fatty acids and a mixture of esters, diesters and hydroxy esters of high molecular weight. The useful effects of the current application of the various extracted wool wax creams have been studied, with increased skin hydration and an enhancement of the skin barrier integrity. Solvent (hexane)-extracted (SE) lanolin in comparison to water-extracted (WE) lanolin has more polar sites, showing the ability to improve skin barrier functions by reinforcing SC lipids. These properties of SE wool wax ensure its suitability for application in the synthesis of cosmetics and dermatological formulations.[72]

11.2.4.3 Beeswax

Beeswax has been used in creams and ointments since ancient times, with various proportions of oils according to the desired consistency. It has zero irritation potential and provides healing, softening, emollient and antiseptic properties to cosmetic products. Beeswax has the following unique properties, making it a perfect ingredient for cosmetics:

- It does not cause allergic reactions.
- Its plasticity and elasticity provide greater permanence on lips and skin.
- It enhances the water-binding ability of creams and ointments.
- It enhances skin elasticity by improving soap functions.

- It is the preferred ingredient in lipstick, as it provides sheen, color stabilization and consistency.

Its applications have been studied in hair conditioners (beeswax 1–3% by weight), hair creams (5–10%), cold creams (8–12%), mascara (6–12%) and eye shadows (6–20%).

Vegetable oils were used traditionally, but they became rancid and decreased the duration in which creams can be used. Emulsifiers such as borax have to be added to mix the beeswax and oils in water. The saponification of the acids in beeswax is a chemical process on which the emulsification is based, resulting in the formation of soap. In this way, in the so-called simple cleansing creams, the associated cleansing effect is exploited. A special process was patented to remove the free radicals from beeswax for mixing it with mineral oils and pigments without the use of emulsifiers.[73]

Recently, beeswax has been used for making nanoparticles with skin-barrier functions, and these beeswax-based nanoparticles (BNs) have been prepared for cosmetic and dermatologic purposes. These BNs are cost-effective, new and commercially viable alternatives in cosmetics with beneficial effects on skin-barrier functions. These can be used for treating particular skin conditions with damaged skin.[74] Beeswax has been used recently to prepare sodium alginate gelatin-based emulsion, which can be used in cosmetic masks.[75]

11.2.5 Mineral Oils

Mineral oils (obtained from petroleum oil) with wax are used in a variety of cosmetic formulations as they regulate the viscosity of formulation and prevent hydration of skin because of possessing lubricating and protective properties and they are well tolerated and stable dermatologically. These oils are incorporated into various cosmetic products from lotions, creams and ointments to hair gels, sunscreens and lip care products. Depending on the type of product, a concentration range of 1–99% mineral oils is used in cosmetics. Mineral oil-based ingredients, which are used in cosmetics, include paraffin, ceresin, isoparaffin, microcrystalline wax, synthetic wax and hydrogenated microcrystalline wax, cera microcrystalline, paraffinum liquidum, ozokerite and petrolatum. Synthetic hydrocarbons may be present in cosmetic products, such as polyethylene and polybutene waxes with polyolefin oligomeric-saturated hydrocarbons (POSH), in addition to mineral oil-saturated hydrocarbons (MOSH), which are structurally identical to MOSH without having polycyclic aromatic hydrocarbons (PAH) as impurities. In lip care products, mixtures of POSH and MOSH are used as raw materials. The quality of these mineral oils is guaranteed by quality affirmation and the introduction of a refining/purification process to

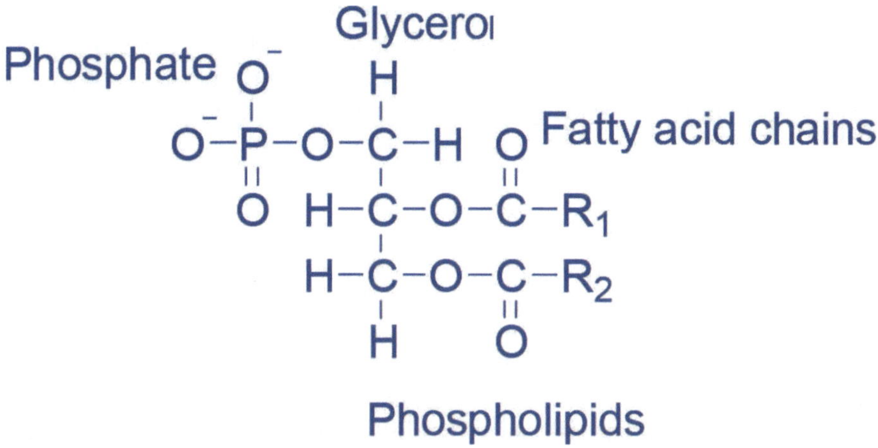

Scheme 11.3 Structure of phospholipids.

decrease the mineral oil aromatic hydrocarbons and remove carcinogenic substances. Human skin does not allow the penetration of mineral oils due to their highly lipophilic properties and, therefore, are not regularly bioavailable in the body. Moreover, the absence of effects on any internal body system and significant alteration in the skin have been reported, which is attributed to the topical application of refined mineral oils. Concerning potential oral exposure of mineral oils from lip care cosmetic products, the personal care industry, the European trade association for cosmetics and Cosmetics Europe have advised for using only mineral oil portion by cosmetic manufacturers due to the application of acceptable daily intake (ADI) values. There should be no toxicological concern because the estimated dose of mineral oils ingested *via* lip care products is <10% of the ADI value.[76] Recently, in cosmetic paraffin wax–mineral oil systems, nanostructured confined oil has been discovered, which decreases transportability and displays solid-like behaviors. This reveals the solid–liquid interaction, which is important for the oil-binding capacity of the material.[77]

11.2.6 Phospholipids

A phospholipid consists of a phosphate group as head and fatty acids as two tails, both head and tail components are connected through glycerol, which is the third component of phospholipid (see Scheme 11.3). It is the main component of the cell membrane. Phospholipids are widely distributed in nature and can be obtained from vegetable sources, such as corn, soybeans, peanut, rape (canola) seed, flaxseeds, wheatgerm and sunflowers, and animal products, such as milk, egg yolk, bovine brain and krill.[78]

The high content of the esterified form of essential fatty acids has credit for extensive applications of vegetable phospholipids in dermatology and cosmetics. These special fatty acids provide specific properties to vegetable phospholipids for topical applications in the cosmetic field and dermatology. After the short period of application of linoleic acid, there is an effective decrease in dehydration, and the function of the skin as a barrier is increased. The surface activity and the special characteristics of making liposomes insist on to use soya phospholipids and vegetable phospholipids in cosmetic products. These phospholipids are valuable in the transportation of linoleic acid into the dermal layers.[79]

In cosmetics, phospholipids are common as friendly natural ingredients. They are used in various classical formulations, such as mixed micelles, suspensions and water-in-oil and oil-in-water emulsions. In cosmetics, three groups of phospholipids are being used: saturated phospholipids, unsaturated phospholipids and lysophospholipids. Saturated phospholipids are used as surface-active compounds and unsaturated phospholipids are used to increase skin penetration.[80] Phospholipids are also used as vehicles for preparing liposomes having other cosmetic ingredients. Liposomes support the inclusion of lipophilic ingredients in formulations and enhance the penetration of these compounds as well as the moisturizing characteristics of phospholipids. The use of phospholipids with argan oil enhances the dermal delivery of their components.

11.2.7 Cellobiose Lipids

Cellobiose lipids are also glycolipids that consist of two glucose molecules bonded by a β-(1-4)-glycosidic bond (see Scheme 11.4). It can be used as

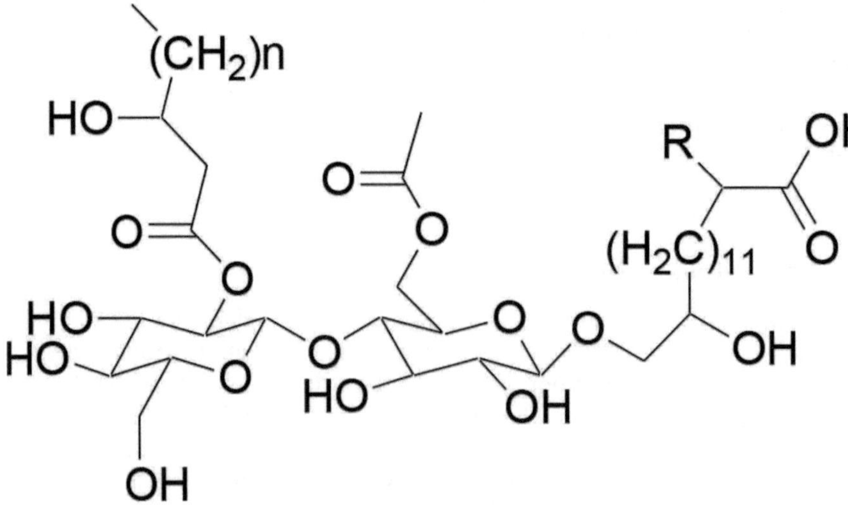

Scheme 11.4 Structure of cellobiose lipid (ustilagic acid).

a surfactant in the production of creams as a gel-forming element, but compared to sophorolipids and mannosyl erythritol lipids (MELs), it is produced in less quantity. A recent study concludes that these lipids can be used as anti-aging agents because they have the ability to activate the fibroblast cells. They can increase the formation of collagen, exhibiting anti-wrinkle and texture conditioner properties in the skin.[81]

11.2.8 Rhamnolipids

Rhamnolipids are a class of glycolipids produced by *Pseudomonas aeruginosa* and have the rhamnose sugar moiety linked to the fatty acid β-hydroxy alkanoic acid grouping (see Scheme 11.5).

Recently, rhamnolipid mixtures have been used in skincare products and cosmetic formulations. Schilling *et al.* (2019) have prepared a mixture of rhamnolipids with surprising physiological compatibility and foaming properties.[82] Rhamnolipids have been used with sophorolipids in various cosmetic formulations, such as body cleansers, shampoos, shower gels, moisturizing agents and anti-dandruff agents.

11.2.9 Mannosylerythritol Lipids

Mannosylerythritol lipids (see Scheme 11.6) belong to glycolipids, which are produced by yeast strains of the genus *Pseudozyma*. These lipids have cosmetical properties such as antioxidant activity, moisturization of dry skin and repair of damaged hair. MEL is used in cosmetics as an anti-aging agent and to remove the roughness of skin. Recently, MEL-A solution (5–10% concentration range) treated with sodium dodecyl sulfate (SDS)

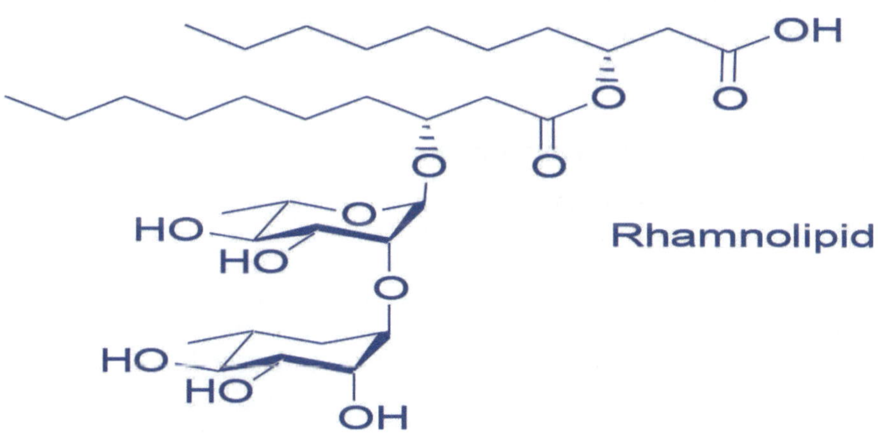

Scheme 11.5 Structure of rhamnolipids.

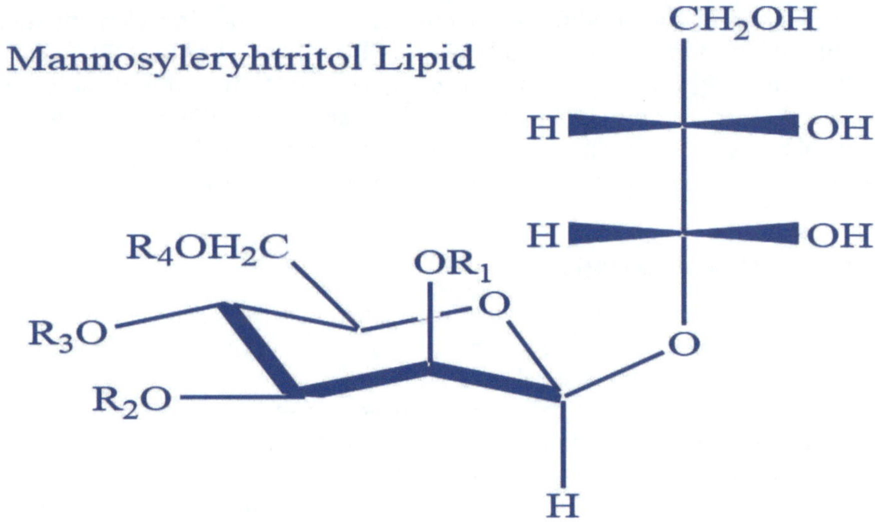

Scheme 11.6 Structure of mannosyl erythritol.

surfactant (1%), which presented moisturizing activity toward human skin and cosmetics, showed anti-wrinkle potential.[83]

The modification of microbial mannosyl erythritol lipids by lipase-catalyzed acylation with uncommon fatty acids from other microbial glycolipids, such as 17-hydroxyoctadecanoic acid from classical sophorolipids and 3-hydroxydecanoic acid from rhamnolipids, produce MEL-A and MEL-B, giving functionalized products of erythritol at the C-1 position. The emulsion of these lipids was prepared by using sunflower oil, and the study of its stability measurement shows that all new glycolipids can stabilize water-in-oil emulsions. These substances may be a useful innovation in the cosmetic industry due to the moisture maintenance of the skin. The surprising moisturizing effect has been studied for mannosyl erythritol lipids, which makes it potentially applicable in skin cream.[84]

11.2.10 Ceramide Nanoemulsions

Ceramides belong to the family of complex lipids known as sphingolipids, and they also have fatty acids (see Scheme 11.7). A high concentration of ceramide is present in the cell membrane of eukaryotes. Ceremide is also present in the cuticle of hair and is known to protect and repair hair fibers. The repairing effect on hair by using MELs is comparable with natural ceremide. However, flexibility, tensile strength and smoothness of hair can be clearly enhanced.[85]

Stable and safe aqueous formulation using oil-in-water nanoemulsions of two hydrophobic actives, ceramide-like molecule

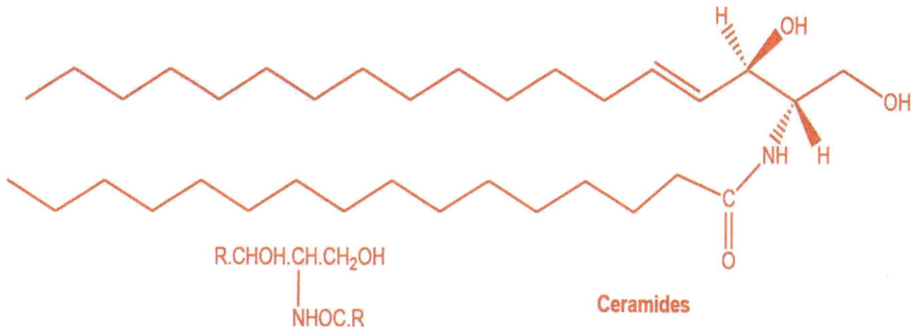

Scheme 11.7 Structure of ceramide.

((9*Z*)-*N*-(1,3-dihydroxyoctadecan-2-yl)octadec-9-enamide) and minoxidil (*d*2,6-diamino-4-(piperidin-1-yl)pyrimidine-1-oxide), was prepared recently. Stable nanoemulsions having a negatively charged surface and nanometric range (around 200 nm) droplet size were prepared. Within the oil droplets, more than 10 w/w% minoxidil and 2 w/w% ceramide-like molecules were assimilated for making formulations. The long-term stability of the emulsions has been studied without any clear tempering for many weeks. It showed that this colloidal dispersion may provide an adaptable implementation in cosmetics for improving the piercing of hydrophobic active components without the need for co-factors and organic solvents.[86] Ceramides are used as raw materials in cosmetics. Galactoceramides and glucosylceramides have been used in skincare products to treat infant acne.

11.2.11 Phosphatidylethanolamine

Phosphatidylethanolamine (PE) (Scheme 11.8) is a phospholipid present in all living organisms, with phosphatidylcholine and phosphatidylserine. It is the backbone of most biological membranes. In mammalian membranes, it is the most abundant phospholipid ranging from 20% to 50%.[87] Lecithin consists of a combination of phosphoric acid, fatty acids and glycerol, and ethanolamine is combined with a phosphate group to form PE (see Scheme 11.8).

In the external shell of cosmetic vesicles, phosphatidylethanolamine is commonly used. The main function of PE is to optimize the lipid composition in skincare products, which plays a supporting role in treating acne in infants.[88]

The modification in thermal transition properties of phosphatidylethanolamine by doping of methylparaben has been studied. The membrane fluidity of PE has been enhanced due to the presence of methylparaben, which is beneficial for cosmetic delivery through the skin membrane. Paraben itself has antimicrobial activity and is less toxic to humans, and, hence, it is used as a preservative in cosmetic toiletries.[89]

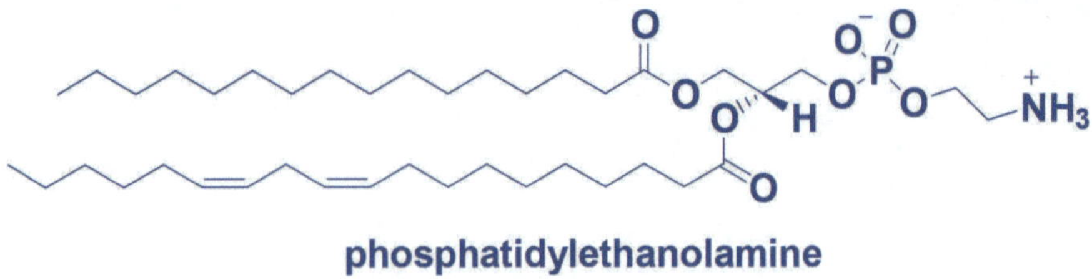

Scheme 11.8 Structure of phosphatidylethanolamine.

11.2.12 Lipid Nanoparticles

11.2.12.1 Liposomes

A liposome is a single lipid bilayer, mostly containing phospholipids. Other lipids are also present in the liposomes. Liposomes are regarded as the most broadly recognized cosmetic transmission system due to their ability to recycle, less toxicity, ability to shield both lipophobic and lipophilic active compounds, biocompatibility and ability to deliver active components to specific sites. The commercial product 'Capture' (anti-aging cream) was the first liposomal cosmetic product by Dior in 1986.[90]

11.2.12.2 Niosomes

Niosomes are formed from the non-ionic surfactants by a self-assembly process. They have structural similarities with liposomes and are prepared as a substitute transporting system for liposomes, as they can tackle the issues related to extensive production, sanitization and physical stability. Moreover, the raw materials for niosomes are comparatively more cost-effective and easily accessible than phospholipids, which are essential for liposomes.

Niosomes were first introduced in the 1970s and 1980s by researchers from L'Oréal (Clichy, France) for cosmetic applications. They have the tendency to enhance the ease of drug transmittance across the skin barrier. They have been studied as a carrier system for cosmetic actives due to their advantages, such as increased stability of entangled active ingredients, bioavailability, improved skin penetration, sustained release properties and improved surface adhesion.

The improved effects of several plant-based active ingredients containing niosomes on the skin have been studied. Curcuminoids present in turmeric have antioxidant and anti-inflammatory activities, and to enhance skin permeation, they are encapsulated in niosomes with 83% efficiency. Niosomes with gallic acid are valuable as skin anti-aging molecules due to their improved skin penetration and chemical stability.

Ellagic acid (EA) is a powerful antioxidant. However, its poor solubility and low permeability limit its application in cosmetics. The synthesis of

EA niosomes using Tween 60 and Span 60 has been studied for transdermal delivery. It showed enhanced skin penetration when used with the niosomal formulation compared to when used alone, by enhancing its perforation through the human skin and dermis layer. The functions of various niosomes are listed in Table 11.2.

11.2.12.3 Ethosomes

Ethosomes are liposomes containing ethanol and are malleable, soft vesicles used for increasing drug delivery. They are actually modified liposomes with a relatively high concentration of alcohol. These are lipid vesicles having ethanol, isopropyl alcohol, phospholipids and water (see Figure 11.3). They have size ranges from nanometers to microns, which enable them to penetrate through the skin more rapidly and have a higher transdermal flux (see Figure 11.4). The purpose of the application of ethosomes in cosmetics is not only to stabilize chemicals but also to minimize the irritation of chemicals and enhance the transdermal penetration of active ingredients, especially in elastic forms. These advantages of elastic ethosomes especially depend on the composition and size of vesicles.[92] The representative structure of a typical ethosome is presented in Figure 11.8.

Ethosomes are modified with propylene glycol in addition to ethanol. Propylene glycol serves as a penetration enhancer, reduces skin irritation, exhibits low toxicity, possesses both hygroscopicity and higher viscosity and offers stability in comparison to ethanol. This modification enhances

Table 11.2 Niosomes used in commercial cosmetic products.[a]

Product name	Functions	Marketed by
Niosomes Plus	Foundation, balance and clear skin tone	Lancôme® (Clichy, France)
Mayu Niosome Base Cream	Whitening and hydrating	Laon Cosmetics® (Seoul, Korea)
Identik Masque Floral Repair	Hair repair masque	Identik® (Paris, France)
Eusu Niosome Makam Pom Whitening Facial Cream	Whitening	Eusu® (Bangkok, Thailand)
Niosome Plus Perfected Age Treatment	Anti-wrinkle	Lancôme® (Clichy, France)
Anti-Age Response Cream	Anti-wrinkle	Nouvelle-HAS Cosmetics® (Varese, Italy)
Indentik Shampooing Floral Repair	Repairing of damaged hair	Identik® (Paris, France)
Anne Möller Anti-Fatigue Eye Contour Roll-On	Anti-puffiness and moisturizing	Anne Möller® (Barcelona, Spain)

[a]Reproduced from ref. 91 with permission from Elsevier, Copyright 2019.

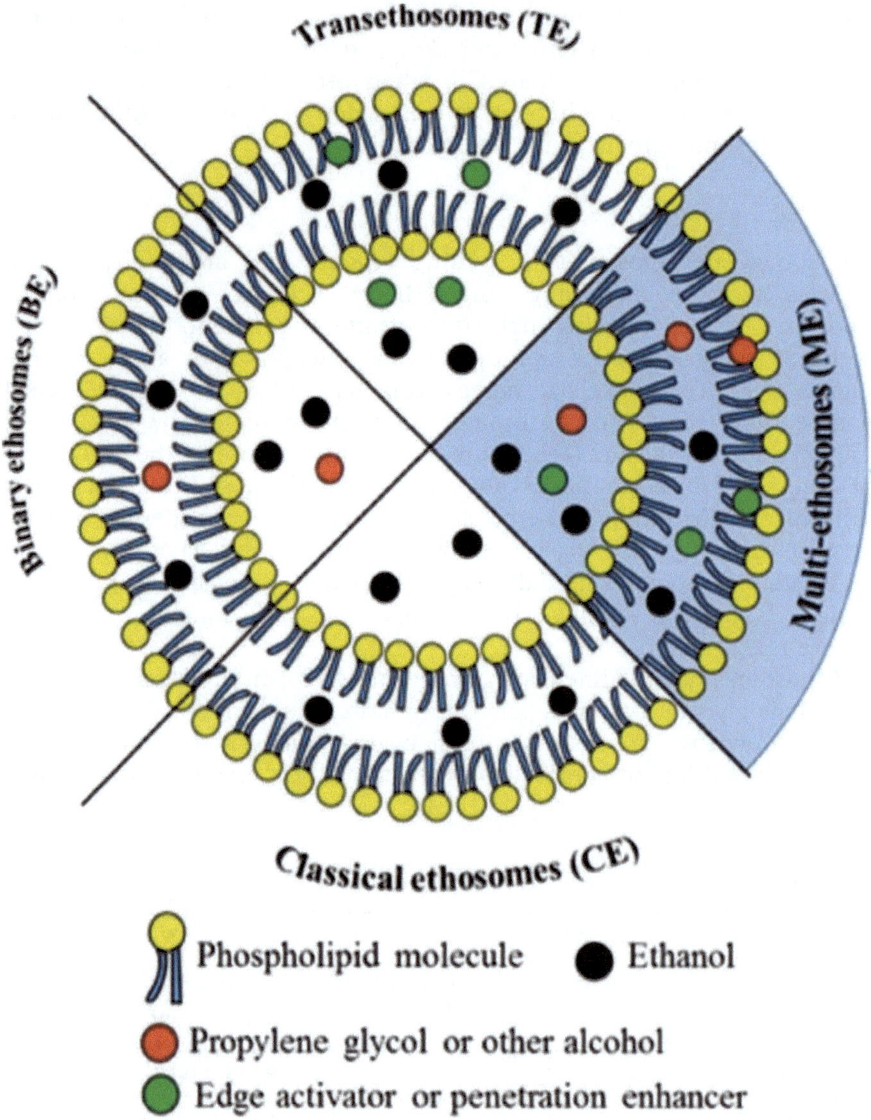

Figure 11.8 Representative structure of typical ethosomes. Reproduced from ref. 93, https://doi.org/10.3390/coatings10040304, under the terms of the CC BY 4.0 license, https://creativecommons.org/licenses/by/4.0/.

the drug retention and the drug's affinity for deeper dermal layers. Binary ethosomes consist of ethanol combined with other alcohols, resulting in smaller vesicle sizes. Due to the mentioned properties, ethosomes have garnered significant interest in cosmetics and dermatology.

11.2.12.4 Solid Lipid Nanoparticles

Solid lipid nanoparticles (SLNs) are used as a substitute carrier system to liposomes, emulsions and polymeric nanoparticles. They have a spherical morphology, and the average size of particles ranges from 40 to 1000 nm. SLNs consist of aqueous phase, having a dispersed solid fat (0.1–30% w/w) (see Figure 11.5). The stability of SLNs is enhanced by using 0.5–5% concentration of surfactants. Lipid parts of SLNs are solid at ambient and room temperatures. Lipids, including waxes, steroids, fatty acids, monoglycerides, diglycerides and triglycerides, are used in the synthesis of SLNs. SLNs have many useful properties, such as low cost, ease of preparation, chemical versatility and physical stability, and their primary role is to make drug delivery easy. A typical form of drug-enriched SLN particle is shown in Figure 11.9.

SC is the outermost lipid bilayer on the skin, which acts as a barrier during the application of active ingredients in cosmetics. SLNs provide lipophilicity to active ingredients for crossing the intact lipid bilayer. SLNs support the controlled-release process by making depots on the skin (see Figure 11.10).

The petrolatum can also be used as an occlusion agent, but it is less esthetic and glossier and affects only a confined portion of the skin, and the extent of occlusion by using solid lipid nanoparticles and nanostructured lipid carriers (NLCs) can be regulated in a controlled manner. The occlusion can be modified by increasing the quantity of particles, which can easily be achieved using nanoparticles. Due to less water loss, this occlusion can lead to the hydration of the skin. The addition of SLN and NLC to the commercially available creams can enhance their moisturizing ability.

Some of the commercially available nanoparticle-based products are the following:

- Nano Repair Q 10 Cream
- Nano Repair Q 10 Serum

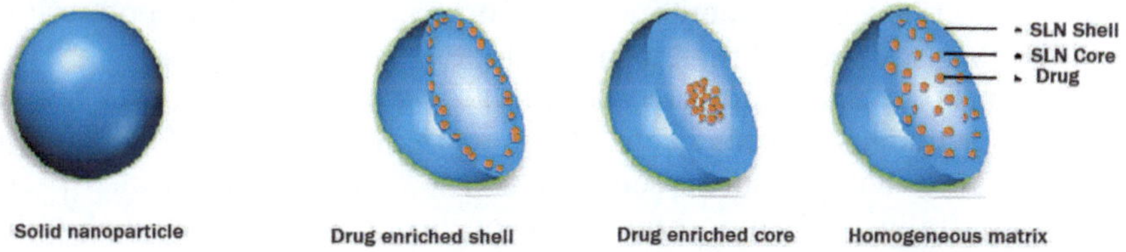

Figure 11.9 Diagrammatic representation of SLNs containing core–shell enriched with the drug. Reproduced from ref. 94, https://doi.org/10.3390/pharmaceutics10040191, under the terms of the CC BY 4.0 license, https://creativecommons.org/licenses/by/4.0/.

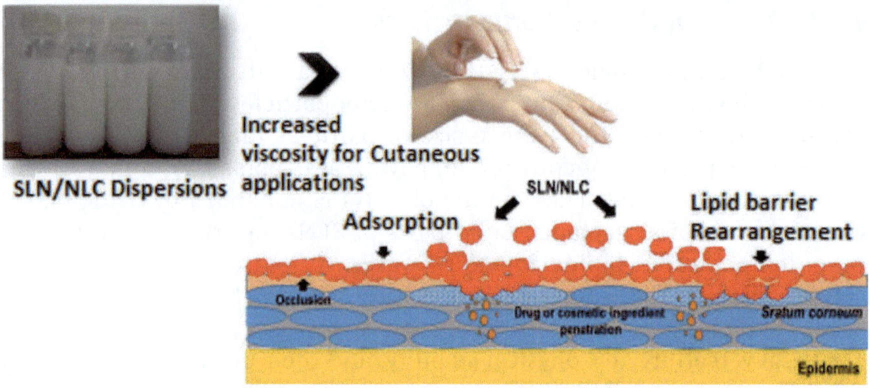

Figure 11.10 Role of SLNs in the penetration of cosmetic ingredients. Reproduced from ref. 95 with permission from Elsevier, Copyright 2018.

- NanoVital (daylight cream)
- Serum, Super Vital eye cream
- Kukui oil[96]

The formulations with lipid nanoparticles have more occlusion than suspensions containing lipid microparticles with similar lipid composition. The occlusion factor of daylight cream can be enhanced by the incorporation of nanoparticles.[97]

Many results have shown that in comparison to traditional aqueous gel, vitamin A and hydroquinone with SLN encapsulation has much more penetration ability. The skin penetration level of hydroquinone when using a hydroquinone–SLN formulation is two times greater than that of an aqueous gel of hydroquinone. This increased penetration is due to occlusion, which in turn causes skin hydration, resulting in a smoothing effect on wrinkles.[98]

SLNs have been prepared for the efficient skin permeation of spironolactone (SP), which is used for treating acne problems. SLN–SP has been synthesized using steric acid (fatty acid), Span 60, Span 80 and Tween 80 (as surfactants) in dichloromethane in the organic phase. Skin penetration of SP from SLNs is twofold than from SP alone.[99]

The use of SLNs in sunscreen formulations has been found to decrease the disadvantages of sunscreens. They also enhance the UV-blocking and protection activity of sunscreens. In addition to this, SLNs also provide better flexibility and smoothness, and occlusion can be obtained without the use of greasy oil and paraffin.[100]

Many SLN sunscreens with some antioxidants have been made, such as silymarin (a natural flavonoid), tocopherol acetate and safranal. They enhance penetration and UV-blocking efficiency and prevent chemical degradation.[101]

Lipid nanoparticles have the property to increase the chemical stability of photosensitive compounds and compounds with sensitivity toward hydrolysis and oxidation. Upon incorporation, they increase the chemical stability of ascorbyl palmitate, retinol (vitamin A), coenzyme Q10 and tocopherol (vitamin E).[102]

11.2.12.5 Nanostructured Lipid Carriers

Nanostructured lipid carriers are advanced forms of solid lipid nanoparticles that improve the carrying capacity and stability and block the drug discharge during storage. The NLC lipid phase consists of both solid and liquid lipids at ambient temperatures. NLCs have three forms: multiple type, formless and imperfect. Improved stability and loading volume of NLCs are because of the formless matrix.[103]

Compared to SLNs, NLCs show a higher carrying capacity for many active compounds; reduced aqueous contents of the suspended particles during storage minimize the potential discharge of active ingredients. Solid lipids are used only to prepare SLNs, and a few portions of particles crystallize after preparation due to excessive energy modification.[104]

The first NLCs containing the cosmetic product Cutanvoa Nanorepair Q10 cream were introduced into the market. It is well tolerated with no skin irritation. Furthermore, after the application of Cutanvoa Nanorepair Q10 cream, a significant increase in skin hydration has been found in comparison to a traditional cream with the same lipid and Q10 coenzyme. Having proven effects on skin hydration, creams with NLC formulations are favorable systems for cosmetics.[105]

Nanostructured lipid carriers containing vitamin E have been prepared for the efficient penetration of vitamin E. The oil phase was prepared by using non-ionic surfactants (Pluronic F-127, Cremophor RH40 and Tween 80), liquid lipids (medium-chain triglycerides, avocado oil and coconut oil) and stearic acid as solid lipids. A high concentration of surfactants made the nanoparticles more homogeneous. Vitamin E possesses photoprotective and antioxidant activities due to which it has been investigated as an active agent in cosmetic applications. Liquid lipids in NLCs enhance lipid matrix properties as well as cosmetic properties.[106]

Tretinoin is a derivative of vitamin A, having anti-acne and anti-aging potential. As an antioxidant, it reduces the effect of free radicals and reactive oxygen species, and its use in the treatment of photodamaged skin has been studied. Tretinoin also has some drawbacks being highly irritant and causing skin peeling and redness. These drawbacks have been overcome by using NLC-containing formulation. NLC formulations of tretinoin have been studied with reduced skin irritation potential, prolonged drug-releasing and increased drug-loading improvements due

Table 11.3 Lipids used in cosmetics.

Name of the lipid	Source	Cosmetic applications	Cosmetic effects	Ref.
Essential oils	Plants	Various cosmetics and perfumes	Fragrance and anti-allergic effects	58–60
Olive oil	Ripe olives	Cream and perfumes	Smoothness and silky effects and aroma	61
Kusum oil	Kusum tree	Hair conditioners, moisturizing lotions and shaving creams	Attractive appearance	62
Jojoba wax	Jojoba seeds	Additives, lubricants and wax polish	Moisturization; reducing acne, wrinkles and psoriasis and restoring skin and hair	63,64
Lanolin wax	Sebaceous secretion of sheep (*Ovis aries*)	Creams, eye products, lotions and lipsticks	Moisturization	65,66
Beeswax	Bees hives	Hair conditioners, mascara, cold creams, hair creams and eye shadows	Skin softening, healing, soothing and improving skin elasticity	66–69
Mineral oils	Crude petroleum oil	Lotions, creams, lip care products and sunscreens	Skin hydration, protection and lubrication	70,71
Phospholipids	Animals and plants	Various cosmetic products	Moisturization	72–74
Cellobiose lipids	Microorganisms (Ustilaginaceae family)	Creams	Anti-wrinkle and anti-aging	75
Rhamnolipids	*Pseudomonas aeruginosa*	Body cleansers, shampoos, shower gels and anti-dandruff and moisturizing agents	Moisturization and anti-dandruff	76,77
Mannosylerythritol lipids (MELs)	Yeast strains of the genus *Pseudozyma*	Skin creams	Moisturization, antioxidant, repairment of damaged skin and anti-aging	75–77

(*continued*)

Table 11.3 *(continued)*

Name of the lipid	Source	Cosmetic applications	Cosmetic effects	Ref.
Ceramides	Cell membrane of eukaryotes	Various cosmetic formulations	Smoothness and hair repairing	77–79
Phosphatidylethanolamine (PE)	Cell membranes of all living organisms	Skincare products	Optimization of lipid composition to treat acne	80,81,83

to the incorporation of NLCs.[107] The sources of a variety of lipids used in cosmetics are given in Table 11.3.

References

1. G. Secchi, *Clin. Dermatol.*, 2008, **26**, 321–325.
2. P. Agache and P. Humbert, in *Measuring the skin*, Springer Science & Business Media, 2004.
3. O. Olatunji, in *Natural polymers: industry techniques and applications*, Springer, 2015.
4. C. Faria-Silva, A. Ascenso, A. M. Costa, J. Marto, M. Carvalheiro, H. M. Ribeiro and S. Simões, *Trends Food Sci. Technol.*, 2020, **95**, 21–32.
5. R. P. Gawade, S. L. Chinke and P. S. Alegaonkar, in *Polymer Science and Innovative Applications*, Elsevier, 2020, pp. 545–565.
6. G. Zocchi, in *Handbook of Cosmetic Science and Technology*, CRC Press, 2009, pp. 375–388.
7. B. Brodsky and A. V. Persikov, *Adv. Protein Chem.*, 2005, **70**, 301–339.
8. D. Meng, H. Tanaka, T. Kobayashi, H. Hatayama, X. Zhang, K. Ura, S. Yunoki and Y. Takagi, *Int. J. Biol. Macromol.*, 2019, **131**, 572–580.
9. L. He, W. Lan, Y. Zhao, S. Chen, S. Liu, L. Cen, S. Cao, L. Dong, R. Jin and Y. Liu, *RSC Adv.*, 2020, **10**, 7170–7180.
10. R. Sriram and V. Gopal, *Res. J. Pharm. Technol.*, 2022, **15**, 1378–1384.
11. S. S. Selvi, M. E. Hasköylü and E. T. Öner, in *Tissue Engineering*, Apple Academic Press, 2022, pp. 159–202.
12. K. Kondej, M. Zawrzykraj, K. Czerwiec, M. Deptuła, A. Tymińska and M. Pikuła, *Int. J. Mol. Sci.*, 2024, **25**, 3702.
13. M. Połomska, L. Kubisz, J. Wolak and D. Hojan-Jezierska, *Appl. Sci.*, 2021, **11**, 8358.
14. P. Zhao, W. Zhao, K. Zhang, H. Lin and X. Zhang, *J. Appl. Polym. Sci.*, 2020, **137**, 48515.
15. S. H. Bentkover, *Facial Plast. Surg.*, 2009, **25**, 073–085.
16. A. Fallacara, S. Manfredini, E. Durini and S. Vertuani, *Facial Plast. Surg.*, 2017, **33**, 087–096.
17. B. Jadach, Z. Mielcarek and T. Osmałek, *Curr. Issues Mol. Biol.*, 2024, **46**, 2043–2070.
18. E. Xhauflaire-Uhoda, K. Fontaine and G. Pierard, *Int. J. Cosmet. Sci.*, 2008, **30**, 131–138.
19. J. Venkatesan, S. Anil, S.-K. Kim and M. S. Shim, *Mar. Drugs*, 2017, **15**, 143.
20. Y. Zhuang, H. Hou, X. Zhao, Z. Zhang and B. Li, *J. Food Sci.*, 2009, **74**, H183–H188.
21. C.-B. Wang, B.-X. Ding, S.-B. Guo, Y.-Z. Wang, Y.-T. Han and Y.-J. Wang, *Acta Pharmacol. Sin.*, 2003, **24**, 692–696.
22. Y.-T. Han, Z.-W. Han, G.-Y. Yu, Y.-J. Wang, R.-Y. Cui and C.-B. Wang, *Pharmacol. Res.*, 2004, **49**, 265–274.

23. N. Jimbo, C. Kawada and Y. Nomura, *Biosci., Biotechnol., Biochem.*, 2016, **80**, 356–359.
24. S. Lo and M. B. Fauzi, *Pharmaceutics*, 2021, **13**, 316.
25. L. Salvatore, M. L. Natali, C. Brunetti, A. Sannino and N. Gallo, *Polymers*, 2023, **15**, 1020.
26. B. Jadach, Z. Mielcarek and T. Osmałek, *Curr. Issues Mol. Biol.*, 2024, **46**, 2043–2070.
27. S. A. Chan, S. Whitehead, D. de Bacq Rose, A. Parys and V. Wong, in *Decision Making in Aesthetic Practice*, CRC Press, 2021, pp. 11–38.
28. N. Kashyap, A. Kumari, N. Raina, F. Zakir and M. Gupta, *Phytomedicine Plus*, 2022, **2**, 100198.
29. G. Li, S. Fukunaga, K. Takenouchi and F. Nakamura, *Int. J. Cosmet. Sci.*, 2005, **27**, 101–106.
30. E. Proksch, M. Schunck, V. Zague, D. Segger, J. Degwert and S. Oesser, *Skin Pharmacol. Physiol.*, 2014, **27**, 113–119.
31. A. Turowski and B. C. Adelmann-Grill, *Int. J. Cosmet. Sci.*, 1985, **7**, 71–84.
32. W. Trilaksani, I. Adnyane, B. Riyanto and N. Safitri, in *IOP Conference Series: Earth and Environmental Science*, IOP Publishing, 2020, **404**, p. 012050.
33. F. Ferrari, R. Striani, D. Fico, M. M. Alam, A. Greco and C. Esposito Corcione, *Polymers*, 2022, **14**, 5519.
34. D. A. Parry and A. North, *J. Struct. Biol.*, 1998, **122**, 67–75.
35. P. K. Dikshit, J. Kumar, A. K. Das, S. Sadhu, S. Sharma, S. Singh, P. K. Gupta and B. S. Kim, *Catalysts*, 2021, **11**, 902.
36. S. Sharma, A. Gupta and A. Kumar, in *Keratin as a Protein Biopolymer: Extraction from Waste Biomass and Applications*, 2019, pp. 1–18.
37. A. Piccioni, C. G. García-Rodrigo, C. Pellegrini, G. Mazzocchetti and M. C. Fargnoli, *J. Cosmet., Dermatol. Sci. Appl.*, 2017, **7**, 48.
38. S. T. Sullivan, C. Tang, A. Kennedy, S. Talwar and S. A. Khan, *Food Hydrocolloids*, 2014, **35**, 36–50.
39. C. G. De Kruif, T. Huppertz, V. S. Urban and A. V. Petukhov, *Adv. Colloid Interface Sci.*, 2012, **171**, 36–52.
40. C. L. Burnett, B. Heldreth, W. F. Bergfeld, D. V. Belsito, R. A. Hill, C. D. Klaassen, D. C. Liebler, J. G. Marks Jr, R. C. Shank and T. J. Slaga, *Int. J. Toxicol.*, 2017, **36**, 17S–56S.
41. Y. Zhao, C. Li, X. Xia, M. Tan, H. Wang, Y. Lv, Y. Cheng, Y. Tao, J. Lu and D. Li, *Chem. Eng. J.*, 2023, **474**, 146013.
42. M. S. P. Abdullah, M. I. Noordin, S. I. M. Ismail, N. M. Mustapha, M. Jasamai, M. F. Danik, W. A. W. Ismail and A. F. Shamsuddin, *Sains Malays.*, 2018, **47**, 323–336.
43. L.-D. Li, P.-W. Mao, K.-D. Shao, X.-H. Bai and X.-W. Zhou, *Appl. Microbiol. Biotechnol.*, 2019, **103**, 9239–9250.
44. A. V. Lyashenko, O. Belova, A. G. Gabdulkhakov, A. A. Lashkov, A. V. Lisov, A. A. Leontievsky and A. b. M. Mikhailov, *Acta Crystallogr., Sect. F: Struct. Biol. Cryst. Commun.*, 2011, **67**, 926–929.

45. C. L. Burnett, B. Heldreth, W. F. Bergfeld, D. V. Belsito, R. A. Hill, C. D. Klaassen, D. C. Liebler, J. G. Marks Jr, R. C. Shank and T. J. Slaga, *Int. J. Toxicol.*, 2017, **36**, 17S–56S.
46. P. Busch, A. Domsch and H. Hensen, in *XVth IFSCC Congress*, London, 1988, p. A23.
47. R. P. Gawade, S. L. Chinke and P. S. Alegaonkar, in *Polymer science and innovative applications*, Elsevier, 2020, pp. 545–565.
48. R. P. Gawade, S. L. Chinke and P. S. Alegaonkar, in *Polymer science and innovative applications*, Elsevier, 2020, pp. 545–565.
49. M. A. Bruusgaard-Mouritsen, J. D. Johansen, C. Zachariae, C. S. Kirkeby and L. H. Garvey, *Contact Dermatitis*, 2020, **83**, 251–270.
50. W. Rungratanawanich, Y. Qu, X. Wang, M. M. Essa and B.-J. Song, *Exp. Mol. Med.*, 2021, **53**, 168–188.
51. R. P. Gawade, S. L. Chinke and P. Alegaonkar, in *Polymer science and innovative applications*, Elsevier, 2020, pp. 545–565.
52. B. Hashemi, M. Varidi and S. M. Jafari, *Food Hydrocolloids*, 2023, **145**, 109082.
53. W. Johnson Jr, W. F. Bergfeld, D. V. Belsito, R. A. Hill, C. D. Klaassen, D. C. Liebler, J. G. Marks Jr, R. C. Shank, T. J. Slaga and P. W. Snyder, *Int. J. Toxicol.*, 2020, **39**, 127S–144S.
54. N. Du, X. Y. Liu, J. Narayanan, L. Li, M. L. M. Lim and D. Li, *Biophys. J.*, 2006, **91**, 4528–4535.
55. L. Alparslan, N. Şekeroğlu and A. Kijjoa, *Curr. Perspect. Med. Aromat. Plants*, 2018, **1**, 53–66.
56. C. S. Hornfeldt, *J. Cosmet. Dermatol.*, 2018, **17**, 209–213.
57. P. P. Nugrahadi, W. L. Hinrichs, H. W. Frijlink, C. Schöneich and C. Avanti, *Pharmaceutics*, 2023, **15**, 935.
58. J. Adler-Nissen, in *Enzymic hydrolysis of food proteins*, Elsevier applied science publishers, 1986.
59. K. G. O. Bezerra, R. D. Rufino, J. M. Luna and L. A. Sarubbo, *Biotechnol. Prog.*, 2018, **34**, 1482–1493.
60. E. Fahy, S. Subramaniam, H. A. Brown, C. K. Glass, A. H. Merrill, R. C. Murphy, C. R. Raetz, D. W. Russell, Y. Seyama and W. Shaw, *J. Lipid Res.*, 2005, **46**, 839–861.
61. H. de Clermont-Gallerande, *OCL: Oilseeds Fats, Crops Lipids*, 2020, **27**, 33.
62. A. Bialek, M. Bialek, M. Jelinska and A. Tokarz, *Int. J. Cosmet Sci.*, 2016, **38**, 382–388.
63. A. Zielińska, K. Wójcicki, D. Klensporf-Pawlik, J. Dias-Ferreira, M. Lucarini, A. Durazzo, G. Lucariello, R. Capasso, A. Santini and E. B. Souto, *J. Spectrosc.*, 2020, **2020**, 1–9.
64. J.-L. Rios, in *Essential oils in food preservation, flavor and Safety*, Elsevier, 2016, pp. 3–10.
65. S. Weber, K. Schrag, G. Mildau, T. Kuballa, S. G. Walch and D. W. Lachenmeier, *Anal. Chem. Insights*, 2018, **13**, 1177390118777757.

66. A. Ahmad and H. Ahsan, *Biomed. Dermatol.*, 2020, **4**, 1–10.
67. M. Radzimierska-Kaźmierczak, K. Śmigielski, M. Sikora, A. Nowak, A. Plucińska, A. Kunicka-Styczyńska and K. H. Czarnecka-Chrebelska, *Molecules*, 2021, **26**, 3074.
68. P. S. Choudhari, M. Gupta and V. Y. Karadbhajne, *Intl. Res. J. Eng. Technol.*, 2018, **5**, 239-242.
69. R. N. Vilas Bôas and H. F. de Castro, *Biotechnol. Bioeng.*, 2022, **119**, 725–742.
70. R. E. Harry-O'kuru, G. Biresaw, S. Gordon and J. Xu, *J. Anal. Methods Chem.*, 2018, **2018**, DOI: 10.1155/2018/7548327.
71. J. Płocica, B. Tal-Figiel and W. Figiel, *Czas. Tech.*, 2013, **2013**, 61–68.
72. C. B. Albanell, V. Carrer, M. Marti, J. Iglesias, J. Iglesias and L. Coderch, *Skin Pharmacol. Physiol.*, 2018, **31**, 198–205.
73. S. Bogdanov, in *Beeswax: Production, Properties, Composition, Control*, Bee Product Science, 2016, pp. 1–18.
74. C. Souza, L. A. P. de Freitas and P. M. B. G. M. Campos, *AAPS PharmSciTech*, 2017, **18**, 2505–2516.
75. W. Prus-Walendziak and J. Kozlowska, *Materials*, 2021, **14**, 745.
76. B. Chuberre, E. Araviiskaia, T. Bieber and A. Barbaud, *J. Eur. Acad. Dermatol. Venereol.*, 2019, **33**, 5–14.
77. F. C. Wang, Y. Miyazaki and A. G. Marangoni, *Cryst. Growth Des.*, 2018, **18**, 2677–2680.
78. C. Allegretti, F. Denuccio, L. Rossato and P. D'Arrigo, *Catalysts*, 2020, **10**, 997.
79. A. M. Rabasco Álvarez and M. L. González Rodríguez, *Grasas Aceites*, 2000, **51**(1–2), 74–96.
80. P. van Hoogevest and A. Fahr, in *Nanocosmetics*, Springer, 2019, pp. 95–140.
81. M. Traversier, T. Gaslondes, S. Milesi, S. Michel and E. Delannay, *Phytochem. Rev.*, 2018, **17**, 1179–1210.
82. M. Schilling, C. Hartung, F. Cabirol, S. Schaffer and P. Allef, *Google Pat.*, 2019.
83. H.-R. Ahmadi-Ashtiani, A. Baldisserotto, E. Cesa, S. Manfredini, H. Sedghi Zadeh, M. Ghafori Gorab, M. Khanahmadi, S. Zakizadeh, P. Buso and S. Vertuani, *Cosmetics*, 2020, **7**, 46.
84. V. K. Recke, C. Beyrle, M. Gerlitzki, R. Hausmann, C. Syldatk, V. Wray, H. Tokuda, N. Suzuki and S. Lang, *Carbohydr. Res.*, 2013, **373**, 82–88.
85. T. Morita, T. Fukuoka, T. Imura and D. Kitamoto, *Appl. Microbiol. Biotechnol.*, 2013, **97**, 4691–4700.
86. E. Guzmán, L. Fernández-Peña, L. Rossi, M. Bouvier, F. Ortega and R. G. Rubio, *Cosmetics*, 2021, **8**, 45.
87. P. Rockenfeller, M. Koska, F. Pietrocola, N. Minois, O. Knittelfelder, V. Sica, J. Franz, D. Carmona-Gutierrez, G. Kroemer and F. Madeo, *Cell Death Differ.*, 2015, **22**, 499–508.

88. M. Yang, M. Zhou, H. Wang, M. Yang, Y. Gao and Y. Jia, *J. Cosmet. Dermatol.*, 2021, **20**, 947–954.
89. L. Panicker, in *AIP Conference Proceedings*, 2013.
90. D. Patel, V. Kumar, R. Kesharwani and B. Mazumdar, *Inven. Rapid Cosmeceuticals*, 2015, 1–6.
91. S. Chen, S. Hanning, J. Falconer, M. Locke and J. Wen, *Eur. J. Pharm. Biopharm.*, 2019, **141**, 18–39.
92. R. Bhosale, A. Osmani and P. Ghodake, *Int. J. Pharm. Med. Res.*, 2013, **1**, 60-69.
93. L. Zhang, X. Li, S. Zhu, T. Zhang, A. Maimaiti, M. Ding and S. Shi, *Coatings*, 2020, **10**, 304.
94. V. Mishra, K. K. Bansal, A. Verma, N. Yadav, S. Thakur, K. Sudhakar and J. M. Rosenholm, *Pharmaceutics*, 2018, **10**, 191.
95. A. Garcês, M. Amaral, J. S. Lobo and A. C. Silva, *Eur. J. Pharm. Sci.*, 2018, **112**, 159–167.
96. R. Müller, R. Petersen, A. Hommoss and J. Pardeike, *Adv. Drug Delivery Rev.*, 2007, **59**, 522–530.
97. S. A. Wissing and R. H. Müller, *Int. J. Pharm.*, 2003, **254**, 65–68.
98. S. Ghanbarzadeh, R. Hariri, M. Kouhsoltani, J. Shokri, Y. Javadzadeh and H. Hamishehkar, *Colloids Surf., B*, 2015, **136**, 1004–1010.
99. H. Kelidari, M. Saeedi, J. Akbari, K. Morteza-Semnani, P. Gill, H. Valizadeh and A. Nokhodchi, *Colloids Surf., B*, 2015, **128**, 473–479.
100. J. Jose and G. Netto, *J. Cosmet. Dermatol.*, 2019, **18**, 315–321.
101. V. Van Tran, T. L. Nguyen, J.-Y. Moon and Y.-C. Lee, *Chem. Eng. J.*, 2019, **368**, 88–114.
102. V. Teeranachaideekul, R. H. Müller and V. B. Junyaprasert, *Int. J. Pharm.*, 2007, **340**, 198–206.
103. N. Naseri, H. Valizadeh and P. Zakeri-Milani, *Adv. Pharm. Bull.*, 2015, **5**, 305.
104. J. Pardeike, A. Hommoss and R. H. Müller, *Int. J. Pharm.*, 2009, **366**, 170–184.
105. J. Pardeike, K. Schwabe and R. H. Müller, *Int. J. Pharm.*, 2010, **396**, 166–173.
106. I. D. de Souza, V. Saez and C. R. Mansur, *Colloids Surf., B*, 2023, 113491.
107. V. M. Ghate, S. A. Lewis, P. Prabhu, A. Dubey and N. Patel, *Eur. J. Pharm. Biopharm.*, 2016, **108**, 253–261.

CHAPTER 12

Protein-derived Materials for Water Purification

IRUM ZAHARA, MUHAMMAD ZUBAIR AND AMAN ULLAH*

Department of Agricultural, Food and Nutritional Science, University of Alberta, Edmonton, Canada
*E-mail: ullah2@ualberta.ca

12.1 Introduction

Water is an essential characteristic of life for organisms on Earth. Paradoxically, water is one of the most abundant resources in the world. However, the global safe water supply is less than 1% for human use. Both inorganic and organic pollutants contaminate freshwater bodies.[1–4] In 2011, according to the data published by the World Health Organization, more than 760 million people had inadequate access to clean drinking water and 50% of the world's population will be living in water-stressed regions by 2025.[5] On the other hand, wherever it is accessible, its cost gradually increases due to growing energy demands, population surges and environmental problems.[6] Unfortunately, water pollution has become a significant global issue due to the rapid increase in anthropogenic contaminants, particularly emerging contaminants from cosmetic or pharmaceutical, oil and gas, mining and other chemical industries.[7–9] According to the United Nations, nearly 70% of industrial wastewater is released into the water bodies without proper treatment in countries with limited resources.[10]

Most traditional water treatment approaches do not efficiently remove pollutants and/or are incapable of eliminating sufficient amounts to

Smart Materials No. 41
Functional Materials from Lipids and Proteins
Edited by Nandika Bandara and Aman Ullah

Published by the Royal Society of Chemistry, www.rsc.org

meet safe water quality standards.[11,12] Surface water pollution also poses a menace to water supplies, as contaminants may enter groundwater resources where they can be transferred to drinking water sources.[13,14] Globally, it is challenging to follow the gradually strict regulation and rising mandate for clean water due to inefficient existing infrastructure for wastewater remediation.[15] Thus, a water treatment method with high removal efficiency and low cost is an urgent need of the time to fulfill the clean water requirement for all of us.[16–18]

Among novel bio-based materials for water treatment processes, proteins are regarded as promising materials for the elimination of inorganic and organic pollutants, as they exhibit excellent attributes, such as good adsorption capacity, recycling, natural abundance, multiple functional groups, ease of modification and preparation at a larger scale, which commend them for use in water remediation.[19–22] Therefore, in this chapter, we discuss the recent advances in protein-based materials for water purification. The processing strategies for proteins to convert into materials for water purification are also discussed. In the end, the chapter concludes with a summary and a future perspective on protein-based materials for water remediation.

12.2 Types of Proteins Used in Water Remediation

Various types of proteins, such as keratin, gelatin, silk and albumin, have been used for water remediation (Figure 12.1).

12.2.1 Keratin Protein

The term "keratin" originated from the Greek word "Keratos" meaning "horn" and emerged in the literature in 1850 to represent the hard tissues (insoluble proteins) categorized as intermediate filaments responsible for developing epithelial and epidermal structures, such as horns, hooves, hairs, nails and wool.[23,24] Keratin is a high-molecular-weight natural polymer with large polypeptide chains[3,4,25] formed by the condensation of amino acids.[26]

Keratins are heterogeneous proteins because of differences in their amino acid structures/chemical compositions.[26] In addition to cysteine, keratins are also abundant with amino acids such as glutamic acid, aspartic acid, threonine, leucine, proline, serine and glycine.[27] Keratins are classified into epithelial (soft) and trichocyte (hard) keratins. Soft keratins, which are found in the stratum corneum of the skin, stabilize the cells in epithelia and have a low cysteine content (<3 mol%). Hard keratins are found in various biological fibrous materials (wool, hairs, feathers, nails and horns of mammals). They have a high content of cysteine (4–17%), which promotes their function as a structural scaffolding, leading to a higher level of protein–protein crosslinking.[28] The keratin protein is chemically

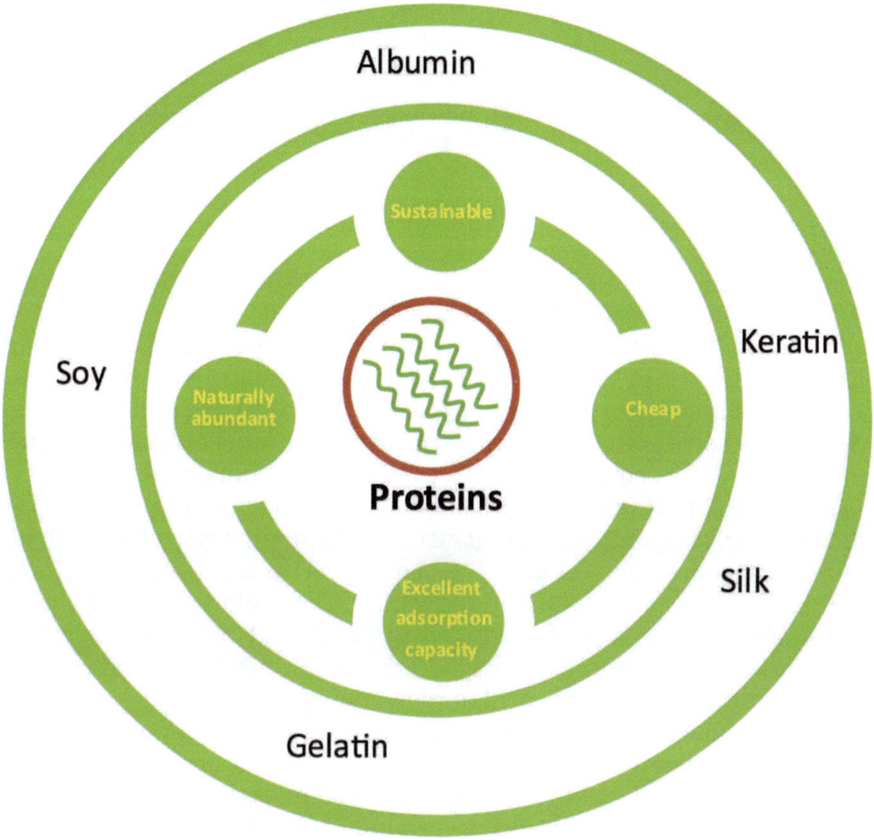

Figure 12.1 Types of proteins.

reactive under favorable reaction conditions due to the presence of several functional groups, mainly peptide backbone, disulfide bond (–S–S), amino group (–NH_2) and carboxylic acids (–COOH).[23,24]

12.2.2 Gelatin

Gelatin is a term commonly used to refer to "animal tissue". Gelatin was first used as a glue around 6000 BCE. With time, the industrial-scale synthesis of gelatin commenced due to its role in several applications.[29,30] Gelatin is a naturally soluble polymer synthesized by the partial hydrolytic degradation of collagen (the main fibrous protein content in bones, cartilages and skins).[31] Commercially, gelatins are obtained from bovine hides, bones or porcine skin.[32,33] But alternatively, the gelatin can also be prepared from poultry and fish by-products.[34] Gelatins derived from collagen fractions usually have molecular weights of more than 30 kDa, as they possess the gel-forming ability.[35] Gelatin consists of 20 amino acids in their polypeptide chains depending on the origin of collagen.[36] The vital amino acids found in gelatin include glycine, proline and hydroxyproline.[37] Gelatin has become important in several applications due to its unique characteristics, such as amphoteric nature, hydrophilic property and specific triple-stranded helical structure.[38]

12.2.3 Soy Protein

For more than 3000 years, soybeans have been cultivated in Asian countries, such as China, Japan and Korea, and later, in the 18th century, France and England tried to grow soybeans using improved strategies. Soybeans have been known to be an essential source of proteins in Asian countries. They are utilized in several types of food, such as tofu (soybean curd), miso (fermented soybean paste), natto (fermented soybeans) and aburage (fried tofu sheets).[39] The soybean proteins contain all the important amino acids that fulfill the nutritional requirements. Furthermore, these proteins lower the cholesterol level in living organisms, reducing the risk of hyperlipidemia and cardiovascular diseases.[40] Soybean proteins also have gelling/emulsifying abilities and oil-holding capacities. The amino acid sequence of soy proteins has not been identified precisely, but the acidic and basic subunits of amino acid composition are known. Acidic subunits contain higher contents of glutamic acid, proline and cysteine, having a molecular weight of about 37 kDa. The basic subunits consist of hydrophobic amino acids, such as alanine, leucine, phenylalanine, tyrosine and valine, having a molecular weight of 20 kDa. The isoelectric point of the protein fraction is pH 4.64. However, the solubility of soy proteins is affected by many factors, such as pH, temperature, relative humidity and storage conditions.[39]

12.2.4 Silk Protein

Silk has been the subject of over 400 million years of research and development, ranging from its use by predators (spider webs) to its applications in housing (honeybees/wasps) and protection (silkworm cocoons).[41,42] Silk is a structural protein typically spun into fibrous materials by living organisms, such as insect/spider species, including ants, fleas and crickets.[43] Silk consists of two main protein contents: fibroin and sericin.[44] Silkworm (*Bombyx mori*) is the source of silk protein (silk fibroin), and its structural composition includes the heavy chain, low chain and glycoprotein. The heavy chain consists of 5263 amino acid residues, has a molecular weight of 391 kDa and has an isoelectric point of pH 4.22. The heavy chain of silk fibroin consists of a block-copolymer-like configuration, which includes 11 hydrophilic regions with 31 amino acids and 12 hydrophobic blocks responsible for 94% of the heavy silk chain. The hydrophobic region has mostly glycine–alanine repeating units (65%) with serine (23%) and tyrosine (9%). The light chain consists of 244 amino acid residues, with a molecular weight of 25 kDa. A single disulfide bond interconnects these chains at the C-terminus.[45] Sericin has a polypeptide linkage with 17–18 different amino acids, with serine, aspartic acid and glycine being the three most abundant amino acids. Sericin has several advantages, such as (1) biodegradability, (2) easy availability and (3) hydrophilicity with polar regions. Hence, sericin can be copolymerized and crosslinked to make several scaffolds that help acquire improved properties for various applications.[46]

12.2.5 Albumin Protein

Studies on albumin have been conducted for five centuries to explore its role and unique properties. The term "albumin" originates from the Latin word "albus" meaning "white", indicating the white-colored precipitates produced by various proteins.[47] Albumin includes the transport proteins (α-fetoprotein), vitamin-D-binding proteins (Gc-globulin) and afamin (α-albumin) found in amphibians and higher vertebrates. Albumin is an aglycosylated protein, containing 585 amino acid residues in a single polypeptide chain with a molecular weight of 66.5 kDa.[48] Albumin contains high cysteine, glutamic/aspartic acid, lysine and arginine contents. Albumin also has lower tryptophan (1–2 residues), methionine (4–6 residues), glycine and isoleucine contents.[49] The mature albumin is organized into an α-helical structure stabilized by 17 disulfide bonds.[48] The nitrogen content of bovine albumin calculated from the amino acid composition is 16.47% (779 nitrogen atoms).[49] The available net charge and ligand-binding site make albumin capable of binding with several substances, such as inorganic ions, vitamins, fatty acids, bilirubin, hormones, thyroxine, steroids and drugs.[50] Based on several benefits, such as (1) availability, (2) stability and (3) ligand-binding characteristics, serum albumin is the most widely studied and applied protein in biochemistry.[51]

12.3 Processing Strategies for Protein-derived Materials for Water Purification

12.3.1 Keratin Protein

Keratin protein is hydrophobic and less reactive apparently, but several factors, such as changes in pH, high temperature and some reducing agents (Na_2SO_3 or Na_2S), increase its solubility in water.[52] Besides protonation/deprotonation of $-NH_2$ and other groups in keratin, chemical reduction breaks down and unfolds the disulfide cross-links into free thiols (–SH), making the keratin's surface reactive, leading to the solubilization of the protein. Due to its distinctive property, keratin protein can be chemically modified and developed into several forms, such as gels, films, beads and nano-/microparticles.[53]

The first step in the processing of keratin protein is its extraction. Over the years, keratin has been extracted from hairs, horns, hooves, beaks, nails and feathers. In 1905, John Hoffmeier from the United States released a commercial method for keratin extraction from animal hooves using lime.[54,55] Then later, oxidative and reductive methods were developed to extract keratin. For the extraction process, the basic requirement is the chemical demolition of intermolecular disulfide bonds while sustaining the covalent bonding of the primary protein structures. It can be attained through reduction, oxidation, sulfitolysis or oxidative sulfitolysis (Figure 12.2) processes. The extracted keratin is then potentially used to develop into powders, films, gels, fibers and foams.[56] Many researchers have used the extracted keratin to develop gels by applying formaldehyde as a hardening agent. Krystayna *et al.* developed fibrous composites from keratins. Ankar synthesized keratin-containing films and coating materials and Kawano prepared keratin-based films.[23,24]

Chemical/covalent modification of keratins can be used to modulate the macromolecular functions of keratin, transforming the highly organized α-helix and β-sheet structures into a random coil structure by disrupting a chemical-bond network. Hence, altering the reactivity of desired functional

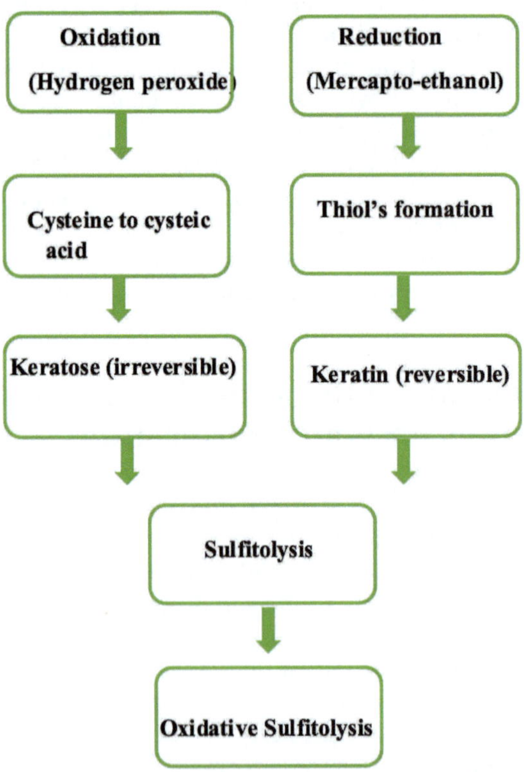

Figure 12.2 Schematic illustration of reaction mechanisms for keratin extraction.

groups can provide more active sites for pollutant adsorption, such as metal ions and volatile organics.[57,58]

In 1935, for the first time, alkylation of cysteine was achieved using electrophiles (iodoacetamide). Later, in 1978, researchers converted the active-site cysteine residues into dehydro-serine, serine and glycine residues.[59] Chemical oxidation of cysteine residues involves disulfide formation, and several simple-to-complex methods were developed for the disulfide formation on keratinous cysteine. The methods include air oxidation, thiol mixing in protein-containing basic buffer and the reaction of 5,5-dithiobis(2-nitrobenzoate) or Ellman's reagent with cysteine.[23,24] Keratin shows promise as an adsorbent for treating inorganic- (heavy metal) and organic-contaminated wastewater. Animal fibrous keratin that is extracted from chicken feathers has a profound effect in removing the inorganic and organic contaminants from wastewater.[60,61]

12.3.2 Gelatin

The collagen is transformed into gelatin by denaturation, which can be achieved by catalyzed hydrolysis using enzymes, acids or alkalis.[62] Collagen can also be thermally denatured under mild acidic/neutral conditions by heating at about 40 °C.[31] Hydrolysis aims to break down the hydrogen and hydrophobic bonds of collagen, which stabilizes helical structure, triggering the fibers and fibrils of collagen to dissociate into tropocollagen units. The next step is to break down the intramolecular bonds between the helical structures.[63]

The gelatin extraction process from porcine and bovine sources includes (1) degreasing: removing fat by soaking and cleaning in hot water for 30 min at 100 °C, (2) acid pre-treatment: soaking in acid (4% hydrochloric acid [HCl], pH <1.5) or alkali pre-treatment: soaking in alkali (potassium/sodium carbonate, Na_2CO_3, pH >7), (3) boiling in distilled water, (4) evaporation: the liquid is filtered to separate the residues and then the liquid is vaporized from solid gelatin and (5) grinding: the solid gelatin is ground into a fine powder.[38]

Gelatin is also developed into gels by establishing the microstructural network. Gelatin gels are characterized by their thermo-reversible behavior, as they are converted into a solution at above 40 °C and solidify when the temperature is decreased to 30 °C. This is due to partially restoring the triple helical structure in the collagen, which is responsible for 3-dimensional networks.[64] Hydrogels produced from gelatin are characterized as water-swollen macromolecules obtained by polymerization of monomers.[65] Hydrogels can hold a significant amount of water within their 3-dimensional network and have a large number of applications in tissue engineering, controlled drug release, biosensors and wound healing in medical sciences.[64]

Similar to keratin, chemical modifications of gelatin can also take place. Several chemicals, such as glyoxal, epoxides, isocyanates, carbodiimides, formaldehyde and glutaraldehyde; natural compounds, such as ferulic and tannin acids, ribose (through Maillard reaction), epigallocatechin gallate (EGCG) and genipin, or enzymes, such as transglutaminase, are being used to cross-link the gelatin.[64] Besides chemical modifications, gelatin can be modified by using physical methods, such as electrospinning and blending with other biomacromolecules (gellan, alginate, chitosan and soy protein isolates).[66] Therefore, gelatin exhibits several functional characteristics, such as film-forming ability by electrospinning, gelling/emulsifying properties, developing into nanocomposites and biodegradability, making it competent for pharmaceutical, food and medical care industries.[64]

12.3.3 Silk Protein

Silk fibroin extraction/regeneration is a three-step process, involving degumming, rehydration and dialysis. In the first step, silkworm cocoons are boiled in a 0.02 M Na_2CO_3, in which the protein partially degrades depending on the duration of degumming time. In the second step, the degummed fibers are liquefied in a 9.3 M lithium bromide (LiBr) solution at 60 °C for 4 h to encourage the rehydration of the proteins, disrupting the hydrogen-bonded crystalline structure and providing the initial silk material fabrication. In the third/final step, the LiBr solution is dialyzed.[67] After extraction, silk protein is processed in aqueous environments to produce several materials, such as films, sponges, electrospun mats and hydrogels.[68] The silk protein sericin is extracted from its source through conventional methods using heat/alkali, which degrades the sericin and produces low-molecular-weight polypeptides that are fragile and highly soluble. After the extraction, the sericin blended with the glutaraldehyde leads to the crosslinking of extracted sericin to form a hydrogel.[46]

Protein nanofibrils have specific surface functionality and a high surface : volume ratio. They also have a repetitive backbone sequence with hydrophilic and hydrophobic amino acid segments enclosed in the twisted structures with a diameter between 5 and 10 nm and a length of up to several micrometers. In addition to the intrinsic properties, nanofibrils have unique features of chirality, polarity and charge, making them dynamic building blocks for environmental applications, such as wastewater remediation.[69]

12.4 Protein-derived Materials for Water Purification

12.4.1 Inorganic Pollutants (Heavy Metals)

Heavy metals are biologically important as they are essential to fulfill the human's nutritional requirements and are considered key components for many enzymes. However, some heavy metals, for example, arsenic (As), cadmium (Cd), lead (Pb), zinc (Zn) and methylated form of mercury (Hg), are recognized as toxic to human health even at very low concentrations.[70,71] When these heavy metals are ingested, they reach the stomach (acidic medium), transform into their soluble oxidation states (Zn^{2+}, Pb^{2+} and Cd^{2+}) and become attached to the biomolecules, such as proteins and enzymes, causing toxic effects by interfering with the metabolic mechanisms.[70,71]

Due to industrial activities and urbanization, water pollution due to heavy metals has been extended to an aberrant level, causing acute toxicities to living organisms and deterioration of the environment. Although there are several methods (reverse osmosis, nanofiltration and disinfection using different chemicals) available to decontaminate water resources, these treatment technologies have several disadvantages, such as some are not cost-efficient and some demand additional maintenance requirements. Among the treatment technologies, adsorbents containing natural polymers (proteins, chitosan and biochar) have gained attention worldwide based on their several benefits (cost-efficiency, biocompatibility and environmentally benign processes).[60]

Over the past few years, researchers started using keratin wool to remove some metals, such as copper (Cu), mercury (Hg), lead (Pb) and cadmium (Cd), from wastewater. Sun *et al.* used modified chicken feathers as an adsorbent to remove chromium (Cr^{VI}) ions from the waste-water stream.[72] Another research conducted by Teixeira and Ciminelli examined a biological route for the adsorption of aqueous arsenic (As^{III}) species by waste biomass with a high fibrous protein (keratin).[73] The examination and comparison of keratin-based adsorbents derived from three sources (chicken feathers, human hairs and animal horns) to remove zinc (Zn) and copper (Cu) ions from wastewater were conducted by Al-Asheh *et al.* They evaluated that animal horn-based adsorbents had greater adsorption efficiency than chicken feathers and human hairs.[74] The same researchers reused the chicken feathers as adsorbents after chemical treatment with NaOH and anionic surfactant dodecyl sulfate to remove copper (Cu) and zinc (Zn) from wastewater.[23,24] The polyurethane–keratin membranes were developed by Manrique-Juárez *et al.* in an assessment of the natural affinity of keratin toward metal adsorption. Keratin was first extracted from chicken feathers, and polyurethane foams were synthesized with two types of isocyanates: diphenyl-methane-diisocyanate (MDI) and toluene-diisocyanate (TDI), which were mixed separately with polyol and keratin to analyze their effects on the membrane microstructure. The influence of pH on the adsorption of contaminants by membranes was evaluated, as it is one of the most important parameters affecting the ionic charges on the adsorbent surface, which is responsible for metal ion attachment.[75] Thakur *et al.* explained the performance of gelatin hydrogels in removing inorganic pollutants.[76]

Li *et al.* discussed the gelatin/silica grafting with benzoquinone for the biotoxicity assay, which also involved the analysis of water pollutants, such as mercury (Hg), copper (Cu) and cadmium (Cd). Researchers developed the novel biosensor by immobilizing the microorganism *E. coli* and redox mediator into a hybrid hydrogel matrix. The 50% inhibiting concentrations were analyzed and found to be 21.2, 44 and 79 mg mL^{-1} for mercury (Hg), copper (Cu) and cadmium (Cd), respectively.[77] Wang *et al.* prepared the dumbbell manganese dioxide/gelatin composites to remove lead and cadmium ions. The maximum adsorption capacities for lead (Pb) and cadmium (Cd) removal were 318.7 and 105.1 mg g^{-1}, respectively.[78] Mahmoud and Mohamed developed the gelatin-based adsorbent to remove chromium (Cr^{VI} and Cr^{III}) from water, explaining that pH is the predominant factor in removing chromium ions. It was evident from the given results that pH = 1 was the optimum pH for the recovery of Cr^{VI}, and pH = 6 was the most favorable pH for the removal of Cr^{III}, corresponding to 97% and 98% removal efficiencies, respectively.[79] Abdellatif *et al.* prepared the cross-linked gelatin/poly(amidoamine) hyperbranched polymer (PAMAM) aerogels for the removal of chromium (Cr^{VI}) and cadmium (Cd^{II}) ions from aqueous media. They assessed the adsorption efficiencies of these metals under the influence of adsorption parameters (contact time and dose–effect). Furthermore, the researchers evaluated the adsorption isotherm data fitted into Langmuir and Freundlich isotherm models, with the maximum adsorption capacities (125.0 and 142.0 mg g^{-1} for Cr^{VI} and Cd^{II} ions, respectively).[80] Wahlström *et al.* developed the dialdehyde–gelatin hydrogels for the removal of heavy metals, such as copper, cobalt, nickel and zinc. The maximum adsorption capacities were 14 mg g^{-1} for copper, 7 mg g^{-1} for cobalt and 6 mg g^{-1} for nickel and zinc, explaining that the hydrogels had a higher affinity for copper than cobalt, nickel and zinc.[81]

Liu *et al.* prepared the composites containing soy protein isolates (SPIs) as a matrix and polyethyleneimine (PEI) as a functional component. The results showed that copper (Cu) was effectively adsorbed by SPI/PEI composites from an aqueous solution.[82] Researchers have discussed the binding of divalent cations, such as calcium (Ca), copper (Cu) and nickel (Ni), to the albumin protein. Albumin has the unique property of having the binding ability to biological materials, including hydrophobic interactions. However, calcium binding was not strong, but copper was adsorbed effectively by albumin. Other metals, such as manganese (Mn), displayed stronger adsorption by the albumin protein, whereas cobalt (Co) and cadmium (Cd) had weaker interactions.[49,50]

12.4.2 Organic Pollutants

Water pollution due to organic contaminants is a threatening issue because of the acute toxicities and carcinogenic nature of the pollutants.[61] Organic pollutants (organic dyes, phenols, hydrocarbons, pesticides, detergents, plasticizers and grease) are being discharged into the environment from chemical, textile, fertilizer and pharmaceutical industries.[28]

Arshad *et al.* developed keratin-derived adsorbents by chemical modification of keratin with polyhedral oligomeric silsesquioxanes (POSS) nanocages and goethite to reveal the functional groups by unfolding the keratin structures. The purpose of these keratin modifications was to remove the organic contaminant (naphthenic acid) from oil sands process-affected water (OSPW).[60,61] Saber-Samandari *et al.* synthesized the nanocomposite material based on the gelatin/iron oxide nanoparticles and carbon nanotube to remove cationic methylene blue (MB) and anionic direct red 80 (DR) dyes. Excellent nanocomposite adsorption performance was observed for both dyes. The nanocomposite displayed the maximum adsorption capacities of 465.5 and 380.7 mg g^{-1} for MB and DR, respectively.[83] In another study, the hydrogels were prepared by encapsulating manganese peroxide on the gelatin matrix using glutaraldehyde as a cross-linker.[84] The purpose of the hydrogels was to decolorize the reactive red 195 A dye from an aqueous solution, and 90% decolorization was achieved within 5 h.[85] In another study, granular gelatin-grafted polyacrylic acid/attapulgite hydrogel was synthesized for the adsorption of orange G and malachite green dyes from the mixed solution. The gelatin/polyacrylic acid and gelatin-grafted polyacrylic acid/attapulgite hydrogel composite displayed the maximum adsorption capacities of 1370 and 1190 mg g^{-1} for malachite green dye.[86]

Silk protein nanofibrils are composed of β-sheets that have well-defined molecular architecture and mechanical stability. These properties make silk protein fibroins a potential candidate for water purification. The supramolecular structure of nanofibrils makes them stable against the adsorption of organic solvents under both acidic and alkaline conditions.[69] Natural organic matter (NOM) is the chemically complex polyelectrolytes that have a greater influence on the adsorption of organic pollutants on biochar. However, Liu *et al.* observed that the adsorption performance of 2,2′,4,4′-tetrabromodiphenyl ether onto biochar was increased after adding bovine serum albumin.[87]

12.4.3 Salts and Other Pollutants

Azadi *et al.* optimized the preparation of gelatin-coated magnetite nanoparticles (MNPs) and utilized the MNPs as a novel draw solute in the forward osmosis process to desalinate brackish water.[88] Jiang *et al.* designed the multifunctional gelatin-based aerogel, with the uniform distribution of TiO_2 nanoparticles, wherein branched polyethyleneimine (PEI) was cross-linked onto it through the effect of glutaraldehyde. The gelatin/TiO_2/PEI showed excellent oil/water separation properties from both oil/water mixtures and emulsions.[89] Pandi *et al.* synthesized the iron oxide-fabricated nano-hydroxyapatite gelatin eco-polymeric composite for defluoridation studies. The results indicated that the iron oxide gel composite displayed a high defluoridation capacity of 5009 mg F^{-1} kg^{-1}. The batch adsorption studies were also conducted to evaluate the influence of process parameters, such as pH, contact time, competitor co-anions and temperature, on adsorption.[90] Karunanidhi *et al.* prepared the keratin-polysulfone blend membranes to treat effluents from the tannery industry. After the treatment, the maximum percentage (53% and 66%) of reduction was observed in effluent's chemical oxygen demand (COD) and biological oxygen demand (BOD), respectively.[91]

12.5 Adsorption Mechanisms

The adsorption mechanisms provide additional insights into the performance of adsorbents for removing the inorganic and organic contaminants from an aqueous environment, hence playing an important role in water remediation.[60] It is difficult to understand the adsorption mechanism, and very few studies are available on the interaction of biopolymers/proteins with inorganic and organic contaminants. It is generally considered that the amino functional group is the most reactive site for the metal ions, whereas the hydroxyl group at the C-position also promotes adsorption. Thus, these two functional groups play a significant role in metal ion interactions. However, several factors, such as system matrix, pH and nature of metal ions, also profoundly impact adsorption significantly. For instance, at a neutral pH or weakly acidic conditions, the two lone pairs of electrons on nitrogen are responsible for binding for divalent metal cations.[92] The protonation of the amino groups takes place in a highly acidic environment, promoting the cationic behavior of biopolymers.[93,94] The functional groups of adsorbents are characterized by surface analyses, such as X-ray photoelectron spectroscopy (XPS), X-ray diffractions (XRDs), scanning electron microscopy (SEM), Fourier-transform infrared (FTIR) spectroscopy and electron-dispersive X-ray spectroscopy (EDX).[89]

The adsorption of heavy metals onto the modified adsorbent's surface involves chemical interactions (chelation/complexation), ion exchange and non-polar interactions (van der Waals forces).[92] Heavy metal ions possess different physical and chemical properties, leading to variations in their affinities for oxygen-, nitrogen- and sulfur-containing ligands. The adsorption of divalent heavy metal cations involves a complexation mechanism with the active sites of keratin-/protein-derived adsorbents. The interactions of the metal ions with the active functional sites of keratin adsorbents are better elucidated with their Lewis acid and base principles. Metal cations are categorized as hard and soft "Lewis" acids depending on their ability to accept an electron pair. The divalent cations Cd^{2+}, Cu^{2+} and Ni^{2+} are categorized in the intermediate class, while Ni^{2+} shifts to the hard class and Cd^{2+} to the soft class. The metal-binding ligands of keratin including oxygen and aliphatic nitrogen are also recognized as hard, aromatic nitrogen as intermediate and sulfur as soft "Lewis" bases, concluding that Ni^{2+} would bind to aromatic nitrogen and Cd^{2+} to S–H groups. Zahara *et al.* demonstrated that the keratin-derived adsorbent modified with

3-(dimethylamino)-1-propylamine exhibited good adsorption for divalent metal cations (Cd^{2+}, Cu^{2+}, Ni^{2+}, Co^{2+} and Zn^{2+}).[60]

Another study by Kuang *et al.* characterized the tetraamine-grafted magnetic chitosan biopolymers before and after the lead (Pb^{2+}) adsorption using the FTIR technique. The results showed that the Pb^{2+} adsorption involves the complexation mechanism due to the presence of many amino, imino and hydroxy groups with the lone pair of electrons on nitrogen and oxygen atoms, which can be donated to Pb^{2+} forming the chitosan–metal complex.[95] In another study, XPS N1s spectra were used to characterize the adsorption of Cu^{2+} using a polystyrene–chitosan (PC–CS) thin film. The researchers concluded that Cu^{2+} coordinates with the N atom of PC–CS. The XPS results (Figure 12.3) showed the two peaks of N1s of Cu^{2+} before adsorption: the first peak has a binding energy of 400.5 eV, which can be designated to N atoms of NH_2 and NH, while the second peak observed at 401.9 eV attributed to the oxidation of N atoms with positive charges of NH^{3+} groups. After Cu^{2+} adsorption, these two peaks shifted to 399.8 and 402.2 eV, respectively, governing the formation of R–NH_2–Cu^{2+} or R–NH–Cu. These differences in binding energies indicate the formation of a complex in which the lone pair of N atoms in the PS–CS surface functional group is shared with the Cu^{2+} ions.[96]

Zahara *et al.* explained that the significant process for oxyanion removal from wastewater could be the chemical reduction of chromium (Cr^{VI}) and vanadium (V^{V}). The oxyanionic species (Cr_2O_4 and H_2VO_4/ HVO_4) could be chemically reduced by keratin-derived adsorbents through exposed amine (–NH_2) and thiol/sulfhydryl (–SH) functional groups. Furthermore, the direct and indirect reduction mechanisms could also be involved in removing Cr^{VI} from wastewater. In the direct reduction mechanism, the Cr^{VI} could directly be reduced to Cr^{III} by interacting with electron-donor groups, and then, the Cr^{III} could lead to the complex formation with adsorbents

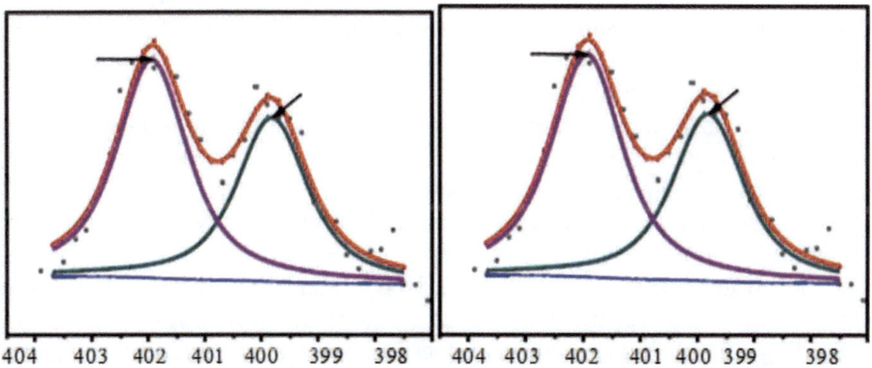

Figure 12.3 XPS N1s spectra before (left) and after (right) the adsorption of Cu. Reproduced from ref. 96 with permission from Elsevier, Copyright 2014.

and remain in the aqueous phase. The indirect mechanism includes the three steps: (1) bond formation of Cr^{VI} with positively charged functional groups on the adsorbent surface, (2) the chemical reduction of Cr^{VI} to Cr^{III} and (3) the complex formation of Cr^{III} with adjacent functional groups. The functional groups that are involved in the indirect reduction mechanism are amino and carboxyl groups.[60] Saber-Samandari *et al.* concluded that the positively charged surface of gelatin-based biopolymers is predominantly responsible for the adsorption of anionic dyes.

Further explanation reveals that gelatin carbon nanotubes can also form a complex with anionic dyes. The carboxylic (–COOH) functional groups electrostatically interact with the cationic dyes.[83] One of the studies explained that the zeta potential explained the malachite green dye adsorption mechanism on hydrogels. Hydrogels displayed negative zeta potential (−20 mV) due to their anionic nature. The researchers mentioned, therefore, that negatively charged orange G dye was repelled by hydrogels.[76]

12.6 Summary and Perspective

Various harmful pollutants, such as metal ions, organics, oils and salts in water, cause a severe ecological problem. The treatment of industrial wastewater before discharge into the water bodies is of utmost priority; however, it is challenging to develop a general material to remove all types of pollutants from contaminated water. In this regard, protein-based materials have exhibited excellent performance in eliminating multiple types of pollutants from contaminated water. Protein-based materials have been considered an excellent replacement as green, renewable and sustainable materials for existing adsorbing materials for healing wastewater contaminants. So far, keratin, soy, albumin, silk and gelatin have been frequently used for water purification. Because of their distinctive physicochemical and structural attributes, natural availability, environmentally benign nature and cost, protein-based materials offer various applications for producing sustainable materials for water purification. The biodegradability and biocompatibility of protein-derived materials can enhance their use in water remediation.

However, although proteins can be used in pollutant removal due to their environment-friendly nature and cost-effective benefits, contaminant removal using proteins is not commercially viable.

Over the last few decades, substantial attempts have been made to design and produce green materials with extremely good performance for wastewater treatment. Protein-derived materials have gained rigorous interest for their applications in the adsorption of various pollutants, including metal ions, such as arsenic, lead, zinc, selenium, copper, chromium, nickel, cadmium and mercury, and organics, such as dyes, phenols, hydrocarbons, pesticides, detergents, plasticizers and grease. One of the significant challenges in the applications of protein-based materials at the industrial level is their recovery after pollutant removal. Therefore, proteins are often modified with other materials to enhance their removal efficiency.

Although proteins are a promising material for wastewater decontamination, they produce harmful secondary environmental waste. It is critical to alleviate this crisis with an environment-friendly and sustainable approach. However, it may be resolved by reusing waste materials for various purposes, *i.e.*, catalysis, sensors and energy applications. In the future, research will be necessary to add value to the waste through recycling and a circular approach. Despite several drawbacks, protein-based materials have shown their efficacy in eliminating toxic impurities from polluted water. The use of proteins to eliminate contaminants, with a focus on the constraints of reuse and separation, has developed into a simple process. Nevertheless, it is in an early stage of development, and several challenges need to be solved relating to the progress of simpler procedures to achieve effective materials in water and wastewater treatment applications. Furthermore, robust collaboration among scientists, industries and government is needed to develop and utilize innovative materials for wastewater treatment.

References

1. V. Kumar, S. K. Dwivedi and S. Oh, *J. Water Process Eng.*, 2022, **45**, 102518.
2. M. M. Pendergast and E. M. V. Hoek, *Energy Environ. Sci.*, 2011, **4**, 1946–1971.
3. S. Rojas and P. Horcajada, *Chem. Rev.*, 2020, **120**, 8378–8415.
4. M. Zubair and A. Ullah, *Biopolym. Their Ind. Appl.*, 2021, 331–349.
5. WHO, *Progress on Drinking Water and Sanitation*, World Health Organization, 2013.
6. A. Haarstrick and M. Bahadir, in *Water and Wastewater Management*, Springer, 2022, pp. 3–14.
7. M. Maharana, M. Manna, M. Sardar and S. Sen, in *Green Adsorbents to Remove Metals, Dyes and Boron from Polluted Water*, Springer, 2021, pp. 245–272.
8. M. Zubair, M. Arshad and A. Ullah, in *Handbook of Chitin and Chitosan*, Elsevier, Amsterdam, The Netherlands, 2020, pp. 773–809.
9. S. Saha, M. Zubair, M. A. Khosa, S. Song and A. Ullah, *J. Polym. Environ.*, 2019, **27**, 1389–1403.
10. UNO, *World Water Development Report*, World Water Assessment Programme (WWAP), 2017.
11. A. Lee, J. W. Elam and S. B. Darling, *Environ. Sci.*, 2016, **2**, 17–42.
12. M. Arshad, M. Zubair, S. S. Rahman and A. Ullah, in *Polymer Science and Nanotechnology, Fundamentals and Applications: Polymers for Advanced Applications*, ed. R. Narain, Elsevier, 2020, pp. 325–340.

13. R. Meffe and I. de Bustamante, *Sci. Total Environ.*, 2014, **481**, 280–295.
14. L. Andrade, J. O'Dwyer, E. O'Neill and P. Hynds, *Environ. Poll.*, 2018, **236**, 540–549.
15. M. A. Shannon, P. W. Bohn, M. Elimelech, J. G. Georgiadis, B. J. Marinas and A. M. Mayes, in *Nanoscience And Technology: A Collection of Reviews from Nature Journals*, World Scientific, 2010, pp. 337–346.
16. M. Varsha, P. S. Kumar and B. S. Rathi, *Chemosphere*, 2022, **287**, 132270.
17. M. Bilal, I. Ihsanullah, M. Younas and M. U. H. Shah, *Sep. Purif. Technol.*, 2021, **278**, 119510.
18. I. Ali, M. Asim and T. A. Khan, *J. Environ. Manage.*, 2012, **113**, 170–183.
19. Q.-T. Zhang, Z.-C. Tu, H. Wang, X.-Q. Huang, L.-L. Fan, Z.-Y. Bao and H. Xiao, *J. Food. Sci. Technol.*, 2015, **52**, 3412–3421.
20. F. Pu, X. Liu, B. Xu, J. Ren and X. Qu, *Chem. - Eur. J.*, 2012, **18**, 4322–4328.
21. A. Aluigi, C. Vineis, C. Tonin, C. Tonetti, A. Varesano and G. Mazzuchetti, *J. Biobased Mat. Bioenergy*, 2009, **3**, 311–319.
22. A. Figoli, C. Ursino, D. O. Sanchez Ramirez, R. A. Carletto, C. Tonetti, A. Varesano, M. P. de Santo, A. Cassano and C. Vineis, *Polym. Eng. Sci.*, 2019, **59**, 1472–1478.
23. M. A. Khosa and A. Ullah, *J. Food Processing & Beverages*, 2013, **1**, 1.
24. J. G. Rouse and M. E. Van Dyke, *Materials*, 2010, **3**, 999–1014.
25. M. Zubair, M. Arshad and A. Ullah, in *Natural Polymers-Based Green Adsorbents for Water Treatment*, Elsevier, 2021, pp. 93–109.
26. M. Feughelman, *J. Applied. Polymer. Sci.*, 2001, **83**, 489–507.
27. G. Rocha Plácido Moore, S. Maria Martelli, C. Gandolfo, P. José do Amaral Sobral and J. Borges Laurindo, *Food Hydrocolloids*, 2006, **20**, 975–982.
28. A. Ghosh and S. R. Collie, *Def. Sci. J.*, 2014, **64**, 209–221.
29. A. Duconseille, T. Astruc, N. Quintana, F. Meersman and V. Sante-Lhoutellier, *Food Hydrocolloids*, 2015, **43**, 360–376.
30. M. C. Gomez-Guillen, B. Gimenez, M. E. Lopez-Caballero and M. P. Montero, *Food Hydrocolloids*, 2011, **25**, 1813–1827.
31. A. A. Mariod and H. F. Adam, *Acta Sci. Pol., Technol. Aliment.*, 2013, **12**, 135–147.
32. S. Sukkwai, K. Kijroongrojana and S. Benjakul, *Int. Food Res. J.*, 2011, **18**, 1129–1134.
33. A. N. Razali, A. M. Amin and N. M. Sarbon, *Int. Food Res. J.*, 2015, **22**, 651–660.
34. S. M. Cho, Y. S. Gu and S. B. Kim, *Food Hydrocolloids*, 2005, **19**, 221–229.
35. M. S. Hoque, S. Benjakul and T. Prodpran, *Food Hydrocolloids*, 2011, **25**, 82–90.
36. S. Singh, K. V. R. Rao, K. Venugopal and R. Manikandan, *Pharm. Technol. North Am.*, 2002, **26**, 36–58.
37. S. Hermanto, L. O. Sumarlin and W. Fatimah, *J. Food Pharm. Sci.*, 2013, **1**, 68–73.

38. J. Alipal, N. A. S. Mohd Pu'ad, T. C. Lee, N. H. M. Nayan, N. Sahari, H. Basri, M. I. Idris and H. Z. Abdullah, *Mater. Today: Proc.*, 2019, **42**, 240–250.
39. K. Nishinari, Y. Fang, S. Guo and G. O. Phillips, *Food Hydrocolloids*, 2014, **39**, 301–318.
40. A. J. Michelfelder, *Am. Fam. Physician*, 2009, **79**, 43–47.
41. B. Mortimer, C. Holland and F. Vollrath, *Biomacromolecules*, 2013, **14**, 3653–3659.
42. A. A. Walker, C. Holland and T. D. Sutherland, *Proc. R. Soc. B*, 2015, DOI: 10.1098/rspb.2015.0259.
43. B. Madsen, Z. Z. Shao and F. Vollrath, *Int. J. Biol. Macromol.*, 1999, **24**, 301–306.
44. B. Panilaitis, G. H. Altman, J. Chen, H. J. Jin, V. Karageorgiou and D. L. Kaplan, *Biomaterials*, 2003, **24**, 3079–3085.
45. C. Z. Zhou, F. Confalonieri, M. Jacquet, R. Perasso, Z. G. Li and J. Janin, *Proteins: Struct., Funct., Genet.*, 2001, **44**, 119–122.
46. Z. Wang, Y. Zhang, J. Zhang, L. Huang, J. Liu, Y. Li, G. Zhang, S. C. Kundu and L. Wang, *Sci. Rep.*, 2014, **4**, 1–11.
47. Z. Liu, D. Trifonova, I. Tulaeva, K. Riabova, A. Karsonova, E. Kozlove, O. Elisyutina, M. Khaitov, M. F. Tejkl, T. H. Chen, A. Karaulov and R. Valenta, *Front. Immunol.*, 2023, **14**, DOI: 10.3389/fimmu.2023.1241518.
48. M. Otagiri and V. T. Giam Chuang, in *Albumin in medicine: Pathological and clinical applications*, Springer, 2016.
49. T. Peters, *Adv. Protein Chem.*, 1985, **37**, 161–245.
50. D. C. Carter and J. X. Ho, *Adv. Protein Chem.*, 1994, **45**, 153–176.
51. S. C. Chien, C. Y. Chen, C. F. Lin and H. I. Yeh, *Biomark. Res.*, 2017, **5**, 1–9.
52. S. Saha, M. Arshad, M. Zubair and A. Ullah, in *Keratin as a Protein Biopolymer*, Springer, 2019, pp. 163–185.
53. S. Alahyaribeik and A. Ullah, *Int. J. Biol. Macromol.*, 2020, **148**, 449–456.
54. K. Gregg and G. E. Rogers, in *Biology of the integument*, Springer, 1986, pp. 666–694.
55. A. Idris, R. Vijayaraghavan, U. A. Rana, D. Fredericks, A. F. Patti and D. R. MacFarlane, *Green Chem.*, 2013, **15**, 525–534.
56. X. C. Yin, F. Y. Li, Y. F. He, Y. Wang and R. M. Wang, *Biomater. Sci.*, 2013, **1**, 528–536.
57. I. S. Carrico, *Chem. Soc. Rev.*, 2008, **37**, 1423–1431.
58. D. Qi, C. M. Tann, D. Haring and M. D. Distefano, *Chem. Rev.*, 2001, **101**, 3081–3111.
59. P. I. Clark and G. Lowe, *Eur. J. Biochem.*, 1978, **84**, 293–299.
60. I. Zahara, M. Arshad, M. A. Naeth, T. Siddique and A. Ullah, *Chemosphere*, 2021, **273**, 128545.
61. M. Arshad, M. A. Khosa, T. Siddique and A. Ullah, *Chemosphere*, 2016, **163**, 334–341.

62. R. A. S. N. Ranasinghe, W. L. I. Wijesekara, P. R. D. Perera, S. A. Senanayake, M. M. Pathmalal and R. A. U. J. Marapana, *Food Rev. Int.*, 2020, **38**, 812–855.
63. M. Nishimoto, R. Sakamoto, S. Mizuta and R. Yoshinaka, *Food Chem.*, 2005, **90**, 151–156.
64. Z. Yang, S. Chaieb and Y. Hemar, *Polym. Rev.*, 2021, **61**, 765–813.
65. J. Kopeček, *Biomaterials*, 2007, **28**, 5185–5192.
66. H. W. Kwak, M. Shin, J. Y. Lee, H. Yun, D. W. Song, Y. Yang, B. S. Shin, Y. H. Park and K. H. Lee, *Int. J. Biol. Macromol.*, 2017, **102**, 1092–1103.
67. D. N. Rockwood, R. C. Preda, T. Yucel, X. Wang, M. L. Lovett and D. L. Kaplan, Materials fabrication from Bombyx mori silk fibroin, *Nat. Protoc.*, 2011, **6**(10), 1612–1631.
68. C. Holland, K. Numata, J. Rnjak-Kovacina and F. P. Seib, *Adv. Healthcare Mater.*, 2018, DOI: 10.1002/adhm.201800465.
69. M. Peydayesh and R. Mezzenga, *Nat. Commun.*, 2021, **12**, 1–17.
70. I. Duruibe, *et al.*, *Int. J. Phys. Sci.*, 2007, **2**, 112–118.
71. P. B. Tchounwou, C. G. Yedjou, A. K. Patlolla and D. J. Sutton, in *Molecular, Clinical and Environmental Toxicology*, 2012, vol. **101**, pp. 133–164.
72. P. Sun, Z. T. Liu and Z. W. Liu, *J. Hazard. Mater.*, 2009, **170**, 786–790.
73. M. C. Teixeira and V. S. T. Ciminelli, *Environ. Sci. Technol.*, 2005, **39**, 895–900.
74. S. Al-Asheh, F. Banat and D. Al-Rousan, *J. Cleaner Prod.*, 2003, **11**, 321–326.
75. M. D. Manrique-Juárez, A. L. Martínez-Hernández, O. F. Olea-Mejía, J. Flores-Estrada, J. L. Rivera-Armenta and C. Velasco-Santos, *Int. J. Polym. Sci.*, 2013, DOI: 10.1155/2013/892547.
76. S. Thakur, P. P. Govender, M. A. Mamo, S. Tamulevicius and V. K. Thakur, *Vacuum*, 2017, **146**, 396–408.
77. J. Li, Y. Yu, J. Qian, Y. Wang, J. Zhang and J. Zhi, *Analyst*, 2014, **139**, 2806–2812.
78. X. Wang, K. Huang, Y. Chen, J. Liu, S. Chen, J. Cao, S. Mei, Y. Zhou and T. Jing, *J. Hazard. Mater.*, 2018, **350**, 46–54.
79. M. E. Mahmoud and R. H. A. Mohamed, *J. Environ. Chem. Eng.*, 2014, **2**, 715–722.
80. M. M. Abdellatif, F. H. H. Abdellatif and S. Ibrahim, *Polym. Bull.*, 2022, DOI: 10.1007/s00289-021-04019-8.
81. N. Wahlström, S. Steinhagen, G. Toth, H. Pavia and U. Edlund, *Carbohydr. Polym.*, 2020, **249**, 116841.
82. J. Liu, D. Su, J. Yao, Y. Huang, Z. Shao and X. Chen, *J. Mater. Chem. A*, 2017, **5**, 4163–4171.
83. S. Saber-Samandari, S. Saber-Samandari, H. Joneidi-Yekta and M. Mohseni, *Chem. Eng. J.*, 2017, **308**, 1133–1144.
84. M. Bilal, M. Asgher, H. Hu and X. Zhang, *Water Sci. Technol.*, 2016, **74**, 1809–1820.

85. A. Zille, T. Tzanov, G. M. Gübitz and A. Cavaco-Paulo, *Biotechnol. Lett.*, 2003, **25**, 1473–1477.
86. Y. Zheng, Y. Zhu, F. Wang and A. Wang, *Water, Air, Soil Pollut.*, 2015, DOI: 10.1007/s11270-015-2624-y.
87. G. Liu, H. Sheng, Y. Fu, Y. Song, M. Redmile-Gordon, Y. Qiao, C. Gu, L. Xiang and F. Wang, *Chemosphere*, 2019, **214**, 176–183.
88. F. Azadi, A. Karimi-Jashni and M. M. Zerafat, *Environ. Technol.*, 2021, **42**, 2885–2895.
89. J. Jiang, Q. Zhang, X. Zhan and F. Chen, *Chem. Eng. J.*, 2019, **358**, 1539–1551.
90. K. Pandi and N. Viswanathan, *J. Chem. Eng. Data*, 2016, **61**, 571–578.
91. A. Karunanidhi, P. S. David and N. N. Fathima, *Water, Air, Soil Pollut.*, 2020, DOI: 10.1007/s11270-020-04682-z.
92. E. Guibal, *Sep. Purif. Technol.*, 2004, **38**, 43–74.
93. S. Qian, G. Huang, J. Jiang, F. He and Y. Wang, *J. Appl. Polym. Sci.*, 2000, **77**, 3216–3219.
94. C. Gerente, V. K. C. Lee, P. Le Cloirec and G. McKay, *Crit. Rev. Environ. Sci. Technol.*, 2007, **37**, 41–127.
95. S. P. Kuang, Z. Z. Wang, J. Liu and Z. C. Wu, *J. Hazard. Mater.*, 2013, **260**, 210–219.
96. W. Jiang, X. Chen, B. Pan, Q. Zhang, L. Teng, Y. Chen and L. Liu, *J. Hazard. Mater.*, 2014, **276**, 295–301.

CHAPTER 13

Lipid- and Protein-based Adhesives

ANUJIT GHOSAL[a,b] AND NANDIKA BANDARA*[a,b]

[a]Department of Food & Human Nutritional Sciences, The University of Manitoba, MB. R3T 2N2, Canada; [b]Richardson Centre for Food Technology and Research (RCFTR), The University of Manitoba, MB. R3T 6C5, Canada
*E-mail: Nandika.Bandara@umanitoba.ca

13.1 Introduction

The term "adhesive" has become common and indispensable in daily life. Adhesives can fall into categories, such as connectors, binding agents, glue, gum, and fixatives, and they are responsible for holding substrates together. "Cement" is an example of adhesives used in construction work, bone healing, and building models. Adhesives are essential in various industrial activities, such as medical devices, household enterprises, and technocrafts, and also in preparing models in schools or businesses, performing electrical fittings, heat insulation, tire manufacturing, floor laminations, phone and laptop production, sensor development, and medical surgeries.

Adhesives can exist in different forms, such as activated or inactivated, and phases, such as solid or liquid. Adhesives can be categorized based on the possibilities of reactions involved during the adhesion process.

1. Reactive and
2. Non-reactive adhesives.

Smart Materials No. 41
Functional Materials from Lipids and Proteins
Edited by Nandika Bandara and Aman Ullah

Published by the Royal Society of Chemistry, www.rsc.org

Figure 13.1 shows a brief classification, limitations, and relative advantages of adhesives. Their adhesive nature is influenced mainly by the available functional groups, hydrocarbon units (chain length), and crosslinking ability. Generally, polymers with longer chain lengths have higher intermolecular cohesion.[1] Adhesives can also be categorized based on the major functionality or active group presented, such as acrylic, polyurethane, epoxy, vinyl, alcohol, styrene, silicone, polycyclic aromatic group, and others. The incorporated functional units in the building block of resins are responsible for anchoring or establishing adhesion with substrates through various interactions, including chemical, ionic, non-covalent, and molecular interactions.[2]

The unique properties of adhesives, such as rapid curing, toughness, mechanical–chemical stability, strain-to-fracture resistance, Young's modulus, delamination, corrosion resistance (moisture, acid, alkali, and salt), sensitivity toward light, temperature, and moisture, can be engineered through chemical/physical/biological modifications.[2–4] The interfacial interactions and cohesion between substrates can be controlled by adjusting the chemical bonds, hydrocarbon structures, and functional additives (nanoparticles, crosslinkers, and oxidants) and establishing an optimum composition.[5,6] For example, the properties of polyurethane (PU)-based adhesives can be altered by varying the molar ratio of isocyanates to the hydroxy groups (hard and soft segments).[7,8] Similarly, in the case of protein- or peptide-based adhesives, the cohesive strength and other functional properties (antimicrobial, UV resistant, self-healing, *etc.*) can be influenced by the type of amino acids present. In other words, the change in primary amines ($-NH_2$ of lysine), hydroxy (–OH), polyamides (PAs) ($-CO-NH_2$), and carboxyl (–COOH) functional groups can alter the adhesiveness

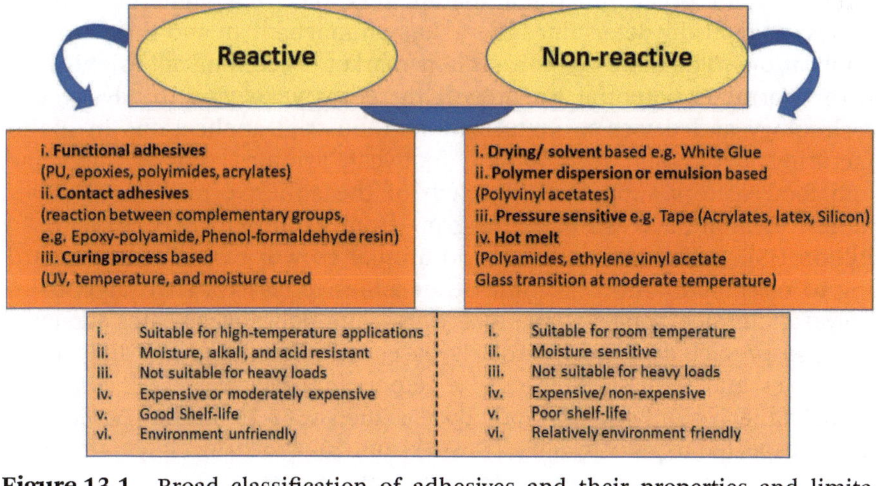

Figure 13.1 Broad classification of adhesives and their properties and limitations.

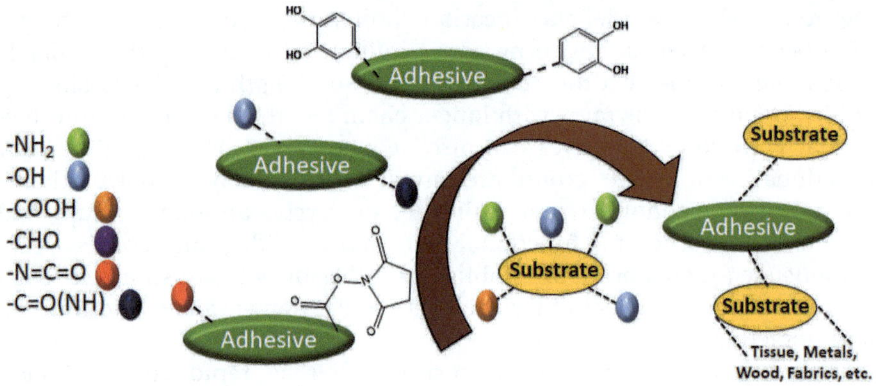

Figure 13.2 Prominent functional groups responsible for strong chemical linkages in adhesives.

of the material.[6,9,10] Figure 13.2 shows some functional units responsible for the material's adhesive, chemical, and functional properties. Adhesives are used in specific places to complement the functionality in substrates (wood, metal, plastics, paper, or concrete), environmental parameters (temperature, moisture, alkaline, or acidic), hydrophobic nature, and available functional/polar tethering groups.

13.2 Economic and Environmental Impact

Adhesives are formulated differently depending on the needs of the industry (such as automotive, aviation, or biomedical), environmental factors (such as temperature, pressure, or moisture), required adhesive strength, and other functional parameters. The raw material used in adhesives typically determines the adhesion mechanism and their range of applications. The adhesive and sealant market is economically significant, with enormous potential for growth in recent years due to advances in technology and device miniaturization. Industrial applications in manufacturing, packaging, automobiles, footwear, avionics, textiles, medicine, and food sectors support the growth of the adhesive market. PU-based adhesives alone are estimated to grow from 7 billion USD (2019) to 9.1 billion USD (2024) (5.6% compound annual growth rate [CAGR]), according to researchandmarkets.com. Other adhesive classes, such as epoxies, polyester amides, acrylates, cyanoacrylates, and polyvinyl acetates/acrylates, are also predicted to grow similarly based on demands. According to the Adhesives and Sealants Market Size Report (grandviewresearch.com), the global adhesive and sealant market was valued at 63.79 billion USD in 2021 and is expected to grow by 5.9% (CAGR) from 2022 to 2030.

Synthetic adhesives derived from petroleum are often used in heavy-duty industrial applications due to their mechanical strength, load-bearing ability, resistance to corrosive environments, and irreversible nature.[11] Most formaldehyde or heterocyclic adhesives are derived from petroleum by-products, which are commonly used in heavy-duty applications. The monomeric reactants required to synthesize adhesives have a low initial cost and chemical versatility.[12] The initial cost of synthesizing adhesives from crude petroleum oil *via* the glycerol route is lower than that of using glycerol from biodiesel manufacturing (0.20 USD kg^{-1} *vs.* 1.04 USD kg^{-1}), as the unit production cost of glycerol from vegetable oil (VO) is 28.98% higher than that from crude oil.[13–15] However, the rising cost of crude oil significantly increases the production cost of glycerol and adhesive prices. Apart from that, the environmental concerns are monumental and should be dealt with. The production of glycerol from VO would reduce greenhouse gas production by 34.5%.[13–15] Although crude oil-based adhesives are efficient for various practical applications, they are non-biocompatible, and their processing (reactive or non-reactive) through hot-melting, baking, multistep curing (two/three/multi-pack), and organic solvents is not environmentally friendly.[16]

During their formulation-processing cycle, the shelf life of adhesives and sealants can be improved by incorporating secondary additives and high/low boiling organic solvents, such as xylene, cyclohexane, acetone, and dimethyl sulfoxide. However, this process results in an increase in the concentration of toxic volatile organic components (VOCs) in adhesives, which mainly consist of benzene, bis-phenols, formaldehyde, other phenolic-rich compounds, propylene carbonates, and other similar chemicals in polymeric resins, which are known for inherent toxicity.[17,18] Figure 13.3 shows the common toxic molecules, solvents, and chemical units used in the adhesive industry during manufacturing. Despite their intrinsic toxicity, these adhesives are still constantly used due to the need for superior flexibility, mobility, tensile properties, fast curability, covalent tethering, and crosslinking, presence of functional long hydrocarbon chains, and lack of alternatives. Overall, crude oil-based adhesives can have direct and indirect harmful effects on human and animal respiratory systems (during production and being in a dump after renovations), constantly polluting the surrounding environment.[19–21] Environmentally friendly alternatives should phase out the use of these adhesives without compromising their physicomechanical or chemical properties.

The COVID-19 pandemic has put additional pressure on the adhesive industry, causing a shift in focus from sustainable and biodegradable adhesives to the packaging industry. Instead, non-biodegradable materials, such as acrylates, polyurethanes, 1,2-dibromo-2,4-dicyanobutane, epichlorohydrin, phthalates, isopropanol, and other solvents, have been increasingly used in the production of masks, medical tape, and other medical products due to their known chemistry and availability. However, these materials cause skin irritations, allergic dermatitis, and other diseases through long-term skin contact with items such as medical masks, hospital

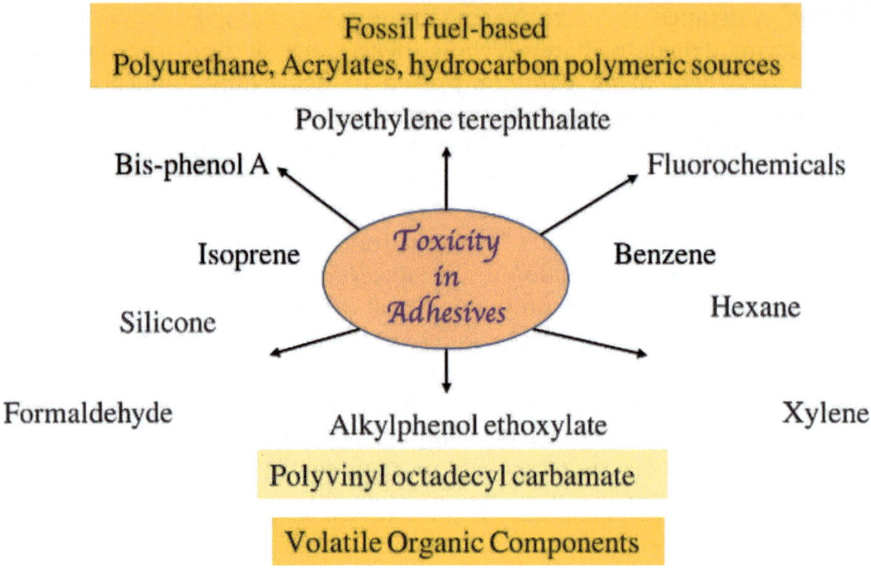

Figure 13.3 Potential pollutants that are responsible for the harmful effects of petroleum-based adhesives.

waste dumps, single-use items, and other similar products.[22,23] Across the globe, these wastes are generating potentially hazardous chemicals and fumes. When subjected to environmental stresses, such as UV exposure, temperature, pressure, and hydrolytic degradation, while on the ground, burial sites, and aquifers, they aggressively pollute the environment. The adhesives from the discarded products will continue to create problems until they are properly processed, and a significant shift toward degradable adhesive materials is achieved. If proper steps are not taken, the resulting environmental imbalance will impact agricultural land, food products, medical supplies, and other essential industrial sectors.[24]

13.3 Alternative and Natural Inspiration for Adhesives

Nature has guided us in our quest to find new and advanced materials. Many unique discoveries have been made possible by mimicking natural structures and compositions, such as spider webs, superhydrophobic lotus leaves, and adhesives from mussels, geckos, and other animals and insects.[25,26] Before the discovery of cement, mud was used as an adhesive to glue the walls and provide insulation against heat, cold, and water. Egg- and plant-based resins, such as gum arabic, gesso, chalk, gypsum, bee wax, honey, wood tar, starch, and other similar natural products, have previously been used in wood panels, canvas, and plaster.[27] These natural products are environmentally benign, biodegradable, and readily available. Even after fulfilling their purpose as adhesives, waste products generated during renovation could be recycled or naturally degraded. However, the need for materials with high mechanical and tensile strength and the discovery of rubbers and plastics revolutionized the adhesive industry. Subsequently, the push for sustainable growth and the adoption of circular economy principles have led to the reuse or modification of suitable biomolecules that can substitute or replace the hydrocarbon units found in petroleum by-product-based adhesives.[28,29] Incorporating biopolymers, either as they are or after undergoing chemical modifications, as substitutes for petroleum-based products in adhesives could strengthen the crosslinked microstructures. The use of sustainable monomers would aid in replacing organic solvents with environmentally friendly solvents, such as vegetable oils or water. A significant effort has been made to ensure a transition from solvent-borne to solvent-free adhesive systems, with a strong emphasis on waterborne adhesives and polymeric nanostructures.[30,31] Biopolymers or biomolecules are degradable and can assist in restoring a sustainable environment more effectively than non-degradable thermosets or other synthetic adhesives. However, the mechanical, physical, and chemical resistance properties of bio-based adhesives need improvement compared to the currently available robust petroleum-based alternatives.[32] Materials that can meet industrial requirements, such as strength and stability, and environmental considerations, such as degradability or recyclability, are the most desirable and sought-after. The current research focuses on a positive outlook toward social, economic, and environmental parameters by developing a three-dimensional, interpenetrating, and networked sustainable polymer-based matrix with bio-additives.[32,33] The processing of biomaterials from agricultural or animal waste/side products, such as seed husks, dry stems, leaves, extracted proteins, starch, cellulose, lignin, and others, is in high demand. For instance, converting seed oils into reactive subunits, such as epoxies, polyurethanes, ester amides, and acrylates, has resulted in the formation of biodegradable adhesive molecules/polymers.[34–36] However, long hydrocarbon chains, low degree of unsaturation, and limited functional groups in virgin seed oil account for the poor mechanical and physicochemical properties of bioadhesives. Smart chemical modification, extraction of reactive biomolecules, fabrication of inorganic–organic nanocomposites, and structural engineering can fabricate biopolymeric systems for broader practical applications in bioadhesives.[37]

In response to the demands of the times, the government, and green initiatives, various companies, such as NatureWorks, DaniMer Scientific, Cargill, Kraton Performance Polymers, Novamont, Innovia, FPI Innovation, Genencor, and others, are working on these issues while meeting the industrial requirements for adhesives. These companies have marketed several bio-based, biodegradable adhesives, such as Ingeo PLA, Nodax™

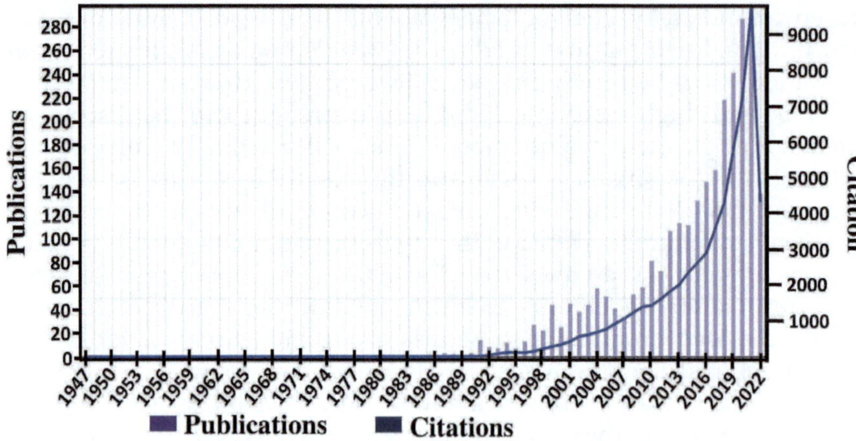

Figure 13.4 Research outputs with "green adhesives" as the keyword on the Web of Science.

PHA, Agri-Pure™ Biosolvent, Anova®, Rosin Ester Tackifier, and others. However, developing adhesives for heavy-duty industries that can completely replace petroleum-based synthetic adhesives requires more insight, time, and initiative. With the subsequent awareness of the possible alternatives, socioeconomic demands, and scientific innovations, industries are working toward greener sources, such as proteins, carbohydrates, starch, cellulose, polysaccharides, and vegetable oil-based polymers as raw materials, *i.e.*, adhesives with low-carbon footprints that can support the concept of the circular economy.[28,29,38] The exponential demand to develop new, flexible, functional, and environmentally friendly adhesives is evident from more than 2500 articles published online with the keyword "green adhesives" since 1999, with around 50 000 citations and a total H-index of 97 (Figure 13.4).

The rapid growth of the economy, technological advancements, miniaturization of devices, and specific requirements in biomedicine and the food industry constantly promote the need for more effective bio/green polymer-based adhesives. In the long run, research inputs and industrial activities while managing the supply chain demand will achieve the final objective of sustainable development in adhesives. The intended large-scale use of bioadhesives in the food, packaging, and biomedical industries is a significant development. There is an urgent need to fabricate bioadhesives with more extended stability, solubility, resistance to humidity, temperature, pH change, and controlled biodigestibility.[37,39] Utilizing proteins, lipids, carbohydrates, and other small green functional molecules, which can transform or contribute to the formulation of suitable polymeric matrices through simple chemical modifications, is ideal and rewarding. In this regard, biodegradable adhesive hydrogels have been developed by mimicking the secretion of geckos, mucus-inspired supramolecular adhesives, and other similar reports, strengthening the possibilities of exciting adhesive systems.[2,25] Many research groups have focused on biomolecules that can partially or fully replace synthetic/petrochemical-based hydrocarbons. Furthermore, developing adhesives from biomaterial/by-products or waste products would simultaneously support the reutilization as per the circular economy and enhance the market value of agricultural waste products, which would support farmers and other associated workers.

13.4 Brief Adhesive Mechanism

Various biomaterials have been explored for their abilities before and after suitable modifications to create a better adhesive profile.[26,40] Biomaterials, such as carbohydrates, starch, lipids, polyphenols, and various proteins, have been studied for applications.[41] However, protein- and lipid-based adhesives are important for applications in food and medicine due to their unique functionality, biodegradability, digestibility, availability, dispersibility, and resistance to gastric conditions and heat. Proteins are composed of amino acid units and have three-dimensional structures. Proteins and lipids may inherit a combination of functional groups, such as polyamide ($-CO-NH_2$), -COOH, -SH, -OH, epoxy, urethane, anhydride, and others. The prominent linkages include disulphide, glycoside, ionic, and hydrogen bonds. The moderate-to-long hydrocarbon chains can serve as the beneficial carbon backbone during the adhesive formulation. Briefly, the adhesive strength of protein-based adhesives could be influenced by protein denaturation (thermal, cold plasma, chemical, or enzymatic treatment), functional modifications of the protein/lipid subunits, blending with reactive polymers or other protein isolates (PIs), and branching with other biopolymers or crosslinking agents. The processing of adhesive substrate and curing temperature/technology used also influence the stability of the material. Figure 13.5 presents the formation of typical crosslinks in any adhesive material, which involves adhesion on the surface and cohesion within the body.

The mechanical properties, such as elongation, tensile strength, and adhesive strength, of adhesives are significantly affected by the presence of polar functionalities in the polymeric subunits. Mostly, the number of complementary groups that can react/interact with each other is responsible for developing an internal crosslinked structure (within the polymer), surface adherence (between the substrate and polymer), and chemical stability of adhesives. Hydrogen bonding is one of the most prominent linkages in adhesives, which is responsible for initiating or strengthening the crosslinks by bringing the functional unit closer together and supporting other linkages simultaneously. The method of curing or final processing of adhesives (hot or cold pressed) can significantly alter the strength and properties of adhesives. The processing technique could

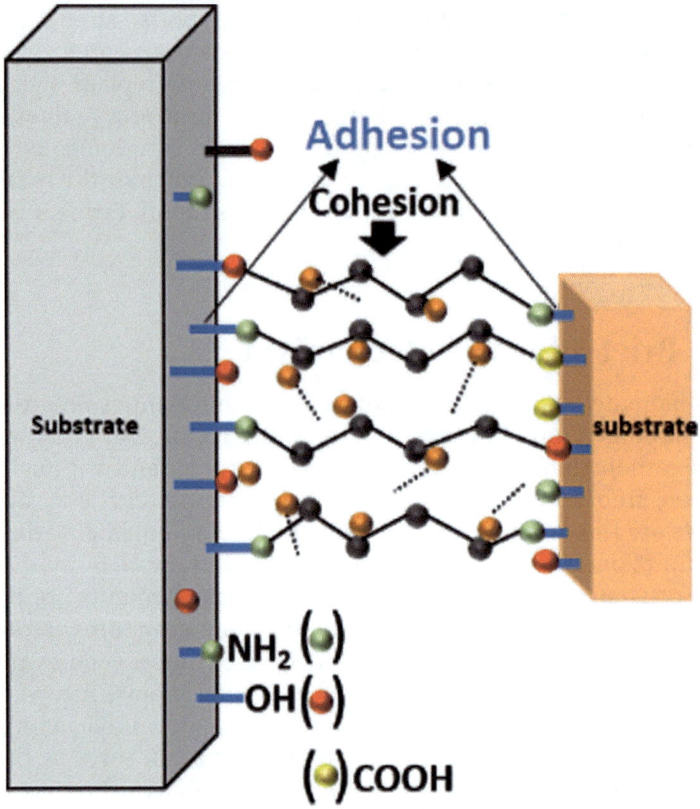

Figure 13.5 Schematic presentation of a crosslinked material with cohesive/adhesive interactions.

assist the chemical reactions in controlling the exposure of hydrophobic structures, solubility, and water-holding capacity. For example, sprayed soy protein (SP) showed higher viscosity [100 to 2200 Centipoise (cP)], hydrophobic content, and partial folding of the molecule around tryptophan over freeze-dried SP adhesives; however, no significant change in adhesive strength was observed.[42]

Apart from the adhesive strength, the chemistry involved in the diffusion of adhesive sol in the wood is also important. The diffusion or distribution of the polymeric structure depends on factors such as the state change (glass transition: rigid or glassy), free volume in the matrix, diffusant dimension, solubility, molecular mass, and motion.[43] The pre-surface treatment and the moisture content of substrates greatly influence the diffusion of adhesives. The processed or pre-treated substrates, such as metallic surfaces (inorganic), etched surfaces (treated with acid or laser), wood (organic), plastics, or hybrid surfaces, can have free hydroxy (–OH) or amine (–NH_2) or amide (–CO–NH_2) groups and grooves to provide anchoring spots on the substrate material. The tethering reaction between the complementary functional groups (–COOH, $-CO_3^{2-}$, $-NH_2$, and others) present in different systems provides additional adhesive ability and strength to the adhesive material. Various analytical techniques, such as X-ray computed tomography (XCT), X-ray fluorescence microscopy (XFM), and computational modeling, could be used to visualize the diffusion process.[44]

13.5 Protein-based Adhesives

Formulating adhesives using proteins, obtained from specific risk materials (proteinaceous waste such as bovine tissues) from slaughterhouses, agricultural by-products, and processing industry wastes, could be highly beneficial and environment-friendly on multiple fronts. These materials could be transformed into adhesives through crosslinking using polyamidoamine-epichlorohydrin, poly(ethylene imine), isocyanates, and reactive polymers, such as epoxy, polyurethane, or anhydride-containing groups.[45] However, this discussion mainly focuses on the proteins obtained from plant sources. Proteins can be extracted profitably from the plant (leaves, routes, stems, and seed husk) and animal (feathers, horns, and nails) by-products and can be used to formulate adhesives in combination with known amounts of starch, carbohydrates, cellulose, lipids, and other biomolecules. Selecting a specific protein for adhesive fabrication can produce a system with more predictable cohesive/adhesive force and physicochemical/mechanical properties. Protein molecules in native form (complexed with carbohydrate, fat, or other biomolecules) may lower the available functional groups and produce an unpredicted adhesive cocktail. Therefore, knowing the average molecular weight, concentration, and purity of protein isolates and other components is beneficial. Compatible copolymers or crosslinking agents are selected according to the requirements of a designed system. Gelatin, casein, keratin, albumin, and fibrin are a few of the proteins extensively used in the adhesive industry. Being biodegradable, they have also found use in the biomedical industry. However, soybean or soy-based adhesives have been prominent in developing formaldehyde-free adhesives. For a long time, they have been the most explored natural adhesive and used in various applications, such as wood, flooring, sealants, paper, coatings, and paint industries. Around 1930, SP- and starch-based adhesives were introduced as paper-coating adhesives.[46] Most of the SP used in the production of adhesives are extracted from the defatted soy flour (SF) or meal, which is a by-product of the vegetable oil industry (soybean oil from seeds). The obtained SP is partially denatured during the process and can be used directly in the adhesive industry without/with minute modifications. Other prominent sources that can be exploited in the production of proteins or active biomolecules for adhesive production are canola protein (CP), fava bean protein (FBP), corn zein protein (ZP), wheat gluten protein (WGP), potato, rice proteins, *etc.*

13.5.1 Alteration of Chemical Structures

Proteins individually may exhibit some degrees of adhesion to specific surfaces but may undergo degradation over time, limiting their applicability. However, modifications in their composition and variations in the functional groups have resulted in bioadhesives with improved properties.[47] Moreover, understanding the regeneration of functional units in protein-based adhesives can assist in the formation of reversible bioadhesives. Transformation of functional groups from active to inactive and *vice versa* through chemical/mechanical/electrical stimulation have been explored to produce innovative reversible biomaterials.[48] Alkali treatment is one of the most efficient and commonly used chemical treatments to improve the adhesive ability of soy protein isolates (SPIs). This treatment allows the possible modification of secondary, tertiary, or quaternary protein structures. Adding acid or salt to defatted SF leads to the denaturation of active functional groups, such as $-NH_2$, –COOH, and –OH, which lower the adhesive ability. However, alkali treatment tends to regenerate the amide groups responsible for crosslinking and water resistance of the bio-product.[49]

Similarly, a combination of alkali, enzymatic, and heat treatment can improve the water resistance of the modified SP adhesive.[50] The increase in pH (8 to 12), temperature (30 °C to 70 °C), and their combination (pH/temperature: 9.0/70 °C, 10.0/50 °C, 11.0/50 °C, and 12.0/40 °C) results in higher adhesive strength of the material. The probable change in surface charge, unfolding of protein structures, exposure of hydrophobic units, and generation of surface reactive groups are responsible for the beneficial properties.[50] These modifications and suitable composition of protein-based derivatives render the improvement that can fit well with the industrial standards in plywood and other industries.

The poor water resistance ability of PI-based adhesives can be improved by introducing hydrophobic groups, thereby enhancing the hydrophobicity of the composition and improving the crosslinking density of the material by the addition of reactants, such as glutaraldehyde (Glu).[51] The complexing agents (Glu, citric acid, anhydrides, epoxy, and PU) react with the available $-NH_2$ and –COOH groups, reducing the material's hydrophilic interactions. The modification of SP by polyethyleneimine (PEI), maleic anhydride (MA), sodium dodecyl sulphonate (SDS), sodium dodecylbenzene sulfonate (SDBS), and other similar additives can overcome the drawbacks of virgin protein adhesives. For example, grafting of maleic acid (10%) on SPI and crosslinking of the modified SPI (80%) using PEI (20%) formulate an adhesive with high dry shear strength and water resistance properties.[52] The interaction between the amine groups of PEI, maleyl esters, and C=C of the maleyl group, *via* the Michael addition reaction, results in a crosslinked structure. Another crosslinking agent, 1,3-dichloro-2-propanol (DCP), has been reported to produce high shear strength in wood adhesives. The adhesive strength was found to be proportional to the increase in the weight of DCP (optimum ratio of 6 : 1), reaction time (1 h), and temperature (85 °C).[53] Similarly, a 35–65% improvement in the water resistance and adhesive strength can be achieved simply by grafting hydrophobic units such as undecylenic acid on the SP chains through complexing functional amines with the carboxyl groups.[54]

Alkaline/enzyme hydrolysis, oxidation, and incorporation of nanoparticles or metallic ions also increase the system's hydrophobic characteristics, viscosity, and water resistance.[55–57] The degree of hydrolysis can determine the adhesive strength of the modified SPI. The literature reveals that a certain degree of hydrolysis can improve the adhesive strength of the materials. However, excessive hydrolysis would destroy the ability of the protein to adhere to the substrate firmly. For example, a hydrolysis time of 1 h is optimum for an SPI-based adhesive with a high adhesive strength of 700 N, while the increase in hydrolysis time (6 h) can reduce the overall ability (280 N).[57] The viscosity is another important parameter for adhesives, and the dispersibility of the material in water can be correlated to the adhesive strength of PI. Low viscous systems are known for easy processing, spreading, better surface tethering or interlocking, and dispersion abilities. In a modified SPI system, the highest adhesive ability was reported for the lowest viscous (70 cP) composition; however, an increase in the concentration of PI (1 to 2 mg cm^{-2}) significantly ($P < 0.05$) improves the adhesiveness of the material.[57] The reduced viscosity may aid the dispersion, but it does not contribute toward the adhesive strength, so a balance in the optimum viscosity and component concentration would be highly beneficial. The strength and viscosity of adhesives can be significantly influenced by altering the degree of disulphide linkages or salt bridges in the chemical structure. Disrupting these linkages using ionic substances, such as NaCl, Na_2SO_4, and Na_2SO_3, ethyl alcohol, and alkaloid reagents, such as tannic acid, thiols, and reducing agents, can disrupt the protein structures. A decrease in the intramolecular H-bonding, ionic interaction, and Vander wall forces typically results in lowered viscosity and adhesive ability. Using 0.1 M NaCl or Na_2SO_4 (28% lower disulphide linkages) reduces the viscosity from >30 000 cP to 6000 or 1050 cP, respectively. The downward trend continues with a subsequent increase in the concentration of salts (0.1 to 0.2 M).[58] However, the impact of salts is limited and can only significantly reduce the viscosity up to a level. In a typical comparison of adhesive strength in wheat gluten and SPI, the use of gluten results in better adhesive abilities and tensile strength; however, ionic salt concentration and particle size of gluten do not influence the adhesive strength significantly.[59] In comparison, introducing free –SH groups in the SPI improves adhesive properties.[60] Introducing free –SH groups by the following reactions can enable the PI to form more disulphide linkages and improve adhesive strength.

1. Acetylation of cysteamine (–SH group, S-acetylcysteamine, 82% acetylation),
2. Conjugation of free –COOH groups in SPI with the –NH group of S-acetylcysteamine in the presence of 1-ethyl-3-(3-dimethylaminopropyl)carbodiimide (EDC), and
3. Alkali (NaOH) treatment to remove the acetyl group (generating free –SH group).

Free –SH or –OH groups (using ethanolamine instead of cysteamine) can be introduced by adding different concentrations of EDC. However, the results revealed that introducing cysteamine (1.42–2.09%) was more efficient in improving the dry shear strength and water resistance of SPI adhesives. Free –SH groups play a significant role in enhancing the adhesion ability up to a certain concentration. Oxidation of PI could increase the formation of disulphide linkages, which supports the formation of a three-dimensional structure and/or covalent interaction of –SH with tyrosine moieties (quinones at pH 10) *via* the Michael addition reaction.[60–62] Oxidizing agents, such as periodate, permanganate, iodate, chlorate, perchlorate, bromate, nitric acid, and others, are also reported in the literature. The oxidants produce more reactive subgroups by oxidizing the functional groups or the unsaturated hydrocarbons in plant-based proteins, carbohydrates, lipids, and phenolics. The treatment of SF with periodate and permanganate improves the wet adhesive strength; interestingly, the extracted SP after similar modification shows relatively better properties.[63] The mechanism includes converting adjacent hydroxyl groups into dialdehydes on treatment with periodate. This is followed by a reaction with amino acids containing amines, such as lysine, to form Schiff bases and secondary amines. The stable chemical structures are mostly responsible for the material's higher adhesion strength (2.89 MPa).[63] Chemical phosphorylation using $POCl_3$ influences the wet bond strength of protein-based adhesives. Incorporating the phosphate group in the protein and carbohydrate chain by treating the SF with $POCl_3$ (SF : $POCl_3$ = 0.15 g g^{-1}) enhances the adhesive strength (2.6 MPa). The phosphate group enables better chemical tethering of the SF to the wood surface *via*, esterification reactions with surface –OH groups and ionic/hydrogen bonding reactions.[64]

Many researchers have also focused on the modification of secondary structures of proteins through chemical and other stimulations to improve the adhesive nature of protein-based materials, particularly the changes that generate more polar functional groups and chemically stable structures. The disruption of the quaternary structures of polypeptides in the proteins can also enhance the surface area of the material and the number of available functional groups, *i.e.*, it supports stronger adhesive strength. However, higher polar groups may increase the hydrophilicity and lower the water resistance of the adhesive. Figure 13.6 presents a schematic route to modify SPI that alters the chemical structure systematically, lowers the molecular weight, and influences the overall adhesive strength of the protein isolates.[65] However, the optimum degree of quaternary structures is also important as they are partially responsible for ascertaining the mechanical strength of these adhesives.[66] Therefore, maintaining an optimum quaternary structure with suitable modification of secondary structures and exposure of more hydrophobic subunits and sufficient polar groups would render a functional material with improved adhesion and mechanical stability. Introducing phenolic compounds obtained naturally could improve the hydrophobicity, toughness, chemical resistivity, mechanical strength, and adhesive strength. The phenolic molecules, such as catechol, juglone, quercetin, caffeic acid, gallic acid, and a few others, could trigger crosslinking of the protein chains.[67,68] For example, zein protein extracted from the corn was crosslinked using a small amount of tannic acid at 120 °C for 3 to 24 h.[69] The final product was a super glue-like adhesive for different substrates, including metals and plastics. Figure 13.7 shows a pictorial diagram of the crosslinked structure of a typical PI using tannic acid as the model phenolic additive. The adhesive

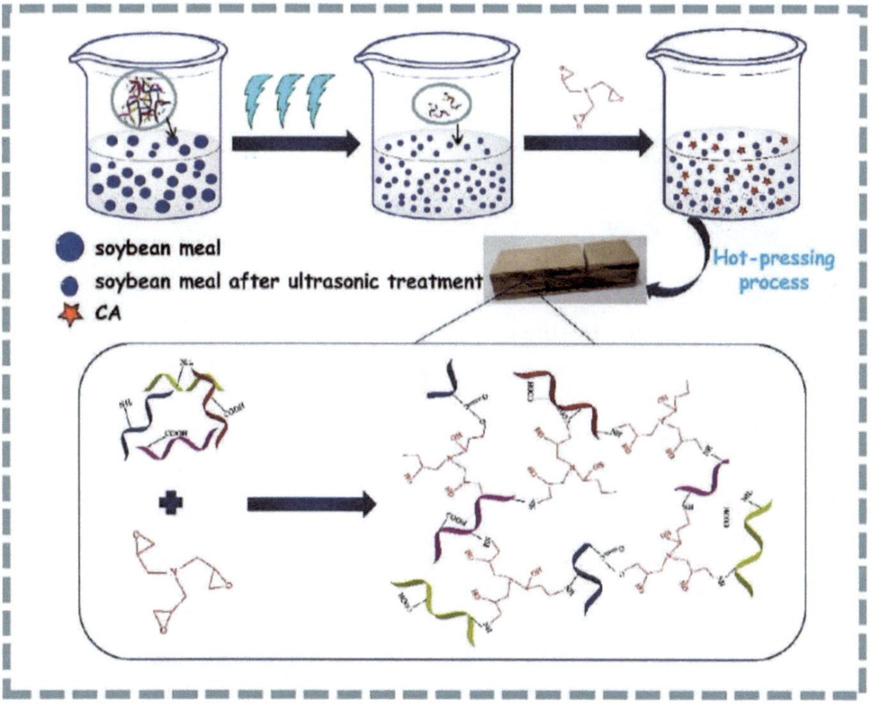

Figure 13.6 Representation of the chemical modification of soy proteins that breaks the quaternary structures and reduces the molecular weight and particle size of the adhesive molecules. Reproduced from ref. 65 with permission from Elsevier, Copyright 2020.

strength of proteins can be influenced by crosslinking using metallic salts (mono/bimetallic/trimetallic) or chemical crosslinking through transglutaminase activity.

The environment or curing technique used to process the adhesive material greatly impacts the overall properties. Higher solubility, uniform dispersion, and optimum crosslinking in the adhesive material would result in sufficient strength. Spray-drying of proteins improves the material's molecular interactions, solubility, surface hydrophobicity, and foaming stability. The enhancement of the molecular interaction was associated with the cleavage of disulphide linkages, unfolding of the protein chains due to weakening or disrupting the intramolecular interactions, and better interaction with the polar water molecules.[70] Furthermore, using protein dispersions in water or a suitable solvent system is more effective in producing an effective adhesive material over dry powder of proteins, even with the exact moisture content.[71] Most synthetic adhesives are processed using the hot-pressing technique, which is interesting because it can improve the viscosity, dispersibility, and fast curing of the material, thus altering the adhesion process. However, protein adhesives are negatively impacted by longer or higher heating conditions, as this would denature

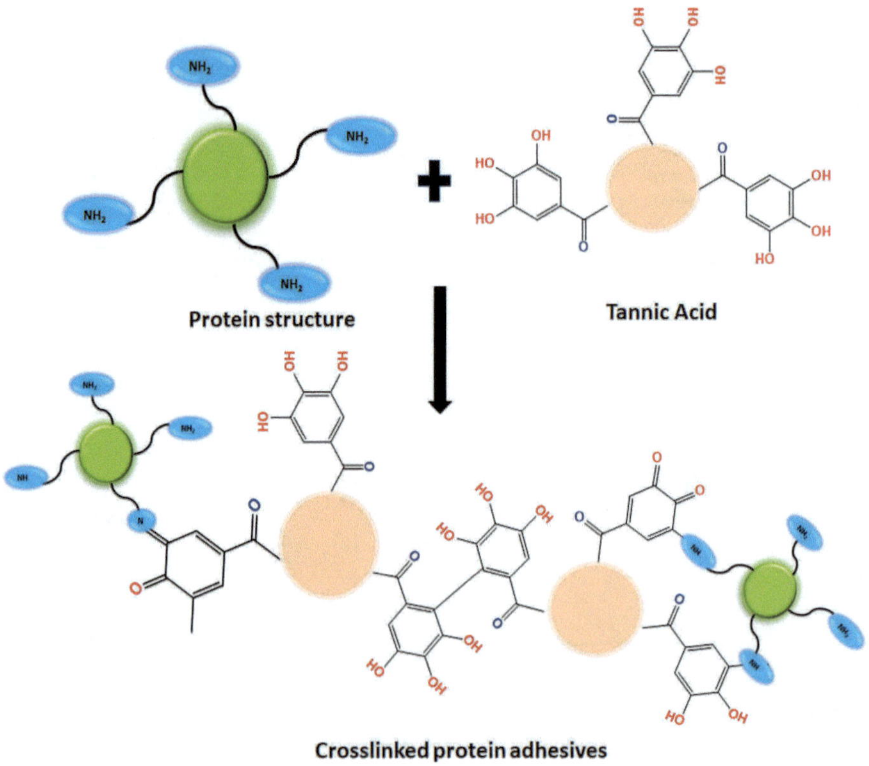

Figure 13.7 Crosslinking between the –NH_2 and –COOH groups.

most of the proteins and deteriorate the physicomechanical and chemical properties of the adhesive system. The system's pH may or may not influence the adhesive properties of the proteins.[50] Higher pH levels can influence the surface charges and hydrophobicity of the PI but may not affect the adhesive strength of the material. This may be due to the small change in the hydrophobic nature of the material resulting from its already exposed or denatured state. The water resistance of the adhesive largely depends on the exposure of large hydrophobic units in the proteins and surface functionalities that interact with the substrate material, such as wood or paper surface. A balanced structure, with sufficient exposure of hydrophobic units while simultaneously having reactive surface groups, would result in a superior moisture-resistant adhesive material.

13.5.2 Enzymatic Treatment

Enzymatic activity can improve the properties of protein-based adhesives, such as stability, water resistance, and adhesive strength. Usually, an alkaline solution of protein isolates is formulated for adhesive applications. However, alkali treatment increases the number of polar groups, which can reduce the water resistance properties of the material. On the other hand, enzymatic treatment can hydrolyze protein isolates and produce better adhesives after a certain degree of hydrolysis. Enzymes, such as bromelain, trypsin, papain, chymotrypsin, pepsin, urease, and others, break down proteins into smaller units, possibly by acting on the polypeptide ternary structures in the protein isolates. The extent of hydrolysis depends on the concentration and time of enzymatic action, and the resulting peptide chains generate new secondary structures that influence the viscosity, residual rate, functional subunits, thermal properties, and fracture strength of the adhesives.[72]

However, not all enzymes act in the same way. Some may improve the hydrophobic characteristics, while others may increase viscosity or decrease the adhesive ability of protein molecules.[73] The amount of enzyme used and its activity duration also affect the adhesive's fate. Enzymatic action on protein isolates or peptide-based adhesives can convert peptide chains into amino acids, such as L-tyrosine and L-lysine, into 3,4-dihydroxyphenylalanine (DOPA), or form co-polypeptide units. Researchers have used this knowledge to mimic natural adhesives and generate similar peptide units, creating green, sustainable adhesive materials.[74] However, the economic aspect of using enzymes and peptide units must be considered. More exploration of better, more economical solutions through the hybridization of different effects is necessary to ensure the sustainability of these modifications.

13.5.3 Biomimetic Modifications

An extensive analysis of natural adhesives in plants and animals has led to a significant understanding of the active molecules responsible for their unique adhesion abilities. The role of proteins in these natural adhesives has been explored and exploited to develop biomimetic adhesives. The biomolecules responsible for strong attachment to underwater rocks and sea surfaces are also stable under turbulent water and other harsh environments (temperature variation, water pressure, saline environment, *etc.*). The average force between 250 N and 300 N per mussel is required to remove a California mussel (*Mytilus californianus*).[75,76] Therefore, the insight into the functionalities, available reactive groups, and the possible mechanism is supposed to inspire the development of synthetic but biomimetic adhesives using bioresources. The analysis of proteins in the byssal thread and adhesive plaques reveals the presence of around six proteins with different concentrations and molecular weights, depending on the species. They mostly have basic isoelectronic points and contain cationic amino acid residues along with catechol subunits of DOPA, chelating peptide chains, and various amino acids with the possibility to form disulphide bonds. A different grouping of functional units would affect the adhesive strength and stability of the adhesive system. Table 13.1 shows a few such functional units and their roles in biomimetic adhesive systems. Extracting appropriate biomolecules from wastes/by-products and chemically converting them into natural bioactive molecules with structures/molecules, such as natural adhesives, have resulted in the fabrication of biomimetic adhesives. The chemical transformation of SPI to incorporate natural marine mussel-like functionality improves the adhesive strength, even in wet environments.[26] The tethering of DOPA groups within SPI results in better adhesion. The concentration of DOPA molecules, peptide units, such as lysine and tyrosine, and molecules with several –OH, –COOH, $-NH_2$, and –SH units have shown the ability to influence the adhesive strength of the modified protein adhesives. Superior adhesive and water resistance abilities have been developed using natural proteins compared to some synthetic adhesive molecules. A series of gelatin–DOPA conjugates with different degrees of substitution levels have been prepared and studied to verify the impact of DOPA substitution. The adhesive strength of these modified proteins is impacted by the degree of substitution.[77]

These biomimetic adhesives are primarily used for tissue engineering, nerve sealing, healing, and sealing bleeding tissues and could replace sutures.[78] Various techniques, such as spin coating, electrospinning, and solution casting, can be used to develop three-dimensional structures using protein conjugates. A few examples of biomimetic adhesives are shown in Table 13.2.

Table 13.1 A few of the functional units present in biomimetic protein-based adhesives and their influence on the adhesive properties.

Functional unit	Structure	Properties
Catechol side group in DOPA	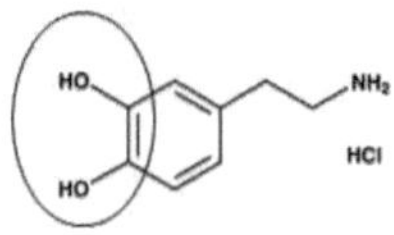	Strong adhesion, hydrophilic compatibility, curing, and interfacial binding
Cysteine residues	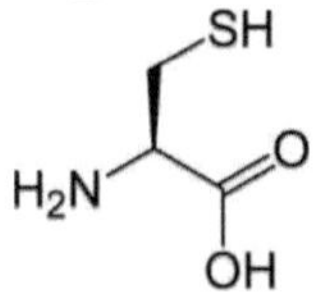	Adhesive and cohesive forces, disulphide bonds, crosslinking, and integrity of adhesive lining
Histidine residues	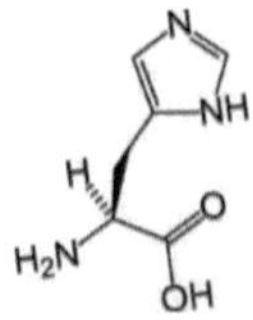	Coordinated bonds with transition metal ions and curing of the liquid form of adhesives
4-Hydroxyarginine	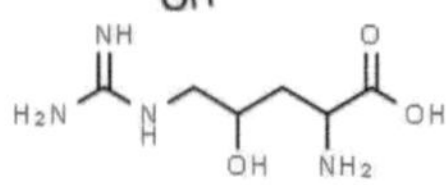	Hydrophilic compatibility
Poly-lysine	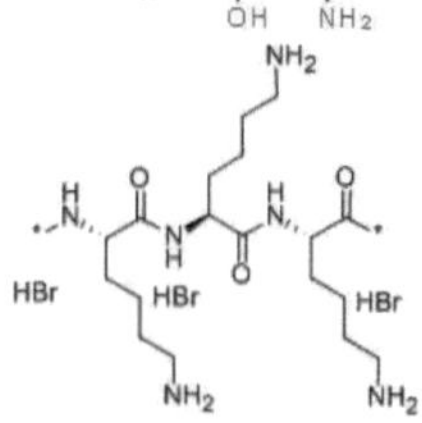	Water solubility, degradability, biocompatibility, and stimuli-responsiveness
Hydrogen bonding	---H----X-H (X = N, O, F)	Adhesive and cohesive strength and hydrophilic compatibility
Pi–Pi interactions	—	Cohesive interactions

13.5.4 Nano-reinforced Proteins as Adhesives

Many industries have been influenced significantly by nanotechnology. The change in nano-regime or the interaction with the nanomaterial influences the material's chemical, biological, and physical aspects, thereby the properties and applications. Nanochemistry has played a significant role in enhancing the adhesive properties of protein-based materials. The nanoparticles, nanofibers, nanosheets, and other metal/metal oxide/polymeric nanostructures have a higher surface-area-to-charge ratio (aspect ratio). As a result, they can either react strongly with the dispersion medium (protein or polymeric units) or act as reinforcing agents/nanofillers for the final adhesive material. They help create additional crosslinking, tethering, mechanical stability, and resistance toward chemical/corrosive ions. In the case of protein-based adhesives, the nanoparticles can form crosslinks with the polypeptide units through covalent or noncovalent

Table 13.2 Examples of biomimetic adhesives developed after certain modifications in the chemical structure of the proteins.

Biomimicking	Active adhesive compositions and mechanism	Properties	Ref.
Mussel foot proteins	Poly[(3,4-dihydroxystyrene)-*co*-styrene] along with pendant catechol groups (*i.e.*, 3,4-dihydroxystyrene)	Adhesion obtained was similar to that of cyanoacrylate "krazy glue"	79
Sandcastle worm	Coacervates of polyelectrolytes triggered by a change in the pH or temperature	Water-borne underwater adhesives	80
Sandcastle worm	Copolymer poly(*N*-isopropyl acrylamide)-poly(acrylic acid)-poly(2-methoxyethyl acrylate) and sodium aspartate; metal ion (Ca^{2+}/Mg^{2+}/Cu^{2+}) chelation effect and temperature-strengthened dehydration effect; and ion coordination-induced coacervation mechanism	Wet adhesives, suitable for substrates, including glass, Si, PTFE, and animal bone, with no oxidation problem due to no DOPA chemistry	81
Mussel foot proteins	DOPA derivatives, polyethylene glycol, and polycaprolactone	Soft tissue engineering; adhesion was better than fibrin glue	82
Insect cuticles	Mineral–organic system and phenol–amine synergy; gallic acid, hydroxyapatite, Ca^{2+}, and polylysine	Mildew-resistant antibacterial soy protein derivatives	83
Catechol chemistry	Soybean meal, tannic acid, and polyethylenimine; natural polyphenol-inspired chemistry	High wet bonding strength for plywood, better network crosslinking density, and thermal stability	84
Lobster cuticular sclerotization	Catechol-rich condensed tannin-functionalized boron nitride nanosheets, chitosan, and soy meal; phenol–amine synergy along with reinforcement with dispersed boron nitride nanosheets	Water resistant, with better mechanical strength, and high cohesive strength	85
Catechol and isocyanate chemistry	Soy protein, isophorone diisocyanate, poly-(tetrahydrofuran), with DOPA and sodium bisulphite units; dual-crosslinking network; and hot pressing technique	Waterborne polyurethane-based soy protein elastomers as wood adhesives; thermally stimulating	86

(*continued*)

Table 13.2 *(continued)*

Biomimicking	Active adhesive compositions and mechanism	Properties	Ref.
Catechol chemistry	Corn-based zein protein and plant-based phenolics, including catechol, tannic acid, juglone, caffeic acid, quercetin, and gallic acid	High bond strength, improved water resistance, and non-toxic biomedical applications	68
Mussel foot proteins	Catecholic and polyelectrolyte system; coacervation-induced adhesion; polyamidoamine-epichlorohydrin, tannic acid, and pyrogallol	Underwater adhesive gel, reusable adhesiveness, and higher strength in basic conditions	87

interactions, developing cohesive strength and stiffness. Formaldehyde or other toxic chemicals used for curing commercial adhesives can be replaced by functionalized nanoparticles.[85] The surface functionalization of nanoparticles, using bioactive molecules, such as DOPA, can impart value-added properties to the nanocomposite adhesive material. The dispersion of biocompatible nanoparticles in protein-based adhesive materials can significantly improve several properties, including curing time and temperature, mechanical strength, antibacterial activity, and UV, moisture, and temperature resistance.[88–90] The nanoparticles, such as SiO_2, ferrites, graphene oxide, clay, calcium phosphate, and others, have been known to influence the desirable properties of protein-based adhesives. The selection of appropriate nanoparticles depends on the intended application (dentistry, sealant, and tissue engineering) of the adhesives.[91–93] Incorporating nanoparticles can readily modify the adhesive properties, making them suitable for specific applications.[94–96] Overall, the strategic utilization of nanotechnology in adhesives has the potential to revolutionize the industry further.

13.5.5 Protein-based Composite and Nanocomposite Adhesives

Protein-based adhesives are green and more environment-friendly, but resorcinol, phenol–resorcinol, melamine–urea–formaldehyde, PU, and other petrochemical-based synthetic adhesives have better physicomechanical and chemical properties. However, suitable modifications of proteins can produce adhesives with comparable or better functional properties. One of the most sought-after methods is to blend them with synthetic or established adhesive formulations in the market. As the main objective is to shift the dependency on petroleum oil derivatives in adhesive industries, such modification can replace 20% to 50% of petroleum derivatives while maintaining the commercial requirements of adhesives. Although not a permanent remediation of petroleum derivatives, it can make a significant difference in supplying relatively greener alternatives and maintaining the commercial requirements.

Crosslinking large hydrocarbon chains of proteins in SPI with diglycidyl ether of bisphenol A (DGEBA) and PEI results in a system (120 min of pot-life) that meets the industrial standards of wood adhesives.[97] The order of reaction among the respective components is crucial in finalizing the adhesive properties. Due to competing reactions, adding DGEBA and PEI to SPI results in a system with zero or very low concentration of SPI. However, pre-mixing DGEBA with SPI, followed by the addition of PEI, can incorporate most of the protein into the backbone of the adhesive.[97] Mixing PI with Kymene 557H, commonly used in wood composites, results in a composite material with similar or even higher shear adhesive strength than commercial phenol–formaldehyde resins.[98] The key is to ascertain the compatibility, percentage of PI, and possibility of crosslinking within the system. In some instances, the lower viscosity of commercial adhesives restricts the processing and handling of the final material. Therefore, additives, such as triglycidylamine, can improve the internal crosslinking in PI, ensuring the better performance of the adhesive. Incorporating molecules with low molecular weight results in zero leaching and homogeneous phase formation while increasing the viscosity, residual rate, polar functional groups, thermal stability, and fracture strength of the adhesives.

Moreover, adding sugar, small amounts of metal ions, and other complexing agents can further enhance the adhesion of composite protein-based adhesives.[99] An additional approach to developing environment-friendly adhesives is to formulate dispersion of waterborne polyurethanes (WPU) in SP or SPI, which can enhance the adhesiveness of the composite by 65%.[100] A smaller but uniform distribution of WPU ensures a stronger interaction, higher viscosity, and bond strength. Although these adhesives have great potential in the wood industry, the use of diisocyanate during the preparation of PU limits the biomedical applications. Among other small molecules, adding lignin can significantly improve the adhesive and mechanical properties of the material. Introducing laccase/3,3,5,5-tetramethyl-1-pyrroline-*N*-oxide (TEMPO)-modified lignin in SP improved the wet shear strength by 106% (1.429 MPa).[100] The improvement is associated with the oxidation of kraft lignin by the laccase enzyme, which produces more reactive functional groups on phenolic and non-phenolic components. It strengthens the interaction with the SPI while lowering the spreading ability due to the high viscosity of the final product.

The cost and dependence on SP or SF in developing adhesives for industrial applications can be reduced by formulating composite structures by mixing low-cost proteins. A mixture of sorghum, canola, and SP in the ratio of 3 : 2 : 5 has been reported to produce a composite with high dry adhesion (62%), wet adhesion (103%), and soaked adhesion (63%) strength. The well-distributed molecular structure, higher viscosity, and hydrophobicity of the composite protein system were responsible for the improved adhesive strength.[101] Apart from SP, Camelina protein (biodiesel residue) can generate bioadhesives by modifying kraft lignin and its oxidation using hydrogen peroxide, assisted by ultrasound treatment to lower the particle size. The interaction of lignin with the protein improves due to increased hydrophobicity and functionalization of lignin particles, but the wet shear strength remains similar to the control lignin–SP-based adhesive (1.43 Mpa).[102] The modification of lignin with pH can result in unfolding, denaturization, and refolding of the protein structures, enhancing the interaction between the lignin–protein structure (pH 8.5 to 4.5), high wet adhesion, resistance to water (40%), and thermal stability.[103] Triticale has significantly more protein among the grains and can be extracted using standard procedures with minor modifications. The extracted protein from the grains can be modified using alkali or urea and crosslinked using glutaraldehyde. The modifications increased the α-helical conformation and molecular weight, improving adhesive strength. As per the report, the protein extracted *via* the acetic acid route showed the best adhesive performance.[104]

Canola, the second most produced oilseed crop after soy, can be another alternative to petrochemical-based adhesives.[105] Along with the chemical modifications of the CP, the exfoliation of nanoparticles significantly improves the adhesive ability of the resulting material. The exfoliation of graphite oxide, bentonite clay, and nanocrystalline cellulose and surface modification of the nanoparticles can significantly improve the adhesive properties of CP. For instance, the addition of graphite oxide and nanocrystalline cellulose boosts the dry/wet adhesive strength (11.67 and 4.85 MPa) of CP. Although the dispersion of nano-clay in the CP also elevates the adhesion strength, it is not comparable to graphite oxide and nanocrystalline cellulose. The poor dispersibility, agglomeration, and a higher concentration (>5%) can deteriorate the adhesion strength of most of the composite materials.[106,107] However, uniform and proper exfoliation of nanoparticles can produce sufficient tensile strength and foster crosslinking of proteins. A pilot-scale study demonstrated the potential of CP as a substitute for commercial petrochemical-based liquid phenol–formaldehyde (PF) in producing oriented strand board (OSB). The CP was chemically modified with ammonium persulfate (APS), followed by the exfoliation of graphite oxide promoted the modified CP's effectiveness in preparing OSB. Even though 100% CP adhesive can be used to prepare OSB, a 40% substitution of PF was found to be the optimum composition, where the OSB produced with 40% canola protein adhesive showed a comparable strength of 100% PF adhesive.[108] The improved strength could be associated with the improvement in the fluidity of the overall resin, uniform in-depth dispersion of the adhesive, and increased or similar crosslinking (cohesive and adhesive) ability. A pictorial diagram presenting the development of CP adhesive and nanocomposite board for wood application has been shown in Figure 13.8. Such formaldehyde-free/low-formaldehyde protein-based adhesives are highly sustainable and green and can generate zero waste systems. The resulting composite material shows better mechanical, physicochemical, and adhesive strength.

Adding other functional biomolecules, such as lipids, modified lipids, lipid nanoparticles, and other polymeric structures, can also enhance the adhesive strength of the proteins. Using lipids as an additive in protein-based adhesives improves the material's overall performance. In some instances, lipids are also used as a solvent for the adhesive precursor, reducing the amount of solvent required during the processing of adhesives. The utilization of lipids is determined by the presence of specific functional groups or the possibility of generating similar functionalities, such as polyester, polyhydroxy, and others.[109]

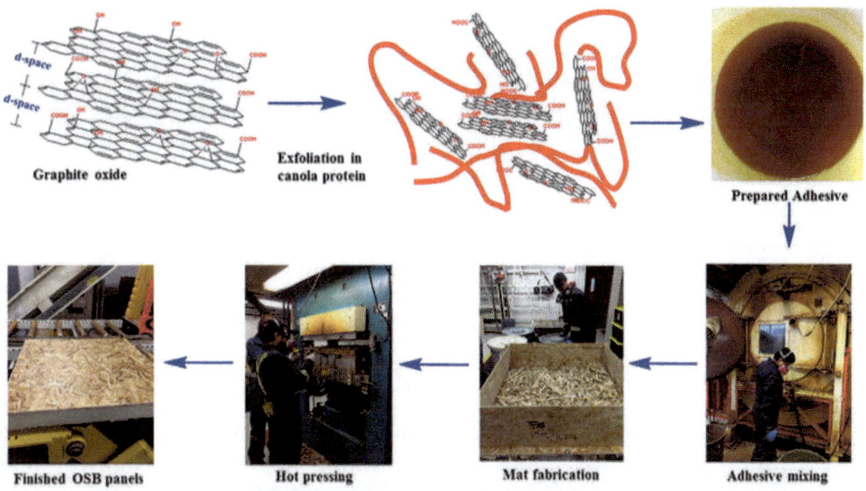

Figure 13.8 The pilot-scale development process of nanocomposite boards for wood application using a mixture of PF and engineered CP adhesives. Reproduced from ref. 108 with permission from the American Chemical Society, Copyright 2018.

13.6 Lipid-based Adhesives

Lipids are highly functional and can be another biomaterial source to fabricate bioadhesives. Lipids are in abundance around us and have numerous valuable properties. The lipid-based adhesive can accommodate commercial adhesives and can potentially substitute petroleum by-products. Long hydrocarbon chains, unsaturated carbons, available functional groups, and high decomposition temperatures compared to proteins make them highly suitable reactive precursors for fabricating bioadhesives. Lipids are sustainable and economical biopolymers that can replace toxic precursors in commercial adhesives. Lipid components, such as oleic acid (OA), linoleic acid, linolenic acid, lecithin, oils from plants/seeds (flaxseed, soybeans, canola seed, and others), and typical triglyceride units, are readily used in the development of adhesives. For example, urushiol lacquer, a naturally obtained oleoresin containing 60–70% of the catechol lipid of urushiol, was modified by incorporating Ag nanoparticles to develop a conducting adhesive paste with massive potential in electronic materials.[110] Chemical modifications of generic lipids to generate functional groups, such as epoxy, urethane, esteramides, acrylates, hydroxy, amide, and others, make them valuable in the adhesive industry.[111–113] Although bioinspired resins/adhesives are fabricated with certain chemicals (oxidants, reducing agents, and catalysts) or organic solvents, the final product is mostly compatible and highly sustainable. Furthermore, new technological advances, such as developing waterborne resins or using precise chemical units (atom economy) and microfluidizers, can reduce the use of chemicals and organic solvents. Nano-enabling of lipid molecules improves the mechanical and adhesive strength of the material. For instance, the formulation of lipid-based nanotubes from a specific custom lipid molecule, AQua ($C_{25}H_{29}NO_4$), has shown efficient adhesion between two poly-dimethylacrylamide (PDMA) hydrogels.[114] The developed lipid structure can be used for adhesion and as a source for sustainable drug release. The vast potential of lipid-based adhesives could further be understood through discussions in the following subsections.

13.6.1 Chemical Modification of Lipids

The long hydrocarbon chain and the available unsaturated carbon and/or other functional groups make lipids suitable for several chemical reactions. However, for the sustainable development of the adhesive industry, it is essential to focus on modifications that use green chemistry principles, facile routes, and other environment-friendly approaches. Modifying lipids, such as tung oil (TO), with diols (with different chain lengths) and maleic anhydride, has been used to produce pressure-sensitive adhesives (PSAs). The adhesives showed high shear (>168 h) and peel (0.3–4.5 Ncm^{-1}) strength, loop track (1–13 N), and good aging resistance.[115] The formulation utilizes diols, such as propylene glycol (PG), di(ethylene glycol) (DEG), poly(ethylene glycol) (PEG), poly(propylene glycol) (PPG), and PEG-block-PPG-block-PEG, in different concentrations. The mechanism involves the modification of the lipid using the Diels Alder reaction, followed by an esterification reaction in the presence of 1,8-diazabicyclo[5.4.0]undec-7-ene to form viscous branched polyesters.[115] Long-chain diols should be preferred in formulating such adhesives, as they produce stronger adhesives than short-chain diols. However, the increase in molecular weight could negatively impact the adhesive strength and should be optimized accordingly. Similarly, renewable methyl oleates (MOs) were developed from partially hydrolyzed epoxidized OA and polyfunctional epoxides (epoxidized soybean oil [ESO] or trimethylolpropane triglycidyl ether) in the presence of different catalysts (dimethylbenzylamine or *N*-methylmorpholine or others).[116] Epoxidation, partial oxidation of double bonds to produce -OH groups, and successive reactions between the carboxyl–hydroxyl groups in OA *via* step-growth polymerization reaction resulted in an active polymeric structure, and conversion into PU resin transforms it into a PSA.[117] The conversion was assisted by chromium(III) tris(acetylacetonate) (CTAA) and dibutyltin dilaurate (DBTDL) catalysts, and polymeric methylene diphenyl diisocyanate (MDPI) was used as the copolymer to formulate the PU. The OA-based green PSA showed good adhesive abilities; however, an alternate reaction that can avoid using isocyanates and heavy metal catalysts would make the process more environmentally friendly. If the major component is biopolymers/bioprecursors (≥80%), petroleum-based reactants, such as DGEBA and PEI, can be justified in any formulation. When DGEBA is added to 80% soy epoxy, it provides chain rigidity and acts as a chain extender and connector. The bis-phenol (hydrophobic) and reactive epoxy units enable the preparation of moisture-resistant, mechanically stable adhesives, whereas PEI (with primary and secondary amino units), known for the extensive crosslinking ability, ensures the formation of a crosslinked product.[111,118] The remote or stimuli-responsive adhesive is another critical class; they can undergo crosslinking as per the demand or directions set by the required chemical reaction. For example, polymerization reaction using UV triggers, without photoactivation by irradiation, will limit the reaction, thus controlling the crosslinking reaction. However, upon initiation, the reaction is completed within a short time frame. Castor oil-based PU can be rapidly formed through UV irradiation to initiate the release of protected cadaverine in a 2-nitrobenzyl-conjugated system.[119] These reactions are highly beneficial in systems where post-curing is preferred over pre-cured adhesives. Therefore, specific reactions that can partially oxidize or generate reactive functional groups in the lipid precursor can generate a series of adhesive formulations.

13.6.2 Epoxidized Adhesives

Epoxy is one of the most versatile, reactive, and widely exploited functional structures that can be converted into the –OH group by simply increasing the temperature. The highly reactive epoxy group enables the chemical reactions at the epoxy carbon. Mostly, molecules with –OH, $-NH_2$, –COOH, and other electronegative groups, under acidic/basic conditions, can undergo free-radical polymerization or nucleophilic addition of primary/secondary amines. The curing mechanism of epoxy adhesives is ring opening, conversion into –OH, and polycondensation or crosslinking reaction (Figure 13.9). The adhesive nature and inherent moisture resistance ability of epoxies are also explored in fabricating polymeric films or coatings, which can be used in surface protection, packaging, and food storage applications.[111,113,120] However, the long hydrocarbon chain and flexibility of lipid-based epoxies prevent them from becoming standalone adhesive systems. Generally, they are mixed with commercial epoxies, such as DGEBA, to regain the required stiffness and mechanical strength. Such epoxy composites partially replace the use of petroleum products and result in the formation of green oleo epoxy adhesives. The adhesive ability and mechanical stability are further enhanced by co-reactants or copolymers (crosslinking agents) containing amide/polyamide/polyhydroxy units. Permanent adhesives are obtained through ring opening, cohesion, surface adherence, and intermolecular crosslinking. Epoxies are excellent glass adhesives with superior mechanical and transparency.

Lipid epoxies can have tuneable glass transition temperatures that ensure better dispersion and adhesion. A simple mixture of epoxidized

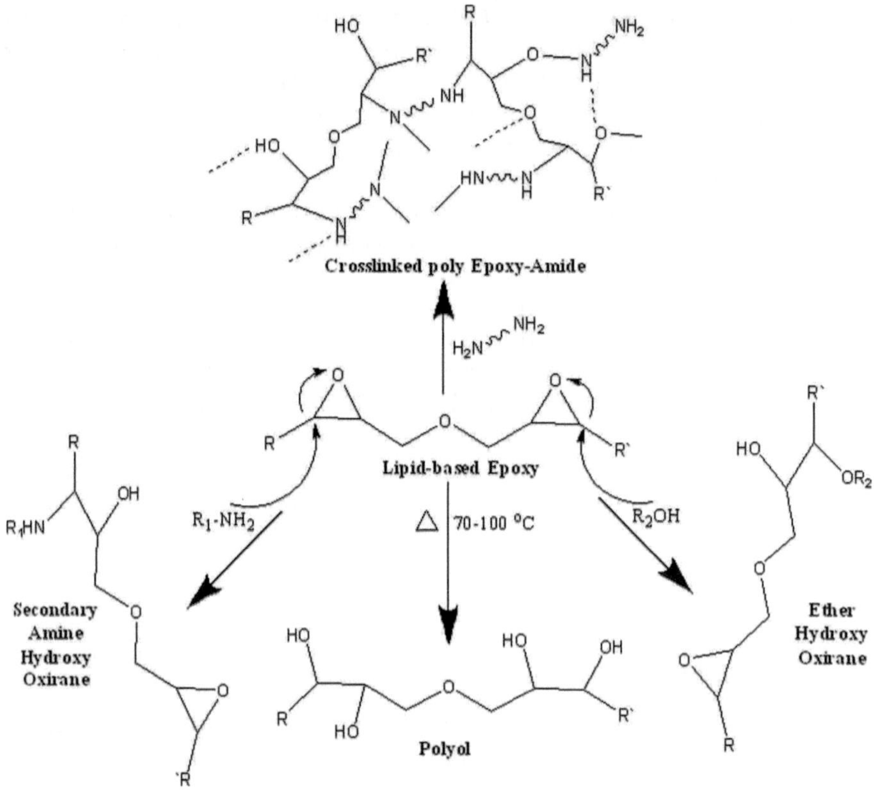

Figure 13.9 Reaction between lipid epoxy molecule and polyamide crosslinker.

cardanol, diglycidyl ether of vanillin, and triglycidyl ether of phloroglucinol, followed by dicyandiamide, can result in bio-based epoxy adhesives.[121] Similarly, *n*-alkyl esters of diphenolic acid (DGEDP) were first modified by TEMPO-oxidized cellulose nanocrystals and cured using **1** (bis-(furfurylamine)).[122] Depending on the curing and degree of epoxidations, they can be used as either lipid-based PSA or permanently cured adhesives. Thermosets for PSA application have been developed by self-condensation/substituting epoxidized oleic acid, which utilizes the available epoxy, unsaturated carbon, and carboxyl groups.[123] Incorporating cyclic siloxanes, nanoparticles, phenolic groups, and complexing agents helps improve the mechanical strength, stability, and adhesive ability of the lipid epoxy.[120] Indeed, addressing the drawbacks of poor shelf life, crack generation tendencies, temperature, and UV sensitivity of epoxy adhesives can improve the material's overall performance.

13.6.3 PU-based Adhesives

PU resins have the most extensive market presence among adhesives. They are two-pack systems known for their high strength and long shelf life. The moisture resistance ability of PU is phenomenal compared to other lipid-based resins. They contain soft (lipid) and hard (–NCO and isocyanate group) segments in the final PU structure. The reaction between –OH groups in lipid-based reactants (polyols, glycerol, PEG, esteramides, and –COOH) with –NCO-containing molecules (toluene diisocyanate [TDI], 4,4-methylenebis(cyclohexyl isocyanate) [HDMI], isophorone diisocyanate [IPDI], *etc.*) results in the formation of PU resin.[124,125] This exothermic process can be achieved at low or room temperature in the presence or absence of catalysts such as dibutyltin dilaurate. Mixing lipid and –NCO-containing molecules **112** can develop PU resin quickly. However, the optimum concentration of –NCO, the type of –NCO (aliphatic or aromatic), and the ratio of –NCO with respect to the –OH groups must be determined. Excess –NCO may lead to the formation of a brittle structure, whereas lower concentration may not produce adhesives or mechanical stability. Vegetable oils, natural lipids, and unsaturated hydrocarbons could be used as starting precursors; however, using isocyanate as such may have environmental impacts. Although the complete consumption of the –NCO group (100% reaction) in any system lowers the associated toxicity, environmental demands suggest that isocyanate should be avoided. PU is restricted in the biomedical or food industry due to the inherent toxic nature of isocyanates. However, replacing isocyanates in PU adhesives is hard due to their high mechanical, thermal, and chemical resistance ability. Better technological facilities and a thrust toward green technology have promoted the exclusion of isocyanates. Alternatively, cyclic carbonate, lignin, and tannin acid assist in developing non-isocyanate PU adhesives.[126,127] Waterborne PU has also received enormous attention, and several reports can be seen in the literature and market.[127] Complexing agents, such as cyanuric chloride, also have huge potential, as they can replace TDI or HDMI while maintaining physicomechanical properties.[128] The dispersion of nanoparticles, clays, fibers, and functional copolymers[129,130] can further improve the stability, strength, and durability of PU resin/adhesive.

13.6.4 Esteramide-based Adhesives

Polyester-amides are the prime reactant in hot-melt adhesives. They are primarily synthesized by reacting acids (acid dimers/acid chlorides/acid anhydrides) with hydroxy amide. They contain outstanding strength, low-temperature flexibility, and chemical stability. Ethylene and vinyl acetate (EVA) copolymers, polyethylene (PE), amorphous polypropylene (PP), block copolymers (styrene butadiene rubber [SBR]), and PAs are a few examples of reactive polymeric units used in hot-melt adhesives.[131] The good adhesive strength and compatibility with other hot polymeric adhesive resins make them suitable bio-additives or diluents. The reactive polymeric units are solid at low or room temperature but melt at high temperatures (120–180 °C). Traditionally, such adhesives do not undergo crosslinking; however, they can be chemically crosslinked at high temperatures to attain high adhesion stability. Studies have been reported to understand the variation in melt viscosity, softening point, glass transition temperature, tensile strength, elongation, and T-peel strength of lipid-based polyester-amides.[131–133]

Additionally, no solvent is required for the application, and it can be beneficial over solvent-borne adhesives or even water-borne adhesives. Quick set time, biodegradable nature, and good thermal stability give them an edge over other adhesives. Seed oil, soy protein derivatives, starch esters, and functional biomolecules can be utilized to formulate green polyester amides. They have shown comparable strength to commercial hot-melt adhesives.[134] With the help of dimer acid and ethylenediamine, modified jatropha seed oil, castor oil, fatty acid components, and other seed oil derivatives have been converted into hot-melt adhesives. The reported polyester amides have better low-temperature flexibility, adhesion, and melting temperatures.[131–133] Including metal ions and secondary curing agents, such as poly(styrene-co-maleic anhydride) and isocyanates, can facilitate the development of room-temperature cured polyester-amides from seed oil and other lipid derivatives.[135–138] Fabrication of hyperbranched polyester-amide and modifications, including DOPA or functional biomimetic groups, can enhance the adhesiveness of polymers and the potential of the lipid-based resin.[139] The conversion of lipid-based polyester amides into films for packaging and coatings makes them beneficial for the biomedical/food industry.[130]

13.7 Blended Adhesives

Petroleum-based epoxies, being robust and mechanically stable, are preferred for most industrial applications. However, with the recent demand for green alternatives and a better understanding of biomaterials, replacing petroleum-based adhesives through blending with biopolymers and other reactive lipid- or protein-based adhesives has been extensively explored. Blending helps in substitution while maintaining the mechanical properties and lowers the impact of non-degradable adhesives, creating better and renewable alternatives that would be available in the market. The most common drawback of blending is the untimely separation due to incompatibility among the resins or differences in the stability of the polymeric subunits. Suitable modification derivatization or chemical branching of biopolymers can reduce the separation issue; it improves the overall properties of the adhesive system. For example, mixing lignin with synthetic epoxy reduces the mechanical properties and lowers the crosslinking density; however, the epoxidation of lignin before mixing enhances the crosslinking density, impact, and adhesive strength of the blend.[140] Table 13.3 presents a few such blended adhesives with improved properties for different applications. The blended systems show beneficial properties, particularly reduced viscosity, enabling easy processing and intermixing, improved crosslinking, and enhanced hydrophobic content, resulting in improved wet adhesion strength and moisture resistance.

Among different bio-additives, lipids play an essential synergistic role in improving adhesive properties. Lipid- and protein-based conjugated adhesives are a prominent class of adhesives among blended systems. The presence of lipids and proteins in certain proportions improves the adhesive strength, wet adhesion, moisture resistance, and biocompatibility of materials. The combination of proteins and lipids results in an equilibrium system with different functionalities, with one being polar and the other being lipophilic.[149] Lipid components, such as oleic, linoleic, linolenic acid, and other derivatives with long hydrocarbon chains, can improve the hydrophobic content and branching/interaction of adhesives. The viscosity of the final product is important as it determines the adhesive's uniform and in-depth dispersion over the substrate. For example, the wet adhesion strength of soy protein (3.2 MPa) was improved by adding 3.5% 2-octen-1-ylsuccinic anhydride, which is primarily attributed to the reactive nature of the anhydride functional group supporting the crosslinking with the protein. The hydrophobic oily nature of the lipid contributes toward the water resistance.[150] The substitution of amine groups in soy glycinin protein by hydrophobic units, such as $-CH_3$ (6%, methylation reaction), ensures higher adhesive strength over the $-COOH$ group (8%, succinylation) at pH 7.6.[151] Nature has provided several examples of such combination, and exploration of these natural adhesives provides huge insight into the molecular mechanism of the adhesives. A better understanding of the chemistry of natural adhesives helps design unique bioadhesives.

The inspiration to develop lipid- and protein-based bioadhesives can be drawn from natural resources. Thoracian barnacles can maximize adhesion on various surfaces under hostile environments. The adhesion is achieved through a biphasic system involving lipids and phosphoproteins in different granules. The secreted lipids create a conducive environment, simultaneously modulating the spreading of phosphoproteins and protecting the nascent adhesive plaque.[149] The presence of lipid derivatives (lipid esters) in the mussel byssal thread adhesive interface has also been confirmed by attenuated Total Reflectance Fourier-Transform Infrared (ATR-FTIR).[152] Similarly, terrestrial arthropod adhesive secretions confirm the presence of lipid and their derivatives along with the protein phase.[153,154] Starfish, similar to other echinoderms, adhere to the surface *via* protein-based glue (20.6%), lipids (5.6%), and carbohydrates.[155] All the natural occurrences of lipid–protein combinations provide directions during the formulation of effective bio-based adhesives. The goal is to

Table 13.3 Examples of blended adhesives.

Final product	Biomaterial	Blended material	Impacted properties	Ref.
Bio-phenol–formaldehyde adhesive	Bio-crude derived from white birch bark and white spruce bark	50% and 25% phenol–formaldehyde adhesive	Low curing temperature and higher thermal stability when formulated at lower temperatures and optimum concentrations; free formaldehyde emission level decreases; and tension shear strength	141
Wood adhesive	Soy protein	Polyacrylate emulsions (methyl methacrylate, butyl acrylate, and acrylic acid)	Improved shear strength and thermal stability	142
Bio-based wood adhesive	Soy protein adhesives	Commercial synthetic latex adhesives	Improved water resistance and lower viscosity	143
Bio-phenol–formaldehyde adhesives for plywood	Pyrolyzed pinewood oil	Phenol–formaldehyde resins	Improved bonding performance	144
Coating application	Castor oil	Bisphenol A diglycidyl ether and polyamide	Improved thermal, physicomechanical, and adhesion properties	145
Photo-crosslinkable tissue adhesives	Castor oil	Polycaprolactone with 2-isocyanatoethyl methacrylate	Better adhesion capacity, lower water sorption capacity, and increased hydrophobicity	146
Hybrid adhesives and sealants	Molecularly optimized lignin	Modified silicone (MS) sealants	Improved thermal- and UV-blocking properties	147
Bio-wood adhesive	Soy protein isolate	Soy-oil-based waterborne polyurethane	Improve wet strength	100
Flexible electrospun scaffolds	Genetically redesigned fusion mussel adhesive proteins	Polycaprolactone, polydioxanone, poly(L-lactide), poly(DL-lactide-co-glycolide), polyethylene oxide, and polyvinyl alcohol	Tissue engineering; better cell adhesion and basic adhesive function of mussel protein enables better surface bioactive functionalization	148

achieve higher adhesive strength, water-resisting ability, improved physico-mechanical properties, and ability to maintain industrial standards.

13.8 Application of Protein- and Lipid-based Bioadhesives

Many natural adhesives have strong adhesive strength, particularly in wet environments. It is beneficial to develop such adhesives from bio-components, as they are effective in different environments. Mimicking natural compositions would be possible, and the product can be used in various applications. Particularly in the field of biomedical and surgical applications, biodegradable and moisture-resistant adhesives have immense importance. Biomimetic hydrogels are formulated for their thermal gelation and tissue adhesion abilities. The mussel chemistry inspires the hydrogel development; they benefit from the strong adhesive action of the DOPA molecule and its crosslinking ability under oxidizing conditions. The phospholipid-based liposomes, filled with an oxidizing agent ($NaIO_4$) and with a bilayer melting transition temperature of 37 °C (physiological pH), release loaded cargo in the presence of DOPA or DOPA-functionalized PEG polymer. The release of periodate led to the oxidation of DOPA and *in-situ* crosslinking of the polymer, resulting in a biomimetic hydrogel system for tissue adhesion.[156]

In addition to the adhesive properties, the presence of lipids with glycoprotein-regulatory nucleic acid results in the development of suitable extracellular matrixes. The adhesive ability of the lipid- and protein-based systems has also been used to formulate drug-delivery vehicles, hydrogels, micelles, liposomes, and films.[157] Grafting of an anticancer drug (paclitaxel)-loaded solid lipid nanoparticles with protein (wheat germ agglutinin lectin) using EDC/*N*-hydroxysuccinimide (NHS) reaction resulted in an enhanced internalization, biodistribution, targetability, and anticancer efficacy.[158] Monodispersed lipid–protein nanoglobules obtained from the slime of *E. rowelli* have been converted into fibers with protein-rich core and lipid-enriched coating. Interestingly, these mechanically drawn fibers could be reversibly formed depending on the condition, thereby proving the idea for developing recyclable adhesive polymers or fibers.[159] Furthermore, membranes, 3D cell culture units, degradable boards, and foams can be prepared from the biowaste products containing proteins, amino acids, lipids, carbohydrates, and others after a few chemical modifications.[160]

Gelatin has been used in numerous research and commercial platforms, from drug-delivery vehicles to adhesive materials. The fast gelation ability of gelatin has been exploited to develop fast-curing hydrogels using a chemically crosslinked hydroxyphenyl propionic acid–gelatin/thiolated gelatin (GH/GS) system.[161] The stability and enhanced adhesive strength could be attributed to the disulphide and phenol–phenol interactions, whereas fibroblast cell compatibility promotes the potential of such adhesives in biomedical applications. In a similar study, injectable *in-situ* gelatin hydrogel-based adhesives were formulated through dual enzymatic (horseradish peroxidase [HRP] and tyrosinase) crosslinking of the polymeric chain after internal chemical modification. The enzymatic conversion of phenol groups into *o*-quinones, phenol tyrosine linkages, and their chemical tethering with amines and thiols on the tissue surface make them a better bioadhesive. The gelation time decreases with the concentration of enzymes, and the adhesive strength (34 kPa) was primarily improved over the commercial fibrin glue (7 kPa).[162] An elastin-like polypeptide was developed using *Escherichia coli*, mimicking the natural peptide-based adhesives. The high tyrosine-rich structure was converted into DOPA, one of the main components to follow the adhesion principles of marine organisms, such as mussels and sandcastle worms. The polypeptide adhesives (mELY16) provide good adhesive strength (~240 kPa) in humid conditions, significantly higher than that of the fibrin sealants.[163]

Furthermore, in gelatin-based hydrogel adhesives, the use of gelatin-methacryloyl (GelMa) has been reported, showing huge potential in biomedical applications.[164–166] In nerve regeneration, the use of peptide-incorporated (Arg–Gly–Asp–Ser [RGDS], Lys–Arg–Asp–Ser [KRDS], and Gly–His–Lys [GHK]) collagen prosthesis has proved the benefit of amino acid-based materials, similar to proteins.[167] However, for biomedical applications, the fouling of proteins may lead to the failure of the structure, and hence, for biomedical applications, various strategies could be followed to avoid such fouling incidences.[168] The solubility of silk fibroin and its biomedical application were enhanced by incorporating catechol functionality (DOPA) into the moiety after adding polyethylene glycol (6%, PEG). Interestingly, the PEG–silk fibroin conjugates were able to form a β-sheet secondary structure that gives the required mechanical stability to the chemical structure and controls the swelling. The successful proliferation of human mesenchymal stem cells (hMSCs) makes this protein adhesive compatible with biological applications.[169]

13.9 Outlook and Scope of Protein–Lipid Biomolecules as Adhesives

The increasing global temperature, pollution, and cost of fossil fuels have created a need for greener alternatives. However, achieving a complete shift from petroleum-based adhesive materials to bio-based alternatives with similar or superior properties will require more effort. Despite this, many alternatives, such as Ingeo PLA, Nodax™ PHA, Agri-Pure™ Biosolvent, Anova®, and Rosin Ester Tackifier, are being actively used in the market, and techno-giants are constantly working toward further advancements for the generalized application of bioadhesives.

Designing bioadhesives is another hot avenue of interest, particularly in mimicking natural adhesives, such as adhesives from mussels and geckos and resins from other insects, and following designs such as spider webs and lotus leaves. This has provided possible combinations and variations for developing better alternative adhesive materials. An optimum composition containing easily obtained proteins, lipid fractions, carbohydrates, and others would benefit society. Regarding the source of materials, using crops or sources not utilized in major food industries would be preferred, for example utilizing chicken feathers for keratin, canola meal for the extraction of canola protein (which has limited utilization due to the presence of allergenic compounds), and other similar alternatives.

Lipid- and protein-based adhesive materials could be used in various applications and have potential in numerous industries. They can be largely used in the food industry (packaging), biomedical industry (drug delivery and wound-healing materials), and construction industry (polyurethane foams). Figure 13.10 presents some areas where such adhesive materials would be greatly valued. Overall, the conjugation of lipids and proteins in the same system improves biocompatibility, aqueous solubility, and targeting or adhesive abilities. However, extensive research is still required

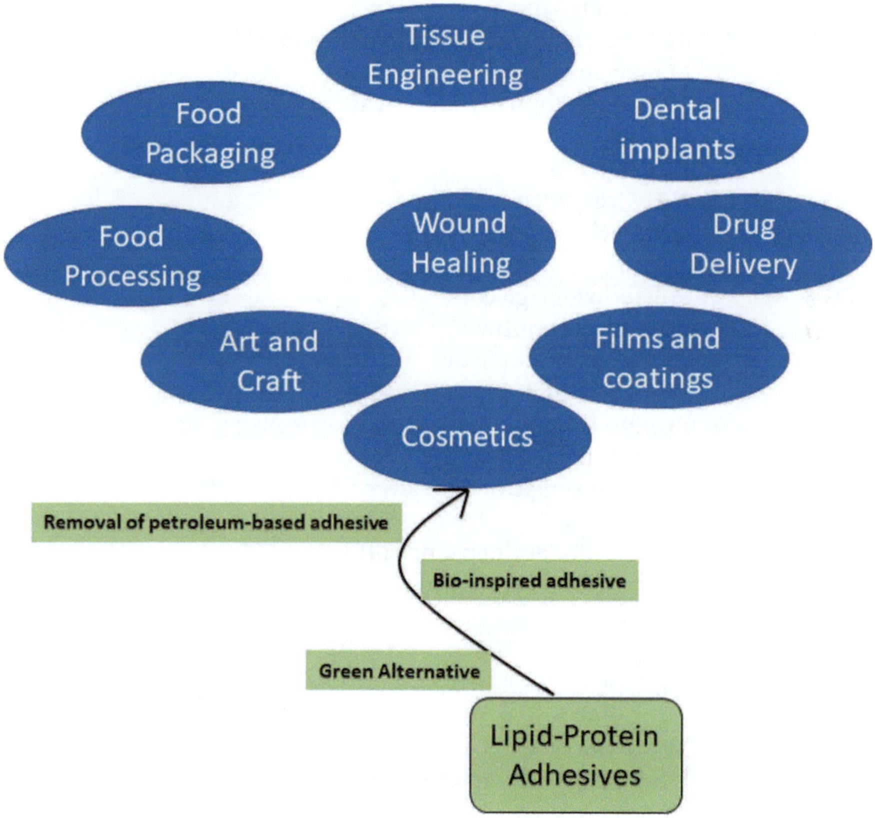

Figure 13.10 Scope and application of lipid- and protein-based adhesives.

to improve the mechanical and physicochemical strength of such adhesives by incorporating nanofibers, particles, metal ions, and other natural or bio-derived additives.

Abbreviations

APS	Ammonium persulfate
ATR-FTIR	Attenuated total reflectance Fourier-transform infrared
CAGR	Compound annual growth rate
–CO–NH	Polyamide
–COOH	Carboxyl
CP	Canola protein
DCP	13-Dichloro-2-propanol
DEG	Di(ethylene glycol)
DGEBA	Diglycedyl ether of bisphenol A
DGEDP	Epoxy *n*-alkyl esters of diphenolic acid
EDC	1-Ethyl-3-(3-dimethylaminopropyl)carbodiimide
ESO	Epoxidized soybean oil
Glu	Glutaraldehyde
HDMI	44-Methylenebis(cyclohexyl isocyanate)
MA	Maleic anhydride
MDPI	Methylene diphenyl diisocyanate
MO	Methyl oleates
$–NH_2$	Primary amine
NHS	*N*-Hydroxysuccinimide
OA	Oleic acid
–OH	Hydroxy
PEG	Poly(ethylene glycol)
PEI	Polyethyleneimine
PF	Phenol formaldehyde
PG	Propylene glycol
PI	Protein isolate
PPG	Poly(propylene glycol)
PSA	Pressure-sensitive adhesives
PU	Polyurethane
SDBS	Sodium dodecylbenzene sulfonate
SDS	Sodium dodecyl sulphonates
SF	Soy flour
SP	Soy protein
TDI	Toluene diisocyanate
TO	Tung oil
VO	Vegetable oil
VOCs	Volatile organic components

References

1. J. Glazer, Monolayer studies of some ethoxylin resin adhesives and related compounds, *J. Polym. Sci.*, 1954, **13**, 355.
2. J. Sun, *et al.*, Fabrication and Mechanical Properties of Engineered Protein-Based Adhesives and Fibers, *Adv. Mater.*, 2020, **32**, 1906360.
3. M. Tsige, C. D. Lorenz and M. J. Stevens, Role of network connectivity on the mechanical properties of highly cross-linked polymers, *Macromolecules*, 2004, **37**, 8466.
4. F. H. Gojny, M. H. G. Wichmann, U. Köpke, B. Fiedler and K. Schulte, Carbon nanotube-reinforced epoxy-composites: enhanced stiffness and fracture toughness at low nanotube content, *Compos. Sci. Technol.*, 2004, **64**, 2363.
5. H. Dodiuk, S. Kenig, I. Blinsky, A. Dotan and A. Buchman, Nanotailoring of epoxy adhesives by polyhedral-oligomeric-sil-sesquioxanes (POSS), *Int. J. Adhes. Adhes.*, 2005, **25**, 211.
6. R. Pinnaratip, M. S. A. Bhuiyan, K. Meyers, R. M. Rajachar and B. P. Lee, Multifunctional Biomedical Adhesives, *Adv. Healthcare Mater.*, 2019, **8**, 1801568.
7. M. Malik and R. Kaur, Influence of aliphatic and aromatic isocyanates on the properties of poly(ether ester) polyol based PU adhesive system, *Polym. Eng. Sci.*, 2018, **58**, 112.
8. A. Das and P. Mahanwar, A brief discussion on advances in polyurethane applications, *Adv. Ind. Eng. Polym. Res.*, 2020, **3**, 93.
9. A. P. Duarte, J. F. Coelho, J. C. Bordado, M. T. Cidade and M. H. Gil, Surgical adhesives: Systematic review of the main types and development forecast, *Prog. Polym. Sci.*, 2012, **37**, 1031.
10. Y. Deng, *et al.*, A sunblock based on bioadhesive nanoparticles, *Nat. Mater.*, 2015, **14**, 1278.
11. D. F. Farrar, Bone adhesives for trauma surgery: A review of challenges and developments, *Int. J. Adhes. Adhes.*, 2012, **33**, 89.
12. R. Kumar, V. Choudhary, S. Mishra, I. K. Varma and B. Mattiason, Adhesives and plastics based on soy protein products, *Ind. Crops Prod.*, 2002, **16**, 155.
13. M. Yang and K. A. Rosentrater, Techno-economic analysis of the production process of structural bio-adhesive derived from glycerol, *J. Cleaner Prod.*, 2019, **228**, 388.
14. L. Landress, *ICIS pricing, glycerine (US gulf)*, ICIS, 2014.
15. J. Sheehan, V. Camobreco, J. Duffield, M. Graboski and H. Shapouri, *Life cycle inventory of biodiesel and petroleum diesel for use in an urban bus. Final report*, 1998.
16. A. Arias, S. González-García, S. González-Rodríguez, G. Feijoo and M. T. Moreira, Cradle-to-gate Life Cycle Assessment of bio-adhesives for the wood panel industry. A comparison with petrochemical alternatives, *Sci. Total Environ.*, 2020, **738**, 140357.

17. W. Jiang, A. Kumar and S. Adamopoulos, Liquefaction of lignocellulosic materials and its applications in wood adhesives—A review, *Ind. Crops Prod.*, 2018, **124**, DOI: 10.1016/j.indcrop.2018.07.053.
18. A. Pizzi, Wood products and green chemistry, *Ann. For. Sci.*, 2016, **73**, DOI: 10.1007/s13595-014-0448-3.
19. H. Guo, F. Murray and S. Wilkinson, Evaluation of total volatile organic compound emissions from adhesives based on chamber tests, *J. Air Waste Manage. Assoc.*, 2000, **50**, 199.
20. J. González-Martín, N. J. R. Kraakman, C. Pérez, R. Lebrero and R. Muñoz, A state-of-the-art review on indoor air pollution and strategies for indoor air pollution control, *Chemosphere*, 2021, **262**, DOI: 10.1016/j.chemosphere.2020.128376.
21. S. Jin, *et al.*, Phytic acid-assisted fabrication for soybean meal/nanofiber composite adhesive via bioinspired chelation reinforcement strategy,*J. Hazard. Mater.*, 2020, **399**, 123064.
22. H. C. Lee and C. L. Goh, Occupational dermatoses from Personal Protective Equipment during the COVID-19 pandemic in the tropics - A Review, *J. Eur. Acad. Dermatol. Venereol.*, 2021, **35**, DOI: 10.1111/jdv.16925.
23. P. E. Enterline, Importance of sequential exposure in the production of epichlorohydrin and isopropanol, *Ann. N. Y. Acad. Sci.*, 1982, **381**, 344.
24. F. D. B. de Sousa, Plastic and its consequences during the COVID-19 pandemic, *Environ. Sci. Pollut. Res.*, 2021, **28**, 46067.
25. P. M. Favi, S. Yi, S. C. Lenaghan, L. Xia and M. Zhang, Inspiration from the natural world: from bio-adhesives to bio-inspired adhesives, *J. Adhes. Sci. Technol.*, 2014, **28**, 290.
26. N. Bandara, H. Zeng and J. Wu, Marine mussel adhesion: Biochemistry, mechanisms, and biomimetics, *J. Adhes. Sci. Technol.*, 2013, **27**, 2139.
27. R. D. Adams, in *Adhesive Bonding: Science, Technology and Applications*, 2021, https://doi.org/10.1016/B978-0-12-819954-1.09990-1.
28. L. A. Heinrich, Future opportunities for bio-based adhesives-aadvantages beyond renewability, *Green Chem.*, 2019, **21**, DOI: 10.1039/c8gc03746a.
29. S. Dahiya, R. Katakojwala, S. Ramakrishna and S. V. Mohan, Biobased Products and Life Cycle Assessment in the Context of Circular Economy and Sustainability, *Materials Circular Economy*, 2020, **2**, 1.
30. H. Honarkar, Waterborne polyurethanes: A review, *J. Dispersion Sci. Technol.*, 2018, **39**, DOI: 10.1080/01932691.2017.1327818.
31. F. Alarcia, J. C. de la Cal and J. M. Asua, Continuous production of specialty waterborne adhesives: Tuning the adhesive performance, *Chem. Eng. J.*, 2006, **122**, 117.

32. K. Zheng, Q. Gu, D. Zhou, M. Zhou and L. Zhang, Recent progress in surgical adhesives for biomedical applications, *Smart Mater. Med.*, 2022, **3**, DOI: 10.1016/j.smaim.2021.11.004.
33. C. Ma, *et al.*, Ultra-strong bio-glue from genetically engineered polypeptides, *Nat. Commun.*, 2021, **12**, 3613.
34. S. Chen, *et al.*, Study on Camellia Oleifera Protein based Wood Adhesive by Epoxy Resin and Its Crosslinking Mechanism, *J. Wuhan Univ. Technol., Mater. Sci. Ed.*, 2021, **36**, 607.
35. A. Tenorio-Alfonso, M. C. Sánchez and J. M. Franco, Impact of the processing method on the properties of castor oil/cellulose acetate polyurethane adhesives for bonding wood, *Int. J. Adhes. Adhes.*, 2022, **116**, 103153.
36. F. Zafar, A. Ghosal, E. Sharmin, R. Chaturvedi and N. Nishat, A review on cleaner production of polymeric and nanocomposite coatings based on waterborne polyurethane dispersions from seed oils, *Prog. Org. Coat.*, 2019, **131**, 259.
37. J. Park, Y. Kim, B. Chun and J. Seo, Rational engineering and applications of functional bioadhesives in biomedical engineering, *Biotechnol. J.*, 2021, **16**, DOI: 10.1002/biot.202100231.
38. N. Bandara and J. Wu, Chemically Modified Canola Protein–Nanomaterial Hybrid Adhesive Shows Improved Adhesion and Water Resistance, *ACS Sustainable Chem. Eng.*, 2018, **6**, 1152.
39. N. Wanasingha, N. K. Dutta and N. R. Choudhury, Emerging bioadhesives: from traditional bioactive and bioinert to a new biomimetic protein-based approach, *Adv. Colloid Interface Sci.*, 2021, **296**, DOI: 10.1016/j.cis.2021.102521.
40. R. Pinnataip and B. P. Lee, Oxidation Chemistry of Catechol Utilized in Designing Stimuli-Responsive Adhesives and Antipathogenic Biomaterials, *ACS Omega*, 2021, **6**, DOI: 10.1021/acsomega.1c00006.
41. S. Gomes, I. B. Leonor, J. F. Mano, R. L. Reis and D. L. Kaplan, Natural and genetically engineered proteins for tissue engineering, *Prog. Polym. Sci.*, 2012, **37**, DOI: 10.1016/j.progpolymsci.2011.07.003.
42. U. Kalapathy, N. S. Hettiarachchy and K. C. Rhee, Effect of drying methods on molecular properties and functionalities of disulfide bond-cleaved soy proteins, *J. Am. Oil Chem. Soc.*, 1997, **74**, 195.
43. J. E. Jakes, C. G. Hunt, S. L. Zelinka, P. N. Ciesielski and N. Z. Plaza, Effects of moisture on diffusion in unmodified wood cell walls: A phenomenological polymer science approach, *Forests*, 2019, **10**, 195.
44. J. E. Jakes, *et al.*, X-ray methods to observe and quantify adhesive penetration into wood, *J. Mater. Sci.*, 2019, **54**, 705.
45. B. B. Adhikari, *et al.*, Enhancing the adhesive strength of a plywood adhesive developed from hydrolyzed specified risk materials, *Polymers*, 2016, **8**, 285.
46. W. M. Bain, *et al.*, in *Protein and synthetic adhesives for paper coating*, Technical Association of the Pulp and Paper Industry, 1952.

47. E. Averina, J. Konnerth, S. D'Amico and H. W. G. van Herwijnen, Protein adhesives: Alkaline hydrolysis of different crop proteins as modification for improved wood bonding performance, *Ind. Crops Prod.*, 2021, **161**, 113187.
48. J. Zhu, *et al.*, Smart bioadhesives for wound healing and closure, *Bioact. Mater.*, 2023, **19**, 360.
49. Q. Lin, N. Chen, L. Bian and M. Fan, Development and mechanism characterization of high performance soy-based bio-adhesives, *Int. J. Adhes. Adhes.*, 2012, **34**, 11.
50. N. S. Hettiarachchy, U. Kalapathy and D. J. Myers, Alkali-modified soy protein with improved adhesive and hydrophobic properties, *J. Am. Oil Chem. Soc.*, 1995, **72**, 1461.
51. Y. Wang, X. Mo, X. S. Sun and D. Wang, Soy protein adhesion enhanced by glutaraldehyde crosslink, *J. Appl. Polym. Sci.*, 2007, **104**, 130.
52. Y. Liu and K. Li, Development and characterization of adhesives from soy protein for bonding wood, *Int. J. Adhes. Adhes.*, 2007, **27**, 59.
53. J. Rogers, X. Geng and K. Li, Soy-based adhesives with 1, 3-dichloro-2-propanol as a curing agent, *Wood Fiber Sci.*, 2004, **36**, 186.
54. H. Liu, C. Li and X. S. Sun, Improved water resistance in undecylenic acid (UA)-modified soy protein isolate (SPI)-based adhesives, *Ind. Crops Prod.*, 2015, **74**, 577.
55. W. Huang and X. Sun, Adhesive properties of soy proteins modified by sodium dodecyl sulfate and sodium dodecylbenzene sulfonate, *J. Am. Oil Chem. Soc.*, 2000, **77**, 705.
56. Y. Wang, D. Wang and X. S. Sun, Thermal properties and adhesiveness of soy protein modified with cationic detergent, *J. Am. Oil Chem. Soc.*, 2005, **82**, 357.
57. U. Kalapathy, N. S. Hettiarachchy, D. Myers and M. A. Hanna, Modification of soy proteins and their adhesive properties on woods, *J. Am. Oil Chem. Soc.*, 1995, **72**, 507.
58. U. Kalapathy, N. S. Hettiarachchy, D. Myers and K. C. Rhee, Alkali-modified soy proteins: Effect of salts and disulfide bond cleavage on adhesion and viscosity, *J. Am. Oil Chem. Soc.*, 1996, **73**, 1063.
59. P. Nordqvist, F. Khabbaz and E. Malmström, Comparing bond strength and water resistance of alkali-modified soy protein isolate and wheat gluten adhesives, *Int. J. Adhes. Adhes.*, 2010, **30**, 72.
60. Y. Liu and K. Li, Modification of soy protein for wood adhesives using mussel protein as a model: The influence of a mercapto group, *Macromol. Rapid Commun.*, 2004, **25**, 1835.
61. S. Takasaki and S. Kawakishi, Formation of Protein-Bound 3,4-Dihydroxyphenylalanine and 5-S-Cysteinyl-3,4-dihydroxyphenylalanine as New Cross-Linkers in Gluten, *J. Agric. Food Chem.*, 1997, **45**, 3472.
62. R. B. Freedman, Proteins: Structures and molecular properties, *Trends Biochem. Sci.*, 1985, **10**, DOI: 10.1016/0968-0004(85)90239-7.

63. C. R. Frihart and L. F. Lorenz, Specific oxidants improve the wood bonding strength of soy and other plant flours, *J. Polym. Sci., Part A: Polym. Chem.*, 2019, **57**, 1017.
64. D. Zhu and S. Damodaran, Chemical phosphorylation improves the moisture resistance of soy flour-based wood adhesive, *J. Appl. Polym. Sci.*, 2014, **131**, DOI: 10.1002/app.40451.
65. L. Yue, *et al.*, A high-performance soybean meal-based plywood adhesive prepared via an ultrasonic process and using significantly lower amounts of chemical additives, *J. Cleaner Prod.*, 2020, **274**, 123017.
66. A. S. Mostaert, M. J. Higgins, T. Fukuma, F. Rindi and S. P. Jarvis, Nanoscale mechanical characterisation of amyloid fibrils discovered in a natural adhesive, *J. Biol. Phys.*, 2006, **32**, 393.
67. G. Schmidt, *et al.*, Strong Adhesives from Corn Protein and Tannic Acid, *Adv. Sustainable Syst.*, 2019, **3**, 1900077.
68. G. Schmidt, B. R. Hamaker and J. J. Wilker, High Strength Adhesives from Catechol Cross-Linking of Zein Protein and Plant Phenolics, *Adv. Sustainable Syst.*, 2018, **2**, 1700159.
69. G. Schmidt, *et al.*, Tunable Tannic Acid–Zein Adhesives for Bonding Different Substrates, *Adv. Sustainable Syst.*, 2022, **6**, 2100392.
70. S. J. Circle, E. W. Meyer and R. W. Whitney, Rheology of Soy Protein Dispersions. Effect of Heat and Other Factors on Gelation, *Cereal Chem.*, 1964, **41**, 157.
71. S. Khosravi, F. Khabbaz, P. Nordqvist and M. Johansson, Protein-based adhesives for particleboards, *Ind. Crops Prod.*, 2010, **32**, 275.
72. Y. Xu, *et al.*, The effect of enzymolysis on performance of soy protein-based adhesive, *Molecules*, 2018, **23**, 2752.
73. R. Kumar, V. Choudhary, S. Mishra and I. K. Varma, Enzymatically-modified soy protein Part 2: Adhesion behaviour, *J. Adhes. Sci. Technol.*, 2004, **18**, 261.
74. K. Numata and P. J. Baker, Synthesis of adhesive peptides similar to those found in blue mussel (Mytilus edulis) using papain and tyrosinase, *Biomacromolecules*, 2014, **15**, 3206.
75. P. Kord Forooshani and B. P. Lee, Recent approaches in designing bioadhesive materials inspired by mussel adhesive protein, *J. Polym. Sci., Part A: Polym. Chem*, 2017, **55**, DOI: 10.1002/pola.28368.
76. J. Witman and T. Suchanek, Mussels in flow: drag and dislodgement by epizoans, *Mar. Ecol.: Prog. Ser.*, 1984, **16**, 259.
77. A. H. J. Gowda, *et al.*, Design of tunable gelatin-dopamine based bioadhesives, *Int. J. Biol. Macromol.*, 2020, **164**, 1384.
78. J. Sun, J. Han, F. Wang, K. Liu and H. Zhang, Bioengineered Protein-based Adhesives for Biomedical Applications, *Chem. - Eur. J.*, 2022, **28**, DOI: 10.1002/chem.202102902.

79. C. R. Matos-Pérez, J. D. White and J. J. Wilker, Polymer composition and substrate influences on the adhesive bonding of a biomimetic, cross-linking polymer, *J. Am. Chem. Soc.*, 2012, **134**, 9498.
80. H. Shao and R. J. Stewart, Biomimetic underwater adhesives with environmentally triggered setting mechanisms, *Adv. Mater.*, 2010, **22**, 729.
81. R. Xu, *et al.*, Synthesis of novel copolymer/Mn+ biomimetic adhesives based on temperature strengthened coacervation effect, *Chem. Eng. J.*, 2021, **425**, 127249.
82. J. L. Murphy, L. Vollenweider, F. Xu and B. P. Lee, Adhesive performance of biomimetic adhesive-coated biologic scaffolds, *Biomacromolecules*, 2010, **11**, 2976.
83. K. Li, *et al.*, Biomimetic development of a strong, mildew-resistant soy protein adhesive via mineral–organic system and phenol-amine synergy, *Ind. Crops Prod.*, 2022, **187**, 115412.
84. Z. Wang, *et al.*, The synergy between natural polyphenol-inspired catechol moieties and plant protein-derived bio-adhesive enhances the wet bonding strength, *Sci. Rep.*, 2017, **7**, 9664.
85. Y. Chen, *et al.*, A biomimetic adhesive with high adhesion strength and toughness comprising soybean meal, chitosan, and condensed tannin-functionalized boron nitride nanosheets, *Int. J. Biol. Macromol.*, 2022, **219**, 611.
86. S. Zhao, *et al.*, Polyurethane as high-functionality crosslinker for constructing thermally driven dual-crosslinking plant protein adhesion system with integrated strength and ductility, *Chem. Eng. J.*, 2021, **422**, 130152.
87. Z. Wang, *et al.*, Facile biomimetic self-coacervation of tannic acid and polycation: Tough and wide pH range of underwater adhesives, *Chem. Eng. J.*, 2021, **404**, 127069.
88. G. Zeng, *et al.*, A bio-based adhesive reinforced with functionalized nanomaterials to build multiple strong and weak cross-linked networks with high strength and excellent mold resistance, *Chem. Eng. J.*, 2023, **453**, 139761.
89. K. Li, *et al.*, Bioinspired interface engineering of soybean meal-based adhesive incorporated with biomineralized cellulose nanofibrils and a functional aminoclay, *Chem. Eng. J.*, 2021, **421**, 129820.
90. X. Huang, *et al.*, Development of a strong soy protein-based adhesive with excellent antibacterial and antimildew properties via biomineralized silver nanoparticles, *Ind. Crops Prod.*, 2022, **188**, 115567.
91. S. Chen, S. Q. Shi, W. Zhou and J. Li, Developments in Bio - Based Soy Protein Adhesives: A Review, *Macromol. Mater. Eng.*, 2022, **307**, 2200277.
92. Z. Wang, *et al.*, Bio-inspired cellulose nanofiber-reinforced soy protein resin adhesives with dopamine-induced codeposition of "water-resistant" interphases, *Appl. Surf. Sci.*, 2019, **478**, 441.

93. D. Gan, *et al.*, Plant-inspired adhesive and tough hydrogel based on Ag-Lignin nanoparticles-triggered dynamic redox catechol chemistry, *Nat. Commun.*, 2019, **10**, 1487.
94. J. Liu, S. Fu, B. Yuan, Y. Li and Z. Deng, Toward a universal 'adhesive nanosheet' for the assembly of multiple nanoparticles based on a protein-induced reduction/decoration of graphene oxide, *J. Am. Chem. Soc.*, 2010, **132**, 7279.
95. D. Ling, *et al.*, Multiple-interaction ligands inspired by mussel adhesive protein: synthesis of highly stable and biocompatible nanoparticles, *Angew. Chem., Int. Ed.*, 2011, **50**, 11360.
96. H. N. Xu, S. Ma, W. Lv and Z. Wang, Soy protein adhesives improved by SiO2 nanoparticles for plywoods, *Pigm. Resin Technol.*, 2011, **40**, 191.
97. S. Y. Mousavi, J. Huang and K. Li, A cold-set wood adhesive based on soy protein, *Int. J. Adhes. Adhes.*, 2021, **106**, 102801.
98. K. Li, S. Peshkova and X. Geng, Investigation of soy protein-kymene® adhesive systems for wood composites, *J. Am. Oil Chem. Soc.*, 2004, **81**, 487.
99. J. K. Román and J. J. Wilker, Cooking Chemistry Transforms Proteins into High-Strength Adhesives, *J. Am. Chem. Soc.*, 2019, **141**, 1359.
100. H. Liu, C. Li and X. S. Sun, Soy-oil-based waterborne polyurethane improved wet strength of soy protein adhesives on wood, *Int. J. Adhes. Adhes.*, 2017, **73**, 66.
101. J. Li, H. Lin, S. R. Bean, X. S. Sun and D. Wang, Evaluation of adhesive performance of a mixture of soy, sorghum and canola proteins, *Ind. Crops Prod.*, 2020, **157**, 112898.
102. X. Zhu, D. Wang, N. Li and X. S. Sun, Bio-Based Wood Adhesive from Camelina Protein (a Biodiesel Residue) and Depolymerized Lignin with Improved Water Resistance, *ACS Omega*, 2017, **2**, 7996.
103. S. Pradyawong, G. Qi, M. Zhang, X. S. Sun and D. Wang, Effect of pH and pH-shifting on adhesion performance and properties of lignin-protein adhesives, *Trans. ASABE*, 2021, **64**, 1141.
104. N. Bandara, L. Chen and J. Wu, Adhesive properties of modified triticale distillers grain proteins, *Int. J. Adhes. Adhes.*, 2013, **44**, 122.
105. N. Bandara, A. Akbari, Y. Esparza and J. Wu, Canola Protein: A Promising Protein Source for Delivery, Adhesive, and Material Applications, *J. Am. Oil Chem. Soc.*, 2018, **95**, 1075.
106. N. Bandara, Y. Esparza and J. Wu, Graphite Oxide Improves Adhesion and Water Resistance of Canola Protein-Graphite Oxide Hybrid Adhesive, *Sci. Rep.*, 2017, **7**, 11538.
107. N. Bandara, Y. Esparza and J. Wu, Exfoliating nanomaterials in canola protein derived adhesive improves strength and water resistance, *RSC Adv.*, 2017, **7**, 6743.
108. N. Bandara and J. Wu, Randomly Oriented Strand Board Composites from Nanoengineered Protein-Based Wood Adhesive, *ACS Sustainable Chem. Eng.*, 2018, **6**, 457.

109. R. Vendamme and W. Eevers, Sweet solution for sticky problems: Chemoreological design of self-adhesive gel materials derived from lipid biofeedstocks and adhesion tailoring via incorporation of isosorbide, *Macromolecules*, 2013, **46**, 3395.
110. B. Lee, *et al.*, Ecofriendly Catechol Lipid Bioresin for Low-Temperature Processed Electrode Patterns with Strong Durability, *ACS Appl. Mater. Interfaces*, 2020, **12**, 16864.
111. A. Ghosal and S. Ahmad, High performance anti-corrosive epoxy–titania hybrid nanocomposite coatings, *New J. Chem.*, 2017, **41**, 4599.
112. A. Ghosal, O. U. Rahman and S. Ahmad, High-Performance Soya Polyurethane Networked Silica Hybrid Nanocomposite Coatings, *Ind. Eng. Chem. Res.*, 2015, **54**, DOI: 10.1021/acs.iecr.5b02098.
113. A. Ghosal, S. Iqbal and S. Ahmad, NiO nanofiller dispersed hybrid Soy epoxy anticorrosive coatings, *Prog. Org. Coat*, 2019, **133**, 61.
114. N. Aydogan, G. D. Kalaycioglu, H. Unsal and M. Ozkan, Custom-made lipid nanotubes as a tissue and hydrogel adhesive, *Colloids Surf., A*, 2021, **608**, 125609.
115. A. Li and K. Li, Pressure-sensitive adhesives based on tung oil, *RSC Adv.*, 2015, **5**, 85264.
116. Y. Wu, A. Li and K. Li, Development and evaluation of pressure sensitive adhesives from a fatty ester, *J. Appl. Polym. Sci.*, 2014, **131**, DOI: 10.1002/app.41143.
117. Y. Wu, A. Li and K. Li, Pressure Sensitive Adhesives Based on Oleic Acid, *J. Am. Oil Chem. Soc.*, 2015, **92**, 111.
118. S. Y. Mousavi, J. Huang and K. Li, Investigation of epoxy-based wood adhesives, *J. Appl. Polym. Sci.*, 2019, **136**, 47741.
119. A. M. Borrero-López, *et al.*, Toward UV-Triggered Curing of Solvent-Free Polyurethane Adhesives Based on Castor Oil, *ACS Sustainable Chem. Eng.*, 2021, **9**, 11032.
120. T. Aziz, *et al.*, Adhesive properties of poly (methyl silsesquioxanes)/bio-based epoxy nanocomposites, *Iran. Polym. J.*, 2020, **29**, 911.
121. F. Ng, L. Bonnet, G. David and S. Caillol, Novel biobased and food contact epoxy coatings for glass toughening applications, *Prog. Org. Coat.*, 2017, **109**, 1.
122. A. Patel, *et al.*, Design and Formulation of a Completely Biobased Epoxy Structural Adhesive, *ACS Sustainable Chem. Eng.*, 2019, **7**, 16382.
123. A. Li and K. Li, Pressure-sensitive adhesives based on epoxidized soybean oil and dicarboxylic acids, *ACS Sustainable Chem. Eng.*, 2014, **2**, 2090.
124. S. Miao, *et al.*, Soybean oil-based polyurethane networks as candidate biomaterials: Synthesis and biocompatibility, *Eur. J. Lipid Sci. Technol.*, 2012, **114**, 1165.
125. K. P. Ang, C. S. Lee, S. F. Cheng and C. H. Chuah, Synthesis of palm oil - based polyester polyol for polyurethane adhesive production, *J. Appl. Polym. Sci.*, 2014, **131**, DOI: 10.1002/app.39967.

126. M. A. Aristri, *et al.*, Recent Developments in Lignin- and Tannin-Based Non-Isocyanate Polyurethane Resins for Wood Adhesives—A Review, *Appl. Sci.*, 2021, **11**, 4242.
127. B. Bizet, É. Grau, H. Cramail and J. M. Asua, Water-based non-isocyanate polyurethane-ureas (NIPUUs), *Polym. Chem.*, 2020, **11**, 3786.
128. A. Sateesh, *et al.*, Surface modification of medical-grade polyurethane by cyanurchloride-activated tetraether lipid (a new approach for bacterial antiadhesion), *J. Biomed. Mater. Res., Part A*, 2008, **84**, 672.
129. X. Tong, X. Luo and Y. Li, Development of blend films from soy meal protein and crude glycerol-based waterborne polyurethane, *Ind. Crops Prod.*, 2015, **67**, 11.
130. F. Zafar, A. Ghosal, E. Sharmin, R. Chaturvedi and N. Nishat, A review on cleaner production of polymeric and nanocomposite coatings based on waterborne polyurethane dispersions from seed oils, *Prog. Org. Coat.*, 2019, **131**, 259.
131. P. Kadam, P. Vaidya and S. Mhaske, Synthesis and characterization of polyesteramide based hot melt adhesive obtained with dimer acid, castor oil and ethylenediamine, *Int. J. Adhes. Adhes.*, 2014, **50**, 151.
132. P. G. Kadam, R. A. Kute and S. T. Mhaske, Synthesis and characterization of a novel polyesteramide from modified jatropha seed oil, dimer acid and ethylenediamine as hot melt adhesive, *J. Adhes. Sci. Technol.*, 2014, **28**, 675.
133. K. Chugh, G. Phalak and S. Mhaske, Fatty acid based novel precursors for polyesteramide hot melt adhesive, *J. Adhes. Sci. Technol.*, 2020, **34**, 1871.
134. P. Utekar, H. Gabale, A. Khandelwal and S. T. Mhaske, Hot-melt adhesives from renewable resources: A critical review, *Rev. Adhes. Adhes.*, 2016, **4**, 104.
135. S. Ahmad, S. M. Ashraf and F. Zafar, Development of linseed oil based polyesteramide without organic solvent at lower temperature, *J. Appl. Polym. Sci.*, 2007, **104**, 1143.
136. F. Zafar, E. Sharmin, S. M. Ashraf and S. Ahmad, Studies on Poly(Styrene-co-Maleic Anhydride)-Modified Polyesteramide-Based Anticorrosive Coatings Synthesized from a Sustainable Resource, *J. Appl. Polym. Sci.*, 2004, **92**, 2538.
137. S. Yadav, F. Zafar, A. Hasnat and S. Ahmad, Poly (urethane fatty amide) resin from linseed oil—A renewable resource, *Prog. Org. Coat.*, 2009, **64**, 27.
138. F. Zafar, H. Zafar, E. Sharmin, S. M. Ashraf and S. Ahmad, Studies on Ambient Cured Biobased Mn(II), Co(II) and Cu(II) Containing Metallo-polyesteramides, *J. Inorg. Organomet. Polym. Mater.*, 2011, **21**, 646.
139. I. Manolakis, B. A. J. Noordover, R. Vendamme and W. Eevers, Novel L-DOPA-derived poly(ester amide)s: monomers, polymers, and the first L-DOPA-functionalized biobased adhesive tape, *Macromol. Rapid Commun.*, 2014, **35**, 71.

140. J. R. Gouveia, G. E. S. Garcia, L. D. Antonino, L. B. Tavares and D. J. Dos Santos, Epoxidation of kraft lignin as a tool for improving the mechanical properties of epoxy adhesive, *Molecules*, 2020, **25**, 2513.
141. S. Feng, Z. Yuan, M. Leitch and C. C. Xu, Adhesives formulated from bark bio-crude and phenol formaldehyde resole, *Ind. Crops Prod.*, 2015, **76**, 258.
142. F. Wang, *et al.*, Combinations of soy protein and polyacrylate emulsions as wood adhesives, *Int. J. Adhes. Adhes.*, 2018, **82**, 160.
143. G. Qi and X. S. Sun, Soy protein adhesive blends with synthetic latex on wood veneer, *J. Am. Oil Chem. Soc.*, 2011, **88**, 271.
144. A. Mao, W. Xu, E. Xi, Q. Li and H. Wan, Evaluation of phenol-formaldehyde resins modified and blended with pyrolysis bio-oil for plywood, *For. Prod. J.*, 2018, **68**, 113.
145. E. Sharmin, *et al.*, Preparation and characterization of nanostructured biohybrid, *Prog. Org. Coat.*, 2011, **72**, 469.
146. T. Cernadas, *et al.*, Preparation of functionalized poly(caprolactone diol)/castor oils blends to be applied as photocrosslinkable tissue adhesives, *J. Appl. Polym. Sci.*, 2020, **137**, 49092.
147. F. Isnard, *et al.*, Stabilization of Hybrid Adhesives and Sealants by Thermodynamic Tuning of Molecularly Optimized Lignin Bio-Additives: Small Changes, Big Effects, *Biomacromolecules*, 2022, **23**, 3174.
148. B. J. Kim, Y. S. Choi and H. J. Cha, Reinforced multifunctionalized nanofibrous scaffolds using mussel adhesive proteins, *Angew. Chem., Int. Ed.*, 2012, **51**, 675.
149. N. V. Gohad, *et al.*, Synergistic roles for lipids and proteins in the permanent adhesive of barnacle larvae, *Nat. Commun.*, 2014, **5**, 4414.
150. G. Qi, N. Li, D. Wang and X. S. Sun, Physicochemical properties of soy protein adhesives modified by 2-octen-1-ylsuccinic anhydride, *Ind. Crops Prod.*, 2013, **46**, 165.
151. L. Zhu and X. S. Sun, Adhesion Performance of Amino Groups (ε -NH 2) of Soybean Glycinin Proteins, *J. Biobased Mater. Bioenergy*, 2020, **1**, 64.
152. Z. Gao, P. J. Bremer, M. F. Barker, E. W. Tan and A. J. McQuillan, Adhesive secretions of live mussels observed in situ by attenuated total reflection-infrared spectroscopy, in *Applied Spectroscopy*, 2007, vol. **61**.
153. S. N. Gorb, The design of the fly adhesive pad: Distal tenent setae are adapted to the delivery of an adhesive secretion, *Proc. R. Soc. B*, 1998, **265**, 747.
154. J. H. Dirks, C. J. Clemente and W. Federle, Insect tricks: Two-phasic foot pad secretion prevents slipping, *J. R. Soc., Interface*, 2010, **7**, 587.
155. P. Flammang, A. Michel, A. van Cauwenberge, H. Alexandre and M. Jangoux, A study of the temporary adhesion of the podia in the sea star asterias rubens (Echinodermata, asteroidea) through their footprints, *J. Exp. Biol.*, 1998, **201**, 2383.

156. S. A. Burke, M. Ritter-Jones, B. P. Lee and P. B. Messersmith, Thermal gelation and tissue adhesion of biomimetic hydrogels, *Biomed. Mater.*, 2007, **2**, 203.
157. T. Dissanayake, X. Sun, L. Abbey and N. Bandara, Recent advances in lipid-protein conjugate-based delivery systems in nutraceutical, drug, and gene delivery, *Food Hydrocolloids Health*, 2022, **2**, 100054.
158. D. Pooja, *et al.*, Improving efficacy, oral bioavailability, and delivery of paclitaxel using protein-grafted solid lipid nanoparticles, *Mol. Pharmaceutics*, 2016, **13**, 3903.
159. A. Baer, *et al.*, Mechanoresponsive lipid-protein nanoglobules facilitate reversible fibre formation in velvet worm slime, *Nat. Commun.*, 2017, **8**, 1.
160. S. A. A. Mohamed, M. El-Sakhawy and M. A. M. El-Sakhawy, Polysaccharides, Protein and Lipid -Based Natural Edible Films in Food Packaging: A Review, *Carbohydr. Polym.*, 2020, **238**, DOI: 10.1016/j.carbpol.2020.116178.
161. T. T. Hoang Thi, Y. Lee, S. B. Ryu, D. H. Nguyen and K. D. Park, Enhanced tissue adhesiveness of injectable gelatin hydrogels through dual catalytic activity of horseradish peroxidase, *Biopolymers*, 2018, **109**, e23077.
162. P. Le Thi, Y. Lee, D. H. Nguyen and K. D. Park, In situ forming gelatin hydrogels by dual-enzymatic cross-linking for enhanced tissue adhesiveness, *J. Mater. Chem. B*, 2017, **5**, 757.
163. M. J. Brennan, B. F. Kilbride, J. J. Wilker and J. C. Liu, A bioinspired elastin-based protein for A cytocompatible underwater adhesive, *Biomaterials*, 2017, **124**, 116.
164. J. R. Soucy, *et al.*, Photocrosslinkable Gelatin/Tropoelastin Hydrogel Adhesives for Peripheral Nerve Repair, *Tissue Eng., Part A*, 2018, **24**, 1393.
165. B. W. Walker, *et al.*, Engineering a naturally-derived adhesive and conductive cardiopatch, *Biomaterials*, 2019, **207**, 89.
166. N. Annabi, *et al.*, Engineering a sprayable and elastic hydrogel adhesive with antimicrobial properties for wound healing, *Biomaterials*, 2017, **139**, 229.
167. A. Pandit, G. Damodaran, R. Collighan and M. Griffin, Tethering a laminin peptide to a crosslinked collagen scaffold for biofunctionality, *J. Biomed. Mater. Res., Part A*, 2009, **89**, 1001.
168. V. B. Damodaran and S. N. Murthy, Bio-inspired strategies for designing antifouling biomaterials, *Biomater. Res.*, 2016, **20**, 18.
169. K. A. Burke, D. C. Roberts and D. L. Kaplan, Silk Fibroin Aqueous-Based Adhesives Inspired by Mussel Adhesive Proteins, *Biomacromolecules*, 2016, **17**, 237.

CHAPTER 14

A Green Biorefinery Concept: Proteins and Lipids as Sustainable Biomass

M. ARSHAD, M. ZUBAIR AND A. ULLAH*

Department of Agricultural, Food and Nutritional Science, University of Alberta, Edmonton, AB T6G 2P5, Canada
*E-mail: amanullah@ualberta.ca

14.1 Introduction

A biorefinery is a facility that transforms a low-value renewable biomass feedstock through multiple processes, such as separations, fractionation, and conversions, into high-value sustainable products, such as feed, food, transportation fuels, fine chemicals, and energy. A biorefinery can work either as a single unit or in conjunction with other facilities for specific purposes while processing products, by-products, or wastes of integrated facilities. As per the International Energy Agency, biorefineries can play a significant role by utilizing biomass resources in a more sustainable and efficient way while producing several products for different markets and industrial sectors.

Biorefinery systems have emerged as an attractive route for energy generation, including combined heat and power and biofuels, as well as chemical production, with great promise for a reduced environmental impact.[1,2] Under this concept, all types of waste produced as a by-product are considered as a new resource. This concept consists of a series of

Smart Materials No. 41
Functional Materials from Lipids and Proteins
Edited by Nandika Bandara and Aman Ullah

Published by the Royal Society of Chemistry, www.rsc.org

unit operations, such as pre-treatment, followed by the extraction and segregation of biomass. The waste produced in each step as a by-product is valorized to generate new high-value-added products.[3] Therefore, the biorefinery concept is valuable, having a very low impact on the environment, and is the best alternative to petrorefinery.

This chapter provides a brief review of biorefinery concepts for lipids and proteins. It covers a brief description of different sources of lipids and proteins and their various methods of extraction from renewable resources. Moreover, the development of bio-derived products from proteins and lipids using various processing strategies and their modifications/functionalization are discussed in detail. Finally, this chapter describes a summary and future perspectives on the use of a biorefinery concept for the valorization of lipids and proteins into value-added products.

14.2 Sources and Extraction of Protein Biomass

Proteins are complex organic molecules and are made up of different amino acids, which have numerous applications in different industrial sectors, such as food, agriculture, and pharmaceutical industries. It is readily available from various sources, such as animals (meat, milk, poultry, fish, and milk products) and plants (legumes, fruits, nuts, seeds, vegetables, and grains). Various methods of protein extraction were reported in the literature.[4–7] Some of the well-known methods are briefly described in this chapter.

14.2.1 Precipitation Methods

Salting out is one of the precipitation methods, where proteins are salted out as a co-precipitate due to the high concentration of ammonium sulfate as a salt. The mechanism of this process is based on preferential solvation, where salt (cosolvent) is excluded from the water layer due to its association with the protein surface (hydration layer).[8]

14.2.2 Iso-ionic Method

Proteins are least soluble and highly precipitable in the iso-ionic state. In the salt-free state (iso-ionic), proteins exist in their most compact and least hydrated conformation, and this phenomenon is much associated with the state of proteins at their iso-electric point. Tanford[9] explained the difference between iso-ionic and iso-electric properties. The use of column chromatography or dialysis for deionization aims to change proteins to an iso-ionic state, resulting in their precipitation. Two key factors influence the solubility of many proteins: the pH of the solution relative to the protein's iso-ionic point and the presence of low salt concentrations (ranging from 0.1 to 0.2 M salt).

14.2.3 Gel Electrophoresis Method

Gel electrophoresis is a technique used to separate charged molecules of DNA or proteins depending on their physical properties when they are pulled through a gel by applying an electrical current. In this technique, the actual sample loaded onto the gel is treated with sodium dodecyl sulfate (SDS) and β-mercaptoethanol. As a result, the protein molecules get denatured, and the SDS binds tightly to the uncoiled protein molecules, making them negatively charged. The separation of proteins through SDS–polyacrylamide gel electrophoresis is associated with their size, as SDS-coated proteins have a balanced charge-to-mass ratio.[10]

14.3 Potential Applications of Proteins in a Biorefinery Concept

Proteins are sustainable and renewable feedstock utilized for food, feed, and bio-based applications (Figure 14.1). Numerous biomass resources can be fractionated into bio-based intermediates, such as proteins. These extracted proteins are further processed using biochemical and thermochemical routes to produce bio-derived products and energy at a commercial level.[11] Proteins are present at more than ten wt.% in biomass resources, mainly used in the food and feed sectors. One of the viable and sustainable opportunities is to separate the functional proteins and add value to a particular food and feed requirement.[12]

Currently, proteins are primarily used in human or pet food and for feeding livestock or aquaculture purposes due to their nutritional value and low degree concentration of less than 50%.[13] Around the globe, about 800 million tons of proteins are produced, with 78% used for feed purposes and 17% for food, including protein isolates and concentrates. Interestingly, less than 1% is used to develop bio-based materials.[14,15] Biomass from the agri- or food by-products will be in high demand due to the global endeavors for low-carbon substitutes.[16] Protein waste is an excellent candidate to meet the growing demand for renewable feedstock.[17] Annually, 100 million tons of protein waste can be produced if the petro-based fuel is shifted to biofuels, which is 10% of the global petroleum demand.

Leaf proteins have been used in food applications, as they have excellent nutritional values. In addition, leaf protein fractions have shown good functional properties. So far, they are used in the food industry for gelation, water-holding capacity, emulsification, and foaming purposes.[18] Alfalfa leaf protein gels exhibited a gel strength twice that of a 15% soy protein isolate.[19] Likewise, purified ribulose-5-bisphosphate carboxylase/oxygenase

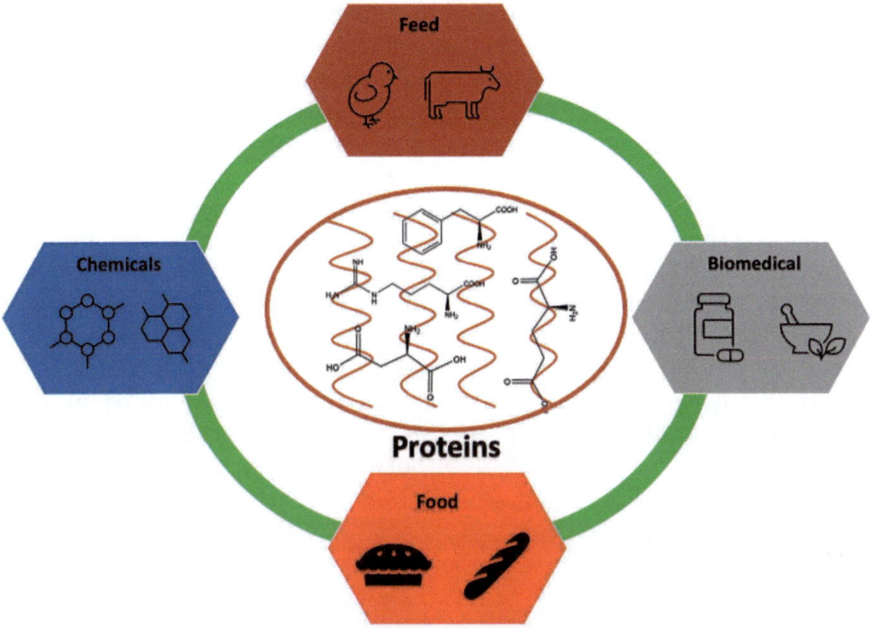

Figure 14.1 Proteins applications in a biorefinery concept.

(RuBisCo) protein from tobacco leaves produced gels stronger than those from soy protein and egg white protein but weaker than those from gelatin proteins.[20] In another study, RuBisCo proteins from sugar beet leaves exhibited stronger gels compared to whey and soy protein isolates even at low concentrations.[21] Lamsal *et al.* studied the gelation capacity of soluble alfalfa proteins at 90 °C. The proteins showed a significant cold-setting and shear-thinning behavior with a low compact structure.[22] In another study by the same group, a combination of soluble alfalfa leaf protein (7%) and whey protein isolate (13%) at pH seven formed standing gels.[23]

Proteins have active surfaces due to their hydrophilic and hydrophobic characteristics in their amphiphilic structure.[17] The surface activity of proteins determines their foaming capacity, which is vital for making aerated foods, such as beverages, bakery, and confectionery. The ability and stability of foaming depend on the ability of protein molecules to unfold and the orientation of the hydrophobic and hydrophilic zones in the air–water interface. They also depend on the formation of a protein film, stabilizing the neighboring foam bubbles.[14,18] The RuBisCo proteins isolated from sugar beet leaves form a whipped foam (the foam overrun capacity of >60%) at pH 4 similar to the whey and soy protein isolates.[21]

Some of the available protein waste contains essential amino acids. It can be used for animal feed, while other amino acids are suitable as starting materials for highly bulk-functionalized petrochemicals (Scheme 14.1). The amino acids can be bio-transformed into chemicals, such as arginine into ornithine, aspartic acid into β-alanine, and glutamic acid into α-ketoglutaric acid and *N*-methyl- and *N*-vinyl-pyrrolidone. At the same time, lysin can be converted into 5-aminovaleric acid and 5-diaminopentane and phenylalanine into cinnamic acid, alcohol, alanine, and L-amino acids, producing primary amines and enantio-compounds, respectively.[24] A good example demonstrates the arginase-catalyzed conversion of arginine into ornithine.[25] The starting material for Nylon-4,6 can be produced by further decarboxylation of ornithine to 1,4-diaminobutane using ornithine decarboxylase.[26] Cinnamic acid is produced from phenylalanine ammonia lyase (PAL)-intermediated reaction of phenylalanine,[27] which can be used as a precursor for the production of styrene. A similar concept can be useful to produce β-alanine by the decarboxylation of aspartate using aspartate α-decarboxylase.[28] β-Alanine has the potential to be used as a starting material for acrylonitrile and acrylamide synthesis. Moreover, oxidation (lysine oxidase) and decarboxylation (lysine decarboxylase) of

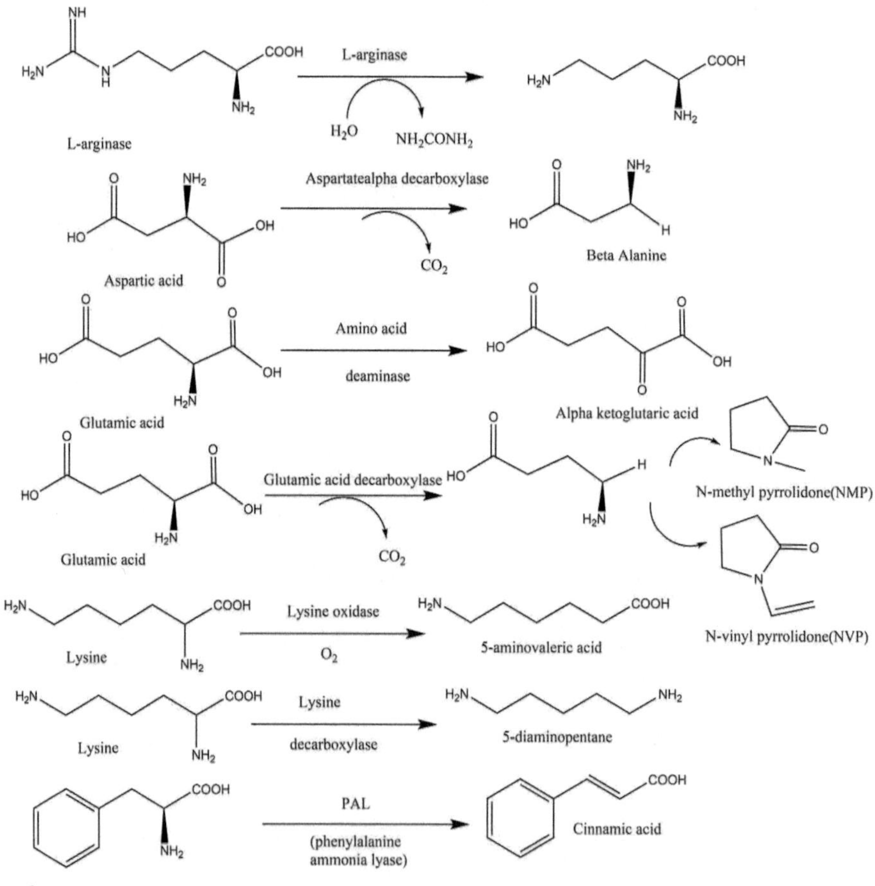

Scheme 14.1 Amino acid transformation into valuable chemicals.

lysine lead to 5-aminovaleric acid and 5-diaminopentane, respectively.[29–31] The potential application of 5-aminovaleric acid is the production of nylon, while 5-diaminopentane is used in the synthesis of polyamide. Another non-essential amino acid, *i.e.*, glutamic acid, is found abundantly in most feedstock proteins. Glutamate can be transformed into γ-aminobutyric acid *via* decarboxylation using the glutamate decarboxylase enzyme.[32] Most importantly, the synthesis of *N*-methylpyrrolidone and *N*-vinylpyrrolidone from γ-aminobutyric acid is an environmentally benign process. Both of these are valuable intermediates for synthesizing many bulk chemicals. On the other hand, glutamate is converted into α-ketoglutaric acid by deamination using glutamate deaminase,[33,34] which can be polymerized into environment-friendly materials, *i.e.*, poly(triol α-ketoglutarate). Alanine and alcohol can produce azo dyes, rubbery products, and antioxidants. They can be obtained from primary alcohols and enzymes, such as alcohol dehydrogenase and L-alanine dehydrogenase.[35]

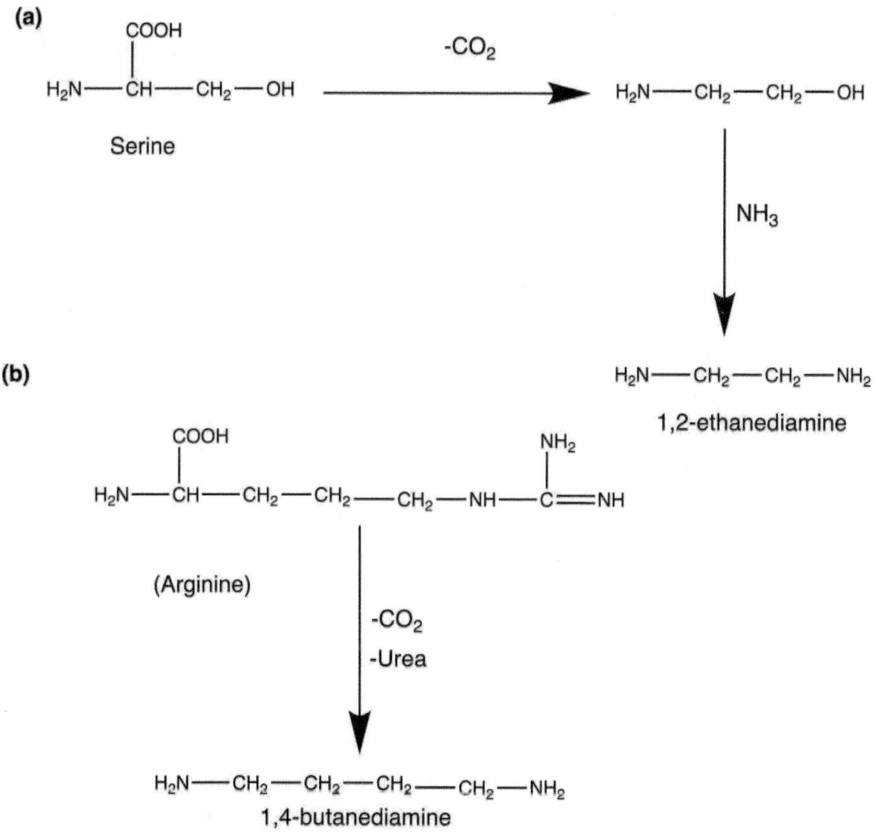

Scheme 14.2 Industrially important diamines: (a) 1,2-ethanediamine and (b) 1,4-butanediamine.

1,2-Ethanediamine is synthesized mostly from ethylene by various routes, including epoxide formation, followed by the amination of ethanolamine to produce 1,2-ethanediamine, chlorination of ethylene, and then substitution of chlorine with ammonia. Similar products can be produced from amino acids, such as serine and arginine, using both enzymatic and chemical reactions (Scheme 14.2). The decarboxylation of serine produced ethanolamine, which is then transformed into a diamine using ammonia. Similarly, arginine is hydrolyzed into urea and ornithine, which form 1,4-butanediamine after the decarboxylation.[36]

Recently, Zubair *et al.* utilized poultry by-products, such as spent hen and chicken feathers, for food packaging and water remediation. They extracted proteins from spent hen and transformed them into films for food packaging applications by nano-reinforcement by either clay, polyhedral oligomeric silsesquioxane (POSS), or cellulose nanocrystals.[37,38] In another study, keratin proteins were isolated from chicken feathers and converted into heavy metal water remediation-adsorbing materials.[39] Zahara *et al.* also modified chicken feather keratin using ionic liquids to remove metal ions from wastewater by energy generation processes.[40] Thus, the economic feasibility or sustainability of the biorefinery platform can be broadened by incorporating the protein-derived materials into the current chemical, feed, and food production. In the future, the supply of renewable and sustainable proteins has opportunities that can be fulfilled by either an optimum valorization of current protein sources or manipulation of novel protein sources.

14.4 Sources and Extraction of Lipids

Lipids are very abundant and economical in nature and, as a result, are considered a potential feedstock for a wide range of industrial applications. Lipids are primarily available from plant and animal sources. They are mainly composed of a class of organic compounds, such as fatty acids (triglycerides), including free fatty acids and their derivatives, sterols, waxes, phospholipids, glycolipids, cholesterol, and sphingolipids, and are soluble in organic solvents due to their amphipathic (hydrophilic and lipophilic) characteristics. Major sources of fats and oils are plants and animals, such as soybean, rice bran, cottonseed, sunflower, palm, palm kernel, groundnut, olive, coconut, linseed, castor, tung, tall oil, milk fat, lard, tallow, and fish.[41–43] At a global scale, about 80% of oils and fats come from plant sources, while 20% are obtained from animal origin. Cottonseed and soybean oils are obtained as by-products from plants primarily grown for their seed meal, which is commonly used as animal feed and fiber. However, the oil crops include groundnut, rape, and sunflower, which provide a residual meal as a by-product. Moreover, trees such as coconut, palm, and olive continue to yield oil for many years once they mature. Non-food vegetable oils, such as tung, castor, jatropha, and linseed, are other sources of plant oils. Camelina oil is also an emerging plant oil, which is mostly used for non-food purposes, though it has received approval for food applications to some extent. Cattle and marine animals are considered major sources of animal oil and fats.[43,44]

Various methods are being used for the extraction of lipids from plant and animal sources. However, the appropriate selection of methods plays an important role in the efficient extraction, which depends on various factors, such as the physical state of the lipid sources (animal or plant), concentration of lipids in the sources, and amount of moisture contents.

14.4.1 Bligh and Dyer and Folch Methods

Bligh and Dyer and Folch methods are considered the most standard methods for lipid extraction from animal sources, where an appropriate ratio of different solvents, such as methanol, chloroform, and water, is used. A solvent ratio of 1 : 2 : 0.75 (methanol : chloroform : water) with twenty times more volume of the sample in the presence of a salt was used in the Folch method, while in the Bligh and Dyer method, a ratio of 1 : 1 : 0.9 of methanol : chloroform : water with 4 times higher volume of the sample in the absence of a salt is utilized for the extraction of lipids.[45] These classic methods use toxic solvents for the extraction of lipids and were modified later by replacing them with *n*-hexane and isopropanol (3 : 2)[46] or cyclohexane, isopropanol, and water (8 : 10 : 11).[47]

14.4.2 Soxhlet Extraction Method

Soxhlet extraction of lipids is a straightforward method, which involves the use of different solvents at various concentrations and provides a high yield of lipids when dried and ground samples are used. However, it has some drawbacks, such as the use of toxic solvents and being a laborious and time-consuming procedure. Moreover, this method is not suitable for samples containing a high amount of water. Typically, in this method, the sample is taken in a porous thimble and placed in a Soxhlet apparatus. In this process, the flask containing the solvents is heated to their boiling points and the evaporated solvents are condensed and collected into the Soxhlet chamber containing the sample. The heated solvents extract the fats from samples and then move them back into the flask. The extraction process usually takes up to 24 h. Once the extraction is completed, the solvent in the flask is evaporated to recover the extracted lipids.[48,49]

14.4.3 Microwave-based Extraction Method

The microwave-assisted method is considered more effective compared to conventional methods because of its advantages, such as less use of solvent, shorter time of extraction, and lower cost. In this method, the sample mixture containing the appropriate solvent is subjected to microwave irradiation for lipid extraction. The solvent containing extracted lipids is separated *via* filtration and evaporated to recover the extracted lipid contents. This method has shown high repeatability and efficiency in comparison to the conventional methods.[50]

14.4.4 Supercritical Fluid Extraction Method

The supercritical fluid extraction method has more advantages, compared to the rest of the methods, such as no use of toxic and flammable organic solvents and sustainable, cheap, and mild extraction conditions. In this method, the sample is slightly heated and pressurized in a chamber followed by the addition of supercritical CO_2 fluid, which is being used as a solvent. The supercritical CO_2 fluid extracts the lipid fraction, which is separated out under reduced pressure and temperature. The samples containing high moisture contents are freeze-dried prior to their lipid extraction to avoid the interaction of water with supercritical fluid, as water acts as a barrier for the diffusion of CO_2 fluid into the sample.[51,52]

14.4.5 Other Extraction Methods

Other extraction methods of lipids are ultrasonic-assisted extraction, solid-phase extraction, and accelerated solvent extraction methods. The ultrasound extraction method is environment-friendly and uses a low amount of solvent. In this method, the sample and an appropriate solvent are placed in a flask in a sonication bath, which uses ultrasound energy for the extraction of lipids from the sample.[53] Solid-phase extraction method is selectively used for the isolation and purification of a specific class of lipids, which involves the use of cartridges or columns packed with different types of stationary phases, such as reversed, normal, or ion exchange sorbents. These cartridges help to separate desirable classes of lipids, such as the normal phase for polar lipids, the reverse phase for hydrophobic lipids, and the ionic phase adsorbs lipids interacting with this phase.[54] Moreover, rapid and efficient extraction of lipids was reported using the accelerated solvent extraction method, which is a commercially available pressurized fluid extraction technique.[55]

14.5 Potential Applications of Lipids in a Biorefinery Concept

Lipids can be converted into high-value products, such as polymers and coatings, lubricants, pharmaceuticals, herbicides/pesticides, solvents, and fuels, using a biorefinery concept (Figure 14.2). These bio-derived industrial products also have great potential to replace petroleum-derived materials. Due to specific characteristics and functional groups, lipids from both plant and animal sources have found numerous industrial-scale applications.

In lipids, plant oils are the most abundant and renewable sources of plant biomass and have found large-scale industrial applications as an alternative fuel/biodiesel. The continuous use of non-renewable and conventional fuels has a high negative impact on the environment due to the high emission of resultant greenhouse gases. As a result, a high rate of interest has been observed in the production of environment-friendly biofuels from renewable resources, which have shown a significant reduction in carbon footprint compared to conventional diesel and petroleum fuel. Considering this, numerous bio-based feedstock, such as plant and animal oils, algae, and various low-value wastes, such as waste cooking oil, soaps, and greases, have been investigated to produce biofuel. However, all of them are not suitable for this purpose, as the desirable

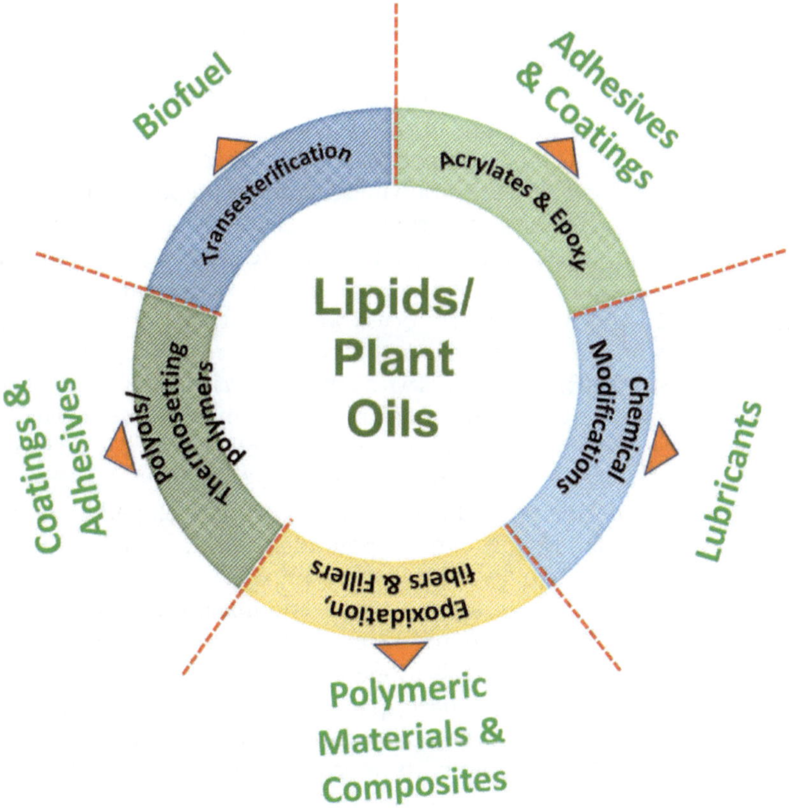

Figure 14.2 Valorization of lipids/plant oils into value-added industrial products.

characteristics of oils for biofuel production include a high oil yield and unique profile of fatty acid components. Plant oils are the most suitable bioresources for biodiesel/fuel production. These oils on treatment with either methanol or ethanol in the presence of acids or bases are converted into methyl or ethyl esters, termed as biodiesel fuel. The developed biodiesel can be used either alone or in a mixture of conventional fuels derived from petroleum resources. The alkyl esters of fatty acids (oils) are widely used as an environment-friendly engine fuel due to their low carbon footprint.

Recently, Rahimi *et al.* (2021)[56] reported the production of biodiesel from camelina oil using the core–shell magnetic nanocatalyst (MgO/Fe_2O_3–SiO_2). The synthesized catalyst has shown the capability to be recycled up to four times. The results from this study revealed that the properties of the produced biodiesel are in the range of specified standards, which makes camelina oil a potential raw material for biofuel production. Moreover, Hsiao *et al.* (2020)[57] studied the production of biodiesel from waste cooking oil using a modified calcium oxide catalyst in the presence of a microwave heating system. A high yield of biodiesel was obtained with the modified catalyst under optimum conditions where the reaction time was greatly reduced.

Utilization of vegetable oils in the production of polymeric materials and composites is highly investigated as a substitute for petroleum-based materials. The development of monomers and polymers from renewable resources is highly emphasized in the recent literature. A complete review covering various routes for the conversion of fatty acids into monomers and polymers *via* a process of radical polymerization was reported by Lomège *et al.* (2019).[58] The preparation of a triblock copolymer as a high-performance pressure-sensitive adhesive from lauryl acrylate (derived from vegetable oil) and triacetic acid lactone *via* reversible addition–fragmentation chain transfer (RAFT) polymerization process was recently reported by Sajjad *et al.* (2020).[59] Another recent contribution includes one-pot copolymerization of high oleic fatty acid methyl esters with ethylene *via* isomerizing copolymerization while investigating their use as a plastic.[60] Moreover, microwave-assisted solvent-free copolymerization of epoxidized fatty acids of waste cooking oil with phthalic anhydride in the presence of (salen)$Cr^{III}Cl$ catalyst and co-catalyst was reported by Shahi *et al.* (2017).[61] The developed biopolymers with good molecular weight can be potentially used as a substitute for petroleum-based products in the polymer industry.

Moreover, vegetable oils after a slight modification have been used with fibers or other fillers to produce composites with enhanced thermal, mechanical, and physical properties, which can potentially find their applications in the aerospace, plastic, or packaging industries. The development of green bio-composites from nano-modified keratin fiber and fatty acid-based copolymer with improved mechanical and flammability properties was reported.[62] Moreover, the preparation of lipid-derived monomers and their corresponding nanocomposites is also published in the literature *via in situ* free radical homo-polymerization of fatty acid monomers and copolymerization of fatty acid-derived monomers and styrene while dispersing different ratios of nanoclay.[63] The developed composites demonstrated enhanced thermal properties and fire retardancy. A recent contribution illustrates the development of bio-composites from cellulose nanocrystals (CNC)-grafted lipids derived from spent hen and reinforcement with nanoclay. The incorporation of nanofillers into lipid-derived composites revealed higher thermal stability and increased tensile strength, making them a potential alternative to conventional composites for the packaging industry.[64]

Considering the concept of biorefinery, lipids and oils or fats are largely reported to serve as a great source of high-value chemicals or monomers, which can be used as precursors for various industrial products. Fatty acid triglyceride structure contains specific functional groups, which can be further modified chemically into fascinating molecular structures. Triglycerides can be converted into fatty acid methyl esters plus glycerol and fatty alcohols and/or by chemical modification of their unsaturated bonds to obtain specific chemicals of interest either by epoxidation, ozonolysis, hydroformylation, thiol–ene coupling, olefin metathesis, dimerization, and radical additions.[41]

Ullah *et al.* (2017 and 2020)[65,66] reported the conversion of oils (fatty acid esters) from different renewable resources into fine chemicals while providing an economical and sustainable solution to transform petrorefinery into a biorefinery. For this purpose, catalytic ethenolysis and alkenolysis of triglycerides and/or fatty acid methyl esters were conducted in the presence of electromagnetic radiation. The results revealed a greener and more rapid microwave-assisted process with high turnover numbers for the conversion of renewable oils and fats into high-value chemicals or monomers as starting materials of various products for the chemical industry. Moreover, during the production of biodiesel from vegetable oils through a process of transesterification, glycerol is produced as a by-product.[67] In recent years, high production of biodiesel in industries has led to an increased production of crude glycerol, which can be further transformed *via* chemical and biological routes into high-value-added chemicals.[68–70]

Thermoset polymers exhibit higher thermal stability and improved chemical and mechanical resistance and are widely used as a coating material for various industries, such as construction and automotive parts.[71,72] For the preparation of these coating materials, resins, such as phenol and epoxy, are mostly used as curing agents. All these petro-based coating components are hazardous and can affect the environment.[73,74] Extensive interest has been given to replacing these petro-based materials with bio-derived alternatives. Among different biomass, plant oils, being versatile in their structures, can be functionalized to prepare a range of thermosetting materials. Acryl-functionalized fatty acid polymeric molecules produced *via* exposure to UV light or auto-oxidation provide high-performance coatings to be used as pressure-sensitive adhesives.[75,76] Meghan *et al.*[71] demonstrated the production of soybean oil-derived thermoset copolymers *via* an emulsion polymerization process from synthesized soybean methacrylate monomer with co-monomers, such as styrene, butyl acrylate, and/or methyl methacrylate. The developed thermoset copolymers were tuned by varying the ratio of soybean methacrylate monomer to obtain good thermal and mechanical properties. The results from this study revealed high improvement in the mechanical properties of thermosets, which make them a potential candidate for coating applications.

Moreover, bio-based polymeric materials from various renewable resources (proteins, rubber, polysaccharides, *etc.*) are extensively being explored in the field of adhesives to fulfill the demand of the sustainable product market.[77] For adhesive application, the utilization of plant-based polymeric material has advantages over others, as their incorporation into adhesive formulations can improve hydrophobicity and, as a result, can provide moisture resistance.[78] Additionally, these plant-derived materials/fragments can also have more advantages, such as crosslinking sites, compatibility between reagents, and plasticizing effect.

A recent study reported the development and utilization of acrylic monomers from different plant oils for adhesive preparations by copolymerization with commercially available styrene, methyl methacrylate, and butyl acrylate. When evaluated as an adhesive for polypropylene (PP), polyethylene terephthalate (PET), uncoated bleached paperboard, and tops coated with a clay mineral paperboard, the resultant ternary latex copolymers were found to be highly effective in their performance.[79] The development of renewable and greener polyurethane adhesives by reacting polyols derived from epoxidized plant oils, such as canola oil, palm oil, and soybean oil, with acids or alcohols is highly studied in the literature. Somani *et al.* (2003)[80] reported the synthesis of polyurethane adhesives by treating castor oil derivatives and diols with diisocyanate adducts while using different ratios of NCO/OH to obtain different compositions. Improved adhesion strength and chemical resistance were observed from the renewable polyurethane adhesives.

The development of environment-friendly lubricant materials is in high demand. Plant oils are also being used to some extent to fulfill the demand for lubricants in the automotive industry. Plant oils are less volatile due to their high molecular weight and exhibit a narrow range of viscosities with the change in temperature. Their use as a lubricant is restricted because of their low thermal oxidation stability and poor cold flow properties.[81] These issues can be resolved by either heating and/or chemically or genetically modifying plant oils or adding additives. The elimination of unsaturated double bonds in the plant oils can help to reduce its oxidation stability, which can be done by the heat modification method (232–330 °C) under an inert atmosphere. As a result, lubricating base oils with a range of viscosities are produced with improved performances.[82–84] In addition, modification of plant oils *via* transesterification, estolide formation, and epoxidation with polyols have demonstrated improvement in their oxidative stability while attaining optimal characteristics to be used for extreme applications as a lubricant.[82]

14.6 Summary and Future Perspective

This chapter focuses on the transformation of renewable biomass, specifically lipids and proteins, into high-value-added products using a biorefinery concept. It covers a brief description of lipid and protein sources and their various well-known extraction methods. Moreover, the utilization of a biorefinery concept for converting lipids and proteins into valuable materials has been discussed in detail, where various routes for the modification or processing of proteins are described for their application as food, feed, agriculture, packaging, and polymeric materials. Furthermore, it describes the utilization of lipids as renewable biomass into biodiesel/fuel, polymer materials and composites, green chemicals and monomers, and lubricants using a biorefinery concept, which includes various chemical modification processes, such as epoxidation, ethenolysis, acrylation, crosslinking, transesterification, and composite preparations by blending with fillers or fibers. The unique inherent functional attributes and biodegradability of proteins and lipids make them a potential candidate to be utilized as a raw material to substitute petroleum-based products for various growing industries, such as food, feed, packaging, biomedical, automotives, and coatings. However, numerous methods and strategies are being used to transform proteins and lipids into high-value-added products, but still, there are some limitations and restrictions in the production of bio-based industrial products in an effective way, such as high curing time and temperature for thermosetting polymers, biofuel production with poor properties due to high unsaturated contents (higher oxidation rate) of plant oils, and rigid or strong amino acid network of proteins. Hence, the emphasis should be placed on developing more effective methods for the development of industrial products from lipids or plant oils and proteins through their functional group modification and/or blending with different fillers and fibers to obtain products of industrial standards for a sustainable future.

References

1. S. Takkellapati, T. Li and M. A. Gonzalez, *Clean Technol. Environ. Policy*, 2018, **20**, 1615–1630.
2. D. Esposito and M. Antonietti, *Chem. Soc. Rev.*, 2015, **44**, 5821.

3. L. Vernès, Y. Li, F. Chemat and M. Abert-Vian, Biorefinery concept as a key for sustainable future to green chemistry—the case of microalgae, in *Plant Based "Green Chemistry 2.0" Moving from Evolutionary to Revolutionary*, 2019, pp. 15–50.
4. G. Franca-Oliveira, T. Fornari and B. Hernández-Ledesma, *Processes*, 2021, **9**, 1626.
5. F. Badii and N. K. Howell, *Food Hydrocolloids*, 2006, **20**, 630–640.
6. J. Nehete, R. Bhambar, M. Narkhede and S. Gawali, *Pharmacogn. Rev.*, 2013, **7**, 107–116.
7. R. B. Freedman, *Trends Biochem. Sci.*, 1985, **10**, 82.
8. H. P. S. Makkar, *J. Agric. Food Chem.*, 1989, **37**, 1197–1202.
9. C. Tanford and L. H. Maurice, Physical chemistry of macromolecules, *J. Electrochem. Soc.*, 1962, **109**(3), 98C.
10. J. Nehete, R. Bhambar, M. Narkhede and S. Gawali, *Pharmacogn. Rev.*, 2013, **7**, 107–116.
11. M. Zubair and A. Ullah, Switchable solvents for bio-refinery applications, in *Green Sustainable Process for Chemical and Environmental Engineering and Science*, Elsevier, 2022, pp. 1–20.
12. W. Mulder, C. van der Peet-Schwering, N.-P. Hua and R. van Ree, *Proteins for food, feed and biobased applications: Biorefining of protein containing biomass*, IEA Bioenergy Task 42, 2016.
13. M. Henchion, M. Hayes, A. M. Mullen, M. Fenelon and B. Tiwari, *Foods*, 2017, **6**, 1–21.
14. M. Hammershøj, A. Prins and K. B. Qvist, *J. Sci. Food Agric.*, 1999, **79**, 859–868.
15. M. L. Amorim, J. Soares, J. S. dos. R. Coimbra, M. de O. Leite, L. F. T. Albino and M. A. Martins, *Crit. Rev. Food Sci. Nutr.*, 2021, **61**, 1976–2002.
16. N. Tripathi, C. D. Hills, R. S. Singh and C. J. Atkinson, *npj Clim. Atmos. Sci.*, 2019, **2**, 35.
17. M. Arshad, M. Zubair, S. S. Rahman and A. Ullah, Polymers for advanced applications, in *Polymer science and nanotechnology: Fundamentals and applications*, ed. R. Narain, 2020, pp. 325–340.
18. A. H. Møller, M. Hammershøj, N. H. M. Dos Passos, H. Tanambell, L. Stødkilde, M. Ambye-Jensen, M. Danielsen, S. K. Jensen and T. K. Dalsgaard, *J. Agric. Food Chem.*, 2021, **69**, 14341–14357.
19. B. E. Knuckles and G. O. Kohler, *J. Agric. Food Chem.*, 1982, **30**, 748–752.
20. S. J. Sheen and V. L. Sheen, *J. Agric. Food Chem.*, 1985, **33**, 79–83.
21. A. H. Martin, O. Castellani, G. A. H. de Jong, L. Bovetto and C. Schmitt, *J. Sci. Food Agric.*, 2019, **99**, 1568–1576.
22. B. P. Lamsal, R. G. Koegel and S. Gunasekaran, Gelation of alfalfa soluble leaf proteins, *Trans. ASAE*, 2005, **48**, 2229–2235.
23. B. P. Lamsal, R. G. Koegel and S. Gunasekaran, *LWT--Food Sci. Technol.*, 2007, **40**, 1520–1526.
24. S. Y. Li, I. S. Ng, P. T. Chen, C. J. Chiang and Y. P. Chao, *Biotechnol. Biofuels*, 2018, **11**, 1–15.

25. P. M. Könst, P. M. C. C. D. Turras, M. C. R. Franssen, E. L. Scott and J. P. M. Sanders, *Adv. Synth. Catal.*, 2010, **352**, 1493–1502.
26. P. M. Könst, M. C. R. Franssen, E. L. Scott and J. P. M. Sanders, *Green Chem.*, 2011, **13**, 1167–1174.
27. A. Ben-Bassat, F. S. Sariaslani, L. L. Huang, R. Patnaik and D. J. Lowe, Methods for the preparation of para-hydroxycinnamic acid and cinnamic acid at alkaline pH, *US Pat.*, 8,003,356, 2011.
28. Y. Shen, L. Zhao, Y. Li, L. Zhang and G. Shi, *Biotechnol. Lett.*, 2014, **36**, 1681–1686.
29. A. V. Pukin, C. G. Boeriu, E. L. Scott, J. P. M. Sanders and M. C. R. Franssen, *J. Mol. Catal. B: Enzym.*, 2010, **65**, 58–62.
30. N. Li, H. Chou and Y. Xu, *Eng. Life Sci.*, 2016, **16**, 299–305.
31. Y. H. Oh, K. H. Kang, M. J. Kwon, J. W. Choi, J. C. Joo, S. H. Lee, Y. H. Yang, B. K. Song, I. K. Kim, K. H. Yoon, K. Park and S. J. Park, *J. Ind. Microbiol. Biotechnol.*, 2015, **42**, 1481–1491.
32. T. M. Lammens, D. De Biase, M. C. R. Franssen, E. L. Scott and J. P. M. Sanders, *Green Chem.*, 2009, **11**, 1562–1567.
33. P. Ödman, W. B. Wellborn and A. S. Bommarius, *Tetrahedron: Asymmetry*, 2004, **15**, 2933–2937.
34. G. S. Hossain, J. Li, H. dong Shin, R. R. Chen, G. Du, L. Liu and J. Chen, *J. Biotechnol.*, 2014, **169**, 112–120.
35. J. H. Sattler, M. Fuchs, K. Tauber, F. G. Mutti, K. Faber, J. Pfeffer, T. Haas and W. Kroutil, *Angew. Chem., Int. Ed.*, 2012, **51**, 9156–9159.
36. J. Sanders, E. Scott, R. Weusthuis and H. Mooibroek, *Macromol. Biosci.*, 2007, **7**, 105–117.
37. M. Zubair, J. Wu and A. Ullah, *ACS Omega*, 2019, **4**, 3772–3781.
38. M. Zubair, J. Wu and A. Ullah, *Biocatal. Agric. Biotechnol.*, 2022, **43**, 102434.
39. M. Zubair, M. S. Roopesh and A. Ullah, *Chemosphere*, 2022, **308**, 136339.
40. I. Zahara, M. Arshad, M. A. Naeth, T. Siddique and A. Ullah, *Chemosphere*, 2021, **273**, 128545.
41. M. Zubair, R. A. Pradhan, M. Arshad and A. Ullah, *Macromol. Mater. Eng.*, 2021, **306**, 2000799.
42. E. Fahy, D. Cotter, M. Sud and S. Subramaniam, *Biochim. Biophys. Acta, Mol. Cell Biol. Lipids*, 2011, **1811**, 637–647.
43. F. D. Gunstone, in *Fatty acid and lipid chemistry*, 1996, pp. 61–86.
44. M. Arshad, A. K. Mohanty, R. Van Acker, R. Riddle, J. Todd, H. Khalil and M. Misra, *RSC Adv.*, 2022, **12**, 27230–27245.
45. E. G. Bligh and W. J. Dyer, *Can. J. Biochem. Physiol.*, 1959, **37**, 911–917.
46. A. Hara and N. S. Radin, *Anal. Biochem.*, 1978, **90**, 420–426.
47. F. Smedes, *Analyst*, 1999, **124**, 1711–1718.
48. K. Ramluckan, K. G. Moodley and F. Bux, *Fuel*, 2014, **116**, 103–108.
49. F. Jalili, S. M. Jafari, Z. Emam-Djomeh, N. Malekjani and V. Farzaneh, *Food Anal. Methods*, 2018, **11**, 598–612.
50. M. Safder, F. Temeli and A. Ullah, *J. Cleaner Prod.*, 2019, **206**, 622–630.

51. M. Safder, F. Temeli and A. Ullah, *J. CO2 Util.*, 2019, **34**, 335–342.
52. M. Solana, C. S. Rizza and A. Bertucco, *J. Supercrit. Fluids*, 2014, **92**, 311–318.
53. H. Teng, L. Chen, Q. Huang, J. Wang, Q. Lin, M. Liu, W. Y. Lee and H. Song, *PLoS One*, 2016, **11**, e0153457.
54. R. K. Saini, P. Prasad, X. Shang and Y. S. Keum, *Int. J. Mol. Sci.*, 2021, **22**, 13643.
55. Y. Tang, Y. Zhang, J. N. Rosenberg, N. Sharif, M. J. Betenbaugh and F. Wang, *RSC Adv.*, 2016, **6**, 29127–29134.
56. T. Rahimi, D. Kahrizi, M. Feyzi, H. R. Ahmadvandi and M. Mostafaei, *Ind. Crops Prod.*, 2021, **159**, 113065.
57. M. C. Hsiao, J. Y. Kuo, S. A. Hsieh, P. H. Hsieh and S. S. Hou, *Fuel*, 2020, **266**, 117114.
58. J. Lomège, V. Lapinte, C. Negrell, J. J. Robin and S. Caillol, *Biomacromolecules*, 2019, **20**, 4–26.
59. H. Sajjad, W. B. Tolman and T. M. Reineke, *ACS Appl. Polym. Mater.*, 2020, **2**, 2719–2728.
60. A. Fleckhaus, P. A. Fokou, G. Klaassen and M. Rüsch gen. Klaas, *Eur. J. Lipid Sci. Technol.*, 2019, **121**, 1700429.
61. M. N. Shahi, M. Arshad and A. Ullah, *Materials*, 2017, **10**, 315.
62. M. Arshad, M. Kaur and A. Ullah, *ACS Sustainable Chem. Eng.*, 2016, **4**, 1785–1793.
63. M. Arshad, L. Huang and A. Ullah, *Polym. Int.*, 2016, **65**, 653–660.
64. M. Kaur, M. Arshad and A. Ullah, *ACS Sustainable Chem. Eng.*, 2018, **6**, 1977–1987.
65. A. Ullah and M. Arshad, *ChemSusChem*, 2017, **10**, 2167–2174.
66. R. A. Pradhan, M. Arshad and A. Ullah, *J. Ind. Eng. Chem.*, 2020, **84**, 42–45.
67. S. Veluturla, N. Archna, D. Subba Rao, N. Hezil, I. S. Indraja and S. Spoorthi, *Biofuels*, 2018, **9**, 305–314.
68. K. Avasthi, A. Bohre, M. Grilc, B. Likozar and B. Saha, *Catal. Sci. Technol.*, 2020, **10**, 5411–5437.
69. P. Sudarsanam, R. Zhong, S. Van Den Bosch, S. M. Coman, V. I. Parvulescu and B. F. Sels, *Chem. Soc. Rev.*, 2018, **47**, 8349–8402.
70. H. W. Tan, A. R. Abdul Aziz and M. K. Aroua, *Renewable Sustainable Energy Rev.*, 2013, **27**, 118–127.
71. M. E. Lamm, P. Li, S. Hankinson, T. Zhu and C. Tang, *Polymer*, 2019, **174**, 170–177.
72. F.-L. Jin, X. Li and S.-J. Park, *J. Ind. Eng. Chem.*, 2015, 1–11.
73. X. Pan, P. Sengupta and D. C. Webster, *Biomacromolecules*, 2011, **12**, 2416–2428.
74. X.-Y. Jian, X.-P. An, Y.-D. Li, J.-H. Chen, M. Wang and J.-B. Zeng, *Macromol.*, 2017, **50**, 5729–5738.
75. C. Zhang, T. F. Garrison, S. A. Madbouly and M. R. Kessler, *Prog. Polym. Sci.*, 2017, **71**, 91–143.

76. I. Tarnavchyk, A. Popadyuk, N. Popadyuk and A. Voronov, *ACS Sustainable Chem. Eng.*, 2015, **3**, 1618–1622.
77. M. N. Islam, F. Rahman, A. K. Das and S. Hiziroglu, *Int. J. Adhes. Adhes.*, 2022, **112**, 102992.
78. Y. Li, S. H. Chou, W. Qian, J. Sung, S. I. Chang and X. S. Sun, *J. Am. Oil Chem. Soc.*, 2017, **94**, 713–721.
79. V. Kirianchuk, B. Domnich, Z. Demchuk, I. Bon, S. Trotsenko, O. Shevchuk, G. Pourhashem and A. Voronov, *Molecules*, 2022, **27**, 5170.
80. K. P. Somani, S. S. Kansara, N. K. Patel and A. K. Rakshit, *Int. J. Adhes. Adhes.*, 2003, **23**, 269–275.
81. A. Campanella, E. Rustoy, A. Baldessari and M. A. Baltanás, *Bioresour. Technol.*, 2010, **101**, 245–254.
82. J. Salimon, N. Salih and E. Yousif, *Arab. J. Chem.*, 2012, **5**, 135–145.
83. S. Z. Erhan, B. K. Sharma and J. M. Perez, *Ind. Crops Prod.*, 2006, **24**, 292–299.
84. N. J. Fox and G. W. Stachowiak, *Tribol. Int.*, 2007, **40**, 1035–1046.

CHAPTER 15

Protein- and Lipid-based Wound-healing Materials

NILAKSHI ABEYSINGHE[a,b] AND NANDIKA BANDARA*[a,b]

[a]Department of Food and Human Nutritional Sciences, Faculty of Agricultural and Food Sciences, University of Manitoba, Winnipeg, Manitoba, R3T 2N2, Canada; [b]Richardson Centre for Food Technology and Research (RCFTR), University of Manitoba, 196 Innovation Drive, Winnipeg, Manitoba, R3T 2N2, Canada
*E-mail: Nandika.Bandara@umanitoba.ca

15.1 Introduction to Biomaterials for Wound Healing

Upsurging numbers of patients suffering from acute and chronic wounds often limit public access to better wound care. Annually, the number of cases reported for leg ulcers, pressure sores, and burn wounds accounts for over 50 million.[1,2] Natural healing of tissues followed by an injury needs a prolonged time.[3,4] Wound dressings accelerate the healing process by providing extra care and reducing the associated risk of infections. Lately, the increased number of incidents of chronic wounds has been predicted to raise the global wound dressing market by 3 billion USD from 2016 to 2022.[2] Technological advances in the research fields and the growing attention on wound dressing products facilitate the necessity to expand this market, fulfilling the patient needs.

Wounds are openings, breaks, or defects mainly formed due to physio-chemical or thermal damages. Wounds are of two types: acute and chronic.

Smart Materials No. 41
Functional Materials from Lipids and Proteins
Edited by Nandika Bandara and Aman Ullah

Published by the Royal Society of Chemistry, www.rsc.org

Burns and cuts, such as acute wounds, take 8–12 weeks to heal. Chronic injuries are fallouts of diseases, such as diabetes, cancers, vascular insufficiency, and pressure necrosis.[5] Failure to access ideal therapeutic materials delays the healing process of wounds. The healing process occurs in four overlapping stages: homeostasis, inflammation, proliferation, and remodeling.[6] Fibrous materials are frequently used in wound-healing applications in biomedical research to facilitate the four stages of healing. These materials are in various physical forms, including films, hydrocolloids, hydrogels, hydrofibers, chitosan, alginates, and foams. In addition, the endogenous biological activities of these materials actively contribute to wound healing. Given the importance of enhancing the healing process, electing a suitable wound dressing is undeniable.[3,6] Therefore, advanced specialized wound dressings are developed and incorporated with polymers, and various other components are designed for superior curative applications.

Selecting primary materials for developing a wound dressing is crucial, as they must meet certain demands within the host tissue. Protein- and lipid-based fibrous materials are attractive as potential wound dressings due to their biocompatibility, reproducibility, excellent affinity, mechanical strength, and processing ability. Applying protein- or lipid-based nanomaterials for biomedical applications has gained interest, as many proteolytic and lipolytic enzymes in the body can degrade natural polymers. The benefits associated are that the degradation causes no toxicity. Furthermore, the body also absorbs the resultant amino acids and steroids. Considering biomedical applications, nontoxic environments and biodegradation of materials are important, as tissue regeneration in injuries requires a toxic-free atmosphere with optimum conditions.[7,8]

Proteins carry integrin-binding domains (cell-recognition sites), such as leucine–aspartic acid–valine and arginine–glycine–aspartic acid (RGD), which promote cell attachment and proliferation. This makes proteins a feasible wound-healing material.[9] Unfortunately, wound healing can be delayed by colonizing and replicating bacteria that cause wound inflammation. However, the use of natural antibiotics extracted from plants, such as essential oils, lipophilic liquids, and semi-liquid compounds, reduces the risk of bacterial growth.[8] Therefore, the ability to store and deliver therapeutic properties of proteins and lipids makes them suitable candidates for fabricating biomaterials.

These bioactive molecules can control the degradation through crosslinking and are available for post-fabrication modifications that blend well with synthetic polymers without threatening bioactivity. Such applications widely open the biomedical field for improved therapeutic applications, compensating for the market demand. This chapter focuses on applying protein- and lipid-based biomaterials for wound healing, their current trends, and future directions.

15.2 Biopolymers as Biomaterials

Biopolymers play increasingly important roles in human life. The research on biopolymers involves fields such as chemistry, biology, and medicine. In terms of source, biopolymers comprise natural biopolymers and synthetic biopolymers. The development of synthetic biopolymers, especially biodegradable polymers, has enriched the structure, properties and varieties of biopolymers, promoting their applications in many areas, such as sutures, bone fixation parts, temporary supporting tubes, controlled drug-release carriers, and tissue engineering scaffolds. The research on synthetic biopolymers has become an attractive subject.

Tissue engineering has evolved over the past few decades. The first historical records of using biomaterials date back to ancient Egyptian times. Ancient history reveals that wound dressings made by animal ligaments successfully heal skin injuries.[10] From the animal sinews, tissue engineering was advanced into incorporating metal joints to facilitate locomotion following the Second World War. The arthritic joints were rebuilt as corrosion-resistant metals to re-establish the motion and facilitate the natural joint function.[11] The concept of introducing biocompatible foreign materials with minimal reactivity came to use about a century ago. These materials had no adverse response on the tissue, enhancing healing properties and actively contributing to the healing process. The toxicity generated by these biomaterials was kept at a minimum.[12] However, the studies addressed the lack of biodegradability of synthetic materials by innovating more biocompatible materials for use as wound dressings.

Recent research interest in biomaterial fabrication has focused on using natural fibers due to the eco-friendliness of the primary materials. When constructing such materials, biopolymeric molecules are gathered to form chain-like structures, which are repetitive chemical blocks that constitute the primary configuration of the biomaterial.[13] The biodegradable matrix of the biomaterials facilitates sustainability, which eventually benefits both wound healing and the environment.[14]

Biopolymers are renewable and extracted from living sources, such as plants, animal body parts/organs, and microbes. Although the polymer market share is small, in 2005, bio-based products, including biopolymers, accounted for 7% of sales worldwide, amounting to US$77 billion in the chemical sector. By 2020, the demand for bio-products had significantly increased, almost doubling the sales of bio-based products. The market is predicted to reach US$250 billion by 2030. There is potential to boost the market share to 90% and replace petroleum-based polymers with biopolymers in the next ten years.[15–17]

The demand for renewable polymeric substances used in biomaterials is increasing due to their unique properties. Biopolymers offer non-toxicity, biocompatibility, and acceptable mechanical properties, making them excellent candidates for fabricating biomaterials.[18,19] Biodegradable properties enable biomaterials to degrade easily and reconstruct new tissue without causing inflammation. The biological and chemical properties of the polymeric matrix enable it to break down either through hydrolytic mechanisms with no enzyme support or through enzymatic degradation. The advantage of using biodegradable materials is that they do not cause permanent chronic reactions to foreign materials. Furthermore, these materials actively facilitate tissue regeneration. The bioactive molecules within these biodegradable assemblies facilitate drug delivery effectively. The compounds that accelerate wound healing and growth factors can be delivered to the wound site through the active conjunctive groups of the polymers.[18–20] Selecting a biopolymer source to fabricate a biomaterial for tissue engineering is crucial, as it requires attention to biocompatibility, favorable cellular interactions, and nontoxicity. Considering the chemical structure of natural polymers, they are classified into major classes, such as polysaccharides, proteins, and polyesters. These categories are based on the nature of the repeating unit of the polymeric chain. The sugars, amino acids, and esters are the monomeric units of polysaccharides, proteins, and polyesters, respectively.[21]

The growth of the biomaterial market depends on the convenience of production and use, low cost of production, and lower rejection rates, which are some of the key driving factors that govern the growth of the biopolymer-based industry. Synthetic biomaterials are toxic and polydisperse, giving an irregular nature to the biomaterials. Synthetic biomaterials are synthesized artificially and are mostly non-degradable.[12,14] Biopolymers are natural biodegradable polymers that can be found simply in the environment within living organisms. The fabrication of biopolymer-inspired biomaterials is known to produce smaller quantities of greenhouse gases, requiring only 60% fewer greenhouse gases and around 50% less non-renewable energy than other polymers. Therefore, biopolymeric biomaterials are easier to use on human skin than synthetic materials.[22–24]

15.2.1 Lipid-based Biomaterials

Lipids from natural resources have extensively been studied in biomedical, pharmaceutical, and health applications of biomaterials. Lipids and lipid derivatives play a significant role in cellular reconstruction and tissue regeneration by reducing inflammations in the local tissue. These lipid-based biopolymers are highly compatible with the human body and have shown low cytotoxicity.[25] Lipid-based biomaterials can be developed from simple, compound, and derived lipids. The abundance of availability, low cost, easy handling, and superior ability of lipids to blend with other materials are some of the key properties of lipids, making them ideal for fabricating wound-healing materials.[26]

Simple lipids (*e.g.*, triglycerides and waxes) from plant-origin oils, such as almond and castor, and animal-origin oils, such as shark liver oil and emu oil, are useful for making healing materials for various reasons, such as having better mechanical strength, and chemical and biological properties. These lipid biopolymers and lipid derivatives can also be used as the base material for biomaterials or as a carrier for drug delivery systems.[25,26] Lipids, such as unsaturated polyester resins (oleic and linolenic), function well with synthetic polymers (polyvinylchloride, polyethylene, and low-density polyethylene), creating robust polymers that improve the characteristics of biomaterials.[26] In addition, blending castor oil with synthetic polymers has improved the flexibility of wound dressings. Lower hydrophilicity, poor thermoplastic properties, brittleness, and hardiness are key challenges associated with starch-based scaffolds during extrusion-based biomaterial fabrication.[27] Therefore, scientists have modified extrusion steps for starch-blended biomaterials by adding sorbitol and glycerol to minimize the use of plasticizers.[28] The resultant bioplastics showed improved thermoplastic properties. However, introducing fatty acids to esterify the starches improved the hydrophilicity, flexibility, brittleness, and permeability and lowered the glass transition temperature, producing starch-based biomaterials.[27–30]

Lipid-coated floating bioadhesive drug delivery systems show promise for the prolonged delivery of drugs to targeted molecules. For example, a fluid multi-layer system incorporated with antibacterial constituents (ofloxacin) and the biofilms coated with waterproof stearic acid layers resulted in retaining drugs in the stomach for more than six hours.[31] Lipids cast off as floating materials in sponges create a prolonged drug delivery characteristic due to their higher contact surface. Drug-incorporated chitosan-based nanocomposites can be coated with lipids to reduce the heaviness of the delivery system, which increases the floatation in gastric fluids. These features enhanced the contact time of the fluids alongside the anti-inflammatory drugs by up to 12 h, eventually decreasing the curing time.[32] Floating ability and integrity of lipids have been useful for these drug delivery matrices to encompass a prolonged retention time.

Furthermore, bioadhesive microsphere matrix systems coupled with glyceryl molecules were used to function as a polymer to extend the retention of drugs in the gut. These microspheres show strong mucoadhesive characteristics that reduce the wound-healing time in the gut wall. Both *in vivo* and *in vitro* studies have proved that glyceryl imparts a high contact surface and slower delivery of drugs over a few hours in the digestive tract. For example, castor oil-entrapped floating microbeads spent 8 h releasing loratadine in the gut.[33]

The formulation of drug delivery systems coated with lipids not only enhances the release of drugs but also improves the immunity where healing occurs. Antibodies blended with calcium, lipids, and mRNA-encoded vaccines that carry anti-tumor effects augment the cellular functions of T cells to improve the immune response. Moreover, lipid-coated cancer cell mimicry vaccines (*e.g.*, αHSP70p-CM-CaP) have been reported to kill and suppress cancer-causing tumor cells by reducing skin inflammation.[34]

Olive oil emulsions integrated into translucent films have been reported to boost mechanical and tensile properties, such as strength, maximum elongation, and Young's modulus. In addition, the lipid layers in these emulsions increase the stability of chitosan-based films to create better adhesion and interaction properties between the biomaterials and the cell matrix. Furthermore, curing materials made with different solid lipids, such as Gelucire® 43/01, cetyl palmitate, Compritol® 888 ATO, Precirol® ATO 5, Softisan® 100, Gelucire® 44/14, Dynasan® 116, Imwitor® 900 k, and Imwitor® 491, have proved to increase the mechanical properties in double-emulsion systems. Furthermore, the authors observed promising results with enhanced drug delivery and healing procedures in *in vivo* systems.[35]

Prevailing improvements in pharmaceutical, biomedical, and tissue engineering fields have shown significant and remarkable applications of lipid-based biopolymer systems, including *in vivo* and *in vitro* biomedical applications on wound healing, drug delivery, immunotherapy, and vaccines. The success of these applications exclusively depends on the versatile characteristics of lipids. Biocompatibility, higher stability, floating ability, prolonged retention of drugs at the application site, target-specific drug delivery, higher contact surface on the local tissue, well-designed control over the release of drugs, and biodegradability are some of the characteristics that advance the use of biomaterials made from lipids. Thus, lipids and lipid-based derivatives provide a broader platform of applications in medical sciences while cultivating the opportunity to incorporate novel techniques and fabricate unique drug delivery matrixes.

15.2.2 Protein-based Biomaterials

Treating wounds is a challenging, multistage process involving homeostasis, inflammation, proliferation, and tissue remodeling.[36,37] However, malfunctioning physiological conditions during recovery could trigger delays or failures in wound healing.[38,39] Therefore, a vast population suffering from unhealed wounds and the complexity of the healing rush the need to develop wound dressings.[40–43] To mimic the structure and function of native tissue, researchers have studied the use of proteins in drug delivery matrices. These matrices/dressings are three-dimensional structures that have proven to perform specific roles in films, hydrogels, and scaffolds.[6] They could be used internally or superficially in the body to enhance the recovery of wounds due to anti-inflammation, cell proliferation-promoting, and rapid homeostasis properties.[44,45] Therefore, protein-based materials address various applications with greater diversity to meet the needs of wound-healing applications.

The use of protein sources as biomaterials is a sustainable approach that involves low-cost techniques. Most natural protein biopolymers are generated from agro wastes, making biomaterial production eco-friendly, economical, and sustainable. Protein sources from both plant and animal materials offer biodegradable biomaterials that grow toward developing green materials. Soy proteins, zein, gliadin-like plant proteins, collagen, gelatin, silk, and keratin-like animal proteins have emerged as novel sustainable materials to produce impurity-free materials. These materials display a minimal risk of disease transmission; additionally, they mitigate the local tissue function more consistently with greater safety for the wound. A few examples of commonly used plant and animal protein-based wound-healing in biomedical fields are given in the subsequent sections.

15.2.2.1 Plant Protein-based Biomaterials

Plant-derived proteins have recently gained popularity as a novel source for developing wound dressers due to their improved cost efficiency, availability, biodegradability, and ability to prevent immunogen transfer.[46–48] Soy protein isolates (SPIs), zein, and wheat proteins are the major plant proteins used to develop wound-healing biomaterials. The characteristics of plant proteins, such as higher net negative charges, help effectively deliver the positively charged drugs to the wound site.[49] Furthermore, the use of plant proteins (*e.g.*, wheat), which are prone to show hydrophilicity, makes them more favorable for attracting cells due to their higher number of polar amino acids.[50] Drug delivery into the body is made easier through the vast range of isoelectric points for plant proteins.[51] However, plant proteins have poor hydrolytic stability.[52] Therefore, researchers use crosslinking and blending to improve the mechanical properties and biocompatibility, making the plant protein-based biomaterials suitable for the controlled release of drugs.

15.2.2.1.1 Soy Protein. Biopolymers, such as soy proteins, carry a vast number of functional groups that are available to be used in biomaterials. These intrinsic abilities are useful when fabricating hydrogels for specific applications.[53] Soy protein isolates are organized into two major globular proteins, namely 7s (β-conglycinin) and 11s (glycine).[54] These globular proteins can be produced into heat-set and cold-set gels, and these techniques generate crosslinks by reaching the isoelectric pH (pI) of the protein.[55] Altered SPI gels enhance their use as adhesives, hydrogels, and films in biomedical applications.[53] Soy protein hydrogels showed various physiochemical properties, making them ideal for wound-healing applications. For instance, they demonstrate a range of water-uptake capacity, swelling behavior, protein release capacity, and drug delivery.[54,56] One of the most critical properties of hydrogels is their water-uptake ability, facilitating them to absorb and diffuse water or body fluids. This feature assists hydrogels in transferring nutrients and metabolites into the wound site.[57] Swelling is necessary for hydrogels to absorb the wound exudates.[56,58] The swelling capacity tested by combining SPI with polyacrylic was reported that swelling reached the maximum capacity of 140% at 20 h and 24 h.

This effect was further enhanced using crosslinking agents *N*,*N*-methylene bisacrylamide and glutaraldehyde.[49] Hydrogels can release nutrients based on the pH medium, a suitable aspect of pH-dependent drug delivery systems. Therefore, tests on hydrogel swelling ratios show higher swelling capacities at higher pH levels. SPI hydrogels are successful delivery systems due to their sound water uptake, swelling capacities, and quick response time to release drugs in manipulative physiological environments. They are pH-responsive but lack stability and show slow release rates at lower pH levels. However, crosslinking could enhance these poor physiochemical properties of SPI-based wound-healing materials.

15.2.2.1.2 Zein. Zein is a plant protein obtained from corn, which can be used as a base material for biomedical applications, such as wound dressings and drug delivery.[59] Zein-based biomaterials promote cell proliferation, cell adhesion, and anti-inflammation. Zein is a highly biocompatible protein, which makes it a suitable candidate for developing hydrogels, films, fiber, and nanoparticle structures as wound-healing materials.[59,60] Zein is used for wound-dressing applications alone or blended with polymers and other nanoparticles. However, despite the benefits of zein, such as its excellent film-forming properties, its application is limited by some mechanical properties. Therefore, research studies have focused on combining zein with biopolymers to improve its less supportive characteristics for cell adhesion, proliferation, and cytotoxicity.[59] Several studies have extensively studied the cell attachment, growth, and proliferation of zein in biomedical applications.[59,60] The cell adhesion of zein hydrogels was evaluated *in vitro* using fibroblast cells. The study tested the cell adhesion by adding fibroblast cells at 105 cells per mat density on the zein-based scaffolds and observed the results through scanning electron microscopy. The study showed that scaffold hydrophilicity increased cell adhesion and proliferation. However, modified zein with ZnO and aloe vera increased the cell adhesion strength compared to pure zein mats. The results confirmed that zein and/or modified zein scaffolds could increase cell adhesion and proliferation without showing any toxic effect on fibroblast cells.[61] Thus, cytocompatibility is a desired feature when making hydrogels that increase the acceptability of new cell environments for biomaterials. Zein hydrogels prepared by ultrasonication (UZCNOs) were found to have higher cell viability for osteoblast cells cultured on their surface. This is due to the high tensile strength, lower degradation, higher hydrophobicity, and crosslinking density of the zein hydrogels. This feature is promising for treating wounds or delivering drugs with minimal toxic effects on the cell.[59]

15.2.2.1.3 Gliadin. Gliadins are alcohol-soluble prolamins generated from wheat and are one of the main proteins found in wheat endosperm. These proteins are versatile materials that can be applied in different platforms of the pharmaceutical industry to enhance elasticity and extensibility. Gliadin was used to fabricate nanofibers that retain various inorganic and organic compounds to deliver hydrophobic active compounds.[62] Drug-loaded gliadin-based nanosystems have proven their ability for site-specific drug release *via* oral administration. Gliadin nanoparticles processed to manipulate the acidic environment of the duodenum and jejunum mucosa have proven to increase the pH level by interacting with the mucosal surface of the intestine. Moreover, these nanoparticles have increased bioavailability by over 40%.[63] Gliadin nanoparticles have also proved helpful in treating *Helicobacter pylori* infection, which triggers gastric cancers and most intestinal-related diseases.[64] Gliadin-based nanostructures conjugated with amoxicillin and omeprazole have created an exploit environment that alters the bacterial function to create optimal conditions inside the colon.[65] Gliadin-based films blended with paracetamol have shown promising results against intestinal degradation. A conventional dosage of paracetamol-loaded gliadin nanostructures has improved the colon environment, sustaining the internal atmosphere to facilitate digestion.[66] Moreover, gliadin does not require crosslinking agents to support the cellular matrix, as it can act as a self-cross-linking agent that increases the strength of the matrix. However, one of the major drawbacks incorporated with gliadin is the site of toxicity imparted into the cells. Further research is required to reduce the cytotoxicity generated from gliadin to make this nanofiber a flourishing agent in biomedical applications.[67] SPI, zein, and gliadin improve cell proliferation, adhesion, drug delivery, strength, and mechanical properties in wound-healing applications. However, their preparation techniques can limit their ability, which needs improvement. Furthermore, many other studies have also focused on the effect of plant proteins as drug delivery matrices and wound dressings, which are summarized in Table 15.1.

15.2.2.2 Animal Protein-based Biomaterials

Animal protein-based biomaterials have been tailored over the past few decades to control cell functions in many different ways. Some of the earliest applications involve silk- and wool-based materials in traditional wound treatment that supported drug delivery and healing. In the context of millennia, the use of animal proteins for biomaterials has extended to elastin-, collagen-, keratin-, and fibrin-like materials, increasing the biomass used to develop biomaterials. The below information highlights a few recent advancements in animal protein applications for wound healing.

15.2.2.2.1 Keratin. Hydrogels made from different sources of keratin respond differently to inflammatory incidents of wounds.[76] The potency of the amount of fibroblast adhered to hydrogels triggers these disparities. Comparatively, the number of fibroblasts that adhere to human hair-based hydrogels is lower than that produced using wool as the keratin source material. The wound size has also been different in such cases. For

Table 15.1 Plant and animal protein-related different approaches for wound care.[a]

Research study group	Protein	Approach	Result
Zhao *et al.* (2018)[68]	Soy protein isolates	Chitosan incorporated into SPI films	An increase in the SPI content reduced the swelling capacity; cell adhesion and proliferation were increased; the chitosan : SPI ratio of 1 : 1 had the maximum healing efficiency
Khabbaz *et al.* (2019)[53]	Soy protein isolates	Electrospinning was used to prepare nanofibers	Exudate absorption was enhanced, and no cytotoxicity was discovered
Akhmetova *et al.* (2020)[69]	Zein	Polyethylene oxide (PEO) or stearic acid was introduced to zein fibers to modify the fiber characteristics	Both the compounds increased the antimicrobial activity against *Staphylococcus aureus* and *Escherichia coli*, and no cytotoxic effect was found on fibroblasts; PEO enhanced the stability
Gunes *et al.* (2020)[70]	Zein	Introduced bioactive compounds from St. John's wort (*Hypericum perforatum*) to the zein films	Water-uptake capacity was reduced; however, the increased permeability eventually helped keep the wound moistened and enhanced tissue regeneration
Zhang *et al.* (2017)[71]	Silk	Preclinical and clinical approach to enhance tissue regeneration	*In vivo* studies on rabbits reduced the wound-healing time, increasing regeneration and tissue repair; in addition, *in vitro* approaches to cure skin defects with silk fibroin films increased the healing process
Lee *et al.* (2014)[72]	Elastin	Elastin-binding proteins increase cell interactions and elasticity	Cellular integrin binds ($\alpha V\beta 3$) at the C-terminal tail of elastin, increasing the cell adhesion and fibroblast attachment on cytoskeletal assembly

(*continued*)

Table 15.1 (*continued*)

Research study group	Protein	Approach	Result
Irmak *et al.* (2019)[73]	Gelatin	Methacrylated gelatin is used in bioprinting applications	Showed more substantial and lower degradation rates of around ~27% over 35 days of application; the cell viability, attachment, proliferation, mineral deposition, and mRNA expression were enhanced in preosteoblastic cells
Yang *et al.* (2010)[74]	Gelatin and chitosan	Radiation-induced crosslinking of gelatin and carboxymethyl chitosan to support cell growth and increase wound healing	Swelling ability, cell attachment, biodegradation, and biocompatibility were increased; in addition, the interconnection between the pores was homogeneous; cell attachment and rapid growth of fibroblasts on the material were observed
Kumar and Singh (2017)[75]	Silk fibroin and casein	Silk fibroin and casein were developed into a physiologically based pharmacokinetic (PBPK) model to test against 100 *in vivo* test subjects.	The bioavailability of the drug was increased by seven-fold compared to non-silk materials with higher retention time.

[a]Different studies on the effects of biomaterials prepared from plant and animal protein have shown promising results over the past years. Novel approaches to incorporating bioactive compounds and chemical cross-linkers, such as citric acid and PEO, have increased the physiochemical characteristics of wound dressers. These results show that proteins are suitable therapeutic agents that can heal wounds and deliver drugs to the wound sites with a minimum cytotoxic effect.

instance, human hair-based hydrogel-treated wounds have had considerably smaller sizes than wool-based hydrogel-treated wounds.[77] Therefore, the choice of wound dressings, especially hydrogels, depends on the purpose of use and the source material of keratin. Chemical treatments can functionalize the active free thiol in the keratin sponges. Iodoacetic acid, 2-bromoethylamine, and iodoacetamide are commonly used to produce different sponges, such as carboxyl-, amino-, and amido-sponges. Chemically modified keratin sponges stimulate extracellular matrix proteins and further hybridization of bioactive molecules.[78] Hybridization involves fabricating calcium and phosphate into sponges. This can be done by chemically binding calcium and phosphate ions or trapping hydroxyapatite particles within the keratin carboxy sponges. Hybridized materials supported osteoblast cultivation, altering their differentiation pattern.[79] Keratin carboxy sponges differentiate the pre-osteoblasts grown with the wound. They can promote internal osteogenesis without harming external heterotopic ossification.[80]

Wool keratin has driven the textile industry to a different level as a renewable raw material. Regenerated keratin films, therefore, have also gained the attention of researchers. Keratin films made from wool are regenerated through wool keratin/ionic liquid solutions dissolving in coagulation solutions, water, methanol, and ethanol. Regeneration creates differences in the wool keratin structures. Ionic solvents make the α-helix arrangements disappear from the original keratin structures, leaving only β-sheets. This new form of wool keratin differs from the natural wool keratin fibers' structure. This molecular chain breakage causes the regenerated wool keratin to lose thermal stability from its original structure, ultimately changing the keratin film structure due to changes in base materials.[81] However, graphene oxide (GO)-induced regeneration produces a different reaction on the keratin structure. GO-modified keratin enhances the mechanical and physical properties of keratin films and improves the thermal stability of films.[82] However, to manifest the maximum possible results, further research is required on regeneration.

15.2.2.2.2 Collagen. Collagen-based biomaterials are useful wound-healing materials to enhance cell migration and proliferation in acellular matrices. A vast number of studies have reviewed the fundamental characteristics of collagen both *in vivo* and *in vitro*. Collagens are widely used to cure vascular diseases, bone reconstruction, and neural migration. In vascular diseases, collagen is advantageous in addressing cardiovascular malfunction and venous or arterial pathologies. They are very useful, as collagen wound-healing materials possess wet strength, allowing a suturing platform for soft tissues. This is a template for new tissues to grow. Heart diseases are prominent due to the high tendency for immunogenic activities and the potential to become calcified. Acellular heart valves derived from collagen are prominent in curing these malfunctions in cardiac and vascular reconstructive applications. The collagen dispersions crosslink with the fibers, ultimately increasing the extra cellular matrix (ECM)'s viscosity, and curing the targeted area.[83] Cryolife is a trademark developed using collagen as the primary material that helps enhance the decellularization and reendothelializing of veins and valves.[84–86] This is a critical application in reconstructing the bone and cartilage in osteochondral applications. They enhance the production of new collagen while bound to an ECM.[85,86] The scaffolds fabricated using collagen support bone tissue engineering by mineralization with calcium phosphate. This is also enhanced by collagen's ability to crosslink with other substances, such as hydroxyapatite. Cartilage regeneration has also recently become an easy task to fulfill with the flexibility conveyed by collagen. Latest cutting-edge technologies have engineered 3-dimensional nervous systems that promote axonal migration and myelination of modern neurons through the formation of connective tissues to improve neuronal function.[87–89]

Collagen is a biopolymer of utmost importance for tissue engineering and regenerative applications, with the ability to provide high immunogenicity, biocompatibility, and biodegradability, with enhanced mechanical strength. Even though these superior characteristics are helpful for wound-healing applications as collagen is a protein material, the difficulty remains to fabricate without making alterations to the original structure. Animal protein-based biomaterial applications in wound healing are a broad aspect that asylums a majority of fields in pharmaceutical and medical industries. Further applications of other animal proteins are also highlighted in Table 15.1.

15.3 Concepts Used Fabricating Biomaterials for Wound Healing

The architecture of the body cells is organized into three-dimensional structures. The biomaterials are designed to mimic the morphology of the body tissues. At the *in-vivo* level of fabricating the wound dressings, 100 to 1000 μm repetitive units are coordinated into a multicellular process to develop a successful 3D structure. The mechanical strength provided by the matrix supports a platform for tissue regeneration. Biochemical, cellular, and physical morphologies further augment cell differentiation, proliferation, migration, and death at a micro level (~10 μm). Therefore, the successful fabrication of functional biomaterials should comprise macro and micro levels to mimic the local cellular environment. The macro-scale level should permit tissue construction and nutrient transportation, while the micro level should safeguard the cell's survival and function.[90]

Conventional fabrication and rapid prototyping are the two major wound-healing material fabrication techniques. Conventional fabrication includes freeze-drying, solvent casting, gas foaming, electrospinning, and thermal-induced phase separation, while rapid prototyping involves stereolithography, selective laser sintering, solvent-based extrusion-free forming, bioprinting, and fused deposition modeling. Several wound dressings have been developed in the context of tissue engineering using conventional techniques. Drug delivery and tissue regeneration are facilitated through these 3D dressings. In addition, the wound-healing materials are provided with a unique shape and pores to facilitate gas and liquid diffusion. However, if produced conventionally, pores may not be evenly disseminated through the 3D structure.[90,91]

Rapid prototyping uses computer-aided design models for the fabrication of wound dressings. The computer sequences directly generate the dressings using a pre-programmed model. The cross-sections are made through computer-aided design (CAD), which overlays a series of sections on top of each other to fabricate the final 3D design. The base materials involve powder, liquid, and sheet materials that are organized layer by layer to fabricate a 3D structure using a laser. The scaffolds, such as wound dressings, are designed from bottom to top by stacking multiple layers. The prototyping methods manufacture the biomaterials using milling and drilling methodologies. Unlike the conventional type, the pore size and shape in rapid prototyping are unique, even, and pre-determined. These features are controlled over the 3D structure through customized computer modeling methods. Each sub-section has unique advantages and disadvantages that are briefly explained in Table 15.2. These techniques serve different purposes, so the manufacturers should be careful when selecting the proper method to fabricate a wound-healing material.[90–92]

15.4 Limitations of Biomaterials for Wound-healing Applications and Future Directions

North America accounts for more than 6.5 million people who require medical attention for wound healing. However, only a few patients can access proper therapeutics due to poor and complex clinical design, inconsistency of therapeutic approaches, lack of knowledge and technological approaches to developing proper healing strategies, and the complexity of the tissue repair process.[16] Due to these reasons and the demand for advanced wound-healing methods over the past few decades, attention has been paid to investing in and investigating unconventional medical therapies. Bioactive molecule production is a cutting-edge technology in biology and pharmaceutical sciences to support tissue regeneration and wound healing. The active compounds and the drug delivery systems are designed to improve the damaged site. Active biomolecules extracted from proteins and lipids are functional composite materials commonly used in wound-healing systems. The proteins, such as keratin, gelatin, silk collagen, natural rubber proteins, and lipid esters, are vastly studied for therapeutic purposes due to their unique characteristics for healing. They are excellent carrier molecules that actively participate in wound healing in the local tissue to fight against inflammations.[25,26,52]

However, the complicated wound-healing process encompasses a series of interactions between the local tissue and extracellular matrix, which are significantly important in regulating the healing process. In this context, the biomaterial and cells work together to heal the cutaneous wounds, and various cellular events take place in the local tissue for skin regeneration. However, certain wound types demand specifically designed wound-healing materials that possibly increase the healing.[93] Even though the healthcare sector is focused on increasing progressive tissue regeneration in the clinical setting, it still needs more validated studies to

Table 15.2 Plant and animal protein-related different approaches for wound care.[a]

Technique	Processing methodology	Advantages	Disadvantages
Conventional method			
Freeze-drying	Fabricates porous scaffolds using sublimation techniques; the polymeric material is dissolved in a chemical solvent to pour into a mold; the solvent-containing molds are then lyophilized to remove the excess water; under a high vacuum provided during freeze-drying, the molds fabricate the substances into scaffolds with high porosity and connectivity	• Silk proteins and lipid polymers are easy to fabricate using this technique • Pore size can be manipulated through freeze-drying, temperature, and pH; for example, fast freeze-drying creates smaller pores and *vice versa* • Controlled solidification creates 3D structures with evenly distributed pores • High temperature or leaching step not required	• Pore sizes are small and irregular • Processing time is long • High-energy-consuming technique • Chemical solvents cause toxicity
Solvent casting	Chemical solvents are evaporated from the scaffolds in two different ways; one method involves dipping the molds in the polymeric solvent and letting it drain for a sufficient time; the polymeric structure is created when the content is drained; in another method, the polymeric solvent is poured into the mold and the solvent is evaporated for an ample time following the 3D structure adheres to the mold	• An easy, simple, and low-cost technique • Does not require large and specific equipment • Membranes are thin and highly porous (50–90%)	• Toxic solvents alter the protein nature by denaturing and retaining the toxicity • Extended time is taken to fabricate the scaffolds

(*continued*)

Table 15.2 (*continued*)

Technique	Processing methodology	Advantages	Disadvantages
Gas foaming	High-pressure (800 psi) carbon dioxide gas passes through the molds to saturate the polymer mix; the amount of gas dissolved in the mix decides the number of pores in the structure; the gas inside the matrix is unstable, separating the phase from the structure; pore nucleation is created through cluster molecules by minimizing the free energy; after the completion of gas diffusion, the pores are made	• Does not require high temperature • Pores can obtain around 80% • Porosity can be controlled by solvents, such as protein, sugar, salts, and wax	• Residues can cause cytotoxicity • If the process is not successful, the pores stay closed • Gas foaming can denature proteins
Electrospinning	This technique utilizes an electrostatic force to fabricate biomaterials; the electrostatic force is controlled and provided between two electrodes with opposite polarity; the polymeric solution is projected from one electrode to the receiving electrode under a high-voltage electric field; a jet of polymer is shot at a droplet form to fabricate the scaffold	• Nanofiber fabrication is facilitated with this technology • The scaffolds made ultra-fine structures with high surface area • Non-invasive technique • Provides a homogeneous structure with a high tensile strength to the final structure • A homogeneous mixture made of fibers with high tensile strength • The 3D matrix is compatible with tissue and cell growth • The structures also facilitate prompt cell adhesion and gas diffusion	• Hard to obtain the desired pore size • The solvents involved can be toxic • Cell seeding is not fully facilitated, which can be overcome by including a cryo-spinning step

(*continued*)

Table 15.2 (*continued*)

Technique	Processing methodology	Advantages	Disadvantages
Thermal-induced phase separation	Temperature fluctuations of this technique separate the polymeric solution dissolved in phenol or naphthalene into two phases with low and high polymer concentrations; first, the liquid–liquid phase is separated when the temperature comes down, followed by a two-phase solid phase is formed; finally, the excessive solvent can be removed by evaporation or sublimation, creating a porous 3D structure	• Low temperatures can be used to integrate the bioactive molecules • Easy to combine with other fabricating techniques • The 3D porous structures can be controlled easily • Creates highly porous structures with 95% pores	• Appropriate liquid–liquid phase separation is required for nanofiber preparation • Limited application, *e.g.*, can only use on the thermoplastics
Rapid prototyping			
Stereolithography	Curable materials are printed layer-by-layer using ultraviolet radiations to create solid, three-dimensional objects; the system comprehends four components: a tank, a transferable built platform, a UV laser for radiating resin, and a dynamic mirror system; a layer of photosensitive liquid resin from the tank is initially ejected and deposited on the platform; the first layer then solidified on the platform; next, the first layer is lowered vertically to receive the second layer; this process is carried out a few times to create the 3D scaffold	• Reduces wastage in subtractive fabrication methods • The method uses high-resolution techniques • The pores created are highly interconnected, and the pore size is uniform • Fabricated scaffolds can be used in bio-sensing, remediation, and drug discovery	• The technique requires massive amounts of monomers and treatments for post-polymerization to improve monomer conversion, which is a limitation of the process

(*continued*)

Table 15.2 (*continued*)

Technique	Processing methodology	Advantages	Disadvantages
			• The beam width limits the size of the scaffold
Selective laser sintering (SLS)	A laser technique is used in this method to sinter the powdered material into a 3D model; the laser source is the power-anchoring component in this technique that asserts the powder layers into thin layers; the SLS method easily fabricates polymers, metals, and plastics into wound dressings	• The use of ultrahigh-molecular-weight polyethylene increases the efficiency of fabricating bio-nanocomposites • Microstructures are controlled well using the SLS technique to adjust the percentage compositions of polymer blends	• Steps after fabricating the scaffolds require additional steps to remove the excess powder
Bioprinting	Two types of 3D bioprinting are available for the manufacturing of wound dressings; they are, namely, acellular and cellular techniques; acellular techniques do not require cells during the printing process, whereas cellular bioprinting uses cells and other bioagents that construct living tissues during the production process; in addition, fabricating biomaterials using bioprinting involves different fabrication methods, such as autonomous self-assembly, biomimicry, inject printing, mini-tissue building block, and microextrusion.	• Bioprinting is a low-cost technique with higher accuracy and provides high cell viability, around 80–90% • This method allows the direct printing of 3D scaffolds without using an aqueous solution • In general, cell growth is supported significantly compared to other methods • Compared to the cellular method, the acellular method	• The major drawback is that the technique can be used only on a limited number of materials • The print head can clog while the machine is running • The primary material should

(*continued*)

Table 15.2 (*continued*)

Technique	Processing methodology	Advantages	Disadvantages
	• Microextrusion is a laser-based technique, and inject printing is a bio-ink technique frequently used in medicine Micro-extrusion is a robot-controlled system with a handling and dispensing system, stage, fiber-optic-based light source, and a video camera. The handling and dispensing system alternates the temperature of the base material; the stage of this system can move along the X, Y, and Z-axes; the light source is used for multiple purposes, including the elimination of deposition area for photoinitiator activation and photographers' activity, and the light source is also used as a piezoelectric humidifier; the video camera commands and controls the X–Y–Z axes; the material deposited on a substrate is extruded through the microextrusion as a continuous flow of material beads deposited in a 2D pattern. The CAD-CAM software moves the microextrusion head alongside the Z-axis, creating the basis for the subsequent 2D layer Inkjet bioprinting is a noncontact technique in which picoliter bio-ink droplets are used to fabricate biomaterials; these structures are either 2D or 3D layers built on a substrate; inkjet techniques are commonly used in thermal ink jetting, acoustic wave jetting, and electro-hydrodynamic jetting	has a higher accuracy with less requirement for cell viability maintenance • Some systems use multiple print heads to dispense several materials that do not require retooling	be prepared and maintained in a liquid state until the fabrication is completed

(*continued*)

Table 15.2 (*continued*)

Technique	Processing methodology	Advantages	Disadvantages
Fused deposition modeling	A polymer layer at its solid is cast through a hot extrusion nozzle on a solid platform; the high temperature melts the solid material, and a computer-controlled 3D arm fabricates the biomaterial; finally, the solid layer is cast in multiple layers of adjacent microfilaments that process thermoplastic biopolymers into wound dressings	• This is a useful technique in scaffold-making that requires low-temperature fabrication • Has the ability to fabricate scaffolds of biodegradable polyesters that help the cell growth • Consistent pore sizes are acquired from this technique	• The technique cannot be applied to biodegradable polymers; therefore, additional steps should be involved to modify the solutions

[a]The manufacturing methods of biomaterials are of two types: conventional type and rapid prototyping. Freeze-drying, solvent casting, gas foaming, electrospinning, and thermal-induced phase separation are conventional techniques of fabrication and stereolithography. On the other hand, selective laser sintering, solvent-based extrusion-free forming, bioprinting, and fused deposition modeling are examples of rapid prototyping. Each technique occupies unique fabrication steps and advantages and disadvantages, as briefly explained in the above table.

fulfill the requirements to cure certain wounds. On the other hand, extensive improvements are required to carry drugs to the wound site to help the physiological sequence of wound healing. The distance from the point of origin to the delivery site greatly impacts the effectiveness of wound healing. The greater distances reduce the drug's effect, which requires substantial improvements through clinical tests. Better transdermal delivery systems that help bypass necrotic tissue also need to improve surgical tissue transplant procedures. Micro-needling can be a potential solution with the recent technological advancements to improve drug delivery efficiency.[94] However, the penetration depth release mechanisms, drug release profile, and the distance to dispatch should be addressed while optimizing the delivery efficiency.

Biomaterials made from biopolymers, such as keratin, silk, chitosan, and lipid derivatives, are giving promising results due to their unique characteristics, such as biocompatibility, biodegradability, ability to promote wound healing while controlling microbial activity, and ability to deliver drugs. Major drawbacks of biopolymers include low stability, lower production feasibility, poor mechanical strength, low hydrophilicity, and higher cost of production.[21,36,45] Conjugating these naturally available biopolymers into composite materials provides a platform for novel wound-healing applications and opens many opportunities to expand the applications to healthcare, pharmaceutical sciences, medicine, and nutritional supplemental production fields. It also enhances the functional properties of the biomaterials. Nanoparticle-incorporated biomaterials support continuous diffusion of gas exchange to accelerate the healing process and maintain a moistened environment. However, depending on the nature of the wound and the physiological stage of the wound-healing process, the ability to speed up the tissue regeneration process could be varied. These pitfalls underline the upsurging need for multifunction biomaterials.[95] Therefore, researchers lately have started researching incorporating active compounds, such as nanoparticles and nanocomposites with antimicrobial, anti-infection, and anti-inflammatory activities. Wound dressings incorporated with natural metabolites are entirely bioactive and biodegradable, promoting the cell generation activity of the extracellular matrix and connective tissue disposition that ramps the wound-healing process.[96] However, more clinical studies are required to prove the potential of using nanoparticles as curing materials in wound dressings because the applications are limited due to the proven toxicity of the nanoparticles, such as silver, gold, zinc, and copper. An attempt to improve the mechanical and biological performances of nanoparticles incorporated into polymeric materials is currently needed with optimum treatment protocols to reduce the side effects.[95,96] Advances in easy and facile techniques to integrate multiple materials toward a single wound dressing hold a greater potential to create multifunctional biomaterials to address several treatment protocols at the time. Further comprehensive studies are needed to explain the interaction of tissue and cellular regeneration at molecular levels to articulate the strategies of several components in an integrated biomaterial. The rationale and effectiveness of translating biopolymers and nanocomposite for effective tissue repair must also be addressed.

A few recent advances in technology have kept the ability of biomaterials to cure wounds on the horizon. Immunomodulation, supramolecular biomaterials, and smart dressings are some 21st century's award-winning strategies to address wound healing. Immunomodulation maintains and adjusts the immune response of a wound at a desired level according to the adjacent environment.[97] These biomaterials are used to confront chronic autoimmune diseases, such as diabetes. The insulin production in the pancreas is no longer destroyed due to impaired immune activity. The effect of synthetic injectable biomaterials to reverse type 1 diabetes in non-obese diabetic mice has created a platform to reduce the effects of chronic disease.[98] However, the effect on *in vivo* studies is yet to be discovered. Supramolecular biomaterials are outstanding medical accessories designed to surpass molecules' constructive and inherent abilities. These molecules are designed to precisely sense and respond to an injury according to the stage of the healing procedure. The potential supermolecule biomaterials use a specific architecture that can mimic the extracellular matrix. These molecules would also be responsive to the correct curing procedure that the wound demands.[99] The smart dressings are flexible electronic skins sensitive to atmospheric changes, such as body temperature and outside pressures. These next-generation smart dressings are equipped with sensors to detect the adjacent environment's changes and respond accordingly. They are also available in the wearable form that detects fluctuations in human health to warn them immediately in an emergency. This method quickens the approach to healing wounds and increases the opportunity for immediate access to medical needs.[98–100] However, these technologies require more clinical approaches to testify against *in vivo* cells to promote safe use. In addition, the large-scale bioprocess engineering of these materials could be challenging due to the high cost of production.[97–100]

Each wound type requires a unique sequence of physiological events to regenerate tissues. Abnormal events during wound healing can cause persistent damage that cannot be reversed. Modern integrated biosensing technologies help overcome abnormal events by addressing peculiar incidents involving healing. Smart materials and automated systems are sophisticated tools in wound-healing applications, capable of self-guiding healing and actively mending dysfunctional cellular sites.[101] Furthermore, automated systems can sense changes in the wound's environment, analyze the injury's requirements, and deliver therapeutic mediators and molecules according to the local tissue's needs. This can be done with or without a medical diagnosis by a professional healthcare practitioner or an automatic diagnosis programmed in the biomaterial.[102] However, more experiments are needed to enhance the loading capacity of biosensors and increase their ability to deliver drugs until the wound is fully cured. Automated systems involve biosensors, wearable devices, and flexible electronics, requiring better sensing ability without providing false-positive or false-negative results. Additionally, the ability to thrive in wound exudates is of greater importance.[101–103] Answering what is lacking in the above cutting-edge technologies will provide superior wound-healing abilities to the biomaterials. The wound-healing materials in the current practice are constantly updated to address the limitations. The integration of interdisciplinary sciences, materials science, tissue engineering, gene therapies, DNA technologies, and nanocomposites fast-track skin repair in biomedical applications. These innovations support better applications of medical treatments, reduction of initial costs, and lowering of the number of experiments on laboratory animals, chemical use, and manpower. While there is much more room to grow and develop, the current initiatives may help grow personalized medicine and rapid healing with less or no pain to leave no scars on the wound site.

15.5 Conclusion

Wound care is an important aspect of clinical care due to the increasing number of patients worldwide suffering from burns and chronic wounds. Various physical forms of wound dressings and drug delivery systems have evolved over the years to meet the emerging necessity for wound dressing. Choosing a material for developing a wound dressing is critical as it requires to be biocompatible within the host organism. However, proteins and lipids display many features that fulfill being a source material for wound dressing. Researchers are investigating, transforming the physical and mechanical properties of proteins and lipids to create different biomaterials compatible with their application of interest. However, certain aspects, such as the ability to cure the wound sites with a minimum impact, should be addressed in future applications. Methods are also required to deliver higher binding ability with minimum resource use. Many studies address target cell accessibility and tissue regeneration with high biocompatibility and biodegradable effects on the local tissue. The current platform and the technologies help fabricate biopolymers into appropriate extracellular environments, boosting tissue regeneration, anti-inflammation, and augmenting healing. These technologies, however, are needed to enhance the structural properties to augment the injury at a narrow time frame with low cytotoxic outcomes. The initial costs for protein and lipid extraction and fabrications using innovative techniques are much needed to bring down, allowing the mass-scale producers to thrive in the industry. To conclude, these techniques give promising results for tissue engineering, taking wound healing to a new level.

References

1. W. Paul and C. P. Sharma, Advanced Biomaterials View project Enteric Coating of Polyclonal Antibody for Colon Delivery View project, 2015, Available from: http://www.polymer-books.com.
2. G. Han and R. Ceilley, Chronic Wound Healing: A Review of Current Management and Treatments, *Adv. Ther.*, 1 March 2017, **34**(3), 599–610.
3. T. G. Sahana and P. D. Rekha, Biopolymers: Applications in wound healing and skin tissue engineering, *Mol. Biol. Rep.*, 2018, **45**, 2857–2867.
4. A. Singh, A. Shakeel, D. Kochhar, S. Jeevanandham, S. Rajput and M. Mukherjee, Polymeric worm-like nanomicellar system for accelerated wound healing, *J. Adv. Pharm. Technol. Res.*, 1 January 2020, **11**(1), 36–43.
5. M. Mir, M. N. Ali, A. Barakullah, A. Gulzar, M. Arshad and S. Fatima, *et al.*, Synthetic polymeric biomaterials for wound healing: a review, *Prog Biomater*, March 2018, **7**(1), 1–21.
6. G. Han and R. Ceilley, Chronic Wound Healing: A Review of Current Management and Treatments, *Adv. Ther.*, 2017, **34**, 599–610.
7. K. G. DeFrates, R. Moore, J. Borgesi, G. Lin, T. Mulderig and V. Beachley, *et al.*, Protein-based fiber materials in medicine: A review, *Nanomaterials*, 2018, **8**, 80–97.
8. K. Khezri, M. R. Farahpour and S. Mounesi Rad, Accelerated infected wound healing by topical application of encapsulated Rosemary essential oil into nanostructured lipid carriers, *Artif. Cells, Nanomed., Biotechnol.*, 4 December 2019, **47**(1), 980–988.
9. K. Yamauchi, H. Hojo, Y. Yamamoto and T. Tanabe, Enhanced cell adhesion on RGDS-carrying keratin film, *Mater. Sci. Eng., C*, June 2003, **23**(4), 467–472.
10. Institute of Biomedical Imaging N. Biomaterials Fact Sheet, Available from: www.nibib.nih.gov.
11. G. Suarato, R. Bertorelli and A. Athanassiou, Borrowing from nature: Biopolymers and biocomposites as smart wound care materials, *Front. Bioeng. Biotechnol.*, 2 October 2018, **6**, 137–148.
12. A. Padinjakkara, in *Biopolymers and Biomaterials*, CRC Press, 2018, pp. 2–3.
13. S. Anil, E. P. Chalisserry, S. Y. Nam and J. Venkatesan, Biomaterials for craniofacial tissue engineering and regenerative dentistry, in *Advanced Dental Biomaterials*, Woodhead Publishing, 2019, pp. 643–674.
14. R. Freeman, J. Boekhoven, M. B. Dickerson, R. R. Naik and S. I. Stupp, Biopolymers and supramolecular polymers as biomaterials for biomedical applications, *MRS Bull.*, 27 November 2015, **40**(12), 1089–1100.
15. M. Han, Synthesis and degradation behavior of poly(ethyl cyanoacrylate), *Polym. Degrad. Stab.*, 2008, **93**(7), 1243–1251.

16. L. Shen, E. Worrell and M. Patel, Present and future development in plastics from biomass, *Biofuels, Bioprod. Biorefin.*, January 2010, **4**(1), 25–40.
17. A. Yadav, S. Mangaraj, R. Singh, N. Kumar and S. Arora, Biopolymers as packaging material in food and allied industry, *Int. J. Chem. Stud.*, 2018, **6**(2), 2411–2418.
18. T. Koltai, S. J. Reshkin and S. Harguindey, Taking advantage of cancer's pH alterations: pH-sensitive nanoparticles, in *An Innovative Approach to Understanding and Treating Cancer: Targeting pH*, Elsevier, 2020, pp. 523–540.
19. S. Shankar and J. W. Rhim, Bionanocomposite Films for Food Packaging Applications, in *Innovative Food Processing Technologies*, Elsevier, 2018, pp. 234–243.
20. M. Hassan, J. Bai and D. Q. Dou, Biopolymers; Definition, classification and applications, *Egypt. J. Chem*, 2019, **62**, 1725–1737.
21. L. P. Datta, S. Manchineella and T. Govindaraju, Biomolecules-derived biomaterials, *Biomaterials*, 2020, **230**(93), 1057–1068.
22. M. Carr, I. Green, T. Brown and C. Murphy, *Biotech Solutions for Climate Report Examining biotechnology's contributions to addressing the climate crisis.*
23. Industrial Biotechnology and Climate Change Organisation for Economic Co-operation and Development (OECD), 2011, Available from: www.oecd.org/sti/biotechnology.
24. R. van Ree, E. de Jong and K. Kwant, Leader of Task42 Co-leader of Task42 Operating Agent Task42, Available from: www.iea-bioenergy.task42-biorefineries.comwww.ieabioenergy.com.
25. P. Basim, S. Gorityala and M. Kurakula, Archives of Gastroenterology Research Review Article Advances in Functionalized Hybrid Biopolymer Augmented Lipid-based Systems: A Spotlight on Their Role in Design of Gastro Retentive Delivery Systems, *Arch. Gastroenterol. Res.*, 2021, **2**(1), 35–47.
26. Qurat-ul-Ain, K. M. Zia, F. Zia, M. Ali, S. Rehman and M. Zuber, Lipid functionalized biopolymers: A review, *Int. J. Biol. Macromol.*, December 2016, **93**, 1057–1068.
27. C. Liu, J. Li, W. Lei and Y. Zhou, Development of biobased unsaturated polyester resin containing highly functionalized castor oil, *Ind. Crops Prod.*, January 2014, **52**, 329–337.
28. K. Martinez Villadiego, M. J. Arias Tapia, J. Useche and D. Escobar Macías, Thermoplastic Starch (TPS)/Polylactic Acid (PLA) Blending Methodologies: A Review, *J. Polym. Environ.*, 17 January 2022, **30**(1), 75–91.
29. R. L. Whistler and G. E. Hilbert, Mechanical Properties of Films from Amylose, Amylopectin, and Whole Starch Triacetates, *Ind. Eng. Chem.*, 1 September 1944, **36**(9), 796–798.

30. H. Winkler, W. Vorwerg and R. Rihm, Thermal and mechanical properties of fatty acid starch esters, *Carbohydr. Polym.*, February 2014, **102**, 941–949.
31. C. Zhang, J. Tang, D. Liu, X. Li, L. Cheng and X. Tang, Design and evaluation of an innovative floating and bioadhesive multiparticulate drug delivery system based on hollow structure, *Int. J. Pharm.*, April 2016, **503**(1–2), 41–55.
32. M. I. Tadros and R. H. Fahmy, Controlled-release triple anti-inflammatory therapy based on novel gastroretentive sponges: characterization and magnetic resonance imaging in healthy volunteers, *Int. J. Pharm.*, September 2014, **472**(1–2), 27–39.
33. S. Mishra and K. Pathak, Formulation and evaluation of oil entrapped gastroretentive floating gel beads of loratadine, *Acta Pharm.*, 1 January 2008, **58**(2), 187–197.
34. T. Kang, Y. Huang, Q. Zhu, H. Cheng, Y. Pei and J. Feng, *et al.*, Necroptotic cancer cells-mimicry nanovaccine boosts anti-tumor immunity with tailored immune-stimulatory modality, *Biomaterials*, May 2018, **164**, 80–97.
35. D. Lopes-de-Campos, M. Pinto, S. A. Costa Lima, T. Santos, B. Sarmento and C. Nunes, Delivering amoxicillin at the infection site - a rational design through lipid nanoparticles, *Int. J. Nanomedicine.*, April 2019, **14**, 2781–2795.
36. T. G. Sahana and P. D. Rekha, Biopolymers: Applications in wound healing and skin tissue engineering, *Mol. Biol. Rep.*, 9 December 2018, **45**(6), 2857–2867.
37. C. O. Chantre, S. P. Hoerstrup and K. K. Parker, Engineering biomimetic and instructive materials for wound healing and regeneration, *Curr. Opin. Biomed. Eng.*, June 2019, **10**, 97–106.
38. M. Mir, M. N. Ali, A. Barakullah, A. Gulzar, M. Arshad and S. Fatima, *et al.*, Synthetic polymeric biomaterials for wound healing: a review, *Prog. Biomater.*, 14 March 2018, **7**(1), 1–21.
39. G. Suarato, R. Bertorelli and A. Athanassiou, Borrowing From Nature: Biopolymers and Biocomposites as Smart Wound Care Materials, *Front. Bioeng. Biotechnol.*, 2 October 2018, **6**, 137–148.
40. S. Y. Wang, H. Kim, G. Kwak, H. Y. Yoon, S. D. Jo and J. E. Lee, *et al.*, Development of Biocompatible HA Hydrogels Embedded with a New Synthetic Peptide Promoting Cellular Migration for Advanced Wound Care Management, *Adv. Sci.*, November 2018, **5**(11), 1800852.
41. X. Mao, R. Cheng, H. Zhang, J. Bae, L. Cheng and L. Zhang, *et al.*, Self-Healing and Injectable Hydrogel for Matching Skin Flap Regeneration, *Adv. Sci.*, 16 February 2019, **6**(3), 1801555.
42. B.-Q. Lai, B. Feng, M.-T. Che, L.-J. Wang, S. Cai and M.-Y. Huang, *et al.*, A Modular Assembly of Spinal Cord-Like Tissue Allows Targeted Tissue Repair in the Transected Spinal Cord, *Adv. Sci.*, September 2018, **5**(9), 1800261.

43. N. J. Darling, E. Sideris, N. Hamada, S. T. Carmichael and T. Segura, Injectable and Spatially Patterned Microporous Annealed Particle (MAP) Hydrogels for Tissue Repair Applications, *Adv. Sci.*, November 2018, **5**(11), 1801046.
44. D. G. Pyun, H. S. Yoon, H. Y. Chung, H. J. Choi, T. Thambi and B. S. Kim, *et al.*, Evaluation of AgHAP-containing polyurethane foam dressing for wound healing: synthesis, characterization, in vitro and in vivo studies, *J. Mater. Chem. B*, 2015, **3**(39), 7752–7763.
45. A. J. Hassiba, M. E. el Zowalaty, G. K. Nasrallah, T. J. Webster, A. S. Luyt and A. M. Abdullah, *et al.*, Review of recent research on biomedical applications of electrospun polymer nanofibers for improved wound healing, *Nanomedicine*, March 2016, **11**(6), 715–737.
46. N. Gjorevski, N. Sachs, A. Manfrin, S. Giger, M. E. Bragina and P. Ordóñez-Morán, *et al.*, Designer matrices for intestinal stem cell and organoid culture, *Nature*, 24 November 2016, **539**(7630), 560–564.
47. S. Ahn, C. O. Chantre, A. R. Gannon, J. U. Lind, P. H. Campbell and T. Grevesse, *et al.*, Soy Protein/Cellulose Nanofiber Scaffolds Mimicking Skin Extracellular Matrix for Enhanced Wound Healing, *Adv. Healthc Mater.*, 23 May 2018, **7**(9), e1701175.
48. S. Ahn, H. A. M. Ardoña, P. H. Campbell, G. M. Gonzalez and K. K. Parker, Alfalfa Nanofibers for Dermal Wound Healing, *ACS Appl. Mater. Interfaces*, 18 September 2019, **11**(37), 33535–33547.
49. N. He, X. Chen, L. Wang, J. Wen, Y. Li and Q. Cao, *et al.*, Fabrication of Composite Hydrogels Based on Soy Protein Isolate and their Controlled Globular Protein Delivery, *Glob. Chall.*, 27 September 2019, **3**(9), 1900030.
50. S. A. El-Lakany, A. I. Abd-Elhamid, E. A. Kamoun, E. M. El-Fakharany, W. M. Samy and N. A. Elgindy, α-Bisabolol-Loaded Cross-Linked Zein Nanofibrous 3D-Scaffolds For Accelerating Wound Healing And Tissue Regeneration In Rats, *Int. J. Nanomed.*, October 2019, **14**, 8251–8270.
51. H. Qi, J. Cao, Y. Xin, X. Mao, D. Xie and J. Luo, *et al.*, Dual responsive zein hydrogel membrane with selective protein adsorption and sustained release property, *Mater. Sci. Eng., C*, January 2017, **70**, 347–356.
52. N. Ni and M. J. Dumont, Protein-Based Hydrogels Derived from Industrial Byproducts Containing Collagen, Keratin, Zein and Soy, *Waste Biomass. Valor.*, 2 March 2017, **8**(2), 285–300.
53. B. Khabbaz, A. Solouk and H. Mirzadeh, Polyvinyl alcohol/soy protein isolate nanofibrous patch for wound-healing applications, *Prog. Biomater.*, 24 September 2019, **8**(3), 185–196.
54. K. Prusty, A. Biswal, S. B. Biswal and S. K. Swain, Synthesis of soy protein/polyacrylamide nanocomposite hydrogels for delivery of ciprofloxacin drug, *Mater. Chem. Phys.*, August 2019, **234**, 378–389.

55. A. Abaee, M. Mohammadian and S. M. Jafari, Whey and soy protein-based hydrogels and nano-hydrogels as bioactive delivery systems, *Trends Food Sci. Technol.*, December 2017, **70**, 69–81.
56. S. Tansaz, R. Singh, I. Cicha and A.Boccaccini, Soy Protein-Based Composite Hydrogels: Physico-Chemical Characterization and In Vitro Cytocompatibility, *Polymers*, 17 October 2018, **10**(10), 1159.
57. T. P. Nguyen and B. T. Lee, Fabrication of oxidized alginate-gelatin-BCP hydrogels and evaluation of the microstructure, material properties and biocompatibility for bone tissue regeneration, *J. Biomater. Appl.*, 16 September 2012, **27**(3), 311–321.
58. R. Silva, R. Singh, B. Sarker, D. G. Papageorgiou, J. A. Juhasz and J. A. Roether, *et al.*, Soft-matrices based on silk fibroin and alginate for tissue engineering, *Int. J. Biol. Macromol.*, December 2016, **93**, 1420–1431.
59. N. Mamidi, A. González-Ortiz, I. Lopez Romo I and E. V. Barrera, Development of Functionalized Carbon Nano-Onions Reinforced Zein Protein Hydrogel Interfaces for Controlled Drug Release, *Pharm.*, 20 November 2019, **11**(12), 621.
60. H. Asadi, A. Ghaee, J. Nourmohammadi and A. Mashak, Electrospun zein/graphene oxide nanosheet composite nanofibers with controlled drug release as antibacterial wound dressing, *Int. J. Polym. Mater. Polym. Biomater.*, 11 February 2020, **69**(3), 173–185.
61. M. Ghorbani, P. Nezhad-Mokhtari and S. Ramazani, Aloe vera-loaded nanofibrous scaffold based on Zein/Polycaprolactone/Collagen for wound healing, *Int. J. Biol. Macromol.*, June 2020, **153**, 921–930.
62. S. Voci, M. Fresta and D. Cosco, Gliadins as versatile biomaterials for drug delivery applications, *J. Controlled Release*, January 2021, **329**, 385–400.
63. M. A. Arangoa, M. A. Campanero, M. J. Renedo, G. Ponchel and J. M. Irache, Gliadin nanoparticles as carriers for the oral administration of lipophilic drugs, *Pharm. Res.*, 2001, **18**(11), 1521–1527.
64. C. Y. Kao, B. S. Sheu and J. J. Wu, Helicobacter pylori infection: An overview of bacterial virulence factors and pathogenesis, *Biomed. J.*, February 2016, **39**(1), 14–23.
65. S. Ramteke, N. Ganesh, S. Bhattacharya and N. K. Jain, Triple therapy-based targeted nanoparticles for the treatment of Helicobacter pylori, *J. Drug Target.*, 3 January 2008, **16**(9), 694–705.
66. V. Stella, Gliadin films. I: Preparation and in vitro evaluation as a carrier for controlled drug release, *Int. J. Pharm.*, 6 July 1995, **121**(1), 117–121.
67. N. Reddy, Q. Jiang and Y. Yang, Novel Wheat Protein Films as Substrates for Tissue Engineering, *J. Biomater. Sci., Polym. Ed.*, 2 January 2011, **22**(15), 2063–2077.
68. Y. Zhao, Z. Wang, Q. Zhang, F. Chen, Z. Yue and T. Zhang, *et al.*, Accelerated skin wound healing by soy protein isolate-modified

hydroxypropyl chitosan composite films, *J. Biol. Macromol.*, October 2018, **118**, 1293–1302.
69. A. Akhmetova, G. M. Lanno, K. Kogermann, M. Malmsten, T. Rades and A. Heinz, Highly Elastic and Water Stable Zein Microfibers as a Potential Drug Delivery System for Wound Healing, *Pharm.*, 18 May 2020, **12**(5), 458.
70. S. Gunes, S. Tamburaci and F. Tihminlioglu, A novel bilayer zein/MMT nanocomposite incorporated with H. perforatum oil for wound healing, *J. Mater. Sci.: Mater. Med.*, 14 January 2020, **31**(1), 7.
71. W. Zhang, L. Chen, J. Chen, L. Wang, X. Gui and J. Ran, *et al.*, Silk Fibroin Biomaterial Shows Safe and Effective Wound Healing in Animal Models and a Randomized Controlled Clinical Trial, *Adv. Healthcare Mater.*, May 2017, **6**(10), 1700121.
72. P. Lee, D. Bax, M. M. M. Bilek and A. S. Weiss, A Novel Cell Adhesion Region in Tropoelastin Mediates Attachment to Integrin αVβ5, *J. Biol. Chem.*, January 2014, **289**(3), 1467–1477.
73. G. Irmak, T. T. Demirtaş and M. Gümüşderelioğlu, Highly Methacrylated Gelatin Bioink for Bone Tissue Engineering, *ACS Biomater. Sci. Eng.*, 11 February 2019, **5**(2), 831–845.
74. C. Yang, L. Xu, Y. Zhou, X. Zhang, X. Huang and M. Wang, *et al.*, A green fabrication approach of gelatin/CM-chitosan hybrid hydrogel for wound healing, *Carbohydr. Polym.*, November 2010, **82**(4), 1297–1305.
75. S. Kumar and S. K. Singh, In silico-in vitro-in vivo studies of experimentally designed carvedilol loaded silk fibroin-casein nanoparticles using physiological based pharmacokinetic model, *Int. J. Biol. Macromol.*, March 2017, **96**, 403–420.
76. P. Kumaran, A. Gupta and S. Sharma, Synthesis of Wound-Healing Keratin Hydrogels Using Chicken Feathers Proteins, and its Properties, *Int. J. Pharm. Pharm. Sci.*, 1 February 2017, **9**(2), 171.
77. M. Park, B. S. Kim, H. K. Shin, S. J. Park and H. Y. Kim, Preparation and characterization of keratin-based biocomposite hydrogels prepared by electron beam irradiation, *Mater. Sci. Eng., C*, December 2013, **33**(8), 5051–5057.
78. J. G. Rouse and M. E. Van Dyke, A Review of Keratin-Based Biomaterials for Biomedical Applications, *Mater.*, 3 February 2010, **3**(2), 999–1014.
79. A. Tachibana, Y. Furuta, H. Takeshima, T. Tanabe and K. Yamauchi, Fabrication of wool keratin sponge scaffolds for long-term cell cultivation, *J. Biotechnol.*, February 2002, **93**(2), 165–170.
80. A. Tachibana, Y. Nishikawa, M. Nishino, S. Kaneko, T. Tanabe and K. Yamauchi, Modified keratin sponge: binding of bone morphogenetic protein-2 and osteoblast differentiation, *J. Biosci. Bioeng.*, November 2006, **102**(5), 425–429.

81. R. Li and D. Wang, Preparation of regenerated wool keratin films from wool keratin–ionic liquid solutions, *J. Appl. Polym. Sci.*, 15 February 2013, **127**(4), 2648–2653.
82. B. Li, J. Yao, J. Niu, J. Liu, L. Wang and M. Feng, *et al.*, Effects of Graphene Oxide on the Structure and Properties of Regenerated Wool Keratin Films, *Polymers*, 28 November 2018, **10**(12), 1318.
83. O. O. Ige, L. E. Umoru and S. Aribo, Natural Products: A Minefield of Biomaterials, *ISRN Mater. Sci.*, 7 May 2012, **2012**, 1–20.
84. O. E. Teebken, C. Puschmann, I. Breitenbach, B. Rohde, K. Burgwitz and A. Haverich, Preclinical development of tissue-engineered vein valves and venous substitutes using re-endothelialised human vein matrix, *Eur. J. Vasc. Endovasc. Surg.*, January 2009, **37**(1), 92–102.
85. M. F. O'Brien, S. Goldstein, S. Walsh, K. S. Black, R. Elkins and D. Clarke, The SynerGraft valve: a new acellular (nonglutaraldehyde-fixed) tissue heart valve for autologous recellularization first experimental studies before clinical implantation, *Semin. Thorac. Cardiovasc. Surg.*, October 1999, **11**(4 Suppl 1), 194–200.
86. T. Konuma, E. J. Devaney, E. L. Bove, S. Gelehrter, J. C. Hirsch and Z. Tavakkol, *et al.*, Performance of CryoValve SG Decellularized Pulmonary Allografts Compared with Standard Cryopreserved Allografts, *Ann. Thorac. Surg.*, September 2009, **88**(3), 849–855.
87. M. Gingras, I. Paradis and F. Berthod, Nerve regeneration in a collagen-chitosan tissue-engineered skin transplanted on nude mice, *Biomaterials*, April 2003, **24**(9), 1653–1661.
88. M. Gingras, M. M. Beaulieu, V. Gagnon, H. D. Durham and F. Berthod, In vitro study of axonal migration and myelination of motor neurons in a three-dimensional tissue-engineered model, *Glia*, February 2008, **56**(3), 354–364.
89. R. Parenteau-Bareil, R. Gauvin and F. Berthod, Collagen-Based Biomaterials for Tissue Engineering Applications, *Mater.*, 16 March 2010, **3**(3), 1863–1887.
90. A. Eltom, G. Zhong and A. Muhammad, Scaffold Techniques and Designs in Tissue Engineering Functions and Purposes: A Review, *Adv. Mater. Sci. Eng.*, 7 March 2019, **2019**, 1–13.
91. B. Subia and J. C. S. Kundu, Biomaterial Scaffold Fabrication Techniques for Potential Tissue Engineering Applications, in *Tissue Engineering*, InTech, 2010.
92. In *3D Printing in Biomedical Engineering*, ed. S. Singh, C. Prakash and R. Singh, Springer, Singapore, 2020.
93. Y. Heun, K. Pogoda, M. Anton, J. Pircher, A. Pfeifer and M. Woernle, *et al.*, HIF-1α Dependent Wound Healing Angiogenesis In Vivo Can Be Controlled by Site-Specific Lentiviral Magnetic Targeting of SHP-2, *Mol. Ther.*, July 2017, **25**(7), 1616–1627.

94. S. Bhatnagar, P. R. Gadeela, P. Thathireddy and V. V. K. Venuganti, Microneedle-based drug delivery: materials of construction, *J. Chem. Sci.*, 1 September 2019, **131**(9), 90.
95. P. Couvreur, Nanoparticles in drug delivery: Past, present and future, *Adv. Drug Delivery Rev.*, January 2013, **65**(1), 21–23.
96. M. J. Mitchell, M. M. Billingsley, R. M. Haley, M. E. Wechsler, N. A. Peppas and R. Langer, Engineering precision nanoparticles for drug delivery, *Nat. Rev. Drug Discovery*, 4 February 2021, **20**(2), 101–124.
97. J. L. Mann, A. C. Yu, G. Agmon and E. A. Appel, Supramolecular polymeric biomaterials, *Biomater. Sci.*, 2018, **6**(1), 10–37.
98. X. Li and X. Su, Multifunctional smart hydrogels: potential in tissue engineering and cancer therapy, *J. Mater. Chem. B*, 2018, **6**(29), 4714–4730.
99. I. Carayon, A. Gaubert, Y. Mousli and B. Philippe, Electro-responsive hydrogels: macromolecular and supramolecular approaches in the biomedical field, *Biomater. Sci.*, 2020, **8**(20), 5589–5600.
100. I. Barros Almeida, L. Garcez Barretto Teixeira, F. Oliveira de Carvalho, É. Ramos Silva, P. Santos Nunes and M. R. Viana Dos Santos, *et al.*, Smart Dressings for Wound Healing: A Review, *Adv. Skin Wound Care*, February 2021, **34**(2), 1–8.
101. S. Takaloo and M. Moghimi Zand, Wearable electrochemical flexible biosensors: With the focus on affinity biosensors, *Sens. Bio-Sens.Res.*, June 2021, **32**, 100403.
102. Y. Yang and W. Gao, Wearable and flexible electronics for continuous molecular monitoring, *Chem. Soc. Rev.*, 2019, **48**(6), 1465–1491.
103. S. P. Lee, G. Ha, D. E. Wright, Y. Ma, E. Sen-Gupta and N. R. Haubrich, *et al.*, Highly flexible, wearable, and disposable cardiac biosensors for remote and ambulatory monitoring, *NPJ Digit. Med.*, 25 December 2018, **1**(1), 2.

Subject Index